Guide to Geometric Algebra in Practice

Leo Dorst · Joan Lasenby
Editors

Guide to Geometric Algebra in Practice

 Springer

Editors

Dr. Leo Dorst
Informatics Institute
University of Amsterdam
Science Park 904
1098 XH Amsterdam
The Netherlands
l.dorst@uva.nl

Dr. Joan Lasenby
Department of Engineering
University of Cambridge
Trumpington Street
CB2 1PZ Cambridge
UK
jl221@cam.ac.uk

ISBN 978-1-4471-5897-4   ISBN 978-0-85729-811-9 (eBook)
DOI 10.1007/978-0-85729-811-9
Springer London Dordrecht Heidelberg New York

British Library Cataloguing in Publication Data
A catalogue record for this book is available from the British Library

Cover design: VTeX UAB, Lithuania

Printed on acid-free paper

Springer is part of Springer Science+Business Media (www.springer.com)

How to Read This Guide to Geometric Algebra in Practice

This book is called a 'Guide to Geometric Algebra in Practice'. It is composed of chapters by experts in the field and was conceived during the AGACSE-2010 conference in Amsterdam. As you scan the contents, you will find that all chapters indeed use geometric algebra but that the term 'practice' means different things to different authors. As we discuss the various *Parts* below, we guide you through them. We will then see that appearances may deceive: some of the more theoretical looking chapters provide useful and practical techniques.

This book is organized in themes of application fields, corresponding to the division into *Parts*. This is sometimes an arbitrary allocation; one of the powers of geometric algebra is that its unified approach permits techniques and representations from one field to be applied to another. In this guide we move, generally, from the description of physical motion and its measurement to the description of objects of a geometrical nature.

Basic geometric algebra, sometimes known as Clifford algebra, is well understood and arguably has been for many years. It is important to realize that it is not just one algebra, but rather a family of algebras, all with the same essential structure. A successful application for geometric algebra involves identifying, among those in this family, an algebra that considerably facilitates a particular task at hand. The current emphasis on rigid body motion (measurement, interpolation, tracking) has focused the attention on a specific geometric algebra, the *conformal model*. This uses an algebra in which such motions are representable as *rotations* in a carefully chosen model space (for 3D, a 5D space denoted $\mathbb{R}^{4,1}$, with a 32D geometric algebra denoted $\mathbb{R}_{4,1}$). Doing so is an innovation over traditionally used representations such as homogeneous coordinates, since geometric algebra has a particularly powerful representation of rotations (as 'rotors'—essentially spinors, with quaternions as a very special case). This conformal model (CGA, for Conformal Geometric Algebra) is used in most of this book. We provide a brief tutorial introduction to its essence in the Appendix (Chap. 21), to make this guide more self-contained, but more extensive accessible introductions may be found elsewhere.

Application of geometric algebra started in physics. Using conformal geometric algebra, we can now update its description of motion in *Part I: Rigid Body Motion*. From there, geometric algebra migrated to applications in engineering and computer science, where motion tracking and image processing were the first to appreciate and apply its techniques. In this book these fields are represented in *Part II: Interpolation and Tracking* and *Part III: Image Processing*. More recently, traditionally combinatorial fields of computer science have begun to employ geometric algebra to great advantage, as *Part IV: Theorem Proving and Combinatorics* demonstrates.

Although prevalent at the moment, the conformal model is not the only kind of geometric algebra we need in applications. It hardly offers more than homogeneous coordinates if your interest is specifically in projective transformations. It takes the geometric algebra of lines (the geometric algebra of a 6D space, for 3D lines), to turn such transformations into rotations (see *Part V: Applications of Line Geometry*), and reap the benefits from their rotor representation. And even if your interest is only rigid body motions, alternative and lower-dimensional algebras may do the job—this is explored in *Part VI: Alternatives to Conformal Geometric Algebra*.

While those parts of this guide show how geometric algebra 'cleans up' applications that would classically use linear algebra (notably in its matrix representation), there are other fields of geometrical mathematics that it can affect similarly. Foremost among those is differential geometry, in which the use of the truly coordinate-free methods of geometric algebra have hardly penetrated; *Part VII: Towards Coordinate-Free Differential Geometry* should offer inspiration for several PhD projects in this direction!

To conclude this introduction, some sobering thoughts. Geometric algebra has been with us in applicable form for about 15 to 20 years now, with general application software available for the last 10 years. There have been tutorial books written for increasingly applied audiences, migrating the results from mathematics to physics, to engineering and to computer science. Still, a conference on applications (like the one in Amsterdam) only draws about 50 people, just like it did 10 years ago. This is not compensated by integrated use and acceptance in other fields such as computer vision (which would obviate the need for such a specialized geometric algebra based gathering; after all, few in computer vision would go to a dedicated linear algebra conference even though everybody uses it in their algorithms). So if the field is growing, it is doing so slowly.

Perhaps a major issue is education. Very few, even among the contributors to this guide, have taught geometric algebra, and in none of their universities is it a compulsory part of the curriculum. Although we all have the feeling that we understand linear algebra much better because we know geometric algebra and that it improves our linear-algebra-based software considerably (in its postponement of coordinate choices till the end), we still have not replaced parts of linear algebra courses by the corresponding clarifying geometric algebra.[1] Most established colleagues may be too set in their ways to change their approach to geometry; but if we do not tell the young minds about this novel and compact structural approach, it may never reach its potential.

Our message to you and them is: 'Go forth and multiply—but use the geometric product!'

---

[1]The textbook *Linear and Geometric Algebra* (by Alan Macdonald, 2010) enables this, and we should all consider using it!

Part I: Rigid Body Motion

The treatment of rigid body motion is the first algebraically advanced topic that the geometry of Nature forces upon us. Since it was the first to be treated, it shaped the field of geometric representation; but now we can repay our debt by using modern geometric representations to provide more effective computation methods for motions. All chapters in this part use conformal geometric algebra to great advantage in compactness and efficiency.

Chapter 1: *Rigid Body Dynamics and Conformal Geometric Algebra* uses conformal geometric algebra to reformulate the Lagrangian expression of the classical physics of combined rotational and translational motion, due to the dynamics of forces and torques. It uses the conformal rotors ('spinors') to produce a covariant formulation and in the process extends some classical concepts such as inertia and Lagrange multipliers to their more encompassing geometric algebra counterparts. In its use of conformal geometric algebra, this chapter updates the use of geometric algebra to classical mechanics that has been explored in textbooks of the past decade. A prototype implementation shows that this approach to dynamics really works, with stability and computational advantages relative to more common methods.

As we process uncertain data using conformal geometric algebra, our ultimate aim is to estimate optimal solutions to noisy problems. Currently, we do not yet have an agreed way to model geometrical noise; but we *can* determine a form of optimal processing for conflicting data. This is done in quite general form for rigid body motions in Chap. 2: *Estimating Motors from a Variety of Geometric Data in 3D Conformal Geometric Algebra*. Polar decomposition is incorporated into conformal geometric algebra to study how motors are embedded in the even subalgebra and what is the best projection to the motor manifold. A general, dot-product-like similarity criterion is designed for a variety of geometrical primitives. Instances of this can be added to give a total similarity criterion to be maximized. Langrangian optimization of the total similarity criterion then reduces motor estimation to a straightforward eigenrotor problem. The chapter provides a very general means of estimation and, despite its theoretical appearance, may be one of the more influential applied chapters in this book.

In robotics, the inverse kinematics problem (of figuring out what angles to give the joints to reach a given position and orientation) is notoriously hard. Chapter 3: *Inverse Kinematics Solutions Using Conformal Geometric Algebra* demonstrates that having spheres and lines as primitives in conformal geometric algebra really helps to design straightforward numerical algorithms for inverse kinematics. Since the geometric primitives are more directly related to the type of geometry one encounters, they lead to realtime solvers, even for a 3D hand with its 25 joints.

Another example of the power of conformal geometric algebra to translate a straightforward geometrical idea directly into an algorithm is given in Chap. 4: *Reconstructing Rotations and Rigid Body Motions from Exact Point Correspondences Through Reflections*. There, rigid body motions are reconstructed from corresponding point pairs by consecutive midplanes of remaining differences. Applying the algorithm to the special case of pure rotation produces a quaternion determination formula that is twice as fast as existing methods. That clearly demonstrates that understanding the natural geometrical embedding of quaternions into geometric algebra pays off.

Part II: Interpolation and Tracking

Conformal geometric algebra can only reach its full potential in applications when middle-level computational operations are provided. This part provides those for recurring aspects of motion interpolation and motion tracking.

Chapter 5: *Square Root and Logarithm of Rotors in 3D Conformal Geometric Algebra Using Polar Decomposition*, gives explicit expressions for the square root and logarithm of rotors in conformal geometric algebra. Not only are these useful for interpolation of motions, but the form of the bivector split reveals the orthogonal orbit structure of conformal rotors. In the course of the chapter, a polar decomposition is developed that may be used to project elements of the algebra to the rotor manifold.

Geometric algebra offers a characterization of rotations through bivectors. Since these form a linear space, they permit more stable numerical techniques than the nonlinear and locally singular classical representations by means of, for instance, Euler angles or direction cosine matrices. Chapter 6: *Attitude and Position Tracking* demonstrates this for attitude estimation in the presence of the notoriously annoying 'coning motion'. It then extends the technique to include position estimation, employing the bivectors of the conformal model.

An important step in the usage of any flexible camera system is calibration relative to targets of unknown location. Chapter 7: *Calibration of Target Positions Using Conformal Geometric Algebra* shows how this problem can be cast and solved fully in conformal geometric algebra, with compact simultaneous treatment of orientational and positional aspects. In the process, some useful conformal geometric algebra nuggets are produced, such as a closed-form formula for the point closest to a set of lines, the inverse of a linear mapping constrained within a subspace, and the derivative with respect to a motor of a scalar measure between an element and a transformed element. It also shows how to convert the coordinate-free conformal geometric algebra expressions into coordinate-based formulations that can be processed by conventional software.

Part III: Image Processing

Apart from the obviously geometrical applications in tracking and 3D reconstruction, geometric algebra finds inroads in image processing at the signal description level. It can provide more symmetrical ways to encode the geometrical properties of the 2D or 3D domain of such signals.

Chapter 8: *Quaternion Atomic Function for Image Processing* deals with 2D and 3D signals and shows us one way of incorporating the rotational structure into a quaternion encoding of the signal, leading to monogenic rather than Hermitian signals. Kernel processing techniques are developed for these signals by means of atomic functions.

The facts that real 2-blades square to −1 and their direct correspondence to complex numbers and quaternions have led people to extend classical Fourier transforms by means of Clifford algebras. The geometry of such an algebraic analogy is not always clear. In the field of color processing, the 3D color space does possess a perceptual geometry that suggests encoding hues as rotations around the axis of grays. For such a domain, this gives a direction to the exploration of Clifford algebra extensions to the complex 1D Fourier transform. Chapter 9: *Color Object Recognition Based on a Clifford Fourier Transform* explores this and evaluates the effectiveness of the resulting encoding of color images in an image retrieval task.

Part IV: Theorem Proving and Combinatorics

A recurrent theme in this book is how the right representation can improve encoding and solving geometrical problems. This also affects traditionally combinatorial fields like theorem proving, constraint satisfaction and even cycle enumeration. The null elements of algebras turn out to be essential!

Chapter 10: *On Geometric Theorem Proving with Null Geometric Algebra* gives a good introduction to the field of automated theorem proving, and a tutorial on the authors' latest results for the use of the null vectors of conformal geometric algebra to make computations much more compact and geometrically interpretable. Especially elegant is the technique of dropping hypotheses from existing theorems to obtain new theorems of extended and quantitative validity.

As a full description of geometric relationships, geometric algebra is potentially useful and unifying for the data structures and computations in Computer Aided Design systems. It is beginning to be noticed, and in Chap. 11: *On the Use of Geometric Algebra in Geometrical Constraint Solving* the structural cleanup conformal geometric algebra could bring is explored in some elementary modeling computations.

Part of the role geometric algebra plays as an embedding of Euclidean geometry is a consistent bookkeeping of composite constructions, of an almost Boolean nature. The Grassmann algebra of the outer product, in particular, eliminates many terms 'internally'. In Chap. 12: *On the Complexity of Cycle Enumeration for Simple Graphs*, that property is used to count cycles in graphs with n nodes, by cleverly representing the edges as 2-blades in a $2n$-dimensional space and their concatenations as outer products. Filling the usual adjacency matrix with such elements and multiplying them in this manner algebraically eliminates repeated visits. It produces compact algorithms to count cycle-based graph properties.

Part V: Applications of Line Geometry

Geometric algebra provides a natural setting for encoding the geometry of 3D lines, unifying and extending earlier representations such as Plücker coordinates. This is immediately applicable to fields in which lines play the role of basic elements of expression, such as projective geometry, inverse kinematics of robots with translational joints, and visibility analysis.

Chapter 13: *Line Geometry in Terms of the Null Geometric Algebra over* $\mathbb{R}^{3,3}$, *and Application to the Inverse Singularity Analysis of Generalized Stewart Platforms* provides a tutorial introduction on how to use the vectors of the 6D space $\mathbb{R}^{3,3}$ to encode lines and then applies this representation effectively to the analysis of singularities of certain parallel manipulators in robotics. Almost incidentally, this chapter also indicates how in the line space $\mathbb{R}^{3,3}$, projective transformations become representable as rotations. Since this enables projective transformations to be encoded as rotors, this is a potentially very important development to the applicability of geometric algebra to computer vision and computer graphics.

In Chap. 14: *A Framework for n-Dimensional Visibility Computations*, the authors solve the long-standing problem of computing exact mutual visibility between shapes, as required in soft shading rendering for computer graphics. It had been known that a Plücker-coordinate-based approach in the manifold of lines offered some representational clarity but did not lead to efficient solutions. However, the authors show that when the full bivector space $\bigwedge^2(\mathbb{R}^{n+1})$ is employed, visibility computations reduce to a convex hull determination, even in n-D. They can then be implemented using standard software for CSG (Computational Solid Geometry).

Part VI: Alternatives to Conformal Geometric Algebra

The 3D conformal geometric algebra $\mathbb{R}_{4,1}$ is five-dimensional and often feels like a slight overkill for the description of rigid body motion and other limited geometries. This part presents several four-dimensional alternatives for the applications we saw in Part I.

Embedding the common homogeneous coordinates into geometric algebra begs the question of what metric properties to assign to the extra representational dimension. Naive use of a nondegenerate metric prevents encoding rigid body motions as orthogonal transformations in a 4D space. Chapter 15: *On the Homogeneous Model of Euclidean Geometry* updates results from classical 19th century work to modern notation and shows that by endowing the dual homogeneous space with a specific degenerate metric (to produce the algebra $\mathbb{R}^*_{3,0,1}$) one can in fact achieve this. The chapter reads like a tutorial introduction to this framework, presented as a complete and compact representation of Euclidean geometry, kinematics and rigid body dynamics.

To some in computer graphics, the 32-dimensional conformal geometric algebra $\mathbb{R}_{4,1}$ is just too forbidding, and they have been looking for simpler geometric algebras to encode their needs. Chapter 16: *A Homogeneous Model for Three-Dimensional Computer Graphics Based on the Clifford Algebra for* $\mathbb{R}^3$ shows that a representation of some operations required for computer graphics (rotations and perspective projections) can be achieved by rather ingenious use of $\mathbb{R}_3$ (the Euclidean geometric algebra of 3-D space) by using its trivector to model mass points.

In Chap. 17: *Rigid-Body Transforms Using Symbolic Infinitesimals*, an alternative 4D geometric algebra is proposed to capture the structure of rigid body motions. It is nonstandard in the sense that one of the basis vectors squares to infinity. The authors show how this models Euclidean isometries. They then apply their algebra to Bezier and B-spline interpolation of rigid body motions, through methods that can be transferred to more standard algebraic models such as conformal geometric algebra.

Chapter 18: *Rigid Body Dynamics in a Constant Curvature Space and the '1D-up' Approach to Conformal Geometric Algebra* proposes yet another way to representing 3D rigid body motion in the geometric algebra of a 4D space. It takes the unusual approach of viewing Euclidean geometry as a somewhat awkward limit case of a constant curvature space and analyzes such spaces first. The Lagrangian dynamics equations take on an elegant form and lead to the surprising view of translational motion in real 3D space as a fast precession in the 4D representational space. The author then compares this approach to that of Chap. 15, after first embedding that into conformal geometric algebra; and the flat-space limit to the Euclidean case is shown to be related to Chap. 17. Thus all those 1D up alternative representations of rigid body motions are shown to be closely related.

Part VII: Towards Coordinate-Free Differential Geometry

Differential geometry is an obvious target for geometric algebra. In its classical description by means of coordinate charts, its structure easily gets hidden in notation, and that limits its applications to specialized fields. Geometric algebra should be able to do better, especially if combined with modern insights in the system of geometrical invariants.

Chapter 19: *The Shape of Differential Geometry in Geometric Calculus* shows how geometric algebra can offer a direct notation in terms of clear concepts such as the tangent volume element ('pseudoscalar'), attached at all locations of a vector manifold, and in terms of its derivative as the codification of 'shape' in all its aspects. The coordinate-free formulation can always be made specific for any chosen coordinates and is hence computational. The chapter ends with open questions, intended as suggestions for research projects. The editors are grateful to have this thought-provoking contribution by David Hestenes, the grandfather of geometric algebra.

The field of "moving frames" has developed rapidly in the past decade, and structured algorithmic methods are emerging to produce invariants and their syzygy relationships for Lie groups. We have invited expert Elisabeth Mansfield, in Chap. 20: *On the Modern Notion of Moving Frames*, to write an introduction to this new field, since we believe that its concretely abstract description should be a quite natural entry to formulate invariants for the Lie groups occurring in geometric algebra. Besides an introductory overview with illustrative examples and detailed pointers to current literature, the chapter contains a first attempt to compute moving frames for $SE(3)$ in conformal geometric algebra.

Part VIII: Tutorial Appendix

In Chap. 21: *Tutorial Appendix: Structure Preserving Representation of Euclidean Motions through Conformal Geometric Algebra*, we provide a self-contained tutorial to the basics of geometric algebra in the conformal model.

Leo Dorst
Joan Lasenby

Contents

Contributors

Nawar Alwesh Industrial Research Limited, Auckland, New Zealand, n.alwesh@irl.cri.nz

Nabil Anwer LURPA, École Normale Supérieure de Cachan, Cachan, France, anwer@lurpa.ens-cachan.fr

Andreas Aristidou Department of Engineering, University of Cambridge, Trumpington Street, Cambridge CB2 1PZ, UK, aa462@cam.ac.uk

Lilian Aveneau XLIM/SIC, CNRS, University of Poitiers, Poitiers, France, lilian.aveneau@xlim.fr

Eduardo Bayro-Corrochano Campus Guadalajara, CINVESTAV, Jalisco, Mexico, edb@gdl.cinvestav.mx

Liam Candy The Council for Scientific and Industrial Research (CSIR), Meiring Naude Rd, Pretoria, South Africa, lcandy@csir.co.za

Yuanhao Cao Key Laboratory of Mathematics Mechanization, Academy of Mathematics and Systems Science, Chinese Academy of Sciences, Beijing 100190, P.R. China, ppxhappy@126.com

Sylvain Charneau XLIM/SIC, CNRS, University of Poitiers, Poitiers, France, sylvain.charneau@xlim.fr

Chris Doran Sidney Sussex College, University of Cambridge and Geomerics Ltd., Cambridge, UK, chris.doran@geomerics.com

Leo Dorst Intelligent Systems Laboratory, University of Amsterdam, Amsterdam, The Netherlands, l.dorst@uva.nl

Daniel Fontijne Euvision Technologies, Amsterdam, The Netherlands, d.fontijne@euvt.eu

Laurent Fuchs XLIM/SIC, CNRS, University of Poitiers, Poitiers, France, laurent.fuchs@xlim.fr

Ron Goldman Department of Computer Science, Rice University, Houston, TX 77005, USA, rng@rice.edu

Charles Gunn DFG-Forschungszentrum Matheon, MA 8-3, Technisches Universität Berlin, Str. des 17. Juni 136, 10623 Berlin, Germany, gunn@math.tu-berlin.de

David Hestenes Arizona State University, Tempe, AZ, USA, hestenes@asu.edu

Anthony Lasenby Cavendish Laboratory and Kavli Institute for Cosmology, University of Cambridge, Cambridge, UK, a.n.lasenby@mrao.cam.ac.uk

Joan Lasenby Department of Engineering, University of Cambridge, Trumpington Street, Cambridge CB2 1PZ, UK, jl221@cam.ac.uk

Robert Lasenby Department of Applied Mathematics and Theoretical Physics, University of Cambridge, Cambridge, UK, robert@lasenby.org

Hongbo Li Key Laboratory of Mathematics Mechanization, Academy of Mathematics and Systems Science, Chinese Academy of Sciences, Beijing 100190, P.R. China, hli@mmrc.iss.ac.cn

Elizabeth Mansfield School of Mathematics, Statistics and Actuarial Science, University of Kent, Canterbury CT2 7NF, UK, E.L.Mansfield@kent.ac.uk

Laurent Mascarilla Laboratory of Mathematics, Images and Applications, University of La Rochelle, La Rochelle, France, laurent.mascarilla@univ-lr.fr

Jose Mennesson Laboratory of Mathematics, Images and Applications, University of La Rochelle, La Rochelle, France, jose.mennesson@univ-lr.fr

Frederic Mora XLIM/SIC, CNRS, University of Limoges, Limoges, France, frederic.mora@xlim.fr

Eduardo Ulises Moya-Sánchez Campus Guadalajara, CINVESTAV, Jalisco, Mexico, emoya@gdl.cinvestav.mx

Glen Mullineux Innovative Design and Manufacturing Research Centre, Department of Mechanical Engineering, University of Bath, Bath BA2 7AY, UK, g.mullineux@bath.ac.uk

Christophe Saint-Jean Laboratory of Mathematics, Images and Applications, University of La Rochelle, La Rochelle, France, christophe.saint-jean@univ-lr.fr

René Schott IECN and LORIA, Nancy Université, Université Henri Poincaré, BP 239, 54506 Vandoeuvre-lès-Nancy, France, schott@loria.fr

Philippe Serré LISMMA, Institut Supérieur de Mécanique de Paris, Paris, France, philippe.serre@supmeca.fr

Leon Simpson Innovative Design and Manufacturing Research Centre, Department of Mechanical Engineering, University of Bath, Bath BA2 7AY, UK, l.c.simpson@bath.ac.uk

G. Stacey Staples Department of Mathematics and Statistics, Southern Illinois University Edwardsville, Edwardsville, IL 62026-1653, USA, sstaple@siue.edu

Robert Valkenburg Industrial Research Limited, Auckland, New Zealand, r.valkenburg@irl.cri.nz

JianXin Yang Robotics and Machine Dynamics Laboratory, Beijing University of Technology, Beijing, P.R. China, yangjx@tsinghua.org.cn

Lixian Zhang Key Laboratory of Mathematics Mechanization, Academy of Mathematics and Systems Science, Chinese Academy of Sciences, Beijing 100190, P.R. China, shadowfly12@126.com

Jun Zhao School of Mathematics, Statistics and Actuarial Science, University of Kent, Canterbury CT2 7NF, UK, J.Zhao-73@kent.ac.uk

Part I
Rigid Body Motion

The treatment of rigid body motion is the first algebraically advanced topic that the geometry of Nature forces upon us. Since it was the first to be treated, it shaped the field of geometric representation; but now we can repay our debt by using modern geometric representations to provide more effective computation methods for motions. All chapters in this part use conformal geometric algebra to great advantage in compactness and efficiency.

Rigid Body Dynamics and Conformal Geometric Algebra

1

Anthony Lasenby, Robert Lasenby, and Chris Doran

Abstract

We discuss a fully covariant Lagrangian-based description of 3D rigid body motion, employing spinors in 5D conformal space. The use of this space enables the translational and rotational degrees of freedom of the body to be expressed via a unified rotor structure, and the equations of motion in terms of a generalised 'moment of inertia tensor' are given. The development includes the effects of external forces and torques on the body. To illustrate its practical applications, we give a brief overview of a prototype multi-rigid-body physics engine implemented using 5D conformal objects as the variables.

1.1 Introduction

Rigid body dynamics is an area which has been worked on for many years and has sparked many new developments in mathematics, but for most people, it is still a conceptually difficult subject, with many features which are non-intuitive.

By using Geometric Algebra, a more intuitive and conceptually appealing framework is possible, and the success of its application to this case is a very good illustration of the power of Geometric Algebra in physical and engineering problems.

A. Lasenby (✉)
Cavendish Laboratory and Kavli Institute for Cosmology, University of Cambridge, Cambridge, UK
e-mail: a.n.lasenby@mrao.cam.ac.uk

R. Lasenby
Department of Applied Mathematics and Theoretical Physics, University of Cambridge, Cambridge, UK
e-mail: robert@lasenby.org

C. Doran
Sidney Sussex College, University of Cambridge and Geomerics Ltd., Cambridge, UK
e-mail: chris.doran@geomerics.com

L. Dorst, J. Lasenby (eds.), *Guide to Geometric Algebra in Practice*,
DOI 10.1007/978-0-85729-811-9_1, © Springer-Verlag London Limited 2011

This extends to both the dynamical side, the Euler equations, which look simpler and more straightforward in the GA approach, and to kinematics, where the introduction of a reference body, which is then rotated and translated to the actual body in space, is a great help to picturing what is going on. The latter might be thought to have nothing intrinsically to do with GA, but in fact there is a close parallel with the geometrical picture which GA gives of quantum mechanics. Here Pauli or Dirac spinors can be viewed as instructions on how to rotate a 'fiducial' frame of vectors to a new frame of vectors (including the particle spin and velocity) attached to the particle itself. The rigid body equivalent of this is of a 'body' frame of vectors which is rotated to give the 'space' frame of vectors defining the orientation of the body at its actual location.

This GA approach is undoubtedly very successful in relation to the rotational degrees of freedom inherent in this process, since these can be described by time-dependent 3D rotors. The aim in the following contribution is to show how the successes of the rotor treatment can be extended to include the translational degrees of freedom as well, via use of the Conformal Geometric Algebra (CGA). In this approach, a single 5D rotor describes both the rotation and translation away from the reference body, resulting in a conceptual unification of these two sets of degrees of freedom. Such unifications have already been described by David Hestenes and others (see e.g. [4]). A novelty here is that we will seek a *covariant* formulation based on a Lagrangian action principle. This should guarantee that all our equations are physically meaningful and mutually consistent.

This development will be taken through to the point of considering interactions of multiple bodies, and we sketch the implementation of a CGA version of the *Fast Frictional Dynamics* (*FFD*) approach [6] in which the equations of motion for large numbers of such bodies, mutually in contact, can be integrated in real time for graphical display.

We start this contribution with a description of the 3D GA approach to rigid body motion already referred to and then show how it may be extended to 5D CGA with the translation and rotational d.o.f. treated on an equal basis. We then discuss some of the issues that must be addressed in implementing this method in a specific algorithm—principally how the rotor update equations are carried out—and look briefly at results obtained in a CGA implementation of the FFD algorithm.

1.2   Rigid Body Dynamics in 3D

We start by looking at the treatment of rigid body motion in ordinary 3D GA. (For a fuller description, see Chap. 3 of [3].)

1.2.1   Kinematics

For a rigid body moving through space, the position $y(t)$ of a particular point can be specified by giving the location x of that point when the body is in some reference configuration (conventionally with origin at the body's centre of mass) and then

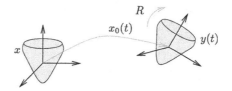

Fig. 1.1 The configuration of a rigid body is described by the location x_0 of its centre of mass, and the orientation of the body relative to some reference configuration, parameterised by the rotor R

describing the transformation from the reference configuration to the actual configuration. This transformation can be expressed in terms of a rotor $R(t)$ which rotates the reference body to the orientation of the actual body, and the position $x_0(t)$ of the body's centre of mass (see Fig. 1.1).

Using this description, $y(t)$ is given by

$$y(t) = R(t)x\tilde{R}(t) + x_0(t) \tag{1.1}$$

Differentiating, the velocity of a point in the body $v(t) = \dot{y}(t)$ is given by

$$\dot{y}(t) = \dot{R}x\tilde{R} + Rx\dot{\tilde{R}} + \dot{x}_0 \tag{1.2}$$

Since $R\tilde{R} = 1$, we have $\dot{R}\tilde{R} + R\dot{\tilde{R}} = 0$, so

$$\dot{y}(t) = (\dot{R}\tilde{R})Rx\tilde{R} + Rx\tilde{R}(R\dot{\tilde{R}}) + \dot{x}_0 \tag{1.3}$$
$$= (\dot{R}\tilde{R})Rx\tilde{R} - Rx\tilde{R}(\dot{R}\tilde{R}) + \dot{x}_0 \tag{1.4}$$
$$= (Rx\tilde{R}) \cdot \Omega + \dot{x}_0 \tag{1.5}$$

where $\Omega = -2\dot{R}\tilde{R}$ is the angular velocity bivector. Defining the body angular velocity $\Omega_B = \tilde{R}\Omega R$, i.e. the spatial-frame angular velocity rotated back to the reference frame, we have the alternative expression

$$\dot{y}(t) = R(x \cdot \Omega_B)\tilde{R} + \dot{x}_0 \tag{1.6}$$

1.2.2   Dynamics

To investigate the dynamics, we need to consider the angular momentum of the body. Letting $\rho(x)$ be the density of the body at location x in the reference configuration, and $y(x, t)$ and $v(x, t)$ be the corresponding spatial-configuration position and velocity, the angular momentum bivector L is defined by

$$L(t) = \int d^3x\, \rho(x)\big(y(x, t) - x_0\big) \wedge v(x, t)$$

$$= \int d^3x\, \rho(x)(Rx\tilde{R}) \wedge (Rx \cdot \Omega_B \tilde{R} + v_0)$$

$$= R\left(\int d^3x\, \rho(x)x \wedge (x \cdot \Omega_B)\right)\tilde{R} \tag{1.7}$$

where $v_0 = \dot{x}_0$, and the v_0 term disappears by the definition of centre of mass.

From this we extract the inertia tensor

$$I(B) = \int d^3x\, \rho(x)x \wedge (x \cdot B) \tag{1.8}$$

and the angular momentum is obtained by rotating the body angular momentum to the space frame:

$$L = RI(\Omega_B)\tilde{R}$$

From its definition we see that $I(B)$ is a linear function mapping bivectors to bivectors.

The dynamical equation of motion is $\dot{L} = T$, where T is the bivector torque in the spatial frame, i.e. a force f applied at location y gives a torque $(y - x_0) \wedge f$. Evaluating this, we have

$$\begin{aligned} \dot{L} &= \dot{R}I(\Omega_B)\tilde{R} + RI(\Omega_B)\dot{\tilde{R}} + RI(\dot{\Omega}_B)\tilde{R} \\ &= R\left[I(\dot{\Omega}_B) - \frac{1}{2}\Omega_B I(\Omega_B) + \frac{1}{2}I(\Omega_B)\Omega_B\right]\tilde{R} \\ &= R\left[I(\dot{\Omega}_B) - \Omega_B \times I(\Omega_B)\right]\tilde{R} \end{aligned} \tag{1.9}$$

Rotating back to the body frame, we have

$$I(\dot{\Omega}_B) - \Omega_B \times I(\Omega_B) = \tilde{R}TR \tag{1.10}$$

which is the GA version of the Euler equations.

As an illustration of the power of the rotor approach, we can look at the solution of these equations in the torque-free case for a symmetric top. We suppose that $i_1 = i_2$ are the two equal moments of inertia, and i_3 is the moment of inertia about the symmetry axis. In this case the solution, which you are asked to verify in the Exercises, is

$$R(t) = \exp\left(-\frac{1}{2}i_1^{-1}Lt\right)R(0)\exp\left(-\frac{1}{2}\omega_3(1 - i_3/i_1)Ie_3 t\right) \tag{1.11}$$

This corresponds to an 'internal' rotation in the $e_1 e_2$ plane (a symmetry of the body), followed by a rotation in the angular-momentum plane. The rotor sandwiched inbetween, $R(0)$, corresponds to the initial orientation of the body.

1.3 Rigid Body Dynamics in 5D CGA

1.3.1 Introduction

We now pass to the Conformal Geometric Algebra (CGA) approach to rigid body dynamics. The idea here is to work in an overall space that is two dimensions higher than the base space, using the usual conformal Euclidean setup. The penalty for doing this, i.e. using a Euclidean setup, is that the number of degrees of freedom is not properly matched to the problem in hand, and we have to introduce additional Lagrange multipliers to cope with this. The alternative, which is also worth exploring and is discussed in Chap. 18, is to use a 1D-up approach, where there is a much better match in terms of degrees of freedom, but where we have to work in a curved space.

So our setup is the standard one with three positive square basis vectors, e_1, e_2, e_3, with two extra vectors adjoined, e and $\bar{e}$, satisfying $e^2 = +1$, $\bar{e}^2 = -1$. From these we form the null vectors $n = e + \bar{e}$ and $\bar{n} = e - \bar{e}$. The representation function we will use has the origin as $-\bar{n}/2$, i.e.[1]

$$X = \frac{1}{2\lambda^2}\left(x^2 n_\infty + 2\lambda x - \lambda^2 \bar{n}\right) \tag{1.12}$$

where x is the ordinary position vector in Euclidean 3-space, and λ is a constant with dimensions of length (included to render X dimensionless). We note that X so defined is covariantly normalised since it satisfies $X \cdot n_\infty = -1$.

The overall idea is to set up a Lagrangian which is covariant with respect to the 5D geometry (that is, which is invariant under appropriate 5D rotor transformations of the 5D vectors and spinors involved) but for which the energy is just the ordinary 3D rigid body energy. Then we should get equations of motion which are correct at the 3D level but covariantly expressed in 5D. Additionally, the way we will express the current configuration of the rigid body is via a combined rotation/translation rotor, so that translations and rotations are integrated as much as possible. Specifically, the configuration is given by an element of the Euclidean group, which corresponds to the subgroup of 5D rotors which preserve n_∞. Any such rotor can be decomposed as

$$R(t) = R_1(t)R_2(t) \tag{1.13}$$

---

[1] *Editorial note*: This notation is used in [3]. Compared to the notation in the tutorial appendix (Chap. 21), Eq. (21.1), by setting $\lambda = 1$, we find that $n = n_\infty$ and $\bar{n} = -2n_o$. Correspondingly, $e = \frac{1}{2}(n + \bar{n}) = -n_o + \frac{1}{2}n_\infty = -\sigma_+$ and $\bar{e} = \frac{1}{2}(n - \bar{n}) = n_o + \frac{1}{2}n_\infty = \sigma_-$. Note that $\bar{n} \cdot n = 2$ corresponds to $n_o \cdot n_\infty = -1$. As a compromise between the notation in this book and [3] that avoids awkward factors, in this chapter we will use $\bar{n}$ but replace n by n_∞.

where $R_2(t)$ is an ordinary 3D rotor, giving the *attitude* of the body, and $R_1(t)$ is a translation rotor, which takes the origin of the reference copy of the body, presumed to be the centre of mass, to where the centre of mass is actually located at time t. Explicitly,

$$R_1 = 1 - \frac{1}{2\lambda} x_c(t) n_\infty \tag{1.14}$$

where $x_c(t)$ is the 3D position of the centre of mass at time t.

If X_{ref} represents the position of a point in the reference copy of the body, then the null vector corresponding to the actual position of this point is

$$X = R X_{\text{ref}} \tilde{R} \tag{1.15}$$

We now proceed in analogy to the usual GA treatment of rigid body dynamics in 3D. The *body angular velocity* bivector is defined by

$$\Omega_B = -2\tilde{R}\dot{R} \tag{1.16}$$

Using this, we get that the time derivative of the 5D null vector of position is

$$\dot{X} = R \, X_{\text{ref}} \cdot \Omega_B \, \tilde{R} \tag{1.17}$$

which means

$$\dot{X}^2 = (X_{\text{ref}} \cdot \Omega_B)^2 \tag{1.18}$$

Now evaluating $\dot{X}$ explicitly, we have

$$\dot{X} = \frac{1}{2\lambda^2} (2x \cdot \dot{x} n_\infty + 2\lambda \dot{x}) \tag{1.19}$$

and therefore

$$\dot{X}^2 = \frac{\dot{x}^2}{\lambda^2} \tag{1.20}$$

This has therefore returned the 3D velocity squared, which is what we need to evaluate the kinetic energy. We can thus write

$$\mathcal{T} = \frac{\lambda^2}{2} \int d^3 x_b \, \rho (X_{\text{ref}} \cdot \Omega_B)^2 \tag{1.21}$$

where the integration is over the 3D reference body, and ρ is the density (in general a function of 3D position). Now

$$(X_{\text{ref}} \cdot \Omega_B)^2 = -\Omega_B \cdot \left(X_{\text{ref}} \wedge (X_{\text{ref}} \cdot \Omega_B) \right) \tag{1.22}$$

and we thus have

$$\mathscr{T} = -\frac{1}{2}\Omega_B \cdot I(\Omega_B) \tag{1.23}$$

where

$$I(\Omega_B) = \lambda^2 \int d^3x_b \, \rho X_{\text{ref}} \wedge (X_{\text{ref}} \cdot \Omega_B) \tag{1.24}$$

This is therefore our version of the moment of inertia tensor. Calculating its components, if B is a spatial bivector, then

$$I(B) = \frac{1}{4\lambda^2} \int d^3x \, \rho \left(x^2 n_\infty + 2\lambda x - \lambda^2 \bar{n}\right)$$
$$\wedge \left(\left(x^2 n_\infty + 2\lambda x - \lambda^2 \bar{n}\right) \cdot B\right) \tag{1.25}$$

$$= \frac{1}{4\lambda^2} \int d^3x \, \rho \left(x^2 n_\infty + 2\lambda x - \lambda^2 \bar{n}\right) \wedge (2\lambda x \cdot B) \tag{1.26}$$

$$= \frac{2}{4\lambda} \int d^3x \, \rho \left(x^2 n_\infty x \cdot B + 2\lambda x \wedge (x \cdot B) - \lambda^2 \bar{n} x \cdot B\right) \tag{1.27}$$

$$= I_{\text{rot}}(B) + \frac{1}{4\lambda^2} n_\infty \int d^3x \, \rho \, x^2 x \cdot B \tag{1.28}$$

where I_{rot} is the 3D rotational moment of intertia, and if a is a spatial vector, then

$$I(n_\infty a) = \frac{1}{4\lambda^2} \int d^3x \, \rho \left(x^2 n_\infty + 2\lambda x - \lambda^2 \bar{n}\right)$$
$$\wedge \left(\left(x^2 n_\infty + 2\lambda x - \lambda^2 \bar{n}\right) \cdot (n_\infty a)\right) \tag{1.29}$$

$$= \frac{1}{4\lambda^2} \int d^3x \, \rho \left(x^2 n_\infty + 2\lambda x - \lambda^2 \bar{n}\right)$$
$$\wedge \left(-2\lambda n_\infty x \cdot a - 2\lambda^2 a\right) \tag{1.30}$$

$$= -\frac{1}{2\lambda} \int d^3x \, \rho \left(\lambda x^2 n_\infty a - 2\lambda n_\infty x(x \cdot a)\right.$$
$$\left. + 4\lambda^2 x \wedge a - \lambda^2 \bar{n} \wedge n_\infty (x \cdot a) - \lambda^3 \bar{n} a\right) \tag{1.31}$$

$$= -\frac{1}{2} M\lambda^2 \bar{n} a - \frac{1}{2} n_\infty a \int d^3x \, \rho x^2 - \frac{1}{2} n_\infty \int d^3x \, \rho x(x \cdot a) \tag{1.32}$$

where $M = \int d^3x \, \rho$ is the total mass of the body.

Note that, since R must be a Euclidean rotor, i.e. $Rn_\infty \tilde{R} = n_\infty$, we have $0 = \frac{d}{dt}n_\infty = \frac{d}{dt}(Rn_\infty \tilde{R}) = Rn_\infty \cdot \Omega_B \tilde{R}$, and thus $n_\infty \cdot \Omega_B = 0$. Therefore Ω_B must be of the form $B + n_\infty a$ for some spatial bivector B and spatial vector a (we could also have come to this conclusion by considering the explicit $R = R_1 R_2$ expression), and consequently the above-evaluated arguments for I are the only ones of interest.

Below, we will see that the main quantity of interest in the equations of motion is $I(\Omega_B) \wedge n_\infty$, which has the simple form

$$I(B + n_\infty a) \wedge n_\infty = I_{\text{rot}}(B)n_\infty + \frac{1}{2}M\lambda^2 a\bar{n} \wedge n_\infty \tag{1.33}$$

1.3.2 Setting up the Lagrangian

We now wish to set up our 5D Lagrangian, working in analogy with the treatment starting on page 425 of [3]. There, we worked with general 3D spinors ψ, which were kept on the rotor manifold via use of a term $\mu(\psi\tilde{\psi} - 1)$ in the Lagrangian. (μ is used instead of the λ in the book, to avoid confusion with our length scale λ.)

Because we are working with 5D spinors, we need some further restrictions to ensure that we are working with a rotor. In addition, we only want to include rotors of a specific type, namely those that preserve the point at infinity n_∞, since these correspond to the rigid body motions we are considering. Our full Lagrangian is thus taken as

$$\mathcal{L} = \left\langle -\frac{1}{2}\Omega_B \cdot I(\Omega_B) - \mu(\psi\tilde{\psi} - 1) - IU\psi\tilde{\psi} - V(\psi n_\infty \tilde{\psi} - n_\infty) \right\rangle \tag{1.34}$$

where μ is a scalar, I the 5D pseudoscalar, and U and V are general 5D vectors.

Here Ω_B is taken to be defined in terms of the (general) spinor ψ as follows:

$$\Omega_B = -\tilde{\psi}\dot{\psi} + \dot{\tilde{\psi}}\psi \tag{1.35}$$

which can be seen to be always a bivector.

The function of the general grade 4 Lagrange multiplier IU is to ensure that the grade 4 part of $\psi\tilde{\psi}$ is zero, and the function of V is to restrict to rigid body motions. With all these restrictions in place, enforced by the multipliers, a general ψ turns into a rotor of the desired form.

1.3.3 The Equations of Motion and Conservation Laws

We employ the usual techniques (see e.g. [8]) to get the equations of motion. These are

$$I(\dot{\Omega}_B) - \Omega_B I(\Omega_B) = \mu + I\tilde{\psi}U\psi + \tilde{\psi}V\psi n_\infty \tag{1.36}$$

Assuming that ψ is of the desired form and restricting to the various grades, we have

$$\begin{array}{ll} -\Omega_B \cdot I(\Omega_B) = \mu + V \cdot n_\infty & \text{scalar part} \\ I(\dot{\Omega}_B) - \Omega_B \times I(\Omega_B) = \tilde{\psi}(V \wedge n_\infty)\psi & \text{bivector part} \\ -\Omega_B \wedge I(\Omega_B) = \tilde{\psi}IU\psi & \text{grade 4 part} \end{array} \tag{1.37}$$

The middle one of these is the direct analogue of the Euler equations. It shows us that the effect of the Lagrange multiplier which is enforcing the restriction to rigid body motions, is to introduce the 'space torque' $V \wedge n_\infty$, which appears in the Euler equation back-rotated to the body frame.

If we employ the techniques in [8] to find the angular momentum and the conservation law it satisfies, then under the change

$$\psi \mapsto e^{\alpha B/2}\psi, \quad V \mapsto e^{\alpha B/2}Ve^{-\alpha B/2}, \quad U \mapsto e^{\alpha B/2}Ue^{-\alpha B/2} \quad (1.38)$$

where α is a scalar, and B is a constant bivector, we find that all the terms in the Lagrangian are invariant except the final one, $V \cdot n_\infty$. This leads to a non-conservation of angular momentum, with the result

$$\frac{dL}{dt} = V \wedge n_\infty \quad (1.39)$$

as already pre-figured in the identification of $V \wedge n_\infty$ as a torque operating on the system. The angular momentum itself is given by

$$L = \psi I(\Omega_B)\tilde{\psi} \quad (1.40)$$

in the usual way.

The conserved Hamiltonian is found to be simply the kinetic energy

$$H = -\frac{1}{2}\Omega_B \cdot I(\Omega_B) \quad (1.41)$$

1.3.4 Solving the Equations

The appearance of V in the main Euler equation is awkward, firstly since we do not have a dynamical equation for it, and secondly since it introduces an explicit ψ in what should be the bivector update equation. A solution of this is to wedge both sides of the bivector equation with n_∞. This yields

$$\left(I(\dot{\Omega}_B) - \Omega_B \times I(\Omega_B)\right) \wedge n_\infty = 0 \quad (1.42)$$

It is this equation which appears to give the required dynamics most quickly, since it contains neither V nor ψ.

1.3.4.1 Counting Arguments

To see that even back at the level of the bivector equation we in fact have all the necessary information for solution, we can use a counting argument.

Let the base space have dimension m. Then Ω_B has $m(m-1)/2$ ordinary rotation bivectors in it, plus m translation bivectors, making $m(m+1)/2$ in total. V meanwhile, though a general vector in $m+2$ dimensions, only has $m+1$ effective components, since we are never going to recover the n_∞ component of V—this is irrelevant to all the dynamics, and without loss of generality we can set it to zero. The number of quantities we have to recover from the bivector equation is thus

$$\frac{m(m+1)}{2} + (m+1) = \frac{(m+1)(m+2)}{2} \tag{1.43}$$

This is equal to the number of bivector components in the overall $(m+2)$-dimensional space, so we do indeed have enough information to separately recover the dynamics we want—encoded in Ω_B—and V. However, Eq. (1.42) provides a more direct way of updating Ω_B, without reference to V or ψ.

1.3.5 An Example—The Dumbbell

To make the above concrete, we consider a simple example, a 2D dumbbell. This is taken to consist of unit mass points at $x = \pm x_0$, $y = z = 0$, and be moving just in the (x, y) plane. The inertia tensor is then just

$$I(\Omega_B) = \lambda^2 \big(X_1 \wedge (X_1 \cdot \Omega_B) + X_2 \wedge (X_2 \cdot \Omega_B) \big) \tag{1.44}$$

where $X_1 = X(x = x_0, y = 0, z = 0)$ and $X_2 = X(x = -x_0, y = 0, z = 0)$.

As an example of how one could proceed, we parameterise Ω_B as

$$\Omega_B = f(t)e_1e_2 + g(t)e_1n_\infty + h(t)e_2n_\infty \tag{1.45}$$

which is the most general angular velocity bivector compatible with the constraints on ψ and the fact the motion is two-dimensional.

Equation (1.42) reads

$$x_0^2 \dot{f} e_1e_2n_\infty - \lambda^2 \big((-\dot{g} + fh)e_1 + (\dot{h} + fg)e_2 \big) e\bar{e} = 0 \tag{1.46}$$

From this we can read off the solution $f = \text{const}$ and

$$g(t) = c_1 \cos(ft) + c_2 \sin(ft), \quad h(t) = c_2 \cos(ft) - c_1 \sin(ft) \tag{1.47}$$

which are the correct 'angular velocities' for a body moving with constant translational velocity $(\lambda c_1, \lambda c_2)$ and rotating at constant rate f.

Going back to find the vector Lagrange multiplier V for this case, one finds that its $\bar{n}$ component is constant and equal to minus the translational kinetic energy. The spatial part carries out a type of helical motion, at twice the underlying frequency f, looking rather like the 'zitterbewegung' which accompanies motion of the electron when considered relativistically (see e.g. [5]). A plot for a case with some illustrative values is shown in Fig. 1.2.

The other vector Lagrange multiplier U is not relevant to this case, since it turns out $\Omega_B \wedge I(\Omega_B)$ is zero here.

Fig. 1.2 Plot of the locus of the spatial part of the Lagrange multiplier V for the 2D dumbbell case. $(V_1(t), V_2(t))$ is plotted against t. The $\bar{n}$ component of V is constant and equal to minus the translational kinetic energy

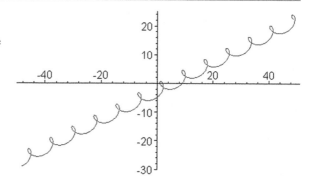

1.3.6 Including Moments and Forces

It is obviously an aim of the method that moments and forces should be included in a unified way. A force acting at a point of contact has two actions: firstly to accelerate the centre of mass—this effect does not depend on the point of contact. Secondly it creates a moment, which does depend on the point of contact. Ideally both these effects would come from a single 'moment' in the higher space, and also whatever expression is involved should be covariant.

The following shows how both these aims can in fact be realised, in a simple way. To demonstrate covariance, we need an extra concept first, however, in order to properly represent forces. This is the concept of *boundary points* in Euclidean space. It will turn out that the wedging with n_∞ that occurs in the final equations renders the distinction between these and ordinary Euclidean 3-vectors somewhat redundant, but the conceptual unification afforded is useful, and the covariance is not properly demonstrated without them, so we will follow this route.

The notion of boundary points in Euclidean space is introduced and discussed in the paper [7]. The idea is basically to be able to represent 'directions', or 'free vectors', in a covariant way. It entails enlarging the class of representative points to include all 5D vectors Y (not just the null vectors X representing position), but with the covariant requirement imposed that $Y \cdot n_\infty = 0$. They are thus the Euclidean analogue of the boundary of the Poincaré sphere, which are the set of points such that $Y \cdot e = 0$. In the same way that in the non-Euclidean space, we found that the boundary points represented momenta; here we also expect them to represent momenta, but in Euclidean space. Taking the derivative with respect to time, our covariant notion of a force is therefore an F satisfying $F \cdot n_\infty = 0$. Specifically we write

$$F = \underline{f} + \alpha n_\infty \tag{1.48}$$

where $\underline{f}$ is the normal Euclidean 3D force vector, and α is a multiplier, which in fact we will never need to determine, due to the final wedging with n_∞. The key point is that under our rotor transformation ψ, we find (taking the back transformation for reasons that will be clear shortly)

$$\tilde{\psi} F \psi = \tilde{R}_2 f R_2 + \beta n_\infty \tag{1.49}$$

where β is a combination of α together with the scalar product of f and the vector through which R_1 translates. Thus this back-transformed F is also a boundary point (and therefore represents a force still), which has merely had its 3D direction rotated, and with a different multiple of n_∞ tacked on. We can see now why we need to add a multiple of n_∞ to the 3-force to make a covariant object—while we could put the multiple to zero for a given force, transforming to a different frame induces a non-zero n_∞ component.

Our fundamental equation is as follows. For a force F acting on a body at the contact position given by X^{cont} (note this is on the surface of the real body, not the reference body), the total effect of this force is to produce the 'moment'

$$M_{\mathrm{space}} = \lambda\, X^{\mathrm{cont}} \wedge F \tag{1.50}$$

Back-rotating this to the body frame, we see that the moment to be added to the right-hand side of the bivector equation in (1.37) is

$$M_{\mathrm{body}} = \lambda\, X^{\mathrm{cont}}_{\mathrm{ref}} \wedge (\tilde{\psi} F \psi) \tag{1.51}$$

Note that if the force is actually specified in the body frame (e.g. at constant angle relative to the body), and of value F_{body}, then the body moment is even simpler and is just equal to

$$M_{\mathrm{body}} = \lambda\, X^{\mathrm{cont}}_{\mathrm{ref}} \wedge F_{\mathrm{body}} \tag{1.52}$$

In terms of Eq. (1.42), we just need to add into the rhs, either of the last two quantities wedged with n_∞. This will of course kill off the n_∞ parts of the force, which is why we do not need to determine them in practice. Thus, e.g. in the case where the force is specified in the body frame, we get

$$M_{\mathrm{body}} \wedge n_\infty = \lambda\, X^{\mathrm{cont}}_{\mathrm{ref}} \wedge f_{\mathrm{body}} \wedge n_\infty \tag{1.53}$$

The above is an assertion; we now need to see why it works. Consider the rhs of the last equation. Decomposing X_{ref} into its constituent parts, we have

$$\lambda\, X^{\mathrm{cont}}_{\mathrm{ref}} \wedge f_{\mathrm{body}} \wedge n_\infty = \frac{\lambda(2\lambda x_{\mathrm{ref}} - \lambda^2 \bar{n})}{2\lambda^2} \wedge f_{\mathrm{body}} \wedge n_\infty$$
$$= x_{\mathrm{ref}} \wedge f_{\mathrm{body}}\, n_\infty + \lambda f_{\mathrm{body}} e\bar{e} \tag{1.54}$$

This is the spatial torque on the body times n_∞, plus the spatial force on the body times $e\bar{e}$. The spatial force in the second term appears multiplied by λ, so that it itself is like a moment and can be added (dimensionally) to the first term.

Rotating back to the space frame and taking the integral with respect to time of this equation, we would conclude (schematically—all in the space frame):

$$\text{(Total 5D angular momentum)} \wedge n_\infty = \text{(spatial angular momentum)} n_\infty$$
$$+ \lambda \text{(spatial momentum)} e\bar{e} \tag{1.55}$$

At this point we realise that what we have ended up with is virtually identical to Eq. (6.19) in the paper on twistors [1].[2] This gives the connection between a line (or ray) in 3D space, represented in the CGA, and the physical observables of the twistor associated with it. So this gives a very interesting connection, in this non-relativistic case, with twistors.

What remains is to establish Eq. (1.55). We need to do this for an object which has non-zero 'spin' angular momentum, and with (reference) centre of mass at the origin. The simplest such object is the dumbbell above. By linearity, the identification is then obliged to work for all objects with the same property (of reference CoM at the origin). For the dumbbell, we indeed find

$$L \wedge n_\infty = \left(\psi I(\Omega_B)\tilde{\psi} \right) \wedge n_\infty$$

$$= (\text{orbital} + \text{spin angular momentum})\, n_\infty$$

$$+ \lambda(\text{spatial momentum})\, e\bar{e} \tag{1.56}$$

thus verifying what we need.

1.3.7   Adding in Gravity

It is pretty obvious how to add in gravity at the level of the equations of motion, but as an exercise in Lagrangian dynamics, we will instead add it into the Lagrangian and derive the extra term in the equation of motion from this.

Let g be the gravitational acceleration vector. Then the gravitational potential for an object of total mass m is given by

$$\mathcal{V} = \text{``}mgh\text{''} = -m\mathbf{g} \cdot \mathbf{x}_c = -\lambda m X_c \cdot \mathbf{g}$$

$$= \left\langle -\lambda m \psi \left(-\frac{\bar{n}}{2} \right) \tilde{\psi} \mathbf{g} \right\rangle \tag{1.57}$$

Since the Lagrangian is meant to be $\mathscr{T} - \mathscr{V}$, we obtain

$$\mathscr{L}_{\text{grav}} = \left\langle -\frac{1}{2}\lambda m \psi \bar{n} \tilde{\psi} \mathbf{g} \right\rangle \tag{1.58}$$

and therefore

$$\partial_\psi \mathscr{L}_{\text{grav}} = -\lambda m \bar{n} \tilde{\psi} \mathbf{g} \tag{1.59}$$

_____

[2]See also [2], which contains further details.

Following through to get the equations of motion reveals an extra bivector part which should appear on the rhs of Eq. (1.37) of

$$\frac{1}{2}\lambda m(\tilde{\psi}g\psi)\wedge\bar{n} \tag{1.60}$$

and wedging this with n_∞, we obtain the following to be added to the rhs of (1.42):

$$\lambda \tilde{R}_2(mg)R_2\,e\bar{e} \tag{1.61}$$

This is in precisely the form we would expect from (1.54), given that the gravitational force does not exert any spatial torque in the body frame.

1.3.8 The Angular Velocity Bivector Update Equation

A central thing that will be needed for a numerical algorithm will be a way of updating the body angular velocity bivector Ω_B at each time step. Equation (1.42), supplemented by a term for the external torques, gives us that

$$I(\dot{\Omega}_B)\wedge n_\infty = \left(\Omega_B \times I(\Omega_B)\right)\wedge n_\infty + T \tag{1.62}$$

where T is the sum of the trivector torques in the body frame. (The general form of these has been discussed above, and the gravitational 'torque' (1.61) is a specific example.) To recover Ω_B, we need to invert the function $F(B) \equiv I(B)\wedge n_\infty$.

Recalling the results of Sect. 1.3.1 that $\Omega_B \cdot n_\infty = 0$, so in general Ω_B is of the form $B + an_\infty$ (B some spatial bivector, a some spatial vector), and from Eq. (1.33) that

$$I(B+n_\infty a)\wedge n_\infty = I_{\text{rot}}(B)n_\infty + \frac{1}{2}M\lambda^2 a\bar{n}\wedge n_\infty \tag{1.63}$$

it is clear that, so long as I_{rot} is non-singular, F will be invertible for bivectors of the form $B \cdot n_\infty = 0$.

To illustrate how things go in the 2D dumbbell example above, we let $E_1 = e_1e_2$, $E_2 = e_1n_\infty$, $E_3 = e_2n_\infty$ be a basis for the appropriate space of bivectors and write $F_i = F(E_i)$. Then,

$$F_1 = 2mx_0^2e_1e_2n_\infty, \quad F_2 = 2m\lambda^2e_1e\bar{e}, \quad F_3 = 2m\lambda^2e_2e\bar{e} \tag{1.64}$$

where each particle is taken as having mass m (they were taken as unit mass previously). The reciprocal frame to this, with $F_i \cdot F^j = \delta_i^j$, is

$$F^1 = -\frac{1}{4mx_0^2}e_1e_2\bar{n}, \quad F^2 = \frac{1}{2m\lambda^2}e_1e\bar{e}, \quad F^3 = \frac{1}{2m\lambda^2}e_2e\bar{e} \tag{1.65}$$

giving

$$F^{-1}(B) = (B \cdot F^i) E_i \tag{1.66}$$

The equation of motion can now be written as

$$\dot{\Omega}_B = F^{-1}\big((\Omega_B \times I(\Omega_B)) \wedge n_\infty + T\big) \tag{1.67}$$

1.3.9 Collisions—One Body

We are now far enough along that we can attempt to take a first look at collisions. The simplest case to consider will be where our object hits a perfectly reflecting smooth surface. Then energy will be conserved, and we only need consider a normal impulse from the surface.

Let T_{imp} be the trivector impulsive torque that this gives rise to (we consider a concrete example below, in case what is meant by this is not fully clear yet). We then input this to the impulsive version of (1.67) to obtain

$$\Delta\Omega_B = T_{\text{imp}} \cdot F^i E_i \tag{1.68}$$

Here $\Delta\Omega_B$ is the sudden jump that will occur in Ω_B due to the impulse. We note that although Ω_B changes, there will be no change in the rotor ψ—it is just the state of motion that alters, not the position or orientation.

To calculate the value of the impulse, we can use energy conservation. Specifically, if Ω_B^{old} is the value of Ω_B just before the impulse, we require

$$-\frac{1}{2}\big(\Omega_B^{\text{old}} + \Delta\Omega_B\big) \cdot I\big(\Omega_B^{\text{old}} + \Delta\Omega_B\big) = -\frac{1}{2}\Omega_B^{\text{old}} \cdot I\big(\Omega_B^{\text{old}}\big) \tag{1.69}$$

This provides an equation which is in general quadratic in T_{imp}. One solution, however, is that $T_{\text{imp}} = 0$, and so it turns out that the value we want is the root of a linear equation, giving a nice simple answer. In a non-perfectly reflecting case, where the energy changes, then we will have a full quadratic to solve.

As a concrete case, let us consider our 2D dumbbell, in a situation where this is projected into the air, with a non-zero angular velocity, and then one of the two point masses (say the one at $x = -x_0$) strikes a perfectly reflecting smooth floor on its descent under gravity, see Fig. 1.3.

The impulsive force will be of the form $\Delta P = p_2 e_2 + p_\infty n_\infty$, i.e. directed upwards, and the trivector torque we need is

$$T_{\text{imp}} = \lambda X_2 \wedge (\tilde{\psi} \Delta P \psi) \wedge n_\infty \tag{1.70}$$

This evaluates to

$$T_{\text{imp}} = -x_0 p_2 \cos\theta e_1 e_2 n_\infty + \lambda p_2 (\sin\theta e_1 + \cos\theta e_2) e\bar{e} \tag{1.71}$$

where θ is the rotation angle in

$$R_2 = \cos(\theta/2) - e_1 e_2 \sin(\theta/2) \tag{1.72}$$

Putting the expression for T_{imp} into (1.68) yields

$$\Delta\Omega_B = \frac{p_2}{2x_0 m\lambda}\left(-\lambda\cos\theta e_1 e_2 + x_0(\sin\theta e_1 + \cos\theta e_2)n_\infty\right) \tag{1.73}$$

Putting this in turn into (1.69) yields a quadratic in p_2, with roots $p_2 = 0$ (no change) and the one we want, which using the parameterisation of Ω_B given in (1.45), is

$$p_2 = -\frac{4m}{1+\cos^2\theta}(\lambda h\cos\theta + \lambda g\sin\theta - x_0 f\cos\theta) \tag{1.74}$$

We can then re-insert this into (1.73) to find the jump in Ω_B in terms of parameters which are all known at the point of contact, thus solving our problem.

To get a feeling for what p_2 is in more conventional terms, we can note that quite generally in 2D, the relation between ψ and Ω_B means that

$$
\begin{aligned}
f(t) &= \dot{\theta}\\
g(t) &= \frac{1}{\lambda}(\dot{x}_c\cos\theta + \dot{y}_c\sin\theta)\\
h(t) &= \frac{1}{\lambda}(-\dot{x}_c\sin\theta + \dot{y}_c\cos\theta)
\end{aligned}
\tag{1.75}
$$

Inserting these into (1.74) yields

$$p_2 = \frac{4m}{1+\cos^2\theta}(x_0\dot{\theta}\cos\theta - \dot{y}_c) \tag{1.76}$$

which is in a form that one can check against the result one would obtain via conventional calculations.

1.3.10 Collisions—Two Bodies

We now move on to consider collisions between two bodies. We again do this in the smooth (non-frictional) case, but it will turn out to be easy to consider the full range from elastic to inelastic collisions.

When two smooth rigid bodies collide at a point, then the resulting impulsive force must be along the joint normal to the two surfaces. That the normals at the point of contact must agree (apart from one being the negative of the other) is obvious, except when one of the bodies has some sharp protuberance, i.e. a discontinuous normal direction at that point. In this case we shall assume that the normal is defined

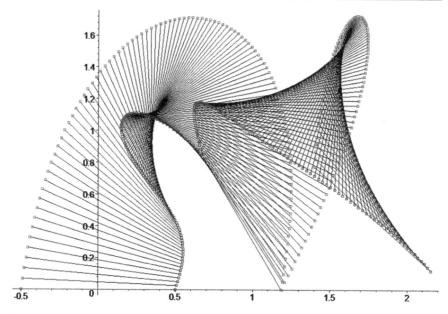

Fig. 1.3 A representation of the motion of the 2D-dumbbell colliding with a frictionless floor, computed in the 5D conformal algebra. The frames of an animation are shown simultaneously, with the dumbbell projected upwards from a horizontal position at the *left side* of the plot. (Online colour version shows different colours for each end of the dumbbell)

by the smooth surface. (If two sharp protuberances collide, we shall need another rule.)

The example we shall work with is when two dumbbells collide. We assume that the light rod joining the two ends of a dumbbell is smooth, and therefore, by the above assumption, if the collision is where the point mass at one end of a dumbbell impacts on the light rod portion of the other, then the impulse is wholly along the direction normal to the light rod at that point.

Thus we can set up a situation, directly analogous to the case where one dumbbell hits a smooth floor, where during its descent under gravity a dumbbell hits the rod part of another dumbbell whilst the second one is temporarily stationary, e.g. by having been launched vertically upwards at some earlier point, in horizontal orientation, and without spin.

The advantage of this setup (which by changing frames is pretty general anyway) is that for the first dumbbell, it allows us to use all the formulae of the previous section, apart from the value for p_2, which will be different (and, we expect, smaller).

We assume that the second dumbbell is hit at a distance b out from the centre towards the $x = +x_0$ mass. The impulse on this will be equal and opposite to that on the first dumbbell, and the impulsive torque equations tell us that

$$\Delta \Omega_B^{(2)} = -\frac{p_2}{2m\lambda x_0^2}\left(\lambda b e_1 e_2 + x_0^2 e_2 n_\infty\right) \tag{1.77}$$

$\Delta\Omega_B^{(1)}$ is still given by the rhs of (1.73).

To fix p_2, we no longer use energy, since we wish to consider inelastic as well as elastic collisions. We parameterise the degree of elasticity in the usual way via a *coefficient of restitution*, which we will call α (to avoid confusion with other uses of e). This is meant to lie in the range 0 to 1, with $\alpha = 0$ being totally inelastic and $\alpha = 1$ perfectly elastic.

The definition of α that we will use, will be the usual one in terms of the relative velocities of the two bodies along the joint normal before and after impact. Specifically, let $\dot{X}_{\text{cont}}^{(i)}$, $i = 1, 2$, be the 5D space velocities of the two points of contact; then if $\boldsymbol{n}$ is the joint normal direction, we define α via

$$\left(\dot{X}_{\text{cont}}^{(1)} - \dot{X}_{\text{cont}}^{(2)}\right) \cdot \boldsymbol{n}\big|_{\text{after}} = -\alpha\left(\dot{X}_{\text{cont}}^{(1)} - \dot{X}_{\text{cont}}^{(2)}\right) \cdot \boldsymbol{n}\big|_{\text{before}} \tag{1.78}$$

It is perhaps not obvious, given that the bodies have spin degrees of freedom, as well as translational, that the value $\alpha = 1$, defined this way, will coincide with no loss of overall KE (translational + rotational), but this is indeed the case. For $\alpha = 0$, one finds that in this case with rotation, then even in the zero momentum frame, the final KE is not zero. However, $\alpha = 0$ is still the proper inelastic limit for rigid bodies. Specifically, values lower than this would imply that interpenetration occurs. The value of p_2 going with $\alpha = 0$ is therefore the minimum impulse that prevents interpenetration, while the value that goes with $\alpha = 1$ is the maximum possible (consistent with energy conservation).

For the two dumbbell case, we can use (1.78) along with (1.17), which gives $\dot{X}$ in terms of Ω_B, to derive p_2, with the result

$$p_2 = -2mx_0^2(1+\alpha)\frac{\lambda h\cos\theta + \lambda g\sin\theta - x_0 f\cos\theta}{x_0^2\cos^2\theta + 2x_0^2 + b^2} \tag{1.79}$$

It is important to note as regards the way we have set things up here, that the f, g and h in this expression are those of the *first* object just before the moment of contact. Substituting in the values of f, g and h from (1.75), we find

$$p_2 = \frac{2mx_0^2(1+\alpha)}{x_0^2\cos^2\theta + 2x_0^2 + b^2}(x_0\dot{\theta}\cos\theta - \dot{y}_c) \tag{1.80}$$

in terms of more standard variables.

We see that p_2 scales directly as $1 + \alpha$. The minimum impulse, needed to prevent interpenetration, is thus half the maximum impulse (corresponding to energy conservation). Note that if we work out the change in the total KE, we find that it is given by a value independent of α times $1 - \alpha^2$. This seems very sensible, and the fixed multiplier corresponds to the amount lost in the inelastic case. Note statements about *changes* in KE are independent of reference frame, so we do not need to worry about whether we are working in the zero momentum frame or not. Note also that total $L \wedge n_\infty$, i.e. angular and linear momentum, is automatically conserved throughout, since the impulse is equal and opposite on each body, and takes place at the same space point (so has the same total moment).

1.4   Implementation

As we have seen, the 5D CGA formulation gives us a new mathematical description of rigid body motion in 3D, which we can use as an alternative to the usual position/rotation representation in numerical simulations. We now discuss the implementation of this approach in a more challenging setting than considered above, namely the development of a prototype multi-body physics engine using CGA operations. The overall simulation scheme used is basically separate from this representation choice, and we used the 'Fast Frictional Dynamics' approach of [6] to develop the physics aspects needed additionally to those already discussed. Basically, these come down to how to handle a potentially large number of simultaneously interacting bodies in a way that scales favourably with the number of bodies involved.

The FFD approach has serious limitations (including non-conservation of energy) but is simple, fast, and provides an adequate demonstration of the underlying CGA representation.

1.4.1   Overview

Our simulation tracks the configuration R and velocity $\dot{R}$ of each rigid body. To update these over a time-step, we go through the following (schematic) routine:
1. Firstly, we update the position and velocity of each body, without regard to any contact forces (but taking into account gravity and any other 'non-contact' forces, e.g. a simulated rocket).
2. We then perform collision detection, seeing whether this update has caused any interpenetration between rigid bodies.
3. If there are any collisions, we update the positions and velocities to fix these (also taking into account the effects of friction).

Examples of operations we need to perform in this process include evolving the position and velocity in the presence of external torques. The following sections go through some of these details in the CGA approach.

1.4.2   Update Equations

1.4.2.1 Velocity Update
Recall that the equation of motion for a rigid body is

$$F(\dot{\Omega}_B) = C + T_B \tag{1.81}$$

where $C = (\Omega_B \times I(\Omega_B)) \wedge n_\infty$ is the 'coriolis torque', $T_B = \tilde{R}TR$ is the body frame external torque, and $F(B) = I(B) \wedge n_\infty$ is the inertia function.

In our simulation, we track the configuration and (generalised) velocity of the body, and want to evolve these over time-steps. The most obvious approach is to parameterise the body's motion by R and Ω_B—we can then use $\dot{\Omega}_B = F^{-1}(C + T_B)$

to do a first-order update for Ω_B and a second-order update for R. Alternatively, we can work directly with the angular momentum $L_n = RF(\Omega_B)\tilde{R}$. This changes as

$$\dot{L}_n = T = RT_B\tilde{R} \tag{1.82}$$

i.e. in response to the spatial-frame trivector torque, so we can update this and then obtain

$$\dot{\Omega}_B = F^{-1}(C + \tilde{R}\dot{L}_n R), \quad \Omega_B = F^{-1}(R\tilde{L}_n R) \tag{1.83}$$

with which we can perform the R update. The advantage of tracking L_n is that, for free motion, L_n is conserved, whereas in general $C \neq 0$, so Ω_B evolves and has the potential to wander off the correct shell. Even in the case of motion under gravity, there is no rotational torque, so the rotational components of L_n do not change, and we retain the most important part of the advantage.

1.4.3 Computer-Level Object Representation

If a software library implementing 5D CGA is available, then all of the quantities referred to above can be used directly, providing a very simple programming experience. However, such libraries are often less efficient than lower-level implementation methods. One compromise is to use existing, optimised vector maths libraries to build the CGA objects we need. For example, configuration rotors are of the form $R = (1 - tn_\infty/2)R_s$, where R_s is a spatial rotor, and t a spatial vector, so can be written as

$$R = q_1 + I_3 n_\infty q_2 \tag{1.84}$$

where $q_i = \alpha_i + B_i$ is a scalar plus spatial bivector object, that is, a quaternion. Thus, configuration rotors can be represented as dual quaternions; this is useful because many optimised vector mathematics libraries contain a quaternion data structure. For example, the multiplication

$$R'R = \left(q_1' + I_3 n_\infty q_2'\right)(q_1 + I_3 n_\infty q_2) = q_1' q_1 + I_3 n_\infty \left(q_1' q_2 + q_2' q_1\right) \tag{1.85}$$

can be implemented in terms of three quaternion multiplications. Extracting the rotation and translation from R, we have $R_s = q_1$, and $t = 2I_3 \langle q_1 \tilde{q}_2 \rangle_2$.

As another example, suppose that we have updated R, but that due to numerical errors, it may not longer be a Euclidean rotor, i.e. $R\tilde{R} \neq 1$ (note that if R is of the form $R = q_1 + I_3 n_\infty q_2$, then $R\tilde{R} = 1$ implies that $Rn_\infty \tilde{R} = n_\infty$). Write ψ as the product of our update step—we want to set $R = \psi/(\psi\tilde{\psi})^{1/2}$, projecting back onto the rotor manifold. Since

$$\psi\tilde{\psi} = (q_1 + I_3 n_\infty q_2)(\tilde{q}_1 + \tilde{q}_2 I_3 n_\infty) = q_1 \tilde{q}_1 + 2I_3 n_\infty \langle q_1 \tilde{q}_2 \rangle$$
$$= \alpha + \beta I_3 n_\infty \tag{1.86}$$

with α, β scalars, we have

$$(\psi \tilde{\psi})^{1/2} = \alpha^{1/2} + \frac{1}{2}\alpha^{-1/2}\beta I_3 n_\infty \tag{1.87}$$

and thus

$$(\psi \tilde{\psi})^{-1/2} = \alpha^{-1/2} - \frac{1}{2}\alpha^{-3/2}\beta I_3 n_\infty \tag{1.88}$$

so

$$R = \frac{\psi}{(\psi \tilde{\psi})^{1/2}} = \alpha^{-1/2}\psi - \frac{1}{2}\beta\alpha^{-3/2}I_3 n_\infty q_1 \tag{1.89}$$

giving us our renormalisation equation.

It should be noted that we could, of course, represent the body configuration as a (rotor, translation vector) pair, in exactly the same way as standard physics libraries use a (normalised quaternion, translation vector) pair. This would be a perfectly sensible choice, and the update equations we get would correspond exactly to those in traditional approaches, though we may choose to derive them through the CGA formalism. In this project, we used the dual quaternion representation since it is slightly closer to the CGA formalism and allows the programming to reflect the CGA equations more closely.

1.4.4 Results

The methods outlined above were used in constructing a prototype real-time rigid-body physics engine, mainly with the goal of demonstrating the use of CGA-based representations and algorithms in such an application. A few short movies of the engine in action can be found at [9]. More complicated situations, especially those involving quasi-static stacking of objects, are not well handled by the simple FFD scheme implemented here, but this issue is orthogonal to the use of CGA, which could equally well be employed alongside a more physically realistic simulation approach.

1.5 Exercises

1.1 Show that for a symmetric top where $i_1 = i_2$ are the two equal moments of inertia, and i_3 is the moment of inertia about the symmetry axis e_3, the inertia tensor can be written

$$I(B) = i_1 B + (i_3 - i_1)(B \wedge e_3)e_3$$

1.2 Verify the rotor solution for a symmetric top given in Eq. (1.11).

1.3 Show that the inertia tensors used in this paper are symmetric (i.e. $I(B) \cdot C = B \cdot I(C)$ for all bivectors B and C) and demonstrate the role this plays in proving conservation of kinetic energy.

References

1. Arcaute, E., Lasenby, A., Doran, C.: Twistors in geometric algebra. Adv. Appl. Clifford Algebras **18**, 373–394 (2008)
2. Arcaute, E., et al.: A representation of twistors within geometric (Clifford) algebra. math-ph/ 0603037v2 (unpublished)
3. Doran, C.J.L., Lasenby, A.N.: Geometric Algebra for Physicists. Cambridge University Press, Cambridge (2003)
4. Hestenes, D.: New tools for computational geometry and rejuvenation of screw theory. In: Bayro-Corrochano, E., Scheuermann, G. (eds.) Geometric Algebra Computing in Engineering and Computer Science, p. 3. Springer, London (2010)
5. Hestenes, D.: Zitterbewegung in quantum mechanics. Found. Phys. **40**, 1–54 (2010)
6. Kaufman, D.M., Edmunds, T., Pai, D.K.: Fast frictional dynamics for rigid bodies. ACM Trans. Graph. **24**(3), 946–956 (2005) (SIGGRAPH 2005)
7. Lasenby, A.N.: Recent applications of conformal geometric algebra. In: Li, H., Olver, P.J., Sommer, G. (eds.) Computer Algebra and Geometric Algebra with Applications. Lecture Notes in Computer Science, p. 298. Springer, Berlin (2005)
8. Lasenby, A.N., Doran, C.J.L., Gull, S.F.: A multivector derivative approach to Lagrangian field theory. Found. Phys. **23**(10), 1295 (1993)
9. http://www.mrao.cam.ac.uk/~anthony/conformal_dynamics

Estimating Motors from a Variety of Geometric Data in 3D Conformal Geometric Algebra

Robert Valkenburg and Leo Dorst

Abstract
The motion rotors, or *motors*, are used to model Euclidean motion in 3D conformal geometric algebra. In this chapter we present a technique for estimating the motor which best transforms one set of noisy geometric objects onto another. The technique reduces to an eigenrotator problem and has some advantages over matrix formulations. It allows motors to be estimated from a variety of geometric data such as points, spheres, circles, lines, planes, directions, and tangents; and the different types of geometric data are combined naturally in a single framework. Also, it excludes the possibility of a reflection unlike some matrix formulations. It returns the motor with the smallest translation and rotation angle when the optimal motor is not unique.

2.1 Introduction

The motion rotors or *motors*, denoted $\mathcal{M}$, are used to model Euclidean motions in 3D conformal geometric algebra (CGA). It is often useful to be able to estimate a motor which best maps one data set onto another in some sense. The canonical problem involves two sets of noisy points where one set is nominally a rotated and translated version of the other. This situation arises frequently, for example when two sets of reconstructed 3D points need to be merged and they share some common points. Several solutions exist to minimise the squared distance between the points, using matrix techniques based on SVD, polar decomposition, and quaternions [3].

R. Valkenburg (✉)
Industrial Research Limited, Auckland, New Zealand
e-mail: r.valkenburg@irl.cri.nz

L. Dorst
Intelligent Systems Laboratory, University of Amsterdam, Amsterdam, The Netherlands
e-mail: l.dorst@uva.nl

L. Dorst, J. Lasenby (eds.), *Guide to Geometric Algebra in Practice*,
DOI 10.1007/978-0-85729-811-9_2, © Springer-Verlag London Limited 2011

In addition to points, many other geometric objects such as lines, directions, and planes provide useful information which can be used to help estimate the rigid body relationship between the data sets.

In this chapter we present a technique for estimating the motor which best transforms one set of noisy geometric objects onto another. The technique reduces to an eigenrotator problem and has some advantages over matrix formulations. It allows motors to be estimated from a wide variety of geometric data such as points, spheres, circles, lines, planes, directions, and tangents; and the different types of geometric data to be combined naturally in a single framework. Also, it does not admit the possibility of a reflection as do some matrix formulations. It returns the motor with the smallest translation and rotation angle when the optimal motor is not unique. To assist the development, we will first examine some useful algebraic and differential properties of the motors.

The following geometric algebra conventions are used in this chapter. The geometric algebra over $\mathbb{R}$ with signature (p, q) (p positive and q negative basis elements) is denoted $\mathbb{R}_{p,q}$. When $q = 0$ we write $\mathbb{R}_p$. A pure Euclidean multivector in $\mathbb{R}_3$ is usually represented in boldface, such as V. The grade-r elements of a geometric algebra $\mathbb{R}_{p,q}$ are denoted $\mathbb{R}_{p,q}^r$. $\mathbb{R}_{p,q}^+$ and $\mathbb{R}_{p,q}^-$ refer to the even and odd elements of $\mathbb{R}_{p,q}$. The conformal geometric algebra (CGA) of the 3D space $\mathbb{R}^3$ is denoted by $\mathbb{R}_{4,1}$. The dual of X is denoted $X^* = X \cdot I^{-1}$. The CGA vector n_o represents the origin and the CGA vector n_∞ represents the point at infinity, with $n_o \cdot n_\infty = -1$. A CGA point or dual sphere s (which is an element of $\mathbb{R}_{4,1}^1$) is normalised if $s \cdot n_\infty = -1$, and a direct sphere (an element of $\mathbb{R}_{4,1}^4$) is normalised if $S \wedge n_\infty = -I_{4,1}$. A round R (including tangents) is normalised if $|R \wedge n_\infty| = 1$, a flat (line or plane) F is normalised if $|F| = 1$, and a direction Δ is normalised if $|n_o \wedge \Delta| = 1$. The notation $\langle X \rangle_{i,j,\dots,k}$ is used as an abbreviation for $\langle X \rangle_i + \langle X \rangle_j + \cdots + \langle X \rangle_k$.

2.2 The Linear Spaces $\mathbb{M}$, $\mathbb{B}$, and $\mathbb{S}$

The 8D linear space $\mathbb{M} = \text{span}\{1, e_{12}, e_{13}, e_{23}, e_1 n_\infty, e_2 n_\infty, e_3 n_\infty, I_3 n_\infty\} \subset \mathbb{R}_{4,1}$ is the smallest linear space in which motors reside. It is convenient to restrict most of the analysis to elements in $\mathbb{M}$ because many simplifications arise. Most of these are consequences of the following split: if $X \in \mathbb{M}$, then $X = R + Q$ where $R \in \mathbb{R}_3^+$ and $Q \in \mathbb{R}_3^- n_\infty = \{V n_\infty : V \in \mathbb{R}_3^-\}$. As $Q\widetilde{Q} = 0$, $\langle X\widetilde{X} \rangle = R\widetilde{R} \geq 0$, so $|X|^2 = |\langle X\widetilde{X} \rangle| = \langle X\widetilde{X} \rangle$, and we can drop the absolute value. We will use the property that $\mathbb{M}$ is closed under multiplication, so if $X, Y \in \mathbb{M}$, and then $XY \in \mathbb{M}$. This is clear by simply multiplying the basis elements. If $X, Y \in \mathbb{M}$, then $\langle XY\widetilde{Y}\widetilde{X} \rangle = \langle X\widetilde{X} \rangle \langle Y\widetilde{Y} \rangle$, so $|XY| = |X||Y|$. In addition, $X \in \mathbb{M}$ is invertible iff $|X| \neq 0$. If X is invertible, then $1 = |X^{-1}X| = |X^{-1}||X|$ and $|X| \neq 0$. Conversely, if $|X| \neq 0$, then

$$X^{-1} = \widetilde{X}\left(\frac{\langle X\widetilde{X} \rangle - \langle X\widetilde{X} \rangle_4}{\langle X\widetilde{X} \rangle^2}\right). \tag{2.1}$$

The denominator of this is simplified because $\langle X\widetilde{X}\rangle_4^2$ vanishes. It is also convenient to split $X \in \mathbb{M}$ into symmetric and antisymmetric parts $X = S + B$ where $S = \frac{1}{2}(X + \widetilde{X}) = \langle X\rangle_{0,4}$ and $B = \frac{1}{2}(X - \widetilde{X}) = \langle X\rangle_2$. The antisymmetric grade-2 elements of $\mathbb{M}$ will be denoted $\mathbb{B} = \text{span}\{e_{12}, e_{13}, e_{23}, e_1 n_\infty, e_2 n_\infty, e_3 n_\infty\}$, and the symmetric grade 0 and 4 elements will be denoted $\mathbb{S} = \text{span}\{1, I_3 n_\infty\}$. The elements of $\mathbb{S}$ are "symmetric" in the sense that for $S \in \mathbb{S}$, we have $S = \widetilde{S}$. $\mathbb{S}$ is closed under multiplication: $S_1, S_2 \in \mathbb{S} \Rightarrow S_1 S_2 \in \mathbb{S}$. Note that if $X \in \mathbb{M}$, then $X\widetilde{X} = \langle X\widetilde{X}\rangle + \langle X\widetilde{X}\rangle_4$. Therefore the condition $X\widetilde{X} = 1$ encodes two constraints: $\langle X\widetilde{X}\rangle = 1$ and $\langle X\widetilde{X}\rangle_4 = 0$ (there is only one grade-4 basis element in $\mathbb{M}$). The following lemma uses these constraints to characterise how the 6D motor manifold $\mathcal{M}$ sits in the 8D linear space $\mathbb{M}$.

Lemma 2.1 $X \in \mathcal{M} \Leftrightarrow X \in \mathbb{M}$ and $X\widetilde{X} = 1$.

Proof Let $X = R + Q \in \mathbb{M}$ where $R \in \mathbb{R}_3^+$ and $Q \in \mathbb{R}_3^- n_\infty$. $X\widetilde{X} = 1$ implies $R\widetilde{R} = 1$ and $Q\widetilde{R} = \langle Q\widetilde{R}\rangle_2$. Thus, R is a rotator, and $X = R + Q\widetilde{R}R = (1 + \langle Q\widetilde{R}\rangle_2)R = TR$ where $T = 1 + \langle Q\widetilde{R}\rangle_2$ is a translator. □

The space $\mathbb{M}$ is incomplete in the sense that, given a basis of $\mathbb{M}$, we cannot find a reciprocal basis that also lies in $\mathbb{M}$. We can enlarge $\mathbb{M}$ to a complete space such as $\mathbb{M} \cup \text{span}\{e_1 n_o, e_2 n_o, e_3 n_o, \widetilde{I}_3 n_o\}$ or $\mathbb{R}_{4,1}^+$ and then construct a reciprocal basis. The subspace spanned by reciprocal vectors associated with elements in $\mathbb{M}$ is denoted $\overline{\mathbb{M}} = \text{span}\{1, \widetilde{e}_{12}, \widetilde{e}_{13}, \widetilde{e}_{23}, e_1 n_o, e_2 n_o, e_3 n_o, \widetilde{I}_3 n_o\}$. Almost every result in $\mathbb{M}$ has a counterpart in $\overline{\mathbb{M}}$. An element $T = 1 + \mathbf{t} n_o = s(1 + \frac{1}{2}\mathbf{t} n_\infty)s$ represents a transversor (reflection in the unit sphere $s = n_o - \frac{1}{2}n_\infty$ followed by a translation $1 + \frac{1}{2}\mathbf{t} n_\infty$ and another reflection in the unit sphere). It is the product of an even number of vectors and satisfies $T\widetilde{T} = 1$, so it is a rotor. Let $\overline{\mathcal{M}}$ denote the rotors of the form $M = TR$ where T is a transversor and R a rotator. The counterpart to Lemma 2.1 takes the form:

Lemma 2.2 $X \in \overline{\mathcal{M}} \Leftrightarrow X \in \overline{\mathbb{M}}$ and $X\widetilde{X} = 1$.

The intersection of $\mathbb{M}$ and $\overline{\mathbb{M}}$ is $\mathbb{R}_3^+$. The rotators $\mathcal{R}$ lie in $\mathbb{R}_3^+$ and are a subset of both $\mathcal{M}$ and $\overline{\mathcal{M}}$. The relationship between the spaces $\mathbb{M}$, $\mathcal{M}$, $\overline{\mathbb{M}}$, $\overline{\mathcal{M}}$, $\mathbb{R}_3^+$, $\mathcal{R}$, and $\mathbb{R}_{4,1}^+$ is shown in Fig. 2.1. We will sometimes want to project an element $X \in \mathbb{R}_{4,1}$ on $\mathbb{M}$ or $\overline{\mathbb{M}}$. Let $\{e_J\}$ be a basis for $\mathbb{M}$, and $\{e^J\}$ be the associated reciprocal basis in $\overline{\mathbb{M}}$. The projection on $\mathbb{M}$ is defined by

$$P_{\mathbb{M}}(X) = \sum_J \langle e^J X\rangle e_J.$$

Fig. 2.1 The relationship between the manifolds of motors $\mathscr{M}$, rotators $\mathscr{R}$, and reciprocal motors $\bar{\mathscr{M}}$ and the linear spaces $\mathbb{M}$, $\mathbb{R}_3^+$, and $\bar{\mathbb{M}}$ they reside in

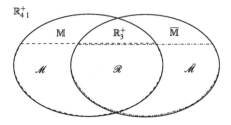

As $\langle \mathrm{P}_\mathbb{M}(X)Y \rangle = \sum_J \langle e^J X \rangle \langle e_J Y \rangle = \langle X\bar{\mathrm{P}}_\mathbb{M}(Y) \rangle$, the adjoint is the projection onto $\bar{\mathbb{M}}$ given by

$$\bar{\mathrm{P}}_\mathbb{M}(Y) = \mathrm{P}_{\bar{\mathbb{M}}}(Y) = \sum_J e^J \langle e_J Y \rangle.$$

This can also be expressed using the multivector derivative $\partial_X = \sum_J e^J \langle e_J \partial_X \rangle$: $\partial_X \langle XY \rangle = \sum_J e^J \langle e_J Y \rangle = \bar{\mathrm{P}}_\mathbb{M}(Y)$. When no ambiguity arises, it is convenient to use the terse notation P_X for the projection onto the basis of the linear space in which the element X resides. For example, if $X \in \mathbb{M}$, then $\mathrm{P}_X = \mathrm{P}_\mathbb{M}$, if $R \in \mathbb{R}_3^+$, then P_R is a projection onto $\mathbb{R}_3^+$, and if $Q \in \mathbb{R}_3^- n_\infty$, then P_Q is the projection onto $\mathbb{R}_3^- n_\infty = \mathrm{span}\{e_1 n_\infty, e_2 n_\infty e_3 n_\infty, I_3 n_\infty\}$. Using the split $X = R + Q \in \mathbb{M}$, where $R \in \mathbb{R}_3^+$ and $Q \in \mathbb{R}_3^- n_\infty$, gives $\mathrm{P}_\mathbb{M} = \mathrm{P}_X = \mathrm{P}_R + \mathrm{P}_Q$, and $\bar{\mathrm{P}}_\mathbb{M} = \bar{\mathrm{P}}_X = \mathrm{P}_R + \bar{\mathrm{P}}_Q$ because $\mathrm{P}_R = \bar{\mathrm{P}}_R$.

2.3 Geometry of the Motors

The following constructions in $\mathbb{M}$ directly parallel constructions in matrix theory, where $\mathscr{M}$ plays the role of the $n \times p$ Stiefel manifold, and $\mathbb{S}$ the symmetric positive definite matrices [4]. Refer to Fig. 2.2 which illustrates some of the concepts introduced in this section. Consider the curve $\psi(t) \in \mathscr{M}$ with $M = \psi(0)$ and $\Delta = \psi'(0)$. Differentiating the constraint $\tilde{\psi}(t)\psi(t) = 1$ and evaluating at $t = 0$ gives $\tilde{M}\Delta = -\tilde{\Delta}M$. As $\Delta \in \mathbb{M}$, $\tilde{M}\Delta \in \mathbb{M}$, and it follows that $\Delta = MB$ where $B \in \mathbb{B}$. We define the *tangent space* of $\mathscr{M}$ at $M \in \mathscr{M}$ by $\mathscr{T}_M = M\mathbb{B} = \{MB : B \in \mathbb{B}\} \subset \mathbb{M}$. Any element $X \in \mathbb{M}$ can be split:

$$X = M(\tilde{M}X) = M\langle \tilde{M}X \rangle_2 + M\langle \tilde{M}X \rangle_{0,4}.$$

The first term in this split is in $\mathscr{T}_M$, while the second term is of the form MS where $S \in \mathbb{S}$. We define the *normal space* of $\mathscr{M}$ at $M \in \mathscr{M}$ (restricted to $\mathbb{M}$) by $\mathscr{N}_M = M\mathbb{S} = \{MS : S \in \mathbb{S}\}$. If $X = MB \in \mathscr{T}_M$ and $Y = MS \in \mathscr{N}_M$, then $\langle X\tilde{Y} \rangle = \langle MBS\tilde{M} \rangle = \langle BS \rangle = 0$, and $\mathscr{T}_M$ is orthogonal to $\mathscr{N}_M$, so $\mathbb{M} = \mathscr{T}_M \oplus \mathscr{N}_M$. From the split, for $X \in \mathbb{M}$ we can define the projection on $\mathscr{T}_M$ along $\mathscr{N}_M$ by

$$\mathrm{P}_{\mathscr{T}_M}(X) = M\langle \tilde{M}X \rangle_2.$$

Fig. 2.2 An intuitive sketch of the geometry of motors $\mathcal{M}$ in $\mathbb{M}$ showing the tangent space $\mathcal{T}_M$ and the normal space $\mathcal{N}_M$ at M, and the projections onto $\mathcal{T}_M$, $\mathcal{N}_M$, and $\mathcal{M}$

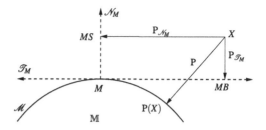

It is clear that $\mathrm{P}_{\mathcal{T}_M}$ is idempotent, onto $\mathcal{T}_M$, and has null-space $\mathcal{N}_M$. Similarly, for $X \in \mathbb{M}$ the projection on $\mathcal{N}_M$ along $\mathcal{T}_M$ is defined by

$$\mathrm{P}_{\mathcal{N}_M}(X) = M \langle \widetilde{M} X \rangle_{0,4}.$$

It is also clear that $\mathrm{P}_{\mathcal{N}_M}$ is idempotent, along $\mathcal{T}_M$, and onto $\mathcal{N}_M$. Closely related to $\mathcal{N}_M$, we can define a polar decomposition for an element in $\mathbb{M}$.

Lemma 2.3 *An element $X \in \mathbb{M}$ with $|X| \neq 0$ has a unique polar decomposition $X = MS = SM$ where $M \in \mathcal{M}$, $S \in \mathbb{S}$, and $\langle S \rangle > 0$.*

Proof Suppose that $MS = M'S'$ are two such decompositions. Then $N = \widetilde{M}'M = S'S^{-1}$ is a symmetric motor ($N = \widetilde{N}$). Hence $N = \alpha + \beta I_3 n_\infty$ and $1 = N^2 = \alpha^2 + 2\alpha\beta I_3 n_\infty$, so $\beta = 0$ and $\alpha = 1$ because $\langle S \rangle > 0$ and $\langle S' \rangle > 0$. As $M I_3 n_\infty \widetilde{M} = I_3 n_\infty$, we have $MS = SM$. The polar decomposition is given by

$$S = |X|\left(1 + \frac{\langle X\widetilde{X}\rangle_4}{2\langle X\widetilde{X}\rangle}\right), \quad M = XS^{-1} = \frac{X}{|X|}\left(1 - \frac{\langle X\widetilde{X}\rangle_4}{2\langle X\widetilde{X}\rangle}\right). \tag{2.2}$$

$\square$

As shown, given $M \in \mathcal{M}$, any $X \in \mathbb{M}$ can be decomposed into components in $\mathcal{T}_M$ and $\mathcal{N}_M$ giving $X = MS + MB$. The polar decomposition can be interpreted as simply choosing M appropriately so that the component in $\mathcal{T}_M$ vanishes leaving $X = MS \in \mathcal{N}_M$. The polar decomposition is applied to more general elements $X \in \mathbb{R}_{4,1}^+$ in [1] (Chap. 5 in this book).

The polar decomposition on $\mathbb{M}$ provides a natural way to define the operation of projection onto $\mathcal{M}$ in the same way as the polar decomposition on $\mathbb{R}^{n \times p}$ defines a projection onto the $n \times p$ orthogonal matrices in matrix theory. If $X \in \mathbb{M}$ has polar decomposition $X = MS$, we define the projection onto $\mathcal{M}$ by

$$\mathrm{P}(X) = XS^{-1} \in \mathcal{M}.$$

The element $S^{-1} \in \mathbb{S}$ nudges X onto $\mathcal{M}$. It is interesting to note that several other situations arise where elements of $\mathbb{S}$ perform some useful transformation. The element $S^{-2} \in \mathbb{S}$ maps $\widetilde{X}$ onto $X^{-1} = \widetilde{X}S^{-2}$ (refer to (2.1)). An element $B \in \mathbb{B}$ can be split into two commuting blades using $S_- = \langle \widetilde{B}B \rangle_4/(2\widetilde{B}B) \in \mathbb{S}$ and $S_+ =$

$1 - S_- \in \mathbb{S}$. If $B_+ = BS_+$ and $B_- = BS_-$, then $B = B_+ + B_-$, and $B_+ B_- = B_- B_+$. This split can be used to factor a motor in accordance with Chasles's decomposition

$$M = \exp\left(-\frac{1}{2}B\right) = \exp\left(-\frac{1}{2}B_+\right)\exp\left(-\frac{1}{2}B_-\right), \qquad (2.3)$$

where $\exp(-B_+/2)$ is a generalized rotator about an axis, and $\exp(-B_-/2)$ is a translator along the axis. Using the polar decomposition, it is a simple matter to show that for any element $Y \in \mathbb{M}$ with $|Y| \neq 0$, we can find an element $X = \log(Y) \in \mathbb{M}$ such that $Y = \exp(X)$. Let Y have a polar decomposition $Y = MS'$. The motor M can be expressed $M = \exp(B)$ where $B \in \mathbb{B}$ (an expression for the motor logarithm may be found in [2]). Also note that if $S = \alpha + Q \in \mathbb{S}$, then $\exp(S) = \exp(\alpha)(1 + Q) = \exp(\alpha) + \exp(\alpha)Q$ because α and Q commute and $Q^2 = 0$. So if $S' = \alpha' + Q' = \exp(S)$, then we take $\alpha = \ln \alpha'$ and $Q = Q'/\alpha'$ giving $S = \ln \alpha' + Q'/\alpha'$. As B and S commute, we can take $X = B + S \in \mathbb{M}$.

There is an equivalent polar decomposition for an element $X \in \overline{\mathbb{M}}$ with $|X| > 0$, of the form $X = MS$, where $M \in \mathscr{\overline{M}}$ models a rotation and transversion, and $S \in \overline{\mathbb{S}} = \mathrm{span}\{1, \tilde{I}_3 n_o\}$.

The rotators $\mathscr{R}$ are used to model rotation about the origin and lie in the linear space $\mathbb{R}_3^+ = \mathrm{span}\{1, e_{12}, e_{13}, e_{23}\}$. All the ideas above simplify when restricted to rotators. If $X \in \mathbb{R}_3^+$, then $X\tilde{X} = \langle X\tilde{X}\rangle$, so the equation $X\tilde{X} = 1$ imposes only one constraint and is equivalent to the statement $\langle X\tilde{X}\rangle = 1$. We will see that the absence of the constraint $\langle X\tilde{X}\rangle_4 = 0$ is an important simplification for rotator estimation. If $R \in \mathscr{R}$, then $\mathscr{N}_R = \{Rs : s \in \mathbb{R}\}$ (i.e. just scalar multiples of R), and the projection on $\mathscr{N}_R$ is given by $\mathrm{P}_{\mathscr{N}_R}(X) = R\langle \tilde{R}X\rangle$. The tangent space $\mathscr{T}_R = \{RB : B \in \mathrm{span}\{e_{12}, e_{13}, e_{23}\}\}$, and $\mathrm{P}_{\mathscr{T}_R}(X) = R\langle \tilde{R}X\rangle_2$. The polar decomposition takes the simple form $X = Rs$ where $R = X/|X|$ and $s = |X| \in \mathbb{R}$.

The translators are used to model translation and lie in the linear space $\mathbb{T} = \mathrm{span}\{1, e_1 n_\infty, e_2 n_\infty, e_3 n_\infty\}$. A translator $T = 1 - \frac{1}{2}\mathbf{t}n_\infty$ has a constant scalar coefficient, so there are only three degrees of freedom, as required. If $X \in \mathbb{T}$, then $\langle X\rangle^2 = X\tilde{X} = \langle X\tilde{X}\rangle$, so the equation $X\tilde{X} = 1$ imposes only one constraint as for rotators. Because $\langle \mathbb{T}\rangle_2$ is made up of null bivectors, significant simplifications arise. If T is a translator, then $\mathscr{N}_T = \{Ts : s \in \mathbb{R}\}$ (i.e. just scalar multiples of T), and the projection on $\mathscr{N}_T$ is given by $\mathrm{P}_{\mathscr{N}_T}(X) = T\langle X\rangle$. The tangent space $\mathscr{T}_T = \mathrm{span}\{e_1 n_\infty, e_2 n_\infty, e_3 n_\infty\}$, and $\mathrm{P}_{\mathscr{T}_T}(X) = \langle \tilde{T}X\rangle_2$. The polar decomposition takes the simple form $X = Ts$ where $T = X/|X|$ and $s = |X| = |\langle X\rangle| \in \mathbb{R}$.

2.4 Estimating Motors

We have two sets of noisy geometric data and wish to estimate the motor that optimally maps one data set onto the other. To solve this problem, we need to be precise about what optimal means, so we will define a measure that is used to determine if two geometric objects are similar. For example, if P and Q are normalised points, then $\langle PQ\rangle = -\frac{1}{2}d^2$ where d is the distance between the points. Two points are considered similar if they are close together. We choose a similarity rather an error measure only because it avoids a sign change for the most common case of points. The inner product between points increases as the points get closer; hence it already has the correct sign. To set the problem up so it has a simple closed-form solution as an eigenrotator problem, we need to restrict the form of the similarity measure as described in the next section. However, even with this restriction, not all possibilities for object representation are admissible into the framework for estimating motors. This is because one of the constraints $\langle X\widetilde{X}\rangle_4 = 0$ for an element $X \in \mathbb{M}$ to be a motor (recall Lemma 2.1) is awkward to handle, and we will only want to consider object representations where it can be dropped so that we can estimate the motor using linear methods. Surprisingly, this occurs quite often as we will see later.

2.4.1 Similarity Measures in CGA

In order to set the problem up as a eigenrotator problem, we need to restrict the similarity measure between objects P and Q to the simple form $\langle P\check{Q}\rangle$, where the *check* operator $\check{Q}$ is a grade-dependent sign change defined by $\check{Q} = \langle Q\rangle_{0,1,3} - \langle Q\rangle_{2,4,5}$. Note that $\check{Q} = \widetilde{Q}$ if $Q = \langle Q\rangle_{0,1,2}$ and $\check{Q} = -\widetilde{Q}$ if $Q = \langle Q\rangle_{3,4,5}$. This operation is motivated by the requirements (i) $\langle p\check{q}\rangle = \langle P\check{Q}\rangle$ where $p = P^*$ and $q = Q^*$ and (ii) $\langle P\check{Q}\rangle = \cos(\theta)$ when P, Q are flats (see below). This simple form is not as much of a restriction as it may first seem. If we carefully consider the object representation, many physically meaningful quantities can be expressed in this way. Consider the following examples:

Points and Spheres   We have already seen that if P and Q are normalised points (grade-1), then

$$\langle P\check{Q}\rangle = \langle PQ\rangle = -\frac{1}{2}d^2 \tag{2.4}$$

where d is the distance between them. Points can be considered as dual spheres with zero radius. When $P = p - \frac{1}{2}\rho_p^2 n_\infty$ and $Q = q - \frac{1}{2}\rho_q^2 n_\infty$ are dual spheres (grade-1), we get

$$\langle P\check{Q}\rangle = \langle PQ\rangle = \langle pq\rangle + \frac{1}{2}\left(\rho_p^2 + \rho_q^2\right)$$

$$= -\frac{1}{2}d^2 + \frac{1}{2}\left(\rho_p^2 + \rho_q^2\right).$$

As the radii ρ_p and ρ_q are constant under rigid body motion, this effectively reduces to the point case, and two spheres are considered similar if their centres are close. When P and Q are normalised spheres (grade-4), we get exactly the same expression because of the way the check operator $\check{}$ is defined. A physical interpretation of $\langle P\check{Q}\rangle$ in terms of a line segment joining the spheres is given in [2, Fig. 14.8, p. 418].

Fig. 2.3 Graph showing $\cos(\theta)$ and $-\frac{1}{2}\theta^2 + 1$. As $\cos(\theta)$ (and $\sin^2(\theta)$) turn up so frequently in geometric calculations, we should embrace there advantages over θ^2

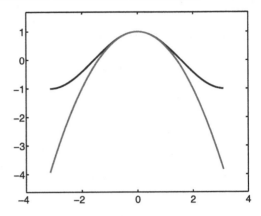

Flats Flats are objects like planes and lines. A flat can be modelled $P = p \wedge V \wedge n_\infty$, where p is a point on the flat and V is a Euclidean blade representing the direction of the flat. If V is a Euclidean vector, then $p \wedge V \wedge n_\infty$ is grade-3 and represents a line. If V is a Euclidean bivector, then $p \wedge V \wedge n_\infty$ is grade-4 and represents a plane. The other cases are less interesting in the current application: if $V = 1$, then $p \wedge n_\infty$ is a flat point, and if V is a Euclidean trivector, then $p \wedge V \wedge n_\infty$ represents a volume but is both translation and rotation invariant. If P and Q are normalised flats so that $|P| = 1$ and $|Q| = 1$, then

$$\langle P\check{Q} \rangle = \cos(\theta) \tag{2.5}$$

where θ is the dihedral angle between them. Two flats are considered similar if the angle between them is small. Note that for small θ, $\cos(\theta) \approx -\theta^2/2! + 1$ as shown in Fig. 2.3. There is no drawback in maximising $\cos(\theta)$ as opposed to $-\theta^2/2!$ for many practical situations. Using $\cos(\theta)$ can even have an added benefit. Because $\cos(\theta) \geq -1$, it restricts the influence of outliers, so we are more likely to get an acceptable solution even with significant outliers. If required, we can then reject outliers and refit until the fit is acceptable. One potential concern with the measure is that it does not capture the distance between lines, only the angle. The distance is usually regarded as the closest distance that the lines pass. It is a simple matter to determine this distance, for example, by forming the motor $P\widetilde{Q}$ and making use of Chasles's decomposition. It is not clear how to do this while keeping the simple form of a scalar product $\langle P\check{Q} \rangle$. This is not so much of a concern with planes as they will always intersect unless they are exactly parallel, so we are often only interested in the angle between them. When there is a specific point of interest on a line or plane, we should consider modelling it as a tangent instead of a flat as discussed below.

Directions Directions are used to model 1D direction and attitude and can be represented in CGA in the form $\Delta = V n_\infty$ where V is a Euclidean blade. They are translation invariant, so for translator T, we have $T\Delta\widetilde{T} = \Delta$. The case where V is grade-1 gives a 1D or line direction, and the case where $\mathbf{V}$ is grade-2 gives a 2D or plane direction. The other cases (scalar and grade-3) are of no practical interest here. For the scalar case, we get a scale multiple of n_∞, and for the grade-3 case, we get a scale multiple of $I_3 n_\infty$, both of which are translation and rotation invariant. If Δ_p and Δ_q are two directions, then $\langle \Delta_p \check{\Delta}_q \rangle = 0$, so we cannot use the directions directly. We can construct a meaningful quantity by representing the directions as flats $n_o \wedge \Delta$, dual flats $n_o \cdot \Delta^*$, or Euclidean directions $n_o \cdot \Delta$. If P and Q are two normalised directions represented in one of the above three forms, then

$$\langle P \check{Q} \rangle = \cos(\theta) \tag{2.6}$$

where θ is the dihedral angle between them. Two directions are considered similar if the angle between them is small.

Tangents Tangents have both location and direction and can be used to model various objects such as tangent planes on a surface, tangent lines on a curve, and rays leaving a camera where the optical centre is the location. A tangent at location p with normalised direction $\Delta = \mathbf{V} n_\infty$ can be represented in CGA as a blade $T' = p \wedge (p \cdot \hat{\Delta})$. If Δ is a bivector, then T' is a tangent line, and if Δ is a trivector, then T' is a tangent plane. When $\Delta = n_\infty$, then $T' = p$, and when $\Delta = I_3 n_\infty$, then $T' = p^*$, and we see that a point can be regarded as a degenerate tangent. Unfortunately, except in the case of points, taking the inner product between two tangents in this form does not give a particularly meaningful quantity. If we are prepared to consider a broader range of representations than blades, then we can construct a meaningful quantity using the measure. To be concrete, we will discuss the case of tangent lines first. Let $T = p + \Lambda$ be a flag (nested sequence of linear spaces) representation of the tangent with grade-1 and 3 parts, where p is the tangent location, and $\Lambda = T' \wedge n_\infty$ is the carrier line with $p \wedge \Lambda = 0$. The representations T and T' are equivalent with $T' = \langle T \rangle_1 \cdot \hat{T}$ and $T = (1 + \tilde{T}')(T' \wedge n_\infty)$. If $P = p + \Lambda_p$ and $Q = q + \Lambda_q$ are two tangent lines, then

$$\langle P \check{Q} \rangle = \langle pq \rangle + \langle \Lambda_p \check{\Lambda}_q \rangle$$
$$= -\frac{1}{2} d^2 + \cos(\theta)$$
$$\approx -\frac{1}{2} \left(d^2 + \theta^2 \right) + 1,$$

where d is the distance between the tangent locations, and θ is the dihedral angle between the tangent carriers. Two tangents are considered similar if their locations are close and the angle between them is small. We can adjust ratio of the locational and angular parts by encoding a weight in the line. For example, if $w = |\Lambda|$, then

$$\langle P \check{Q} \rangle = -\frac{1}{2} d^2 + w^2 \cos(\theta)$$
$$\approx -\frac{1}{2} \left(d^2 + w^2 \theta^2 \right) + w^2.$$

Exactly the same construction works with tangent planes. Here we take $P = p + \Pi_p$ and $Q = q + \Pi_q$ to be two tangent planes where Π_p, Π_q are planes with $p \wedge \Pi_p = 0$ and $q \wedge \Pi_q = 0$.

Rounds Rounds are objects like spheres, circles, and point pairs. We have already discussed spheres above, and we will now generalise this to include the remaining round objects. A direct round can be represented in CGA as a blade of the form $R = s \wedge (s \cdot \hat{\Delta})$ where s is a dual sphere and $\Delta = \mathbf{V}n_\infty$ is the direction. This is the same expression as for tangents, and tangents can simply be regarded as rounds with zero radius. A normalised direct round object R can also be represented as a tangent-like flag object $T = s + F$ where s is a dual sphere and F is a carrier flat with $s \wedge F = 0$. Just as for tangents the two representations are equivalent with $T = (1 + \tilde{R})(R \wedge n_\infty)$ and $R = \langle T \rangle_1 \cdot \hat{T}$. If $P = s_p + F_p$ and $Q = s_q + F_q$ are two rounds represented in this way, with radii ρ_p and ρ_q, respectively, then

$$\langle P\check{Q} \rangle = -\frac{1}{2}d^2 + \cos(\theta) + \frac{1}{2}(\rho_p^2 + \rho_q^2),$$

where d is the distance between the centres of the rounds, and θ is the dihedral angle between the carrier flats. As mentioned when discussing spheres, the radii are invariant under rigid body motion, so this effectively reduces to the tangent case. If P and Q are direct spheres, then $P \wedge n_\infty = -I_{4,1}$ and $Q \wedge n_\infty = -I_{4,1}$ and $\cos(\theta) = 1$, and it reduces further to the point case.

We have associated a physically meaningful measure with the basic objects available in CGA. Some objects, such as points, spheres, and flats, are represented in their basic blade form, and we will refer to these as primitive objects. Other objects, such as rounds and tangents, are represented in flag form and constructed using primitive objects. The directions, on the other hand, are converted to a primitive object representation. Other ways of representing the objects P and Q can be designed to give different measures. The only structural requirement is that they are expressed in the form $\langle P\check{Q} \rangle$.

2.4.2 Motor Estimation Problem Formulation

We are now in a position to formulate the estimation problem. Let P_k, $k = 1, \ldots, n$, be a set of normalised CGA objects before motion, and Q_k, $k = 1, \ldots, n$, be the set of objects after motion, $w_k \in \mathbb{R}$ be scalar weights, and $M \in \mathcal{M}$. The total similarity is given by the weighted sum of the symmetrised similarity between $MP_k\tilde{M}$ and Q_k as follows:

$$E = \frac{1}{2}\sum_{k=1}^{n} w_k \left(\langle M P_k \tilde{M} \check{Q}_k \rangle + \langle \check{Q}_k M \tilde{P}_k \tilde{M} \rangle \right) = \langle \tilde{M}\mathscr{L}M \rangle, \tag{2.7}$$

where

$$\mathscr{L}X = \frac{1}{2}\sum_{k=1}^{n} w_k(\check{Q}_k X P_k + \tilde{\check{Q}}_k X \tilde{P}_k).\tag{2.8}$$

Note that $\mathscr{L}$ satisfies the useful symmetry property $\langle\tilde{A}\mathscr{L}B\rangle = \langle\tilde{B}\mathscr{L}A\rangle$ for all $A, B \in \mathbb{R}_{4,1}$. If P_k and Q_k have the same symmetry and are either both symmetric (i.e. $A = \tilde{A}$) or both antisymmetric (i.e. $A = -\tilde{A}$), then $\mathscr{L}X$ reduces to $\mathscr{L}X = \sum_{k=1}^{n} w_k \check{Q}_k X P_k$. This is clearly true when P_k and Q_k are homogeneous (and the same grade). However, in some mixed grade situations (e.g. for the flags $P = \langle P\rangle_1 + \langle P\rangle_3$ and $Q = \langle Q\rangle_1 + \langle Q\rangle_3$) we require the full form given by (2.8). The data P_k, $k = 1, \dots, n$, need not all be of the same object type but could contain a variety of geometric objects such as points, spheres, flats, and directions. Clearly, for a given k, P_k and Q_k represent the same object type as one is simply a rotated and translated version of the other. The magnitude of the weights w_k can be used to adjust the contribution a data element makes based on its reliability, or to introduce attractive and repulsive contributions. We can now couch the problem of finding an optimal motor more precisely as maximising $\langle\tilde{X}\mathscr{L}X\rangle$ subject to $X \in \mathcal{M}$. Using Lemma 2.1, we can rewrite this as

$$\max_{X \in \mathbb{M}}\langle\tilde{X}\mathscr{L}X\rangle \text{ subject to } \langle X\tilde{X}\rangle = 1 \text{ and } \langle X\tilde{X}\rangle_4 = 0.\tag{2.9}$$

2.4.3 Optimal Rotator and Translator Estimation

First consider the simpler case of rotator estimation so that problem (2.9) reduces to

$$\max_{X \in \mathbb{R}_3^+}\langle\tilde{X}\mathscr{L}X\rangle \text{ subject to } \langle X\tilde{X}\rangle = 1\tag{2.10}$$

This has a simple solution which is captured in the following theorem.

Theorem 2.1 *Let P_k and Q_k, $k = 1, \dots, n$, be two sets of normalised conformal objects in $\mathbb{R}_{4,1}$, $w_k \in \mathbb{R}$ be scalar weights, and $\mathscr{L}$ be defined by*

$$\mathscr{L}X = \frac{1}{2}\sum_{k=1}^{n} w_k(\check{Q}_k X P_k + \tilde{\check{Q}}_k X \tilde{P}_k).$$

Then the maximiser of $\langle\tilde{R}\mathscr{L}R\rangle$ subject to $R \in \mathscr{R}$ is an eigenrotator of $P_R\mathscr{L}$ associated with the largest eigenvalue, where P_R is the projection onto $\mathbb{R}_3^+$.

Proof The Lagrange function associated with problem (2.10) is given by $L(X) = \frac{1}{2}\langle\tilde{X}\mathscr{L}X\rangle - \frac{\alpha}{2}(\langle\tilde{X}X\rangle - 1)$ where $X \in \mathbb{R}_3^+$. Using the first-order optimality condition $\partial_{\tilde{X}}L = 0$ and noting that $P_R X = X$ gives $P_R\mathscr{L}X = \alpha X$ at the maximiser. In addition, $\alpha = \langle \tilde{X} \mathscr{L} X \rangle$, so X is the eigenrotator of $P_R \mathscr{L}$ associated with the largest eigenvalue.　　　　　　　　　　　　　　　　　　　　　　　　　　　　　□

The optimal rotator can be readily obtained by forming the matrix representative of $P_R \mathscr{L}$ as outlined in the following procedure:

1. Form an orthonormal basis $e_k, k = 1, \ldots, 4$, of $\mathbb{R}_3^+$ (e.g. $\{1, e_{12}, e_{13}, e_{23}\}$).
2. Form the 4×4 symmetric matrix $L_{ij} = \langle \tilde{e}_i P_R \mathscr{L} e_j \rangle = \langle \tilde{e}_i \mathscr{L} e_j \rangle$.
3. Calculate $r \in \mathbb{R}^4$, a unit eigenvector of L associated with the largest eigenvalue.
4. Calculate the optimal rotator $R = \sum_k r_k e_k \in \mathscr{R}$.

If the dimension d of the eigenspace associated with the largest eigenvalue is greater than one, then the optimal eigenrotator is not unique. This will happen in degenerate situations such as estimating a rotator from a single pair of planes. The planes will be made parallel, but any additional rotation about an axis normal to the planes is permissible and will not affect the measure. A specific solution can be returned at the expense of a small increase in complexity as follows. Let $V \in \mathbb{R}^{4 \times d}$, $d \leq 4$, be an orthogonal matrix whose range is the eigenspace of L associated with the largest eigenvalue. Any maximum unit eigenvector can be expressed as $r = Vx$ for unit vector $x \in \mathbb{R}^d$. Note $\cos(\frac{\theta}{2}) = \langle R \rangle = \sum_k r_k \langle e_k \rangle = r^T z$ where $z \in \mathbb{R}^4$ with $z_k = \langle e_k \rangle$, and θ is the angle of rotation. With the natural basis above we get $z = (1 \quad 0 \quad 0 \quad 0)^T$. Hence $x^T V^T z$ can be identified with $\cos(\frac{\theta}{2})$. Maximising $x^T V^T z$ subject to $x^T x = 1$ gives the following enhancement to step 3 above:

3'. Calculate $r = \mathrm{unit}(V V^T z) \in \mathbb{R}^4$, the unit eigenvector of L associated with the largest eigenvalue and the smallest angle of rotation, where $z \in \mathbb{R}^4$ with $z_k = \langle e_k \rangle, k = 1, \ldots, 4$.

If $d = 1$, then there is no choice, and $r = V$ or $r = -V$, as expected.

This approach has an advantage over the standard methods of estimating an orthogonal 3×3 matrix using polar decomposition (or SVD) because improper rotations are excluded at the outset rather than removed at the end with a determinant check [3]. This advantage can be achieved with a matrix formulation based on quaternions [3]. However, the rotator formulation is also directly applicable to a wider range of objects than just points, including spheres, flats, and directions, and allows all these objects to be incorporated into a single framework.

The translator case is simpler because we can encode the constraint in the parameterisation of the translator. Let $T = 1 + Q$ where $Q = q_1 e_1 n_\infty + q_2 e_2 n_\infty + q_3 e_3 n_\infty \in \mathbb{R}_3^1 n_\infty$. Let $\mathscr{F}^+$ denote the Moore–Penrose pseudo-inverse of a linear transformation $\mathscr{F}$.

Theorem 2.2 *Let P_k and Q_k, $k = 1, \ldots, n$, be two sets of normalised conformal objects in $\mathbb{R}_{4,1}$, $w_k \in \mathbb{R}$ be scalar weights, and $\mathscr{L}$ be defined by*

$$\mathscr{L} X = \frac{1}{2} \sum_{k=1}^{n} w_k (\check{Q}_k X P_k + \tilde{\check{Q}}_k X \tilde{P}_k).$$

Then the maximiser of $\langle \tilde{T} \mathscr{L} T \rangle$ subject to T being a translator is given by $T = 1 + Q$ where $Q = -(\bar{P}_Q \mathscr{L} P_Q)^+ \mathscr{L} 1$.

Proof The objective function is given by $L(T) = \langle \widetilde{T} \mathscr{L} T \rangle = \langle 1 \mathscr{L} 1 \rangle + 2 \langle \widetilde{Q} \mathscr{L} 1 \rangle + \langle \widetilde{Q} \mathscr{L} Q \rangle$ for $T \in \mathbb{T}$. The first-order optimality condition $\partial_{\widetilde{Q}} L = 0$ gives $\bar{\mathsf{P}}_Q \mathscr{L} Q + \bar{\mathsf{P}}_Q \mathscr{L} 1 = 0$, so $Q = -(\bar{\mathsf{P}}_Q \mathscr{L} \mathsf{P}_Q)^+ \mathscr{L} 1$. $\qquad\qquad\square$

The optimal translator can be obtained by forming the matrix representative of $\bar{\mathsf{P}}_{\mathbb{T}} \mathscr{L}$ as outlined in the following procedure:

1. Form a basis e_k, $k = 1, \ldots, 4$, of $\mathbb{T}$, where the first basis vector is scalar (e.g. $\{1, e_1 n_\infty, e_2 n_\infty, e_3 n_\infty\}$).
2. Form the 4×4 symmetric matrix $L_{ij} = \langle \widetilde{e}_i \bar{\mathsf{P}}_{\mathbb{T}} \mathscr{L} e_j \rangle = \langle \widetilde{e}_i \mathscr{L} e_j \rangle$ and break it into sub-matrices $L = \left(\begin{smallmatrix} L_{rr} & L_{rq} \\ L_{qr} & L_{qq} \end{smallmatrix} \right)$ where $L_{rr} \in \mathbb{R}$ and $L_{qq} \in \mathbb{R}^{3 \times 3}$.
3. Calculate $q = -L_{qq}^+ L_{qr} \in \mathbb{R}^3$.
4. Form the full coefficient vector $t = \left(\begin{smallmatrix} 1 \\ q \end{smallmatrix} \right) \in \mathbb{R}^4$.
5. Calculate the optimal translator $T = \sum_k t_k e_k$.

The use of the Moore–Penrose pseudo-inverse will ensure that the smallest translation q is returned when there is not a unique maximiser of $\langle \widetilde{T} \mathscr{L} T \rangle$. This will happen when no locational information is provided, for example, finding the translator between two sets of directions. In such a case the above procedure will return an identity translator $T = 1$.

2.4.4 Optimal Motor Estimation as an Eigenrotator Problem

It is interesting to see that much of the structure for rotators and translators is preserved when we consider the more complex case of motor estimation. First note that the key difference between the full motor problem (2.9) and the rotator problem (2.10) is the addition of the extra constraint $\langle X \widetilde{X} \rangle_4 = 0$. We will show that by restricting the representation of CGA objects the constraint $\langle X \widetilde{X} \rangle_4 = 0$ can be dropped entirely, leaving a problem no more difficult than the rotator estimation problem. The other difference between problems (2.9) and (2.10) is the linear space involved. The motors lie in $\mathbb{M}$, while the rotators lie in $\mathbb{R}_3^+ \subset \mathbb{M}$. The only implication is that $\mathbb{M}$ is incomplete in the sense discussed previously: we cannot construct a reciprocal basis that also lies in $\mathbb{M}$. The following lemma characterises those elements which are nearly motors, where we have not enforced the constraint $\langle X \widetilde{X} \rangle_4 = 0$.

Lemma 2.4 $X \in \mathbb{M}$ *and* $\langle X \widetilde{X} \rangle = 1 \Leftrightarrow X = M + \beta M I_3 n_\infty$, $M \in \mathscr{M}$, *and* $\beta \in \mathbb{R}$.

Proof Let $X = MS$ be the polar decomposition of $X \in \mathbb{M}$ with $S = \alpha + \beta I_3 n_\infty$. Because $1 = \langle \widetilde{X} X \rangle = \langle S^2 \rangle = \alpha^2$ and $\alpha \geq 0$, we have $\alpha = 1$ and $X = M + \beta M I_3 n_\infty$. If $X = M + \beta M I_3 n_\infty$ where $M \in \mathscr{M} \subset \mathbb{M}$, then X is the sum of products of elements in $\mathbb{M}$, so $X \in \mathbb{M}$, and $\langle X \widetilde{X} \rangle = \langle M \widetilde{M} \rangle = 1$. $\qquad\square$

On the LHS of Lemma 2.4 we have the 8D space $\mathbb{M}$ with one constraint imposed, and on the RHS we have the 6D motor manifold with an extra degree of freedom added through β. It is convenient to use the notation $\Psi = I_3 n_\infty = \widetilde{\Psi}$ for the quad-vector basis element of $\mathbb{M}$ as it is used frequently. In addition, we will denote the set

Fig. 2.4 Sketch showing the
2D normal space $\mathcal{N}_M$ of $\mathcal{M}$
at M (restricted to $\mathbb{M}$).
Imposing the constraint
$\langle X\widetilde{X}\rangle = 1$ restricts us the 1D
subspace of $\mathcal{N}_M$ consisting
of elements of the form
$M(1 + \beta I_3 n_\infty)$

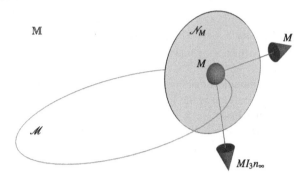

defined in Lemma 2.4 by $\mathcal{M}' = \{M + \beta M\Psi : M \in \mathcal{M}, \beta \in \mathbb{R}\}$. One way to study
the problem is to consider the behaviour of the objective function $\langle \widetilde{X}\mathcal{L}X\rangle$ with elements $X \in \mathcal{M}'$. This allows us to separate out the terms which result from relaxing
the constraint $\langle X\widetilde{X}\rangle_4 = 0$. When $\beta = 0$, X lies on the motor manifold $\mathcal{M}$. As $|\beta|$
increases, X leaves $\mathcal{M}$ along a 1D subspace of $\mathcal{N}_M$. A sketch of the situation is
shown in Fig. 2.4. Note that M is both a point on $\mathcal{M}$ and a direction vector in $\mathcal{N}_M$.
Expanding the objective function at $X = M + \beta M\Psi \in \mathcal{M}'$ gives

$$\langle \widetilde{X}\mathcal{L}X\rangle = \langle \widetilde{M}\mathcal{L}M\rangle + 2\beta\langle \widetilde{M}\mathcal{L}(M\Psi)\rangle + \beta^2\langle \widetilde{\Psi}\widetilde{M}\mathcal{L}(M\Psi)\rangle. \qquad (2.11)$$

For a given $M \in \mathcal{M}$, this is a quadratic in β. We are interested in the cases where
the coefficient of β vanishes and coefficient of β^2 is not positive, independently
of M. When the coefficient of β^2 is negative, leaving $\mathcal{M}$ decreases the objective
function, and maximising $\langle \widetilde{X}\mathcal{L}X\rangle$ subject to $X \in \mathcal{M}'$ will give us the optimal motor
which solves problem (2.9). If the coefficient of β^2 vanishes, then the solution is not
unique, and if $M \in \mathcal{M}$ is a solution, then so is $M(1 + \beta\Psi)$. In such situations we
can maximise $\langle \widetilde{X}\mathcal{L}X\rangle$ to give a solution, and then project the resulting X onto $\mathcal{M}$
to get the optimal motor. We first make some general observations which help to
manipulate (2.11).

1. When $\mathcal{L}X = \frac{1}{2}\sum_{k=1}^{n} w_k(\check{Q}_k X P_k + \widetilde{\check{Q}}_k X \widetilde{P}_k)$ is substituted in the coefficients of
   β and β^2, the term $P_k' = P_k'(M) = \widetilde{M}Q_k M$ turns up which has the same grades
   as Q_k and P_k. (It represents the same kind of object.)
2. The coefficient of β is made up of terms $\langle \Psi(\check{P}_k' P_k + P_k \check{P}_k')\rangle$.
3. If P_k and Q_k have the same symmetry, the coefficient of β reduces to $2\langle \Psi \check{P}_k' P_k\rangle$.
4. The coefficient of β^2 is made up of terms $\langle \Psi \check{P}_k' \Psi P_k\rangle$.
5. $\langle \Psi \check{P}' \Psi P\rangle = -\langle n_\infty \check{P}' n_\infty P\rangle$ for all $P, P' \in \mathbb{R}_{4,1}$.
6. $\langle \widetilde{X}\check{Q}Y P\rangle = \langle \widetilde{X}\check{q}Y p\rangle$ where $p = P^*$ and $q = Q^*$ for all $X, Y, P, Q \in \mathbb{R}_{4,1}$.

We will examine what conditions need to be imposed on P_k, Q_k, and w_k so that we
can ensure that the coefficient of β vanishes and the coefficient of β^2 is not positive.
Let us first consider the cases where P_k and Q_k are homogeneous and then extend
to mixed grade elements. We only need to provide proofs for scalars, vectors, and
bivectors because the trivector, quadvector, and pseudoscalar cases follow by observation 6 above. First examine the case where P_k and Q_k are vectors or quadvectors.

Lemma 2.5 *Let P_k and Q_k, $k = 1, \ldots, n$, be two sets of vectors or quadvectors. Then $\langle \widetilde{X} \mathscr{L} X \rangle = \langle \widetilde{M} \mathscr{L} M \rangle - \beta^2 \langle n_\infty \mathscr{L} n_\infty \rangle$.*

Proof As $\check{P}'_k P_k$ has no grade-4 part, the coefficient of β vanishes. Also note that for vector or quadvector Q_k, we have $n_\infty \widetilde{M} Q_k M n_\infty = \widetilde{M} n_\infty Q_k n_\infty M = n_\infty Q_k n_\infty$ so $\langle n_\infty \check{P}'_k n_\infty P_k \rangle = \langle n_\infty \check{Q}_k n_\infty P_k \rangle$, and the coefficient of β^2 is independent of M. □

Using Lemma 2.5, we can now provide the following useful results for normalised points, spheres, and dual spheres; and planes and dual planes:

Lemma 2.6 *Let P_k and Q_k, $k = 1, \ldots, n$, be two sets of normalised conformal points, spheres, or dual spheres in $\mathbb{R}_{4,1}$, w_k be scalar weights with $\sum_k w_k > 0$, and*

$$\mathscr{L} X = \frac{1}{2} \sum_{k=1}^{n} w_k (\check{Q}_k X P_k + \widetilde{Q}_k X \widetilde{P}_k).$$

Then the maximiser of $\langle \widetilde{X} \mathscr{L} X \rangle$ subject to $X \in \mathbb{M}$ and $\langle \widetilde{X} X \rangle = 1$ is a motor.

Proof Assume that X is not a motor, so $X = M(1 + \beta \Psi)$. For normalised spheres, dual spheres, and points, we have $\langle n_\infty \check{Q}_k n_\infty P_k \rangle = 2$, so $\langle n_\infty \mathscr{L} n_\infty \rangle = 2 \sum_k w_k > 0$. By Lemma 2.5 we have $\langle \widetilde{X} \mathscr{L} X \rangle = \langle \widetilde{M} \mathscr{L} M \rangle - 2\beta^2 \sum_k w_k$, and X cannot be a maximiser. □

It is interesting that only the sum of the weights $\sum_k w_k > 0$ need be positive. Some points can have a repulsive force as long as the sum of the attractive contribution is greater than the repulsive terms. The result for planes is as follows:

Lemma 2.7 *Let P_k and Q_k, $k = 1, \ldots, n$, be two sets of normalised conformal planes or dual planes in $\mathbb{R}_{4,1}$, and $\mathscr{L}$ be defined by (2.8). The maximum value of $\langle \widetilde{X} \mathscr{L} X \rangle$ subject to $X \in \mathbb{M}$ and $\langle \widetilde{X} X \rangle = 1$ is obtained by a motor.*

Proof For planes or dual planes P_k and Q_k, we have $\langle n_\infty \check{Q}_k n_\infty P_k \rangle = 0$ so $\langle n_\infty \mathscr{L} n_\infty \rangle = 0$, and the coefficient of β^2 also vanishes. □

This is a weaker result than for points and spheres since we can only state that the maximum is obtained by a motor because the maximiser is not unique. If $M \in \mathscr{M}$ is a maximiser, then so is $X = M + \beta M \Psi$. The case of bivectors and trivectors is not quite as clean.

Lemma 2.8 *Let P_k and Q_k, $k = 1, \ldots, n$, be two sets of bivectors or trivectors such that $n_\infty P_k n_\infty = n_\infty Q_k n_\infty = 0$. Then $\langle \widetilde{X} \mathscr{L} X \rangle = \langle \widetilde{M} \mathscr{L} M \rangle$.*

Proof $n_\infty P_k n_\infty = n_\infty Q_k n_\infty = 0$ iff Q_k and P_k have no terms of the form $\mathbf{V} n_o$ where $\mathbf{V}$ is a Euclidean blade. This precludes the appearance of a term $\widetilde{I}_3 n_o$ in the product $\check{P}' P$; hence $\langle \Psi \check{P}_k P_k \rangle = 0$, and the coefficient of β vanishes. As $n_\infty Q_k n_\infty = 0$, we have $\langle \Psi \check{P}_k \Psi P_k \rangle = 0$, and the coefficient of β^2 also vanishes. □

While Lemma 2.8 is somewhat restrictive, it is still sufficiently general to allow the following useful result for lines, which is analogous to Lemma 2.7 for planes.

Lemma 2.9 *Let P_k and Q_k, $k = 1, \ldots, n$, be two sets of normalised conformal lines or dual lines in $\mathbb{R}_{4,1}$, and $\mathscr{L}$ be defined by (2.8). The maximum value of $\langle \widetilde{X} \mathscr{L} X \rangle$ subject to $X \in \mathbb{M}$ and $\langle \widetilde{X} X \rangle = 1$ is obtained by a motor.*

Proof If P_k and Q_k are lines, then $n_\infty P_k n_\infty = n_\infty Q_k n_\infty = 0$. □

For completeness, the case for scalars and grade-5 elements is also given but is of limited interest as these elements are invariant to rigid body motion.

Lemma 2.10 *If P_k and Q_k, $k = 1, \ldots, n$, are scalars or grade-5, then $\langle \widetilde{X} \mathscr{L} X \rangle = \langle \widetilde{M} \mathscr{L} M \rangle = \sum_k w_k \langle P_k \check{Q}_k \rangle$.*

Proof If P_k and Q_k are scalar, $\langle \Psi \check{P}_k' P_k \rangle = \check{P}_k' P_k \langle \Psi \rangle = 0$, and the coefficient of β vanishes. Similarly $\langle \Psi \check{P}_k' \Psi P_k \rangle = \check{P}_k' P_k \langle \Psi^2 \rangle = 0$, so the coefficient of β^2 also vanishes. Also $\langle \widetilde{M} \mathscr{L} M \rangle = \sum_{k=1}^n w_k \langle \widetilde{M} \check{Q}_k M P_k \rangle = \sum_{k=1}^n w_k \langle \check{Q}_k P_k \rangle$. □

The cases where P and Q are mixed grade can now be expressed in terms of the homogeneous cases. We will only consider the mixed grade case where $P = \langle P \rangle_1 + \langle P \rangle_r$ and $Q = \langle Q \rangle_1 + \langle Q \rangle_r$ because this is all we currently require. The coefficient of β will have terms of the form

$$\langle \Psi (\check{P}' P + P \check{P}') \rangle = 2 \langle \Psi \langle \check{P}' \rangle_1 \langle P \rangle_1 \rangle + 2 \langle \Psi \langle \check{P}' \rangle_r \langle P \rangle_r \rangle$$
$$+ 2 \langle \Psi (\langle \check{P}' \rangle_1 \cdot \langle P \rangle_r + \langle P \rangle_1 \cdot \langle \check{P}' \rangle_r) \rangle. \tag{2.12}$$

The first two terms involve a single grade and are handled by the homogeneous cases. The last term can only make a contribution when $r = 5$. The coefficient of β^2 will have terms of the form

$$\langle \Psi \check{P}' \Psi P \rangle = - \langle n_\infty \langle \check{P}' \rangle_1 n_\infty \langle P \rangle_1 \rangle - \langle n_\infty \langle \check{P}' \rangle_r n_\infty \langle P \rangle_r \rangle$$
$$- \langle n_\infty \langle \check{P}' \rangle_1 n_\infty \langle P \rangle_r \rangle - \langle n_\infty \langle P \rangle_1 n_\infty \langle \check{P}' \rangle_r \rangle. \tag{2.13}$$

Again the first two terms involve a single grade and are handled by the homogeneous cases. If $v = \langle v \rangle_1$, then $n_\infty v n_\infty = 2(v \cdot n_\infty) n_\infty$ is a scale multiple of n_∞. The last two terms can only make a contribution if $r = 1$, which has already been taken into consideration by the first two terms. Let

$$\mathscr{L}_r X = \sum_{k=1}^n w_k (\langle \check{Q}_k \rangle_r X \langle P_k \rangle_r + \widetilde{\langle \check{Q}_k \rangle_r X \langle P_k \rangle_r})$$

denote the restriction of $\mathscr{L}$ to the grade-r parts of P_k and Q_k. We can summarise the above discussion by stating that for mixed grade objects of the form $P_k = \langle P_k \rangle_1 + \langle P_k \rangle_r$, $Q_k = \langle Q_k \rangle_1 + \langle Q_k \rangle_r$, where $r \neq 5$, we have

$$\langle \widetilde{X} \mathscr{L} X \rangle = \langle \widetilde{X} \mathscr{L}_1 X \rangle + \langle \widetilde{X} \mathscr{L}_r X \rangle. \tag{2.14}$$

Lemma 2.5 to Lemma 2.10, together with the comments of mixed grade cases, tell us for which object representations we can ignore the constraint $\langle \widetilde{X} X \rangle_4 = 0$ during motor estimation. For convenience we, will refer to these objects as *admissible*, and we see immediately that all the objects represented earlier when discussing measures are admissible. We wish to maximise $\langle \widetilde{X} \mathscr{L} X \rangle$ where $X \in \mathscr{M}$ as stated in problem (2.9). For admissible objects, we can neglect the awkward condition $\langle \widetilde{X} X \rangle_4 = 0$ and solve

$$\max_{X \in \mathbb{M}} \langle \widetilde{X} \mathscr{L} X \rangle \text{ subject to } \langle X \widetilde{X} \rangle = 1. \tag{2.15}$$

Thus we can maximise $\langle \widetilde{X} \mathscr{L} X \rangle$ under the more relaxed constraints $X \in \mathbb{M}$ and $\langle X \widetilde{X} \rangle = 1$. This problem can be readily solved, and we can now present the generalisation of Lemma 2.1 and Lemma 2.2 to the case of motors:

Theorem 2.3 *Let P_k and Q_k, $k = 1, \ldots, n$ be two sets of admissible normalised conformal objects in $\mathbb{R}_{4,1}$, $w_k \in \mathbb{R}$ be scalar weights, and $\mathscr{L}$ be defined by*

$$\mathscr{L} X = \frac{1}{2} \sum_{k=1}^{n} w_k (\check{Q}_k X P_k + \tilde{Q}_k X \tilde{P}_k).$$

Then the maximiser of $\langle \widetilde{M} \mathscr{L} M \rangle$ subject to $M \in \mathscr{M}$ is given by $M = R + Q$ where R is an eigenrotor of $P_R \mathscr{L}'$ associated with the largest eigenvalue, $\mathscr{L}' = \mathscr{L} - \mathscr{L}(\bar{P}_Q \mathscr{L} P_Q)^+ \mathscr{L}$, and $Q = -(\bar{P}_Q \mathscr{L} P_Q)^+ \mathscr{L} R$.

Proof The Lagrange function associated with problem (2.15) is given by $L(X) = \frac{1}{2} \langle \widetilde{X} \mathscr{L} X \rangle - \frac{\alpha}{2} (\langle \widetilde{X} X \rangle - 1)$ for $X \in \mathbb{M}$. The first-order optimality condition $\partial_{\widetilde{X}} L = 0$ gives $\bar{P}_M \mathscr{L} X = \alpha \bar{P}_M X$. Let $X = R + Q \in \mathbb{M}$ where $R \in \mathbb{R}_3^+$ and $Q \in \mathbb{R}_3^- n_\infty$. Using $\bar{P}_M = P_R + \bar{P}_Q$, we can separate $\partial_{\widetilde{X}} L = 0$ into R and Q components as follows:

$$P_R \mathscr{L} R + P_R \mathscr{L} Q = \alpha R,$$
$$\bar{P}_Q \mathscr{L} R + \bar{P}_Q \mathscr{L} Q = 0.$$

This is a standard form for quadratic minimisation with a homogeneous quadratic constraint, and we can calculate Q from the second equation and then eliminate Q from the first equation in the usual way. This gives $P_R \mathscr{L}' R = \alpha R$ where $\mathscr{L}' = \mathscr{L} - \mathscr{L}(\bar{P}_Q \mathscr{L} P_Q)^+ \mathscr{L}$ and $Q = -(\bar{P}_Q \mathscr{L} P_Q)^+ \mathscr{L} R$. At the maximum, α equals

$\langle \widetilde{X} \mathscr{L} X \rangle$; therefore R is the eigenrotator of $P_R \mathscr{L}'$ associated with the largest eigenvalue.   □

We see that, as for the rotator case, the problem reduces to an eigenrotator problem. This motor estimation method is easily implemented by forming the matrix representative of $\bar{P}_{\mathbb{M}} \mathscr{L}$ as outlined in the following procedure:

1. Form a basis e_k, $k = 1, \ldots, 8$, of $\mathbb{M}$, where the first four basis vectors are associated with R and orthonormal, and the last four are associated with Q in the split $X = R + Q$ (e.g. $\{1, e_{12}, e_{13}, e_{23}, e_1 n_\infty, e_2 n_\infty, e_3 n_\infty, I_3 n_\infty\}$).
2. Form the 8×8 symmetric matrix $L_{ij} = \langle \widetilde{e}_i \bar{P}_{\mathbb{M}} \mathscr{L} e_j \rangle = \langle \widetilde{e}_i \mathscr{L} e_j \rangle$ and break it into 4×4 sub-matrices $L = \begin{pmatrix} L_{rr} & L_{rq} \\ L_{qr} & L_{qq} \end{pmatrix}$.
3. Form the 4×4 matrix $L' = L_{rr} - L_{rq}(L_{qq}^+ L_{qr})$.
4. Calculate $r = \text{unit}(V V^T z) \in \mathbb{R}^4$, the unit eigenvector of L' associated with the largest eigenvalue and the smallest angle of rotation, where $z \in \mathbb{R}^4$ with $z_k = \langle e_k \rangle$, $k = 1, \ldots, 4$.
5. Calculate $q = -(L_{qq}^+ L_{qr})r \in \mathbb{R}^4$.
6. Form the full coefficient vector $m = \begin{pmatrix} r \\ q \end{pmatrix} \in \mathbb{R}^8$.
7. Calculate the optimal motor $M = \sum_k m_k e_k \in \mathscr{M}$.

The key steps of estimating R and estimating Q are both robust in the sense that a reasonable value will be returned even if insufficient information is provided. Let $M = TR$ where $T = 1 - \frac{1}{2} t n_\infty$ is a translator and t is the Euclidean translation vector. Note that $Q = -\frac{1}{2} t R n_\infty$, so we have $|q| = |n_o \cdot Q| = \frac{1}{2}|t|$. The use of the Moore–Penrose pseudo-inverse will ensure that the smallest translation t is returned when there is not a unique maximiser as discussed after the procedure for estimating translators. The estimated motor will maximise the measure and provide the motor with the smallest translation and rotation angle when there is not a unique maximiser.

2.5   Examples

In this section we provide some illustrations of the algorithm. The data is generated as follows. A random geometric object P_k is generated, such as a point, sphere, line, circle, or tangent. Noise is added to the data P_k by perturbing it with a small random motor $M_k \approx 1$ before applying the general fixed rigid body transformation M_o to give $Q_k = M_o M_k P_k \widetilde{M}_k \widetilde{M}_o$. The noise is sufficient to provide clear delineation between the objects in the figures presented. The motor estimation procedure is then applied to the data pairs (P_k, Q_k), $k = 1, \ldots, K$, to obtain an optimal estimate M of M_o. In the figures presented the dark data is the source data after the action of the estimated motor $Q'_k = M P_k \widetilde{M}$, and the light data is the target data Q_k. The difference between the sets is the error remaining after applying the estimation procedure and is due to the noise on the data. If no noise is present, the fit is perfect, and the data sits exactly on top of each other.

Fig. 2.5 Two pairs of spheres used to estimate the rigid body motion. The centres all lie on a line

First consider the problem of fitting spheres. As discussed earlier, the radius of the spheres plays no role, as it is invariant to rigid body transformations, and the situation is identical to the case of noisy points. With just one pair of spheres, the fit is perfect, and the centres of the spheres coincide. The rotational part vanishes because the smallest angle of rotation is zero and the estimated motor is a pure translator. With two pairs of spheres, the optimal motor makes their centres lie on the same axis with equal separation as shown in Fig. 2.5. This example is a typical situation where there is insufficient information to get a unique maximiser of $\langle \widetilde{M} \mathscr{L} M \rangle$. For the estimated motor, the rotation about the axis is zero, and the rotation is in a plane parallel with the axis through the points before and after motion. A more general situation is shown in Fig. 2.6, where there are five pairs of noisy spheres.

The more complex example in Fig. 2.7 shows the algorithm being applied to five pairs of different objects. We use spheres, lines, circles, and 1D and 2D tangents. We have not included planes simply because, unlike lines, it is hard to visualise the separation between planes in a figure.

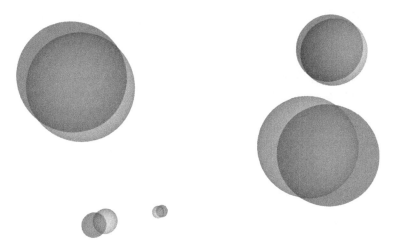

Fig. 2.6 Five pairs of spheres used to estimate the rigid body motion

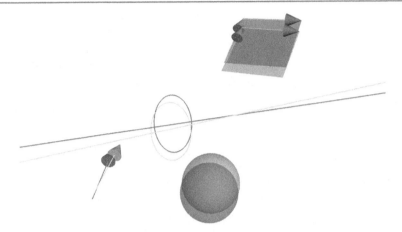

Fig. 2.7 Five pairs of different objects (spheres, lines, circles, 1D and 2D tangents) used to estimate the rigid body motion

2.6 Discussion

We have presented a technique for estimating motors from noisy geometric data. The data may comprise a variety of objects including points, rounds (point pairs, circles, spheres), flats (lines, planes), tangents, and directions. To assist the developments, we first studied the geometry of the motors in the smallest linear embedding space $\mathbb{M}$. The estimation technique reduced to a small eigenrotator problem and allowed the different types of geometric data to be combined naturally in a single framework while excluding reflection. In order to formulate the problem, we restricted the similarity measure between geometric objects P and Q to the simple form $\langle P \check{Q} \rangle$ (with the aid of a grade dependent sign operator). In addition we restricted the representation of objects to what we referred to as the admissible objects. These are representations that allow us to ignore the motor constraint $\langle M \tilde{M} \rangle_4 = 0$ during optimisation. With these restrictions, we are able to associate a physically meaningful measure to the primitive objects: points, spheres, lines, and planes. Other objects such as circles, point pairs, and tangents were incorporated by representing them as flags using sums of primitive objects. Directions were incorporated by representing them as associated flats. The estimation procedure reduced to a standard constrained optimisation problem with a closed-form solution, which could be expressed as an eigenrotator problem.

2.7 Exercises

2.1 Find a representation for spheres so that if P and Q are two spheres we get the following measure $\langle P \check{Q} \rangle = -\frac{1}{2} d^2 - \frac{1}{2}(\rho_p - \rho_q)^2$ where d is the distance between the centres, and $\rho_p - \rho_q$ the difference in radii.

2.2 Show that for $X, Y \in \mathbb{M}$ we have $|XY| = |X||Y|$.

2.3 Consider an object of the form $F = p + \Lambda + \Pi$, where p is a point, Λ a line through p, and Π a plane through p. Show that if $P = p + \Lambda_p + \Pi_p$ and $Q = q + \Lambda_q + \Pi_q$ have this form then $\langle P\check{Q} \rangle = -\frac{1}{2}d^2 + \cos(\theta) + \cos(\phi)$ where d is the distance between the points, θ is the dihedral angle between the lines, and ϕ the dihedral angle between the planes. Are objects of this form admissible? What if Λ and Π are perpendicular?

Acknowledgements This work was supported by the New Zealand Foundation for Research, Science and Technology.

References

1. Dorst, L., Valkenburg, R.: Square root and logarithm of rotors in 3D conformal geometric algebra using polar decomposition. In: Dorst, L., Lasenby, J. (eds.) Guide to Geometric Algebra in Practice. Springer, London (2011), Chap. 5 in this book
2. Dorst, L., Fontijne, D., Mann, S.: Geometric Algebra for Computer Science: An Object-Oriented Approach to Geometry. Morgan Kaufman, San Mateo (2007/2009)
3. Kanatani, K.: Geometric Computation for Machine Vision. Oxford University Press, Oxford (1993)
4. Valkenburg, R.J., Kakarala, R.: Lower bounds for the divergence of orientational estimators. IEEE Trans. Inf. Theory **47**(6) (2001)

Inverse Kinematics Solutions Using Conformal Geometric Algebra

3

Andreas Aristidou and Joan Lasenby

Abstract

This paper describes a novel iterative Inverse Kinematics (IK) solver, FABRIK, that is implemented using Conformal Geometric Algebra (CGA). FABRIK uses a forward and backward iterative approach, finding each joint position via locating a point on a line. We use the IK of a human hand as an example of implementation where a constrained version of FABRIK was employed for pose tracking. The hand is modelled using CGA, taking advantage of CGA's compact and geometrically intuitive framework and that basic entities in CGA, such as spheres, lines, planes and circles, are simply represented by algebraic objects. This approach can be used in a wide range of computer animation applications and is not limited to the specific problem discussed here. The proposed hand pose tracker is real-time implementable and exploits the advantages of CGA for applications in computer vision, graphics and robotics.

3.1 Introduction

This paper describes a fast iterative Inverse Kinematics (IK) solver which is implemented using Conformal Geometric Algebra (CGA). Geometric Algebra (GA) [1] provides a convenient mathematical notation for representing orientations and rotations of objects in three dimensions. The conformal model of GA extends the usefulness of the 3D GA by expanding the class of rotors to include translations, dilations and inversions, as well as being able to express lines, planes, circles and spheres as elements of the algebra. Rotors are more numerically stable and more

A. Aristidou (✉) · J. Lasenby
Department of Engineering, University of Cambridge, Trumpington Street, Cambridge CB2 1PZ, UK
e-mail: aa462@cam.ac.uk

J. Lasenby
e-mail: jl221@cam.ac.uk

L. Dorst, J. Lasenby (eds.), *Guide to Geometric Algebra in Practice*,
DOI 10.1007/978-0-85729-811-9_3, © Springer-Verlag London Limited 2011

efficient than rotation matrices, making GA popular for applications in computer graphics and robotics. A more detailed treatment of GA can be found in [2].

The CGA geometric representation and its algebraic richness offer great flexibility in the process of modelling virtual or mechanical objects. In this paper, a method for solving the IK problem of a 3D human hand, which uses the CGA framework, is presented. The model is highly constrained with both rotational and orientational constraints, allowing motion only within a feasible set. Using data from a markered optical motion capture system, the 3D hand pose was efficiently tracked and reconstructed. It is important to note that this is not a system designed specifically for the task of hand tracking and reconstruction, but rather to provide a framework for many IK applications in computer vision and robotics. Both the IK solver and the hand model are real-time implementable, and the system produces motion which is smooth and natural.

3.2   Background

Inverse Kinematics is defined as the problem of determining a set of appropriate joint configurations for which the end effectors move to desired positions as smoothly, rapidly and as accurately as possible. Several models have been implemented for solving the IK problem from different areas of study. A detailed review of IK solvers is given in [3]. Most of the literature which uses CGA to address the IK problem presents kinematic solutions which focus on the advantages that the CGA model offers, rather than presenting a complete IK solver. For instance, [4] gives a simple framework solution for a robot arm, underlining the generality and the efficiency of the CGA mathematical model for solving the IK problem. Corrochano and Kähler [5] used a language of points, lines and planes (which are later replaced by spheres in [6]) to solve the IK problem of a specific robot arm. Similar solutions were given by [7–10], where CGA was used to deal with forward kinematics, dynamics and projective geometry problems. In [11], a technique for the combination of very efficient algorithms, based on two different optimisation approaches using Gaigen 2 and MAPLE, is presented. CGA therefore appears to be a promising mathematical tool for computing the IK of a robot arm and solving the problem of visually guided grasping. Recently, [12] described an application of CGA to the analysis of a parallel manipulator with limited mobility. [13] gives a brief introduction to CGA and describes basic geometric entities; it also gives a synopsis of different IK framework solutions and grasping processes of a robot arm. Finally, [14] proposed an optimised algorithm to provide IK solutions using reconfigurable hardware, leading to very efficient implementations. In summary, most of these methods are applied to the simple kinematic problems of a robot arm with 5 degrees of freedom (DoF). They mainly describe how to constrain the movement of the arm to a feasible set within a framework rather than describing a solver itself. In this paper we describe a heuristic algorithm that solves the IK problem in an iterative fashion, akin to the popular CCD method [15]. FABRIK (Forward And Backward Reaching Inverse Kinematics) is a reliable iterative algorithm that uses points, lines and spheres to solve the IK problem. It divides the problem into two phases, a forward and a backward reaching approach, and it can treat most of the joint types and supports biomechanical constraints on chains with both single and multiple end effectors. It is fast, computationally efficient and provides visually smooth results.

3.3 FABRIK: An Iterative Inverse Kinematics Solver

FABRIK uses the previously calculated positions of the joints to find the updates in a forward and backward iterative manner. The proposed IK solver starts from the last joint of the chain and works forwards, adjusting each joint along the way. Thereafter, it works backwards in the same way, in order to complete a full iteration.

Therefore, assume that $p_1, \ldots, p_n$ are the joint positions of a manipulator. For the simple case where only a single end effector exists, take p_1 as the root joint and p_n as the end effector. The target is t, and the initial base position is b. First calculate the distances between each joint $d_i = |p_{i+1} - p_i|$ for $i = 1, \ldots, n - 1$. Then, to check whether the target is reachable or not, find the distance between the root and the target, $dist$, and if this distance is smaller than the total sum of all the inter-joint distances, $dist < \sum_1^{n-1} d_i$, the target is within reach; otherwise, it is unreachable. If the target is within reach, a full iteration is constituted by two stages. In the first stage, the algorithm estimates each joint position starting from the end-effector, p_n, moving inwards to the manipulator base, p_1. So, let the new position of the end-effector be the target position, $p_n' = t$. The new position of the $(n - 1)$th joint, p_{n-1}', is assigned as the nearest point on the sphere Σ_{n-1}, with centre the joint position p_n' and radii the distance d_{n-1} from the joint position p_{n-1}. Similarly, the new position of the $(n - 2)$th joint, p_{n-2}', is selected as the nearest point on sphere Σ_{n-2}, with centre the joint position p_{n-1}' and radii the distance d_{n-2} from the joint p_{n-2}. The algorithm continues until all new joint positions are calculated, including the root, p_1'. The nearest point on a sphere from a point in space is clearly found by simply taking a point along the line joining the centre of the sphere to the point, which has distance from the centre equal to the sphere radius. An entirely CGA solution is also given in Sect. 3.4.2.1.

A full iteration is completed when the same procedure is repeated but this time starting from the root joint and moving outwards to the end effector. Thus, let the new position for the first joint, p_1'', be its initial position b. Then, the new joint position p_2'' is assigned as the nearest point on the sphere Σ_1, with centre the p_1'' and radii the distance d_1 from the joint p_2'. This procedure is repeated for all the remaining joints, including the end effector. FABRIK is illustrated in pseudo-code in Algorithm 1, and a graphical representation of a full iteration of the algorithm is demonstrated in Fig. 3.1.

The forward and backward procedure is then repeated for as many iterations as needed, until the end effector is identical or close enough (to be defined) to the desired target. FABRIK can easily handle end effector orientations and supports, to the best of our knowledge, all chain classes. It can also cope with cases where the model has multiple chains and end effectors and is applicable to problems with

---

Algorithm 1: A full iteration of the FABRIK algorithm using CGA.

Input: The joint positions $\mathbf{p}_i$ for $i = 1, \ldots, n$, the target position $\mathbf{t}$ and the distances between each joint $d_i = |\mathbf{p}_{i+1} - \mathbf{p}_i|$ for $i = 1, \ldots, n-1$

Output: The new joint positions $\mathbf{p}_i$ for $i = 1, \ldots, n$.

1.1   % The distance between root and target
1.2 $dist = |\mathbf{p}_1 - \mathbf{t}|$
1.3   % Check whether the target is within reach
1.4   **if** $dist > d_1 + d_2 + \cdots + d_{n-1}$ **then**
1.5   % The target is unreachable
1.6   **for** $i = 1, \ldots, n-1$ **do**
1.7   % Find the nearest point on sphere, with centre the joint position $\mathbf{p}_i$ and radius the distance d_i, from a point is space, $\mathbf{t}$
1.8 $\mathbf{p}_{i+1} = \textbf{NearestPointSphere}(\mathbf{p}_i, d_i, \mathbf{t})$;
1.9   **end**
1.10 **else**
1.11   % The target is reachable; thus, set as $\mathbf{b}$ the initial position of the joint $\mathbf{p}_1$
1.12 $\mathbf{b} = \mathbf{p}_1$
1.13   % Check whether the distance between the end effector $\mathbf{p}_n$ and the target $\mathbf{t}$ is greater than a tolerance.
1.14 $dif_A = |\mathbf{p}_n - \mathbf{t}|$
1.15   **while** $dif_A > tol$ **do**
1.16   % STAGE 1: FORWARD REACHING
1.17   % Set the end effector $\mathbf{p}_n$ as target $\mathbf{t}$
1.18 $\mathbf{p}_n = \mathbf{t}$
1.19   **for** $i = n-1, \ldots, 1$ **do**
1.20   % Find the nearest point on sphere, with centre the joint position $\mathbf{p}_{i+1}$ and radius the distance d_i, from a point is space, $\mathbf{p}_i$
1.21 $\mathbf{p}_i = \textbf{NearestPointSphere}(\mathbf{p}_{i+1}, d_i, \mathbf{p}_i)$;
1.22   **end**
1.23   % STAGE 2: BACKWARD REACHING
1.24   % Set the root $\mathbf{p}_1$ its initial position.
1.25 $\mathbf{p}_1 = \mathbf{b}$
1.26   **for** $i = 1, \ldots, n-1$ **do**
1.27   % Find the nearest point on sphere, with centre the joint position $\mathbf{p}_i$ and radius the distance d_i, from a point is space, $\mathbf{p}_{i+1}$
1.28 $\mathbf{p}_i = \textbf{NearestPointSphere}(\mathbf{p}_i, d_i, \mathbf{p}_{i+1})$;
1.29   **end**
1.30 $dif_A = |\mathbf{p}_n - \mathbf{t}|$
1.31   **end**
1.32 **end**
1.33
1.34 % Where the function **NearestPointSphere**(X, Y, Z), finds the nearest point on a sphere from a point in space. X is the sphere's centre, Y is the sphere's radii and Z is the point in space.

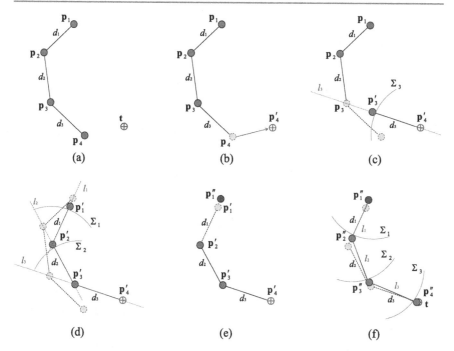

Fig. 3.1 A full iteration of FABRIK for the case of a single target and 4 joints using CGA. (**a**) The initial position of the manipulator and the target, (**b**) move the end effector $\mathbf{p}_4$ to the target, (**c**) find the joint $\mathbf{p}'_3$ which is the intersection of the sphere Σ_3 and the line l_3 which passes through the points $\mathbf{p}'_4$ and $\mathbf{p}_3$, (**d**) continue the algorithm for the rest of the joints, (**e**) the second stage of the algorithm: move the root joint $\mathbf{p}'_1$ to its initial position, (**f**) repeat the same procedure but this time start from the base and move outwards to the end effector. The algorithm is repeated until the position of the end effector reaches the target or gets sufficiently close

closed loops. A reliable method for incorporating constraints is presented in [16]; the main idea is the repositioning and reorientation of the target to be within the allowed range of motion. In this paper we give details of how these constraints can be applied to a hand model, restricting the hand motion to a feasible set. It is worth noting that FABRIK, as described in Fig. 3.1, can be implemented simply by taking distances along lines rather than intersecting with spheres [16]. However, when we wish to incorporate constraints, we often need the sphere-line information; so we choose to work entirely in this unified framework. Also, the CGA framework offers several algorithm optimisations such as for cases where the 'end effector' is not positioned at the end of the chain (i.e. it is a leaf). For instance, assume that the joint positions $\mathbf{p}_i$ and $\mathbf{p}_{i-2}$ are known and that we want to estimate the joint position $\mathbf{p}_{i-1}$. This can be done by finding the intersection of the spheres Σ_1 and Σ_2 with centres the known joint positions $\mathbf{p}_i$ and $\mathbf{p}_{i-2}$ and radii the distances $d_i = |\mathbf{p}_i - \mathbf{p}_{i-1}|$ and $d_{i-2} = |\mathbf{p}_{i-2} - \mathbf{p}_{i-1}|$, respectively. If the intersection is a circle, then the estimated joint position can be assigned as the nearest point on that circle from its previous position (as described in Sect. 3.4.2.2). If the intersection is a single

Fig. 3.2 The hand's model geometry used in our implementation

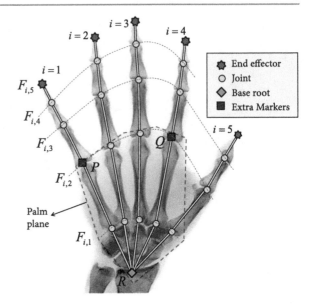

point, the estimated joint position is assumed to be that point; otherwise, if the two spheres do not intersect, the estimated joint position is equal to $\mathbf{p}_{i-1} = \frac{\mathbf{p}_i + \mathbf{p}_{i-2}}{2}$. Another simple optimisation is the direct construction of a line pointing towards the target, when the latter is unreachable. In that case, each joint $\mathbf{p}_i$ is assigned to be the nearest point on the sphere, with centre the previous joint $\mathbf{p}_{i-1}$ and radii the distance d_{i-1} from the target.

3.4 Using FABRIK for Hand Pose Tracking

In this section, an example of FABRIK implementation and how it performs on hand pose tracking is presented. The hand rotational and orientational limitations have been incorporated using CGA. The proposed approach is an example of object modelling for kinematic solutions, and we note that it can be adjusted to solve a variety of different modelling problems.

3.4.1 The Hand Geometry

It is assumed that the hand geometry, meaning the initial joint configuration of the hand, is known a priori. An example of a hand model is graphically represented in Fig. 3.2. The proposed hand model consists of 25 joints and has in total 25 DoFs. The end effector positions are captured using an optical motion capture rig, such as the Phasespace Impulse System [17]. Since our hand model does not have a mesh which defines its external shape, constraints such as self collisions are not considered here. The markers are identified (e.g. in the Phasespace system, each

LED marker is pulsed at a different frequency) so that it is known a priori on which finger each marker is placed. It is also important to know the orientation of the hand in order to efficiently incorporate constraints. This can be achieved by attaching two extra markers at specific positions, p and q, on the back of the hand (reverse palm). Assuming that the palm is always flat, we can find the plane describing the orientation of the hand using p, q and the position of the base root, r, which also lies on the palm plane. For simplicity, markers p and q can be placed at the joint positions $F_{1,2}$ and $F_{4,2}$ respectively, as shown in Fig. 3.2.

Before employing the IK solver, it is crucial to find the fingers' orientations, the chain roots and the end effectors for each chain; the target positions are assumed to be known since they are tracked by the motion capture system. The procedure is simple. Firstly, we estimate the hand orientation; thereafter, we calculate the palm joints and the finger orientations at each time step. When each finger orientation is known, the finger joints at the previous time step are translated and rotated in such a way that all joints belong to the current finger plane. Finally, a constrained version of FABRIK, with rotational limitations, is incorporated to fit the joints of each finger. This procedure is given in detail in the following paragraphs.

The first step is to find the hand orientation; hence, by accepting that the hand plane Φ_x is similar to the palm plane and that the markers p, q and r are lying on that plane, the hand orientation, meaning the plane Φ_x, can be estimated. Therefore,

$$
P = \frac{1}{2}\left(p^2 n_\infty + 2p - \bar{n}\right)
$$
$$
Q = \frac{1}{2}\left(q^2 n_\infty + 2q - \bar{n}\right) \tag{3.1}
$$
$$
R = \frac{1}{2}\left(r^2 n_\infty + 2r - \bar{n}\right)
$$

where P, Q and R are the 5D null vectors representing points p, q and r, respectively, and n_∞ and $\bar{n}$ are the null vectors in CGA.[1] The plane Φ_x is given by

$$
\Phi_x = P \wedge Q \wedge R \wedge n_\infty = \langle\langle\langle PQ\rangle_2 R\rangle_3 n_\infty\rangle_4 \tag{3.2}
$$

Note that the form given on the right-hand side of (3.2), and other relevant equations, is useful for implementation purposes and so is included here.

Calculating the Palm Joints   The next step is to incorporate constraints to obtain other palm joints. Thus, by assuming that the inter-joint distances (for the joints $F_{i,1}$ where $i = 1, \ldots, 5$ and $F_{j,2}$ where $j = 1, \ldots, 4$) are fixed over time and that all these joints lie on the palm plane, we can easily locate them using basic geometric entities such as planes, circles and spheres. An example of palm constraints is given

---

[1]*Editorial note*: Note here that in this chapter CGA equations are given in terms of n_∞ and $\bar{n}$, where $\bar{n} = -2n_o$.

Fig. 3.3 The palm plane
constraints: the hand plane
can be calculated using the
marker positions P, Q and R,
accepting that the markers lie
on that plane and that the
hand and palm planes are
similar. The rest of the palm
joints can be estimated,
assuming that the inter-joint
distances remain constant
over time, by intersecting the
spheres Σ_p and Σ_q with
centres at the marker
positions P and Q and radii
of the distance between their
centre and the joint position
we are looking for

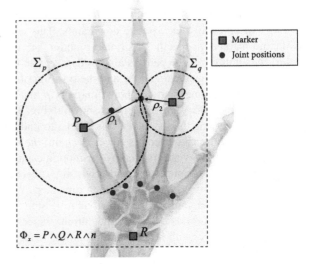

in Fig. 3.3. For instance, the joint position we are working on can be estimated
by intersecting the spheres with centres being the marker positions p and q and
radii being the distance between the marker and that joint position (taken from the
model). Therefore, find the sphere with its centre at the marker position P and radius
equal to the distance between the marker P and the joint we are working on

$$\Sigma_p = \left(P - \frac{1}{2}\rho_1^2 n_\infty \right) I \tag{3.3}$$

where ρ is the sphere radius. Similarly, find the sphere with centre the marker position Q and radius equal to the distance between the marker Q and the joint we are
working on

$$\Sigma_q = \left(Q - \frac{1}{2}\rho_2^2 n_\infty \right) I \tag{3.4}$$

The intersection of the two spheres gives a circle or a single point or no intersection.
The meet between the two spheres is given by

$$C = \Sigma_p \vee \Sigma_q = \left[\langle \Sigma_p \Sigma_q \rangle_2 \right]^* \tag{3.5}$$

- If $C^2 > 0$, then C is a circle. In that case, the possible solutions are given by
  intersecting the circle C and the palm plane Φ_x:

$$B = C \vee \Phi_x = \left[\langle C\Phi_x \rangle_3 \right]^* \tag{3.6}$$

- If $B^2 > 0$, the meet between C and Φ_x gives two points which can be extracted via projectors, as described in [18]. The new joint position is assigned as the point that is closer to the previous joint position (at time $k - 1$).
- If $B^2 = 0$, the intersection is a single point $X = Bn_\infty B$.
- If $B^2 < 0$, the intersection does not exist. For that instance, the new joint position is then taken as the nearest point on circle, C, from the previous joint position (at time $k - 1$, see Sect. 3.4.2.2).
• If $C^2 = 0$, the intersection is a single point $X = Cn_\infty C$.
• If $C^2 < 0$, the two spheres do not intersect. In that case, the final joint position is given by averaging the distance between the two markers, $x = (p + q)/2$.

Calculating the Finger Joints In order to estimate the finger joints, we need to find the finger planes Φ_i for $i = 1, \ldots, 4$. Each Φ_i can be calculated using the known joint positions $F_{i,2}$, the marker positions $F_{i,5}$ and by assuming that they are perpendicular to the palm plane Φ_x (note that this does not hold for the thumb plane Φ_5). Since both points from each finger are known (the motion capture system tracks the end effector positions $F_{i,5}$, and the finger roots $F_{i,2}$ lie on the palm plane with constant distance from the attached markers p and q, as explained in previous paragraphs), each finger plane can be estimated at the current time frame. The vector that is perpendicular to the hand plane Φ_x is given by

$$\hat{n} = \Phi_x^* - \frac{1}{2}\left(\Phi_x^* \cdot \bar{n}\right)n_\infty \qquad (3.7)$$

as explained in [18]. The finger planes can then be calculated as

$$\Phi_i = F_{i,2} \wedge F_{i,5} \wedge \hat{n} \wedge n_\infty = \left\langle\!\left\langle\left\langle(F_{i,2}F_{i,5})2\hat{n}\right\rangle_3 n_\infty\right\rangle_4\right. \quad \text{for } i = 1, \ldots, 4 \qquad (3.8)$$

The thumb orientation Φ_5 can be estimated using the marker position $F_{5,4}$ and the joint positions $F_{1,2}$ and $F_{5,2}$ that lie on the palm, assuming that when the thumb bends to the ventral side of the palm, it always points at the joint $F_{1,2}$ (approximately true in practice).

The next step is to estimate the rotation between the previous and the current frame of each finger plane. This can be done using rotors; the rotor R which expresses the rotation between the plane in the previous frame and the plane in the current frame, for each finger, can be found using the closed-form expression given in [19].[2] Then each finger joint at time $k - 1$ is translated and rotated in such a way that all joints of a given finger lie on the plane of the current frame k, as demonstrated in Figs. 3.4 and 3.5. Hence,

$$\hat{F}_{i,j}^k = RF_{i,j}^{k-1}\tilde{R} \qquad (3.9)$$

---

[2]*Editorial note*: This is essentially (4.2) for $n = 3$.

Fig. 3.4 The joint positions at times $k-1$ and k. Each finger joint at time $k-1$ needs to be rotated in such a way that all joints of that finger lie on the plane of the current frame k

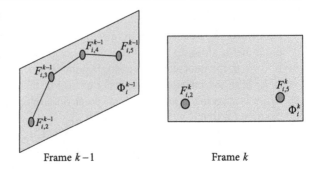

Frame $k-1$ Frame k

where $i = 1, \ldots, 4$ and $j = 3, 4, 5$ (except for the thumb where $i = 5$ and $j = 2, 3, 4$).

All joints now lie on plane Φ_i^k. Lastly, FABRIK is applied to each finger chain, assuming that the root of the chain is $F_{i,2}^k$, the end effector is the rotated point $\hat{F}_{i,5}^k$, and the target is the current marker position $F_{i,5}^k$, as shown in Fig. 3.5. The inter-joint distances are constant over time; thus, for computational efficiency, they can be calculated and stored at the first frame. It is important here to note that the marker occlusion problem is considered solved using constrained prediction algorithms, such as [20].

The resulting posture can be further improved in accuracy and naturalness by incorporating properties of the fingers, muscles, skin and individual joints via constraints [21].

3.4.2 Trigonometric Solutions

This section presents trigonometric solutions, using CGA, to problems which appear during the implementation of the proposed methodology.

3.4.2.1 Nearest Point on a Sphere from a Point in Space

This section shows how to calculate the nearest point on a sphere from a point in space using CGA. Assume that a sphere has centre $\underline{c}$ and radius ρ. The sphere Σ_1 can be expressed as a blade in CGA as follows:

$$\Sigma_1 = \left(c - \frac{1}{2}\rho^2 n_\infty\right) I \tag{3.10}$$

where $c = \frac{1}{2}\underline{c}^2 n_\infty + \underline{c} - \frac{\bar{n}}{2}$.

Assume a point in space q. In order to find the nearest point on the sphere from that point, we need to find the intersection of the line L_1 that passes through the point q and the sphere centre $\underline{c}$. Thus,

$$L_1 = Q \wedge c \wedge n_\infty = \langle\langle Qc\rangle_2 n_\infty\rangle_3 \tag{3.11}$$

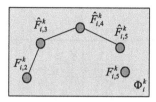

Frame k

Fig. 3.5 The current joint positions, after rotating them in order to lie on the current finger plane Φ_i^k. The problem of orientation is therefore solved, and FABRIK can then be utilised assuming that the root of the chain is $F_{i,2}^k$, the end effector is the point $\hat{F}_{i,5}^k$, and the target is the current marker position $F_{i,5}^k$

where $Q = H(q)$ is the Hestenes mapping of q. The intersection between the line L_1 and the sphere Σ_1 always returns two possible solutions, which are given by the bivector $X_1 \wedge X_2$.

$$X_1 \wedge X_2 = L_1 \vee \Sigma_1 \qquad (3.12)$$

Finally, the vectors X_1 and X_2 can be extracted from $X_1 \wedge X_2$ using projectors. Then, the nearest point on the sphere is assigned as the point that returns the minimum distance from the point in space.

We note here that although the nearest point on a sphere from a point in space can be found very easily using distance along lines, because we are working entirely in the CGA framework (in order to easily incorporate constraints), it is generally more computationally efficient to do all calculations in CGA.

$$X = \arg\big(\max(X_1 \cdot X, X_2 \cdot X)\big) \qquad (3.13)$$

3.4.2.2 Nearest Point on a Circle from a Point in Space

This section describes how to find the nearest point on a circle from a point in space. In particular, the minimum distance on a circle from a point in space is related to the projection of that point onto the plane Φ of the circle. This can be achieved by reflecting the point in the plane and finding the mid-point of the reflected and the original point. Hence, let the circle $C = H(b) \wedge H(c) \wedge H(d)$, where b, c and d are points that lie on the circle. The centre c of the circle C can be calculated as

$$c = Cn_\infty C \qquad (3.14)$$

and the plane Φ of the circle can be formulated as

$$\Phi = C \wedge n_\infty = \langle Cn_\infty \rangle_4 \qquad (3.15)$$

Having the plane Φ and the point $X = H(x)$ in space, the nearest point on the circle can be found by reflecting that point in the plane Φ.

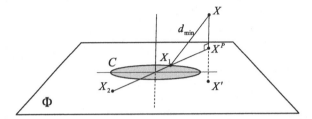

Fig. 3.6 The nearest point on circle to point in space. The point X is projected on the circle's plane Φ. A line is then formed through the midpoint of X and its projected counterpart and the centre of the circle. The intersection between the line and the circle returns two possible solutions; the one that is shorter to the point X is chosen

$$X' = \Phi X \Phi \tag{3.16}$$

The mid-point X^P is then calculated as

$$X'^P = X^P + \alpha n_\infty = H\left(\frac{1}{2}\left(H^{-1}(X') + x\right)\right) \tag{3.17}$$

Then, a line, L, is formed through this midpoint and the centre of the circle,

$$L = X^P \wedge c \wedge n_\infty \tag{3.18}$$

The intersection between line L and circle C will return a bivector, $A \wedge B$, which represents the shortest and longest distances on the circle from the point in space. The vectors X_1 and X_2 can be extracted from $X_1 \wedge X_2$ using projectors. The nearest point is then selected using a simple distance comparison method. This method is also illustrated in Fig. 3.6.

$$(X_1, X_2) = L \vee C$$
$$X = \arg\left(\max(X_1 \cdot X, X_2 \cdot X)\right) \tag{3.19}$$

3.5 Experimental Results

Experiments were carried out using a 10 camera Phasespace motion capture system, capturing data at 100 Hz [17]. The implemented methodology was able to process up to 70 frames per second, using MATLAB [22]. Our dataset comprises markered motion capture data; data captured using colour video cameras are also used to compare the reconstruction quality between the estimated and the true hand postures. The reconstructed hand postures were visualised using a mesh deformation algorithm.

The proposed method is real-time implementable, requiring only 1.43 ms per frame for tracking and fitting 25 joints. FABRIK is able to fit the joints and reconstruct the hand accurately; the rotational and orientational constraints ensure that

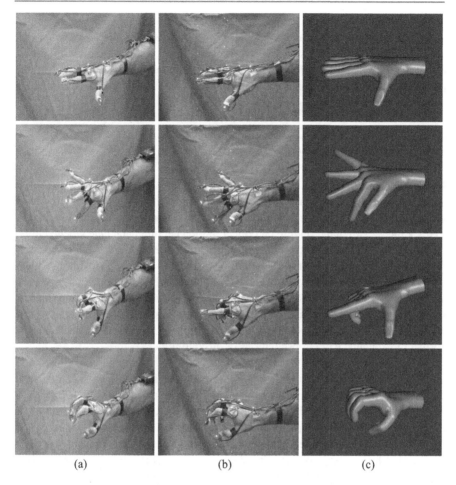

(a)   (b)   (c)

Fig. 3.7 An example of hand reconstruction using our methodology at a frame rate of 100 Hz. (**a**) View of the hand from RGB camera 1, (**b**) a different view of the hand from RGB camera 2 and (**c**) the final visualised posture. The resulting poses are visually natural and biomechanically correct

each finger movement remains normal without showing asymmetries, or irregular bends and rotations.

The implemented system can smoothly track the hand movements. The reconstruction quality can be checked visually by comparing the generated 3D hand animations with the data captured using a colour video camera, as seen in Fig. 3.7. It is difficult to illustrate the reconstruction quality in still images, but the resulting motion does not suffer from oscillation or discontinuities, and each finger smoothly moves to the target.

Despite the accuracy in performance, the resulting postures of our approach are not unique; several possible poses could result from the 3D articulated hand tracking. However, the advantages of this method are its efficiency and ability to return natural and feasible motion, which meets the user constraints, with low computational cost. It is also important to note here that FABRIK results in poses which are closely related to previous states. Therefore, the final joint configuration might be different when the IK problem is solved with the end effectors in different initial positions but with similar final states. Nevertheless, these differences are minimal causing only a small decrease in performance.

3.6   Conclusions and Future Work

In this paper, we presented an iterative Inverse Kinematics solver that was implemented using CGA. Rotational and orientational constraints were then incorporated for hand modelling; using a minimum number of available markers, we were able to track the 25 DoF hand relying on optical motion capture data. One labelled optical marker was attached to the end of each finger, treated as an end effector, and 3 more markers were placed at strategic positions on the hand reverse-palm to help us identify the root and orientation of the hand. The proposed methodology produced smooth and natural hand postures over time; the required processing time remained low enabling an effective real-time hand motion tracking and reconstruction system. The results were precise, producing visually natural, smooth and biomechanically correct movements, without oscillations or discontinuities.

This application exploits the advantages of CGA for incorporation of constraints in IK problems and proves that it is a useful mathematical tool which can be successfully used for applications in computer vision, graphics and robotics. In general, CGA gives us the ability to describe algorithms in a geometrically intuitive and compact manner. More particularly, it simplifies the mathematical model of the IK solver, since basic entities, such as spheres, lines, planes and circles, are simply represented by algebraic objects. In addition, the structure and elegance of CGA leads to low computational cost and real-time performance.

In future work, a more sophisticated model will be implemented which takes into consideration, in addition to the joint rotational and orientational restrictions, physiological constraints such the flexion, inertia, abduction, the finger's intradigital and transdigital correlation, the rigidity and the friction of the hand, as described in [21].

3.7   Exercises

3.1 What is the complexity of a simple unconstrained version of FABRIK for a six-joint kinematic chain?

3.2 Similarly to finding the nearest point on a circle from a point in space, find trigonometric solutions for: (a) the nearest point on a line from a sphere, (b) the nearest point on a line from a circle.

3.3 Describe a model with joint constraints for a human arm using the FABRIK Inverse Kinematics algorithm and CGA as the mathematical framework (hint: assume that the shoulder and the joint connecting the palm with the arm are ball and socket joints with rotational and orientation limits, and that elbow is a hinge joint. For simplicity, the hand should be considered as a solid limb segment).

References

1. Hestens, D., Sobczyk, G.: Clifford Algebra to Geometric Calculus: A Unified Language for Mathematics and Physics. Reidel, Dordrecht (1984)
2. Doran, C., Lasenby, A.: Geometric Algebra for Physicists. Cambridge University Press, Cambridge (2003)
3. Aristidou, A., Lasenby, J.: Inverse kinematics: a review of existing techniques and introduction of a new fast iterative solver. Cambridge University Department of Engineering Technical Report, CUED/F-INFENG/TR-632 (2009)
4. Dorst, L., Fontijne, D., Mann, S.: Geometric Algebra for Computer Science: An Object-Oriented Approach to Geometry. Morgan Kaufmann, San Mateo (2009)
5. Bayro-Corrochano, E., Kähler, D.: Motor algebra approach for computing the kinematics of robot manipulators. J. Robot. Syst. **17**(9), 495–516 (2000)
6. Bayro-Corrochano, E.: Robot perception and action using conformal geometric algebra. In: Handbook of Geometric Computing, pp. 405–458, Chap. 13. Springer, Berlin (2005)
7. Zamora, J., Bayro-Corrochano, E.: Inverse kinematics, fixation and grasping using conformal geometric algebra. In: Proceedings of the IEEE International Conference on Intelligent Robots and Systems (IROS '04), vol. 4, pp. 3841–3846 (2004). doi:10.1109/IROS.2004.1390013
8. Hildenbrand, D.: Tutorial: Geometric computing in computer graphics using conformal geometric algebra. Comput. Graph. **29**(5), 795–803 (2005)
9. Zamora, J., Bayro-Corrochano, E.: Kinematics and grasping using conformal geometric algebra. In: Lenarčič, J., Roth, B. (eds.) Advances in Robot Kinematics, pp. 473–480. Springer, Berlin (2006)
10. Hildenbrand, D., Zamora, J., Bayro-Corrochano, E.: Inverse kinematics computation in computer graphics and robotics using conformal geometric algebra. Adv. Appl. Clifford Algebras **18**, 699–713 (2008)
11. Hildenbrand, D., Fontijne, D., Wang, Y., Alexa, M., Dorst, L.: Competitive runtime performance for inverse kinematics algorithms using conformal geometric algebra. In: Proceedings of Eurographics Conference, 2006
12. Tanev, T.K.: Geometric algebra approach to singularity of parallel manipulators with limited mobility. In Lenarcic, J., Wenger, P. (eds.) Advances in Robot Kinematics: Analysis and Design, pp. 39–48. Springer, Dordrecht (2008)
13. Bayro-Corrochano, E., Zamora, J.: Differential and inverse kinematics of robot devices using conformal geometric algebra. Robotica **25**(1), 43–61 (2007)
14. Hildenbrand, D., Lange, H., Stock, F., Koch, A.: Efficient inverse kinematics algorithm based on conformal geometric algebra (using reconfigurable hardware). In: Proceedings of the 3rd International Conference on Computer Graphics Theory and Applications, Madeira, Portugal, 2008
15. Wang, L.-C.T., Chen, C.C.: A combined optimization method for solving the inverse kinematics problems of mechanical manipulators. IEEE Trans. Robot. Autom. **7**(4), 489–499 (1991)
16. Aristidou, A., Lasenby, J.: FABRIK: a fast, iterative solver for the inverse kinematics problem. Graph. Models **73**(5), 243–260 (2011)
17. PhaseSpace Inc: Optical motion capture systems. http://www.phasespace.com
18. Lasenby, A.N., Lasenby, J., Wareham, R.: A covariant approach to geometry using geometric algebra. Cambridge University Department of Engineering Technical Report, CUED/F-INFENG/TR-483 (2004)

19. Lasenby, J., Fitzgerald, W.J., Lasenby, A.N., Doran, C.J.L.: New geometric methods for computer vision: an application to structure and motion estimation. Int. J. Comput. Vis. **26**(3), 191–213 (1998)
20. Aristidou, A.: Tracking and modelling motion for biomechanical analysis. PhD Thesis, University of Cambridge, Cambridge, UK (October 2010)
21. Kaimakis, P., Lasenby, J.: Physiological modelling for improved reliability in silhouette-driven gradient-based hand tracking. In: Proceedings of the International Conference on Computer Vision and Pattern Recognition, Miami, USA, 25 June 2009, pp. 19–26
22. The Mathworks—MATLAB and Simulink for technical computing. http://www.mathworks.com

Reconstructing Rotations and Rigid Body Motions from Exact Point Correspondences Through Reflections

4

Daniel Fontijne and Leo Dorst

Abstract

We describe a new algorithm to reconstruct a rigid body motion from point correspondences. The algorithm works by constructing a series of reflections which align the points with their correspondences one by one. This is naturally and efficiently implemented in the conformal model of geometric algebra, where the resulting transformation is represented by a versor. As a direct result of this algorithm, we also present a very compact and fast formula to compute a quaternion from two vector correspondences, a surprisingly elementary result which appears to be new.

4.1 Introduction

Reconstructing rigid body motions from correspondences is a common problem in geometry. The correspondences can be directions (when only looking for rotation) or locations (when translation is also used). The applications range from satellite tracking to registration of point clouds. Our own need is fast evaluation of the rigid body motion of groups of features in a marker-less motion capture application.

In classical vector algebra, a straightforward method is to first compute the translation of the centroid (the average of all the points) and then to compute the rotation. The rotation can be computed by solving a matrix equation $A = R B$, where A and B are the matrices of the data vector correspondences, and R is the unknown 3×3

This work was performed while the first author was at the University of Amsterdam

D. Fontijne (✉)
Euvision Technologies, Amsterdam, The Netherlands
e-mail: d.fontijne@euvt.eu

L. Dorst
University of Amsterdam, Amsterdam, The Netherlands
e-mail: L.Dorst@uva.nl

L. Dorst, J. Lasenby (eds.), *Guide to Geometric Algebra in Practice*,
DOI 10.1007/978-0-85729-811-9_4, © Springer-Verlag London Limited 2011

63

rotation matrix. If the input is overdetermined or noisy, the Procrustes method [7] can be used. This involves computing the singular value decomposition of a 3×3 matrix.

In homogeneous coordinate approaches, rotation and translations are in principle unified in a single rigid body motion matrix. Yet the computation of the motion from data is based on a split of the translational and rotational parts, as in the vector algebra method.

Geometric algebra contains all these algebras and hence all these solution methods. Until recently, rigid body motion reconstruction would proceed as in the homogeneous coordinate approach: compute the translation using the centroid method and then use one of several compact geometric algebra methods to compute the rotation. The advantage over homogeneous coordinates is then the natural incorporation of quaternions (as ratios of real vectors) and the extendibility to n-D.

To compute the shortest rotation from one unit vector **a** to another unit vector **b** in n-D, one would compute (see [9])

$$\mathbf{Q} = 1 + \mathbf{ba}, \tag{4.1}$$

which can be normalized through dividing by $\sqrt{2(1 + \mathbf{b} \cdot \mathbf{a})}$. This **Q** is an even *versor*, the motion representation that replaces matrices in geometric algebra since it can transform any element X of the algebra through $X \mapsto \mathbf{Q}X\mathbf{Q}^{-1}$. Unfortunately, the equation has a singularity at 180 degrees (it returns the result $\mathbf{Q} = 0$ when a versor for a 180-degree rotation in an arbitrary plane might have been preferable). Equation (4.1) also appears in quaternion literature, see e.g. [8].

Geometric algebra also has a formula to determine the rotation of a given *frame of vectors* $\{\mathbf{e}_i\}$ to a given rotated frame $\{\mathbf{f}_i\}$ in n-D. One determines the reciprocal frame (a.k.a. the dual basis) $\{\mathbf{f}^i\}$ and the forms the rotation versor **V** from

$$\mathbf{V} = 4 - n + \sum_{i=1}^{3} \mathbf{f}^i \mathbf{e}_i, \tag{4.2}$$

where n is the dimension of the vector space (see e.g. [5, p. 257]). Equation (4.2) is a generalization of (4.1). It has the same issues as (4.1): it is not normalized (though it easily can be) and suffers from a singularity for 180-degree rotations. The frames do not need to be orthonormal but then require the explicit computation of a reciprocal frame, which is relatively expensive.[1]

Since our interest is specifically rigid body motions, we may expect an efficient method in a geometric algebra specifically designed to represent such motions, the *conformal model* [10] (referred to as *CGA*, for Conformal Geometric Algebra). In that model, there is a much stronger unification of translations and rotations than in the homogeneous coordinate approach (in fact, a unification of all conformal

---

[1] When applied within the conformal model, (4.2) can find pure rotations and pure translations, but no general rigid body motions which require a grade-4 part in their versor.

transformations, explaining the name). All such transformations in n-D turn out to be representable as *orthogonal transformations* of an $(n + 2)$-dimensional space. One would hope that a more compact method could therefore be designed using CGA.

Recently, there have been efforts to find a unified and preferably closed-form solution to rigid motions computation by conformal geometric algebra methods. One result is the general method of [11] which can handle noisy data but is rather expensive computationally as it involves finding a least-squares solution to a system of 2^n linear equations.[2] There is an n-D method for a closed solution in [4], based on assembling linear equations for the various grades of the versor, but collecting the data on the higher grades involves $O(2^n)$ combinations of the original data points, and the method cannot guarantee to return a versor from noisy data. Another exact method is [2], which can even handle general conformal transformations from unweighted point data (none of the others can); when the data is somewhat noisy, the method will still return a versor, probably close to the noiseless result. The downside to this method is that it only works for the conformal model and, for rigid body motions, is much less efficient than the method presented in this paper (which it actually contains twice, to solve two subproblems).

This paper presents a new method for the retrieval of rigid body motions, as the application of a more general algorithm for the determination of orthogonal transformations from vector data. We present the algorithm in a geometric algebra formulation independent of the CGA usage, making it applicable to efficient rotation estimation as well (resulting in a new and compact quaternion formula).

Our algorithm is related to the Cartan–Dieudonné theorem, which states that in n-D any orthogonal transformation can be represented as at most n planar reflections. A proof in terms of geometric algebra may be found in [1], which decomposes a known orthogonal transformation into reflections. By contrast, we use reflections to reconstruct an unknown n-D orthogonal transformation from vector correspondences.

The input to the method must specify at least the minimal set of correspondences which fully specifies the transformation. For rotations in 3D, this means that at least two vector correspondences are required; for rigid body motions in 3D, at least three point correspondences are required. In principle, the method is closed-form, but checks need to be made to determine whether correspondences were degenerate; thus the end result is an algorithm instead of a closed-form equation. When the correspondences are imprecise (e.g., due to noise), the algorithm will still give a sensible result which is guaranteed to be a versor but not a certifiably optimal solution. This result may still be very useful as a seed for optimization or as a minimal model for RANSAC. The algorithm can handle over-specified input but does not benefit from it in a sense that the solution becomes neither more stable nor more precise. The recent algorithm [12] is better suited for such situations.

---

[2]An extension of this method to n-D (with $n > 6$) may moreover be problematic since the Lagrangian constraints for the versor manifold are not yet known in general.

The new method presented here can reconstruct several classes of transformations. First of all, it can reconstruct all orthogonal transformations in Euclidean spaces of any dimension. Secondly, it can be used to find rigid body motions in Euclidean spaces of any dimension using the conformal model. Finally, it can reconstruct any orthogonal transformation in spaces with arbitrary metric in any dimension, though with a caveat on degeneracy of the input data, which is discussed later in Sect. 4.6.

The paper proceeds as follows: first we present the principle behind the method using only elementary geometry. We then formalize the algorithm using geometric algebra in Sect. 4.3. This is followed by the derivation of a new quaternion formula for the 3D Euclidean rotation case in Sect. 4.4. Section 4.5 presents benchmarks which compare the performance of some classical linear algebra and quaternion approaches to our new geometric algebra approach. Section 4.6 primarily discusses issues in the extended application of the algorithm.

4.2   Method for Reconstructing Rigid Body Transformations from Point Correspondences Through Plane Reflections

In this section we illustrate the geometry behind our algorithm before we formalize it using geometric algebra in the next section. We do so to allow the readers who are less familiar with geometric algebra to understand the resulting procedure. Note that for the purpose of illustration, we talk of '3D points', 'reflections in planes' and 'distances' in this section, but the same principles apply to 'vectors', 'reflections in hyper-planes', and 'dot products', respectively. For simplicity, degenerate cases are ignored in this section.

Suppose that we have n points denoted by their position vectors $\mathbf{p}_i$ and their correspondences $\mathbf{p}_i'$. We want to find the orthogonal transformation F which relates them,

$$\mathbf{p}_i' = F(\mathbf{p}_i).$$

Let us denote reflection in a plane $\mathbf{R}$ as $\mathbf{R}(\cdot)$.

The algorithm works by finding successive reflections which align the points one by one. The first reflection $\mathbf{R}_1$ aligns $\mathbf{p}_1$ with $\mathbf{p}_1'$, the second reflection $\mathbf{R}_2$ aligns $\mathbf{p}_2$ with $\mathbf{p}_2'$ without affecting the already aligned $\mathbf{p}_1$, and so on, until all points have been aligned. How many reflections are required depends on the dimension of the vector space and the type of orthogonal transformation that has actually taken place. The algorithm terminates when it detects that all points have been aligned.

It turns out that these reflections $\mathbf{R}_i$ can be found as *bisector planes* related to the data. A bisector plane is defined as the plane consisting of points with equal distance to two specified points. A trivial fact we need is that a point contained in a plane is not affected by reflection in that plane.

In our algorithm, the first reflection plane $\mathbf{R}_1$ is the bisector plane of $\mathbf{p}_1$ and $\mathbf{p}_1'$, see Fig. 4.1. For the second reflection plane $\mathbf{R}_2$, the bisector plane of $\mathbf{R}_1(\mathbf{p}_2)$ and $\mathbf{p}_2'$ is used. Perhaps surprisingly, $\mathbf{R}_2$ does not move the already aligned point $\mathbf{R}_1(\mathbf{p}_1)$,

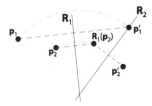

Fig. 4.1 Aligning points using plane-reflections. Note that $\mathbf{p}_1' = \mathbf{R}_1(\mathbf{p}_1) = \mathbf{R}_2(\mathbf{R}_1(\mathbf{p}_1))$, $\mathbf{p}_2' = \mathbf{R}_2(\mathbf{R}_1(\mathbf{p}_2))$. The total transformation $\mathbf{R}_2(\mathbf{R}_1(\cdot))$ is a rotation about the intersection of the planes $\mathbf{R}_1$ and $\mathbf{R}_2$

for we can show that $\mathbf{R}_1(\mathbf{p}_1)$ is contained in $\mathbf{R}_2$. We use the properties of the distance preservation of the various reflections in the setup to establish the sequence:

$$
\begin{aligned}
\text{distance}(\mathbf{p}_1, \mathbf{p}_2) &\overset{\text{correspondence}}{=} \text{distance}(\mathbf{p}_1', \mathbf{p}_2') \\
&\overset{\mathbf{R}_2 \text{ dist. pres.}}{=} \text{distance}(\mathbf{R}_2(\mathbf{p}_1'), \mathbf{R}_2(\mathbf{p}_2')) \\
&\overset{\text{definition of } \mathbf{R}_2}{=} \text{distance}(\mathbf{R}_2(\mathbf{p}_1'), \mathbf{R}_1(\mathbf{p}_2)) \\
&\overset{\mathbf{R}_1 \text{ dist. pres.}}{=} \text{distance}(\mathbf{R}_1(\mathbf{R}_2(\mathbf{p}_1')), \mathbf{p}_2).
\end{aligned}
$$

This should hold for any point $\mathbf{p}_2$; that is only possible if $\mathbf{p}_1$ equals $\mathbf{R}_1(\mathbf{R}_2(\mathbf{p}_1'))$. Therefore $\mathbf{p}_1' = \mathbf{R}_1(\mathbf{p}_1) = \mathbf{R}_1(\mathbf{R}_1(\mathbf{R}_2(\mathbf{p}_1'))) = \mathbf{R}_2(\mathbf{p}_1')$. Hence $\mathbf{p}_1'$ is not affected by a reflection in the plane $\mathbf{R}_2$, and so it lies on it.

If more reflections would be required to align more points, the principle of this proof still holds: earlier aligned points are not moved by subsequent reflections. After processing of all data points, the total orthogonal transformation is effectively represented as the series of reflection planes.

4.3   Formalization Using Geometric Algebra

Formalization of the ideas from the previous section is straightforward using (conformal) geometric algebra. In CGA, points of $\mathbb{R}^n$ are homogeneously represented as null vectors of $\mathbb{R}^{n+1,1}$. The difference between two such point representatives $\mathbf{p}$ and $\mathbf{q}$ is a vector representing the (dual) bisector plane of the two points (when they have equal homogeneous weight, and a dual sphere when not). This plane (or sphere) vector can be used as a reflection operator $\mathbf{R}$. Applying this 'versor' $\mathbf{R}$ to a vector $\mathbf{a}$ is done using the well-known 'sandwiching' operation $-\mathbf{R}\mathbf{a}\mathbf{R}^{-1}$. The versor $\mathbf{R} = \mathbf{p} - \mathbf{q}$ can be used to reflect $\mathbf{p}$ into $\mathbf{q}$, or vice versa, as we will show below.

However, the resulting algorithm works not only for points in CGA, but can be applied to reconstruct orthogonal transformations in other spaces as well. See Algorithm 1.

Algorithm 1: Reconstruct Orthogonal Transformation.

Input:
- A set of n exact correspondences between vectors $\mathbf{p}_i$ and $\mathbf{p}'_i$.
- A small threshold ε.

Input requirement: The vectors must be related by an orthogonal transformation, i.e. $\mathbf{p}'_i = (-1)^{\mathrm{grade}(\mathbf{V})}\mathbf{V}\mathbf{p}_i\mathbf{V}^{-1}$.

Output: the versor $\mathbf{V}$.

1. Initialize a set $\mathscr{A}$ of active vectors $\mathbf{a}_i \leftarrow \mathbf{p}_i$ for $i = 1,\ldots,n$.
2. Set $\mathbf{V} = 1$.
3. While $(|(\mathbf{a}_i - \mathbf{p}'_i)^2| \geq \varepsilon)$ for some $\mathbf{a}_i$ in $\mathscr{A}$ do
   a. Find in $\mathscr{A}$ a vector $\mathbf{a}_j$ for which $|(\mathbf{a}_j - \mathbf{p}'_j)^2| \geq \varepsilon$. Preferably, this should be the $\mathbf{a}_j$ with the largest $|(\mathbf{a}_j - \mathbf{p}'_j)^2|$.
   b. Set reflector $\mathbf{R} \leftarrow (\mathbf{a}_j - \mathbf{p}'_j)$.
   c. Append $\mathbf{R}$ to current transformation $\mathbf{V} \leftarrow \mathbf{R}\mathbf{V}$.
   d. Remove $\mathbf{a}_j$ from $\mathscr{A}$.
   e. Reflect all remaining active vectors:
      $$\mathbf{a}_k \leftarrow -\mathbf{R}\,\mathbf{a}_k\,\mathbf{R}^{-1} \text{ for every } \mathbf{a}_k \text{ in } \mathscr{A}.$$

The 'preferable' part of Step 3a is important for numerical stability. By purposely searching for the vector which has a large distance to its correspondent, we avoid accidentally using a vector which is very close (in terms of floating point precision) to its correspondent, which would make the algorithm numerically unstable.

When one is sure that reflectors with $\mathbf{R} \cdot \mathbf{R} < 0$ will never occur, it pays off to normalize all reflectors in Step 3b. The advantage is that the inverse in Step 3e can be avoided, and the algorithm will automatically return unit versors. This optimization can be used in Euclidean metrics and for example in the conformal model when all input vectors are unit-weight representations of points.

4.3.1 Proof of Correctness

To prove that the algorithm works correctly, we have to demonstrate

1. that the reflection computed in Step 3b properly aligns the active vector under consideration with its correspondent;
2. that non-active vectors are indeed not moved from their final position by later reflections, so that they can be excluded from transformation in Step 3e; and
3. that the test in Step 3 does not terminate the algorithm before $\mathbf{V}$ has been found.

Proof of 1 In any geometric algebra, the reflection of the vector $\mathbf{a}$ in the non-null vector $(\mathbf{a} - \mathbf{p}')$ gives:

$$-(\mathbf{a} - \mathbf{p}')\mathbf{a}(\mathbf{a} - \mathbf{p}')^{-1} = \left(\mathbf{p}'(\mathbf{a} - \mathbf{p}') - (\mathbf{a}'^2 - \mathbf{p}'^2)\right)(\mathbf{a} - \mathbf{p}')^{-1}$$

$$= \mathbf{p}' - \left(\mathbf{a}^2 - \mathbf{p}'^2\right)\left(\mathbf{a} - \mathbf{p}'\right)^{-1} \tag{4.3}$$

$$\overset{*}{=} \mathbf{p}', \tag{4.4}$$

where the final step marked with * is only valid if $\mathbf{a}^2 = \mathbf{p}'^2$. But this precisely happens when $\mathbf{p}'$ is obtained from $\mathbf{a}$ by some orthogonal transformation. This holds for all active vectors, both initially and still after the repeated orthogonal transformations of Step 3e. $\qquad\square$

Proof of 2 In Step 3e of the algorithm, we only reflect the remaining active vectors. We show that this is permitted, since the vectors $\mathbf{a}_i$ that were already aligned with their correspondents $\mathbf{p}'_i$ by the current $\mathbf{V}$ (so that $\mathbf{a}_i = \mathbf{V}\mathbf{p}_i\mathbf{V}^{-1} = \mathbf{p}'_i$) will not be affected by the new reflection $\mathbf{R}$. (This was essentially the intuitive motivation of Sect. 4.2.)

Let $\mathbf{a}$ be the new vector from $\mathscr{A}$ selected by Step 3a of the current loop; note that it is related to its original vector $\mathbf{p}$ by $\mathbf{a} = \mathbf{V}\mathbf{p}\mathbf{V}^{-1}$, since all vectors in $\mathscr{A}$ have effectively been reflected by $\mathbf{V}$ through application of Step 3e in every passage through the loop thus far. Then $\mathbf{R}$ is the reflection $\mathbf{R} = \mathbf{a} - \mathbf{p}'$, which is non-null (or the algorithm would have terminated).

If we would use this $\mathbf{R}$ to reflect one of the previously aligned vectors $\mathbf{a}_i$, we would find:

$$-\mathbf{R}\mathbf{a}_i\mathbf{R}^{-1} = \mathbf{a}_i - 2(\mathbf{a}_i \cdot \mathbf{R})\mathbf{R}^{-1},$$

so this would not affect $\mathbf{a}_i$ if and only if $\mathbf{a}_i \cdot \mathbf{R} = 0$ (in conformal geometric algebra, this is interpretable as the condition that point $\mathbf{a}_i$ lies on the plane $\mathbf{R}$). We compute:

$$\begin{aligned}
\mathbf{a}_i \cdot \mathbf{R} &= \mathbf{p}'_i \cdot \left(\mathbf{a} - \mathbf{p}'\right) \\
&= \mathbf{p}'_i \cdot \mathbf{a} - \mathbf{p}'_i \cdot \mathbf{p}' \\
&= \left(\mathbf{V}\mathbf{p}_i\mathbf{V}^{-1}\right) \cdot \left(\mathbf{V}\mathbf{p}\mathbf{V}^{-1}\right) - \mathbf{p}'_i \cdot \mathbf{p}' \\
&= \mathbf{p}_i \cdot \mathbf{p} - \mathbf{p}'_i \cdot \mathbf{p}' \\
&= 0.
\end{aligned}$$

In the last two steps we use the inner product preservation of orthogonal transformations twice: once for $\mathbf{V}$ (being constructed of multiple reflections, it is an orthogonal transformation) and once for the original transformation converting all $\mathbf{p}_j$ into their correspondents $\mathbf{p}'_j$. So indeed, the previously aligned vectors would remain aligned by subsequent reflections. $\qquad\square$

Proof of 3 Step 3 checks if there are still valid reflectors to apply. It does so by checking the distance between the current value of all active vectors $\mathbf{a}_i$ and their respective target values $\mathbf{p}'_i$. Although this test may look sensible, its validity limits the algorithm from being fully generally applicable to geometric algebras of arbitrary metric signatures.

When using a Euclidean metric, the test $|(\mathbf{a}_i - \mathbf{p}'_i)^2| = \|\mathbf{a}_i - \mathbf{p}'_i\|^2 \stackrel{?}{=} 0$ is indeed equivalent to $\mathbf{a}_i \stackrel{?}{=} \mathbf{p}'_i$. But when using a non-Euclidean metric, it may be true that $|(\mathbf{a}_i - \mathbf{p}'_i)^2| = 0$ while $\mathbf{a}_i$ is not equal to $\mathbf{p}'_i$. This happens when $(\mathbf{a}_i - \mathbf{p}'_i)$ is a null vector (i.e. a non-zero vector with zero norm). Since the conformal model has a non-Euclidean metric, it may be affected. For the application that motivated us: the determination of rigid body motions based on point data, the test is still valid (the norm of the difference of two vectors representing points which are related by a rigid body motion is proportional to their squared distance; so indeed only zero when the points are identical). But already for dilations in the conformal model, exceptional data can be constructed that make the test fail, see Sect. 4.6.   □

4.3.2 Reconstructing 3D Rigid Body Motions from Three Point Correspondences

One would expect that three point correspondences are sufficient to reconstruct a general 3D rigid body motion using conformal geometric algebra. However, in general our algorithm reconstructs a rigid body motion with a negative determinant using only three points, since three reflections are enough to map the points to their images.

Fortunately, it is not required to use four points to avoid this situation. Applying one extra reflection in the plane spanned by the three points results in the correct rigid body motion. This is what the *4planes* method benchmarked in Sect. 4.5.2 does.

4.4 Reconstructing a Quaternion from Two Vector Correspondences

In the case of 3D Euclidean vector space, our algorithm reduces to a very simple and efficient formula when reconstructing a rotor from two properly selected vectors $\mathbf{x}$ and $\mathbf{y}$ and their transforms $\mathbf{x}'$ and $\mathbf{y}'$:

$$V = (\mathbf{y}' + \mathbf{y}) \cdot (\mathbf{x}' - \mathbf{x}) + (\mathbf{y}' - \mathbf{y}) \wedge (\mathbf{x}' - \mathbf{x}), \tag{4.5}$$

where $\mathbf{x}$ and $\mathbf{y}$ are the original vectors, and $\mathbf{x}'$ and $\mathbf{y}'$ their rotated counterparts. As a quaternion Q, this would be written in terms of its scalar and vector components as

$$Q = ((\mathbf{y}' + \mathbf{y}) \cdot (\mathbf{x}' - \mathbf{x}), (\mathbf{y}' - \mathbf{y}) \times (\mathbf{x}' - \mathbf{x})). \tag{4.6}$$

Computing this quaternion takes 9 multiplications and 14 additions. By comparison, even for an orthonormal frame in 3D, (4.2) takes 27 multiplications and 19 additions. The resulting quaternion is not necessarily a unit quaternion in general.

Fig. 4.2 An example of a degenerate configuration for (4.5)

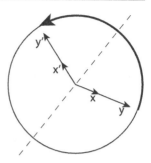

Equation (4.5) is derived as follows. Starting with the two reflection vectors

$$\mathbf{R}_1 = \mathbf{x} - \mathbf{x}',$$
$$\mathbf{R}_2 = -\mathbf{R}_1\mathbf{y}\mathbf{R}_1^{-1} - \mathbf{y}',$$

multiply them into a single versor, leading to

$$\begin{aligned}
\mathbf{V} = \mathbf{R}_2\mathbf{R}_1 &= \left(-\mathbf{R}_1\mathbf{y}\mathbf{R}_1^{-1} - \mathbf{y}'\right)\mathbf{R}_1 \\
&= -\mathbf{R}_1\mathbf{y} - \mathbf{y}'\mathbf{R}_1 \\
&= -\left(\mathbf{x} - \mathbf{x}'\right)\mathbf{y} - \mathbf{y}'\left(\mathbf{x} - \mathbf{x}'\right) \\
&= -\left(\mathbf{y}' + \mathbf{y}\right) \cdot \left(\mathbf{x} - \mathbf{x}'\right) - \left(\mathbf{y}' - \mathbf{y}\right) \wedge \left(\mathbf{x} - \mathbf{x}'\right).
\end{aligned} \tag{4.7}$$

Unfortunately, there is a singularity where (4.5) fails [3]. When $\mathbf{x} \wedge \mathbf{y}$ is orthogonal to the plane of rotation, $\mathbf{x} - \mathbf{x}'$ is parallel to $\mathbf{y} - \mathbf{y}'$, and the grade 2 part of the rotor will be 0 no matter what the rotation angle is.

The situation is illustrated in Fig. 4.2. The figure shows the top-view of a rotation plane (shown as a disc) and the projections of vectors $\mathbf{x}$ and $\mathbf{y}$ and their images $\mathbf{x}'$ and $\mathbf{y}'$. $\mathbf{x}$ and $\mathbf{y}$ are in a plane perpendicular to the rotation plane. $\mathbf{x}$ and $\mathbf{y}$ do not have to be collinear for the singularity to occur, but their projections onto the rotation plane should be.

Note that the vectors $\mathbf{x} - \mathbf{x}'$ and $\mathbf{y} - \mathbf{y}'$ are parallel, so the grade 2 part of (4.5) is zero. But the fundamental problem is that a single reflection (the dashed line in the figure) can map $\mathbf{x}$ to $\mathbf{x}'$ and $\mathbf{y}$ to $\mathbf{y}'$. That is, the input data is degenerate, and the general algorithm in Sect. 4.3 would return a grade 1 reflection versor. Thus, the singularity occurs because the second reflection $\mathbf{R}_2 = -\mathbf{R}_1\,\mathbf{y}\,\mathbf{R}_1{}^{-1} - \mathbf{y}'$ is zero because it is unnecessary to map the vectors to their correspondences.

However, using our prior knowledge that the returned answer should be a rotation and not a reflection, a fix can be applied, suggested by Clawson [3]. The vector $\mathbf{z} = (\mathbf{x} \wedge \mathbf{y})^*$ orthogonal to both $\mathbf{x}$ and $\mathbf{y}$ will never be in the $\mathbf{x} \wedge \mathbf{y}$-plane in this degenerate situation. The vector $\mathbf{z}$ and its transformed correspondence $\mathbf{z}' = (\mathbf{x}' \wedge \mathbf{y}')^*$ can therefore be used in place of $\mathbf{x}$ or $\mathbf{y}$ in (4.5) to avoid the singularity. Thus, to be able to handle degenerate input like the one in Fig. 4.2, we suggest Algorithm 2.

This is the algorithm that is benchmarked in the next section under the label *Ours*.

Algorithm 2: Reconstruct 3-D Rotation.

Input:
- Two 3D vectors $\mathbf{x}$ and $\mathbf{y}$ and their correspondences $\mathbf{x}'$ and $\mathbf{y}'$.
- A small threshold ε.

Input requirement: The vectors must be related by a 3D rotation $\mathbf{V}$.

Output: the versor $\mathbf{V}$.

1. Compute $\mathbf{V} = (\mathbf{y}' + \mathbf{y}) \cdot (\mathbf{x}' - \mathbf{x}) + (\mathbf{y}' - \mathbf{y}) \wedge (\mathbf{x}' - \mathbf{x})$.
2. If $(\|\mathbf{V}\| < \varepsilon \|\mathbf{x}\| \|\mathbf{y}\|)$, apply the singularity fix:
   a. Compute vectors $\mathbf{z} = \mathbf{x} \wedge \mathbf{y}^*$ and $\mathbf{z}' = \mathbf{x}' \wedge \mathbf{y}'^*$.
   b. Compute $\mathbf{V}_{xz} = (\mathbf{z}' + \mathbf{z}) \cdot (\mathbf{x}' - \mathbf{x}) + (\mathbf{z}' - \mathbf{z}) \wedge (\mathbf{x}' - \mathbf{x})$.
   c. Compute $\mathbf{V}_{zy} = (\mathbf{y}' + \mathbf{y}) \cdot (\mathbf{z}' - \mathbf{z}) + (\mathbf{y}' - \mathbf{y}) \wedge (\mathbf{z}' - \mathbf{z})$.
   d. Select from $\mathbf{V}_{xz}$ and $\mathbf{V}_{zy}$ the versor with the largest norm. Set $\mathbf{V}$ to that versor.

4.5   Benchmarks

We performed benchmarks in order to compare the efficiency of our algorithm with existing approaches. We did so for the two practical applications, reconstructing rotations and reconstructing rigid body motions.

4.5.1   Performance for 3D Rotations

Reconstructing a rotation of a 3D frame is a common task in geometry. Two vectors $\mathbf{x}$ and $\mathbf{y}$ and their rotated images $\mathbf{x}'$ and $\mathbf{y}'$ contain enough information to do so. We benchmarked four different methods:

Ours Equation (4.5) using Algorithm 2.

Quat Applying quaternion equation (4.1) ($\mathbf{Q} = 1 + \mathbf{ba}$) twice. The first rotation aligns $\mathbf{x}$ with $\mathbf{x}'$. The second rotation should be about $\mathbf{x}'$ axis, and align $\mathbf{y}$ with $\mathbf{y}'$. To enforce this, we project $\mathbf{y}$ and $\mathbf{y}'$ onto the plane orthogonal to $\mathbf{x}'$ before applying (4.1) the second time. By concatenating both rotations we reconstruct the full rotation.

Mat Solving a system of equations $A = RB$ where R is the desired rotation matrix. The first column of A is $\mathbf{x}$, the second column is $\mathbf{y}$, and the third is set to $\mathbf{x} \times \mathbf{y}$. Likewise the columns of B are set to the $\mathbf{x}'$, $\mathbf{y}'$ and $\mathbf{x}' \times \mathbf{y}'$. The system is solved by inverting B using Cramer's rule (which is cheaper than Gaussian elimination for this 3D case) and then computing $R = AB^{-1}$.

Proc Using the Procrustes [7] method (essentially, using the SVD of the matrix AB^T to determine an optimal rotation matrix as $R = UV^T$ with $U\Sigma V^T = \mathrm{SVD}(AB^T)$).

The first two methods are most natural in a quaternion or geometric algebra environment, and the last two methods are natural in a matrix environment. We allow each method to return the rotation in its own natural format. That is, *Ours* and *Quat* return a rotor (quaternion), and *Mat* and *Proc* return a 3×3 matrix.

The Procrustes method involves computing a relatively expensive singular value decomposition (SVD), but its advantage is that it computes the least-squares solution in the presence of noise. An implementation of the SVD was written specifically for 3×3 matrices following [6]. It is more than 10 times faster than Intel's LAPack (MKL) version which is optimized for large matrices.

As input to the benchmark, we created 10 million pairs of random 3-vectors that are transformed by 10 million random rotations. The coordinates of the random 3-vectors are uniformly distributed in the interval $[-1, 1]$. The random rotations are created by assigning random coordinates to the rotors and normalizing them.

We then measured how long each method took to reconstruct 10 million rotations from the pairs of vectors and their transformed correspondences. The results are, ordered from fast to slow:[3]

Method	Time (s)	Frequency (M/s)
Ours	0.79	12.7
Mat	1.29	7.6
Quat	3.46	2.9
Proc	7.50	1.3

Our new equation (4.5) *Ours* is the clear winner. The matrix-based method *Mat* comes in second but is almost two times slower. What was most surprising is that the quaternion method (based on applying (4.1) twice) is more than four times slower than (4.5). This is due to the required normalizations and projections. It is very likely that someone who is not aware of (4.5) would invent and apply this approach, guided by the consensus that quaternions are efficient in treating rotations.

4.5.2   Performance for 3D Rigid Body Motions

One of the most interesting applications of our algorithm is reconstructing rigid body motions (RBM) from measured points. In fact, this application was the reason to develop the new algorithm in the first place.

As input to the RBM benchmark, we created 10 million triplets of points at random locations and 10 million random rigid body motions. The coordinates of the points were uniformly distributed in the interval $[-1, 1]$, i.e. they were in a cube of size 2 around the origin. The random rigid body motions were created as the geometric product of two or four random reflection planes, following Cartan–Dieudonné. In 20% of the cases, two planes were used, in all other cases, four planes were used. The coordinates of these planes were uniformly distributed in the interval $[-1, 1]$.

---

[3]All benchmarks were performed on a 2.8-GHz Intel Core2Duo processor using 64-bit (double) floating point precision.

The planes were directly used to apply the random rigid body motions to the random points.

We benchmarked five different methods to recover the rigid body motions. One is our new algorithm as discussed in Sect. 4.3. This method is labeled *4planes* below. It finds four (or two, in degenerate cases) reflection planes. The planes are multiplied (using the geometric product) to form a conformal versor which has 8 coordinates.

The four other methods differ only in the way they compute the rotation, as they all determine the translation by computing the translation of the centroid of the points. The different methods to compute the rotations are exactly the four methods already benchmarked in Sect. 4.5.1 above.

Again each method returns the results in its own natural format. That is, *Ours* and *Quat* return a quaternion and a translation vector, *Mat* and *Proc* return a 3×3 rotation matrix and a translation vector, and *4planes* returns a conformal rigid body motion versor.

The results of the benchmark are, ordered from fast to slow:

Method	Time (s)	Frequency (M/s)
Centroid + Ours	1.11	9.0
Centroid + Mat	1.39	7.2
4planes	2.12	4.7
Centroid + Quat	3.74	2.7
Centroid + Proc	8.23	1.2

It is perhaps somewhat disappointing that the general version of our algorithm *4planes* is about 50% slower than the matrix method *Centroid + Mat*. In return for this performance loss, our algorithm solves for both rotation and translation in one go. If you are willing to give up this stylistic advantage, then the *Centroid + Ours* method easily beats the matrix method. *Centroid + Ours* uses the centroid to compute the translation and our (4.5) to compute the rotation. The translation and rotation are then combined into a single conformal versor, the same type of representation that the *4planes* method returns.

4.6   Discussion

Our algorithm has some limitations which make it unsuitable for certain applications. These are rather subtle, and not recognizing them may lead to unexpected failure.

4.6.1 Null Reflectors

The first limitation is the fact that the difference between two non-equal vectors may be a null vector when using a non-Euclidean metric (like the conformal model). This can cause the algorithm to terminate due to the test in Step 3 (i.e. $\mathbf{a}_i - \mathbf{p}'_i$ may be a null vector for all valid i). This problem cannot be fixed or worked around as a null vector cannot be used as a reflector (it is never a factor of a versor). Hence the algorithm has to terminate even though it can in principle detect (by using a Euclidean norm) that it is not done yet.

The question is whether this limitation hampers practical application of the algorithm. When using a Euclidean metric, there is no problem since no null vectors exist in those metrics. Rigid body motions in the conformal model can be reconstructed from point data without problems because the difference between two non-equal unit-weight point representations is never a null vector. In other cases, one should do an analysis of whether null vectors could arise as the difference of two vectors.

An example of a versor/data combination where our algorithm fails to retrieve the versor in the conformal model is the following combination of a dilation (scaling) versor $\mathbf{S}$ and sample vectors at the origin and infinity:

$$\mathbf{S} = e^{n_o \wedge n_\infty},$$
$$\mathbf{p} = n_o, \quad \mathbf{p}' = \mathbf{S}\mathbf{p}\mathbf{S}^{-1} \approx 0.14 n_o,$$
$$\mathbf{q} = n_\infty, \quad \mathbf{q}' = \mathbf{S}\mathbf{q}\mathbf{S}^{-1} \approx 7.39 n_\infty.$$

Both potential reflectors $\mathbf{p} - \mathbf{p}'$ and $\mathbf{q} - \mathbf{q}'$ produced by the algorithm are null vectors and thus cannot be used. This is a rather contrived example since the dilation center and infinity are the only two points which are not moved by the dilation, but it shows that care has to be taken to be sure that this situation can never arise in practice.

4.6.2 The Scaling of Vectors

Our algorithm determines an orthogonal transformation of vectors; this may only be indirectly related to the transformation of the input data. An example is a dilation versor in the conformal model, which works by scaling the n_o and n_∞ coordinates of points such that after normalization (i.e. dividing by the n_o coordinate), we interpret the point as dilated. To reconstruct a dilation in the conformal model, one has to feed points with correct homogeneous weights to the algorithm, which then automatically produces the correct sphere reflectors to align the point representatives. But to know these weights, one has to know the amount of dilation itself. So far, only the algorithm in [2] can solve this chicken-and-egg problem, by using data other than points as input (namely, a frame and a point). The limitation of that algorithm is that it is designed only for the conformal model and that it is considerably less efficient than the one described in this paper for correspondences with known weights.

4.6.3   The Determinant of the Reconstructed Versor

In many applications, the determinant of the reconstructed versor ($+1$ or -1) will be known beforehand. That is, one knows whether a reflection or a rotation should be reconstructed. As shown at the end of Sect. 4.4, there are input configurations where there are multiple solutions: either a rotation or a reflection could be used to align the data. The algorithm will then return the versor with the lowest number of factors. In that case, one could force the reconstructed versor to the required determinant by applying one extra reflection in the (hyper-)plane containing all data points $\mathbf{p}'_i$.

Also, in the case of noisy overdetermined data, the algorithm may compute an extra reflection. This can be solved either by applying yet an extra reflection beyond that or adjusting the algorithm to terminate at the required determinant. In any case, more research is required to determine the effect of noisy data on the algorithm. Rather than adapt our exact algorithm to the noisy input, one may prefer to use the recent algorithm of [12] which was designed to process noisy and redundant data.[4]

4.7   Conclusion

We have presented a new algorithm to reconstruct orthogonal transformations from exact vector correspondences. It is suitable for many applications, including reconstructing rigid body motion and rotations.

Our benchmarks show that our new (4.5) is by far the most efficient method to reconstruct 3D rotations, being $1.67\times$ faster than the next best method. In the case of rigid body motions, the algorithm is not the fastest possible, but it is still acceptable ($1.53\times$ slower than a classical rotation matrix+centroid solution). The combination of centroid+our rotation method beats that classical method by 25%.

For other potential applications beyond rigid body motions and rotations, the algorithm has two main limitations, one fundamental and another practical. The fundamental limitation is that in some metrics, with ill-chosen data, the algorithm may abort early due to null reflectors. Fortunately, this does not affect its use for the determination, from point data, of Euclidean rotations, or rigid body motions in the conformal model. The practical limitation is that the algorithm needs vector correspondences in the model algebra, including their weights, and that these cannot always be retrieved from the actual input data. An example is the determination of dilations in the conformal model, from point data. Again, this does not affect the application to rotations and rigid body motions.

The algorithm can work with noisy input data and always returns an orthogonal transformation (though not a least squares solution). In the case of noisy overdetermined data, special care must be taken to ensure the correct sign of the determinant of the transformation.

---

[4] *Editorial note*: Chapter 7 gives a related method designed for redundant data.

4.8 Exercises

4.1 Using (4.7), do a hand computation of the rotation transforming e_1 into e_2 and e_3 into $(e_1 + e_3)/\sqrt{2}$.

4.2 Some point configurations are degenerate and do not require the full orthogonal transformation of the space they reside in. Investigate what happens when the algorithm is applied to the point pair $e_1 \mapsto e_2$ and $e_3 \mapsto e_3$.

4.3 If it was known that the point pairs in the previous problem were related by a rotation (rather than a reflection), what more should you do to find its versor? (Hint: see Sect. 4.6.3.)

4.4 If you use the above configuration $e_1 \mapsto e_2$ and $e_3 \mapsto e_3$ in (4.5), it fails to produce the rotation versor. Invoke the Clawson correction of Sect. 4.4 to fix this.

4.5 As stated in the introduction, some methods have problems with 180-degree rotations. Compute the results of our algorithm to determine the rotation with point pairs $e_1 \mapsto -e_1$ and $e_2 \mapsto -e_2$.

Acknowledgements   We acknowledge the support of NWO in the DASIS project (Discovery of Articulated Structures in Image Sequences) for funding this work. We are indebted to Richard Clawson who discovered and fixed the singularity in quaternion (4.5). For a more detailed description of the singularity, please refer to his write-up [3].

References

1. Aragon-Gonzalez, G., Aragon, J., Rodriguez-Andrade, M., Verde-Star, L.: Reflections, rotations, and Pythagorean numbers. Adv. Appl. Clifford Algebras **19**, 1–14 (2009)
2. Cibura, C., Dorst, L.: From exact correspondence data to conformal transformations in closed form using Vahlen matrices. In: Proceedings of GraVisMa 2009, Plzen
3. Clawson, R.: Exceptional cases for Fontijne's quaternion rotation formula. Personal communication. Available at http://gmsv.kaust.edu.sa/people/postdoctoral_fellows/clawson/clawson.html#clawson-publications
4. Dorst, L.: Determining a versor in n-D geometric algebra from the known transformation of n vectors. In: Proceedings of GraVisMa 2009, Plzen
5. Dorst, L., Fontijne, D., Mann, S.: Geometric Algebra for Computer Science: An Object Oriented Approach to Geometry, revised edn. Morgan Kaufmann, San Mateo (2009)
6. Garcia, E.: Singular Value Decomposition (SVD): A Fast Track Tutorial. Retrieved 2009-11-02 from http://www.miislita.com/information-retrieval-tutorial/singular-value-decomposition-fast-track-tutorial.pdf
7. Golub, G., Van Loan, C.: Matrix Computations, 3rd edn. Johns Hopkins University Press, Baltimore (1997)
8. Hanson, A.J.: Visualizing Quaternions. Morgan Kaufmann, San Mateo (2006)
9. Lasenby, J., Fitzgerald, W.J., Doran, C.J.L., Lasenby, A.N.: New geometric methods for computer vision. Int. J. Comput. Vis. 191–213 (1998)
10. Li, H., Hestenes, D., Rockwood, A.: Generalized homogeneous coordinates for computational geometry. In: Geometric Computing with Clifford Algebras. Springer Series in Information Science (2001)

11. Perwass, C.: Geometric Algebra with Applications in Engineering. Springer, Berlin (2008)
12. Richard, A., Fuchs, L., Charneau, S.: An algorithm to decompose n-dimensional rotations into planar rotations. In: Computational Modeling of Objects Represented in Images. Lecture Notes in Computer Science, vol. 6026, pp. 60–71. Spinger, Berlin (2010)

Part II
Interpolation and Tracking

Conformal geometric algebra can only reach its full potential in applications when middle-level computational operations are provided. This part provides those for recurring aspects of motion interpolation and motion tracking.

Square Root and Logarithm of Rotors in 3D Conformal Geometric Algebra Using Polar Decomposition

5

Leo Dorst and Robert Valkenburg

Abstract

Conformal transformations are described by rotors in the conformal model of geometric algebra (CGA). In applications there is a need for interpolation of such transformations, especially for the subclass of 3D rigid body motions. This chapter gives explicit formulas for the square root and the logarithm of rotors in 3D CGA. It also classifies the types of conformal transformations and their orbits. To derive the results, we employ a novel polar decomposition for the even subalgebra of 3D CGA and an associated norm-like expression.

5.1 Rotor Interpolation for Conformal Motions

Euclidean rigid body motions and similarities are special cases of conformal transformations (since they preserve angles). In applications like computer graphics, we typically want to interpolate them, to perform a specified motion gradually.

In principle, geometric algebra offers a natural way to do this for all conformal transformations. They are represented in conformal geometric algebra (CGA) as rotors (even-graded versors that satisfy $R\tilde{R} = 1$). Any rotor can be represented as the exponential of a bivector $R = e^{B}$, and a simple form of interpolation would split this element into n equal parts through $R^{1/n} = \sqrt[n]{R} = e^{B/n}$. This ability to take nth roots of rotors would enable the interpolation between two poses characterized by rotors R_1 and R_2 by means of the intermediate rotors $R_i = (R_2/R_1)^{i/n} R_1$. However, the n-th root of a rotor in this procedure is not easy to extract in general geometric

L. Dorst (✉)
Intelligent Systems Laboratory, University of Amsterdam, Amsterdam, The Netherlands
e-mail: L.Dorst@uva.nl

R. Valkenburg
Industrial Research Limited, Auckland, New Zealand
e-mail: r.valkenburg@irl.cri.nz

L. Dorst, J. Lasenby (eds.), *Guide to Geometric Algebra in Practice*,
DOI 10.1007/978-0-85729-811-9_5, © Springer-Verlag London Limited 2011

algebras, or even merely in the conformal model. To retrieve the required bivector B, one needs the logarithm $B = \log(R)$ of the rotor R, but the noncommutative nature of rotors makes this nontrivial.

An alternative to this general nth-root interpolation would be 'repeated halving' of the transformations; this requires the square root $\sqrt{R}$ of a rotor R. Such a square root of a rotor could be simpler to extract than a logarithm, and it is also useful when constructing rotors as the square root of ratios of suitably chosen geometric elements.

In many applications we are specifically interested in rigid body motions in 3D; their rotors in the conformal model $\mathbb{R}_{4,1}$ are sometimes called 'motors'. For these motors, the explicit logarithm was given in [2], but no square root. Moreover, for a general rotor of $\mathbb{R}_{4,1}$, both logarithms and square roots appear to be lacking in the literature. In this chapter, we provide both:

- A *closed-form formula to compute the square root* $\sqrt{R}$ *of a 3D conformal rotor* R. This is new and based on the development of a new polar decomposition for certain elements of the even subalgebra $\mathbb{R}_{4,1}^{+}$ which may be useful in its own right. It employs a new normlike expression for multivectors of $\mathbb{R}_{4,1}$.
- A *closed-form solution for the principal logarithm* $\mathrm{Log}(R)$ *of a 3D conformal rotor* R. Our solution here essentially extends a method from [3] to conformal rotors and brings it into more compact form by means of the normlike expression. We also give a geometrical interpretation of the Log in terms of locally perpendicular orbits of simple rotors.

The results can be given in closed form because we restrict ourselves to spaces of less than six dimensions; many of the principles will generalize, but this is not explored in this chapter.

5.2   Rotor Roots Through Polar Decomposition

We first remark that the square root of a rotor always exists and is a rotor. This follows simply from the exponential representation of a rotor: halving the bivector exponent gives a square root. The square root is often not unique, even apart from the obvious $\pm$ ambiguity.

We will also find that it is convenient to restrict our treatment to rotors for which $\langle R \rangle \geq 0$. Since $-R$ acts the same as R, this does not lose any power of expression.

5.2.1   Sketch: How to Take the Square Root of a Rotor

When one wants to halve an angle in the plane, one forms an equal-sided parallelogram on the two legs; the diagonal of that parallelogram is at the half angle. The same method can easily be applied to planar Euclidean rotors (see Fig. 5.1) and gives the rotor square root when used between the identity rotor 1 and the target rotor R. (This gives the 'principal square root': taking its negative, and or adding

Fig. 5.1 Motivational justification of the formula $\sqrt{R} \propto (1 + R)$

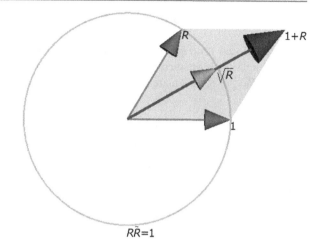

multiples of 2π to the rotor angle will lead to other square roots.) A simple computation shows that this principle may be extended to more general rotors; for the rotor $\sqrt{R}$, a scaled version of the even multivector $1 + R$ is very much like:

$$1 + R = \sqrt{R}(\sqrt{R}^{\sim} + \sqrt{R}). \tag{5.1}$$

The right-hand side is the product of the rotor $\sqrt{R}$ with a quantity that is its own reverse. In 3D CGA, that self-reverse element must be of the form 'scalar + quadvector'. For a simple rotor $\sqrt{R}$ (i.e., the exponential of a 2-blade), there is no grade higher than 2 in $\sqrt{R}$, and the self-reverse factor is scalar. Denoting the scalar part of R by $\langle R \rangle$, normalization of $(1 + R)$ by the scalar $\|1 + R\| = \sqrt{(1 + \tilde{R})(1 + R)} = \sqrt{2(1 + \langle R \rangle)}$ (which exists since $\langle R \rangle \geq 0$ by assumption) produces the rotor of the principal square root $\sqrt{R}$:

$$R \text{ simple rotor}: \quad \sqrt{R} = \text{normalize}[1 + R] = \frac{1 + R}{\sqrt{2(1 + \langle R \rangle)}}. \tag{5.2}$$

Many interesting rotors in 3D CGA, such as the motors, have a nonzero grade-4 part. For such rotors, we propose to extend the concept of normalization of elements of the even subalgebra, so that we may still write

$$\sqrt{R} = \text{normalize}[1 + R]. \tag{5.3}$$

Ultimately, we do this through defining a *polar decomposition* of an element of the even subalgebra $\mathbb{R}^{+}_{4,1}$ into a rotor and a symmetric (self-reverse) element and defining the rotor part as the outcome of the normalization. However, we can give the result without knowing this background.

5.2.2 The Square Root of Rotors in 3D CGA

We give a closed-form expression for the square root of a rotor R, in terms of its i-grade components denoted by R_i. There are two cases:

- If $(1 + R_0)^2 \neq R_4^2$, a square root is:

$$\sqrt{R} = (1 + R) \frac{1 + R_0 - R_4}{2((1 + R_0)^2 - R_4^2)} \frac{1 + R_0 + R_4 + \sqrt{(1 + R_0)^2 - R_4^2}}{\sqrt{1 + R_0 + \sqrt{(1 + R_0)^2 - R_4^2}}}.$$

(5.4)

- If $(1 + R_0)^2 = R_4^2$, there is an infinity of square roots:

$$\sqrt{R} = \left(\frac{1 + R}{2} + B \frac{1 - R}{2} \right),$$

(5.5)

where the 2-blade B should satisfy $B^2 = -1$ and should commute with R_4 (but is otherwise arbitrary).

- Always $\pm \sqrt{R}$ are also square roots, and when $R_4^2 > 0$, so are $\pm \sqrt{R} R_4 / \|R_4\|$. When $R = 1$, any quadvector Q with square 1 is a square root.

Equation (5.4) is the regular case; (5.5) is the exception. It is actually only required in the truly exceptional cases '$R = -1$' and '$R = R_4$ with $R_4^2 = 1$'. The case $R = -1$ may be excluded without loss of applicability by treating only rotors with nonnegative scalar part.

One may verify the correctness of these rather intimidating equations by direct computation of their squares. But more insight in their construction and the nature of the exceptions is obtained by relating them to a polar decomposition of elements of $\mathbb{R}_{4,1}^+$. We do so in the next, rather technical, sections. The reader who merely wants to program the results may proceed to Sect. 5.2.8, where we treat some practical issues followed by special cases.

5.2.3 Polar Decomposition of Invertible Even Elements in 3D CGA

As we have seen in (5.1), the even multivector $1 + R$ is proportional to $\sqrt{R}$ by a factor $(\sqrt{R}^{\sim} + \sqrt{R})$, which is in general non-scalar. Since $\sqrt{R}$ must be a rotor, we could retrieve it from $1 + R$ by somehow enforcing a multiplicative 'projection' onto the rotor manifold. This is highly reminiscent of a standard technique in matrix algebra called polar decomposition, in which a matrix is factored into an orthogonal matrix and a symmetric matrix. The orthogonal matrices correspond to our rotors; apparently, we should interpret 'symmetry' as 'self-reverse', at least in the even subalgebra $\mathbb{R}_{4,1}^+$ of 3D CGA. Once we can make this decomposition, we can define the 'normalize' operation in (5.3) as 'taking the rotor part of the polar decomposition'. The justification for using the term 'normalization' is that in the case of a non-null element X without 4-grade part in $\widetilde{X}X$ (such as versors), this operation reduces to dividing X by $\|X\| = \sqrt{|\langle \widetilde{X}X \rangle|}$, which agrees with the classical normalization procedure.

For our purposes in the even subalgebra of $\mathbb{R}_{4,1}^+$ (or a subalgebra), we therefore define the *polar decomposition* of an element X as the multiplicative decomposition into a rotor U and a self-reverse element S:

$$\text{polar decomposition:} \quad X = US \quad \text{with } \widetilde{U}U = 1 \text{ and } \widetilde{S} = S. \tag{5.6}$$

The existence and conditions for uniqueness of the polar decomposition for arbitrary elements of $\mathbb{R}_{4,1}^+$ are rather involved, and we are preparing a paper on that.[1] In this chapter we only need to apply the decomposition to elements X of the form $X = 1 + R$, with R a rotor, and this simplifies such issues. For instance, since (with two exceptions detailed below) such elements are invertible, we may limit ourselves to the polar decomposition of invertible elements. In the following sections, we first treat some general properties of the symmetric element in the polar decomposition, to provide an explicit formula for the invertible case. We then apply that to $X = 1 + R$ with R a rotor. Finally, we resolve some exceptional cases.

5.2.4 Determining the Square Root of a Symmetric Element

If an invertible element X has a polar decomposition as $X = US$, with U a rotor and S a self-reverse element, then the quantity $\widetilde{X}X$ contains information on S only:

$$\widetilde{X}X = (US)^\sim US = \widetilde{S}\widetilde{U}US = \widetilde{S}S = S^2.$$

To establish formulas for the polar decomposition, we retrieve S from S^2 by taking a square root; this will turn out to be non-unique in some cases. Since S is invertible because X is, for each S we can then uniquely find the associated U as $U = XS^{-1}$.

A square root always has an ambiguity of sign. Let us call a square root of S^2 with nonnegative scalar part a *principal square root*. We will treat square roots that are pure quadvectors separately below.

Lemma 5.1 *Let $\mathbb{R}_{4,1}^{0,4}$ be the subalgebra of $\mathbb{R}_{4,1}^+$ with symmetric (self-reverse) elements of the form $\langle \Sigma \rangle + \langle \Sigma \rangle_4$.*

Consider an element $\Sigma \in \mathbb{R}_{4,1}^{0,4}$ for which $\langle \Sigma \rangle^2 \geq \langle \Sigma \rangle_4^2$ and define

$$[[\Sigma]] = \sqrt{\langle \Sigma \rangle^2 - \langle \Sigma \rangle_4^2}. \tag{5.7}$$

Then Σ has

---

[1] *Editorial note*: Some more details on the polar decomposition for the linear space of motors in 3D CGA may be found in Sect. 2.3 of Chap. 2 in this volume.

- *a unique principal square root in $\mathbb{R}_{4,1}^{0,4}$ when $\langle \Sigma \rangle + [[\Sigma]] > 0$, namely*

$$\sqrt{\Sigma} = \frac{\Sigma + [[\Sigma]]}{\sqrt{2}\sqrt{\langle \Sigma \rangle + [[\Sigma]]}};$$

- *exactly two distinct principal square roots in $\mathbb{R}_{4,1}^{0,4}$ when $\langle \Sigma \rangle - [[\Sigma]] > 0$, namely*

$$\sqrt{\Sigma} = \frac{\Sigma + [[\Sigma]]}{\sqrt{2}\sqrt{\langle \Sigma \rangle + [[\Sigma]]}} \quad \text{and} \quad \sqrt{\Sigma} = \frac{\Sigma - [[\Sigma]]}{\sqrt{2}\sqrt{\langle \Sigma \rangle - [[\Sigma]]}};$$

- *an infinite number of principal square roots when $\langle \Sigma \rangle + [[\Sigma]] = 0$, namely any quadvector Q such that $Q^2 = \Sigma \leq 0$.*

For all other elements of $\mathbb{R}_{4,1}^{0,4}$, no square root exists in $\mathbb{R}_{4,1}^{0,4}$.

Proof Since quadvectors square to scalars in $\mathbb{R}_{4,1}^{0,4}$, the square of Σ has to be a linear combination of Σ and 1:

$$\Sigma^2 = (\Sigma_0 + \Sigma_4)^2 = \left(2\Sigma_0 - (\Sigma_0 - \Sigma_4)\right)\Sigma$$
$$= 2\Sigma_0\Sigma - \left(\Sigma_0^2 - \Sigma_4^2\right) = 2\Sigma_0\Sigma - [[\Sigma]]^2. \tag{5.8}$$

Under the assumption of the lemma, $[[\Sigma]] = \sqrt{\Sigma_0^2 - \Sigma_4^2}$ is well-defined. Since the square root $\sqrt{\Sigma}$ is requested to be an element of $\mathbb{R}_{4,1}^{0,4}$, it must be a linear combination of 1 and Σ_4 (no other quadvector works due to their scalar squares), or rather of 1 and Σ. Solving $\sqrt{\Sigma}^2 = \Sigma$ with $\sqrt{\Sigma} = \alpha + \beta\Sigma$ for α and β gives two potential solutions, conditional on whether $\sqrt{\langle \Sigma \rangle \pm [[\Sigma]]}$ is well defined (i.e., if the arguments of the square roots are positive):

$$\sqrt{\Sigma} = \frac{\Sigma \pm [[\Sigma]]}{\sqrt{2}\sqrt{\langle \Sigma \rangle \pm [[\Sigma]]}}. \tag{5.9}$$

Both are principal square roots, since $\langle \sqrt{\Sigma} \rangle = \sqrt{(\langle \Sigma \rangle \pm [[\Sigma]])/2}$ is nonnegative.

To verify the domain of validity of the main case of the lemma, let us study when Σ can actually be the square of another element of $\mathbb{R}_{4,1}^{0,4}$. To be specific, take $S = S_0 + S_4$, the indices again denoting the grades. Then

$$S = S_0 + S_4 \quad \Rightarrow \quad S^2 = \left(S_0^2 + S_4^2\right) + 2S_0S_4. \tag{5.10}$$

Since

$$[[\Sigma]]^2 = \Sigma_0^2 - \Sigma_4^2 = \left(S_0^2 + S_4^2\right)^2 - 4S_0^2S_4^2 = \left(S_0^2 - S_4^2\right)^2 \geq 0, \tag{5.11}$$

$[[\Sigma]]$ is indeed well defined for such obviously square elements. We now verify the conditions of the principal roots in the lemma:

$$\langle \Sigma \rangle + [[\Sigma]] = S_0^2 + S_4^2 + \left| S_0^2 - S_4^2 \right|$$

$$= \begin{cases} 2S_0^2 \geq 0 & \text{if } S_0^2 \geq S_4^2 \quad \text{case (a)} \\ 2S_4^2 \geq 2S_0^2 \geq 0 & \text{if } S_0^2 \leq S_4^2 \quad \text{case (b),} \end{cases} \tag{5.12}$$

while

$$\langle \Sigma \rangle - [[\Sigma]] = S_0^2 + S_4^2 - \left| S_0^2 - S_4^2 \right|$$

$$= \begin{cases} 2S_0^2 \geq 0 & \text{if } S_0^2 \leq S_4^2 \quad \text{case (c)} \\ 2S_4^2 & \text{if } S_0^2 \geq S_4^2 \quad \text{case (d).} \end{cases} \tag{5.13}$$

So $\langle \Sigma \rangle + [[\Sigma]] \geq 0$ for all $\Sigma = S^2$, with the zero value occurring only when both $S_0 = 0$ and $S_4^2 \leq 0$ (then Σ is a nonpositive scalar). In contrast, $\langle \Sigma \rangle - [[\Sigma]]$ can only be guaranteed to be non-negative when $S_0^2 \geq S_4^2 \geq 0$. When $\langle \Sigma \rangle - [[\Sigma]] = 0$, we have that $\langle \Sigma \rangle + [[\Sigma]] = 2[[\Sigma]] \geq 0$, so that devolves to the $+-$root except when $\langle \Sigma \rangle = [[\Sigma]] = 0$ (when it devolves to the exception of the $+-$root).

The only case for which a square root cannot be determined by either of the above, even though we know it exists, is therefore when Σ is a non-positive scalar. In that case, any quadvector Q for which $Q^2 = \Sigma$ is a permissible square root; the original S is among the possibilities, but no longer the only one.

There are elements in $\mathbb{R}_{4,1}^{0,4}$ for which the above does not provide any square root, namely those for which $\langle \Sigma \rangle + [[\Sigma]] < 0$ (which implies $\Sigma_4^2 > 0$ and $\Sigma_0 \leq 0$), and elements for which $\Sigma_4^2 > \Sigma_0^2$ (so that $[[\Sigma]]$ does not exist). Neither are in fact the square of any element of $\mathbb{R}_{4,1}^{0,4}$: substitution of $\Sigma = S^2$ gives equivalence with the impossible '$S_0^2 + S_4^2 < 0$ while $S_4^2 > 0$' and '$0 > (S_0^2 - S_4^2)^2$', respectively.   □

In the cases (a) and (c) of (5.12) and (5.13), the square root $\sqrt{S^2}$ of (5.9) evaluates to $(S_0 + S_4)S_0/|S_0|$; in the cases (b) and (d), to $(S_0 + S_4)S_4/\|S_4\|$ (if $S_4^2 > 0$ and additionally $S_0^2 \geq S_4^2$ in case (d)). In the cases (a) and (d), this therefore retrieves the original S (making it a principal root by possibly flipping a sign on the scalar part). In cases (b) and (d), it retrieves another root differing by a quadvector $Q = S_4/\|S_4\|$ squaring to 1. Such a quadvector Q is actually a rotor (since $\widetilde{Q}Q = Q^2 = 1$).

5.2.5 The Polar Decomposition of Invertible Elements of $\mathbb{R}_{4,1}^{+}$

If $\widetilde{X}X = S^2$ is not a square in $\mathbb{R}_{4,1}^{0,4}$, then X does not have a polar decomposition. When it is a square, the norm of (5.7) is properly defined, and the formulas of the lemma can be applied to retrieve a candidate S. For each invertible S, one then finds $U = XS^{-1}$.

Clearly, there is a sign ambiguity in any polar decomposition definition: both (U, S) and $(-U, -S)$ are legitimate polar decompositions of X. By choosing a principal square root for S, we can resolve this. But when there are two distinct principal square roots to $\widetilde{X}X$, there would still be two distinct polar decompositions. However, as we have seen above, those two principal roots differ by a rotor $S_4/\|S_4\|$; therefore we can choose only one root and have the rotor part of the decomposition absorb or supply such a rotor factor. That has the advantage of making the polar decomposition unique. We therefore define the 'principal polar decomposition' by choosing the $+$ case of the square root in (5.9) as follows.

Let X be an invertible element of $\mathbb{R}_{4,1}^+$. If $\widetilde{X}X$ is such that $\langle\widetilde{X}X\rangle^2 \geq \langle\widetilde{X}X\rangle_4^2$, define

$$\|X\| = \sqrt[4]{\langle\widetilde{X}X\rangle^2 - \langle\widetilde{X}X\rangle_4^2}. \tag{5.14}$$

Note that this is related to (5.7) through $\|X\|^2 = [[\widetilde{X}X]]$. For X for which $\langle\widetilde{X}X\rangle + \|X\|^2 > 0$, the *principal polar decomposition* of X is defined as:

$$U = \frac{X}{\widetilde{X}X}\frac{\widetilde{X}X + \|X\|^2}{\sqrt{2}\sqrt{\langle\widetilde{X}X\rangle + \|X\|^2}}, \quad \text{and} \quad S = \frac{\widetilde{X}X + \|X\|^2}{\sqrt{2}\sqrt{\langle\widetilde{X}X\rangle + \|X\|^2}}. \tag{5.15}$$

For other elements $X \in \mathbb{R}_{4,1}^+$, we do not define a polar decomposition in this chapter.

Proof To prove the formula for U when X and hence S are invertible, it is enough to verify that $X = US$. This is left as an exercise for the reader.   □

5.2.6 The Square Root of a Rotor

Equation (5.1) shows that a factorization of $1 + R$ is possible, splitting off a symmetric element $S = \sqrt{R}^{\,\sim} + \sqrt{R}$ from a rotor part $\sqrt{R}$. The proof of Lemma 5.1 shows that for such an S, the principal square root of S^2 in the case (5.12) always exists. It sometimes returns $SS_0/|S_0|$ (namely in case (a): $S_0^2 > S_4^2$) and sometimes $SS_4/\|S_4\|$ (namely in case (b): $S_0^2 < S_4^2$). A problem appears to be that we cannot tell from our data S^2 which of these is the case: $S^2 = (S_0^2 + S_4^2) + 2S_0S_4$ is symmetrical in S_0 and S_4 when S_4^2 is just as positive as S_0^2 can be.

However, precisely in that case, the ambiguous factor $Q = S_4/\|S_4\|$ is a rotor, and a symmetry of our original problem. Since S_4 in the polar decomposition can only be proportional to R_4, Q commutes with R, with $1 + R$, and with $\sqrt{R}$. So if $\sqrt{R}$ is a square root of R, so is $\pm Q\sqrt{R}$. When $R_4^2 > 0$, the rotor R therefore has at least four square roots (and even an infinity of them when moreover $R_0 = 0$; see the next section).

We remark that in the very special case $R = 1$, any quadvector Q with $Q^2 = 1$ is a square root. With those considerations, we have shown (5.4):

Lemma 5.2 *A rotor R of $\mathbb{R}_{4,1}^+$ for which $1 + R$ is invertible has a principal square root that can be determined from the principal polar decomposition as*

$$\sqrt{R} = \frac{1+R}{2+R+\tilde{R}} \frac{2+R+\tilde{R} - \|1+R\|^2}{\sqrt{2}\sqrt{2(1+\langle R \rangle) - \|1+R\|^2}}$$

$$= \frac{1+R}{2(1+R_0+R_4)} \frac{1+R_0+R_4+\sqrt{(1+R_0)^2 - R_4^2}}{\sqrt{1+R_0+\sqrt{(1+R_0)^2 - R_4^2}}}$$

$$= (1+R)\frac{1+R_0-R_4}{2((1+R_0)^2 - R_4^2)} \frac{1+R_0+R_4+\sqrt{(1+R_0)^2 - R_4^2}}{\sqrt{1+R_0+\sqrt{(1+R_0)^2 - R_4^2}}},$$

where $R_i = \langle R \rangle_i$. Always $-\sqrt{R}$ is also a square root. If $R_4^2 > 0$, so are $\pm\sqrt{R} \times R_4/\|R_4\|$. If $R = 1$, any quadvector Q for which $Q^2 = 1$ is also a square root.

When $1 + R$ is noninvertible, the polar decomposition approach cannot be applied. The following section investigates and resolves the square root in that case.

5.2.7 Roots When $1 + R$ Is Noninvertible

By (5.1), the quantity $1 + R$ always can be written in the form US of a rotor times a symmetric element, and the rotor part of each of such factorizations is the square root of $\sqrt{R}$. When $1 + R$ is noninvertible, our treatment of the polar decomposition does not apply to retrieve $\sqrt{R}$. This happens precisely when $\|1 + R\| = \sqrt{2}\sqrt[4]{(1 + R_0)^2 - R_4^2} = 0$, so when $(1 + R_0)^2 = R_4^2$. Examples can be found of 3D CGA rotors satisfying this, for instance $R = -1$ or $R = \mathbf{e}_1\mathbf{e}_2\mathbf{e}_3\mathbf{e}_+$. Actually, those examples are prototypical, as we now prove.

We begin with a useful characterization of a rotor in terms of its grades.

Lemma 5.3 *The element R of $\mathbb{R}_{4,1}^+$ is a rotor iff $R_2^2 = (R_0 + R_4)^2 - 1$ and $R_2 R_4$ is a bivector (where R_i is shorthand for $\langle R \rangle_i$).*

Proof The rotor R should satisfy

$$1 = R\tilde{R} = (R_0 + R_2 + R_4)(R_0 - R_2 + R_4)$$
$$= (R_0 + R_4)^2 - R_2^2 + (R_2 R_4 - R_4 R_2).$$

The final term in brackets is a commutator of a bivector and a quadvector, and therefore potentially a quadvector (it is self-reverse). We show that it is actually zero. The rotor R can be written as the exponential of some bivector B. Since $B^2 = B \cdot B + B \wedge B$, when we form the Taylor expansion $R = \exp(B)$, we can only generate a grade-2 part R_2 proportional to a linear combination of B and $B(B \wedge B)$, and a grade-4 part R_4 proportional to $B \wedge B$. Then $R_2 R_4 - R_4 R_2$ is proportional to a linear combination of the commutators $B \times (B \wedge B)$ and $B \times (B(B \wedge B))$; and both of these are easily seen to evaluate to zero using $B \wedge B = B^2 - B \cdot B$.

The statement $R_2 R_4 = R_4 R_2$ is equivalent to '$R_2 R_4$ is a bivector', since it leads to $(R_2 R_4)^{\sim} = R_4 \tilde{R}_2 = -R_4 R_2 = -R_2 R_4$, and vice versa.

Conversely, when the grade relationships hold, it is immediate that R is even and $R\tilde{R} = 1$, so that R is a rotor.   □

Lemma 5.4 *The element* $1 + R$, *with* R *a rotor of* $\mathbb{R}_{4,1}^+$, *is noninvertible iff* '$R = -1$' *or* '$R = \langle R \rangle_4$ *with* $R^2 = 1$'.

Proof Any rotor U satisfies $U\tilde{U} = 1$, which in less than 6D is equivalent to $U^2 = 2U(U_0 + U_4) - 1$. Then stipulating $U^2 = R$ leads to a simple relationship between S and some components of U: since for invertible U we have $US = 1 + R = 1 + U^2 = 2U(U_0 + U_4)$, it follows that $S = 2(U_0 + U_4)$. As a consequence, $U_2^2 = (U_0 + U_4)^2 - 1 = \frac{1}{4}S^2 - 1$. It also easily follows that $R_2 = U_2 S$.

For noninvertible $1 + R$, the symmetric factor $S = S_0 + S_4$ is noninvertible. This is equivalent to $\|S\| = 0$, which gives $S_0^2 = S_4^2$. Substituting this gives $S^2 = 2S_0 S$.

The case $R = -1$ is a trivial case of the lemma. Excluding it from here on, in the noninvertible case we can still determine the principal root of $S^2 = 2(1 + R_0 + R_4)$ using (5.9) since $\langle S^2 \rangle = 2S_0^2 > 0$. It gives $S = (1 + R_0 + R_4)/\sqrt{1 + R_0}$.

We must have $R_2^2 = (R_0 + R_4)^2 - 1$ since R is a rotor, see Lemma 5.3. With the expressions above, we can develop both sides in terms of R_0 and R_4. We find, on the one hand, $R_2^2 = U_2^2 S^2 = (\frac{1}{4}S^2 - 1)S^2 = 2S_0(S_0^2 - 1)S = 2R_0(1 + R_0 + R_4)$, while, on the other hand, $(R_0 + R_4)^2 - 1 = R_0^2 + R_4^2 + 2R_0 R_4 - 1 = 2R_0^2 + 2R_0 R_4 = 2R_0(R_0 + R_4)$. Equating the two expressions results in $R_0 = 0$. That in turn gives $R_4^2 = 1$, so that $R_2^2 = (R_0 + R_4)^2 - 1 = 0$. It follows that R_2 is a null blade that should produce a bivector with the non-null quadvector R_4. This is only possible when $R_2 = 0$. This noninvertible situation therefore reduces to a rotor that is a pure quadvector $R = R_4$ squaring to 1. □   □

In these noninvertible cases, the square root is nonunique.

Lemma 5.5 *A rotor* R *of* $\mathbb{R}_{4,1}^+$ *for which* $1 + R$ *is noninvertible has a principal square root of*

$$\sqrt{R} = \left(\frac{1 + R}{2} + B \frac{1 - R}{2} \right), \tag{5.16}$$

where B *is any 2-blade commuting with* $\langle R \rangle_4$ *such that* $B^2 = -1$. *Moreover,* $-\sqrt{R}$ *and* (*when* $R \neq -1$) *also* $\pm R\sqrt{R}$ *are square roots.*

Proof With Lemma 5.4 circumscribing the invertible cases, verification of (5.16) is a matter of straightforward computation by squaring and substituting $R^2 = 1$. □

Since BR gives the same result as B, a 3-parameter family of square roots results. The multiplication by $R_4/\|R_4\|$ resulting in extra square roots here amounts to $R\sqrt{R}$, which has the same effect as replacing B by $-B$, so the B-roots come in conjugated pairs.

As we remarked before, we can limit ourselves to rotors with nonnegative scalar part and treat $R = -1$ as $R = 1$, using Lemma 5.2 to find its square roots.

5.2.8 Interpolation

Since some rotors have multiple square roots, one may wonder what happens in subsequent interpolation. Fortunately, the cases do not expand exponentially. Where $\sqrt{R}$, $-\sqrt{R}$ and, under the condition $R_4^2 > 0$, also $\sqrt{R}(R_4/\|R_4\|)$ and $-\sqrt{R}(R_4/\|R_4\|)$ are square roots of R, the fourth roots still only number two or four, for the twofold or fourfold ambiguity of the various cases overlap.

In the case that R has infinitely many roots, only in the first interpolation step does one need to break the symmetry by an arbitrary bivector B. Once that has been done, the square roots have become regular rotors, with their regular multiple square roots.

5.2.9 Special Cases of Rotor Roots

We treat some important special cases of rotor roots.
- *Special Case Motors: Rigid Body Motions.*

  For the useful case of a motor M in $\mathbb{R}_{4,1}$, we can restrict ourselves to the eight-dimensional linear space $\mathbb{M}$ from which the motors take their multivector components [5] (Chap. 2 in this volume), with basis $\{1, \mathbf{e}_i \wedge \mathbf{e}_j, \mathbf{e}_i n_\infty, \mathbf{I}_3 n_\infty\}$. It follows that all rotors $M \in \mathbb{M}$ have a null 4-grade part. We then compute $\|1 + M\| = \sqrt{\langle(1 + \widetilde{M})(1 + M)\rangle} = \sqrt{2(1 + \langle M \rangle)}$ (which is well defined since $\langle M \rangle \geq -1$). This ultimately gives:

square root of motor M: $\sqrt{M} = \dfrac{1 + M}{\sqrt{2(1 + \langle M \rangle)}}\left(1 - \dfrac{\langle M \rangle_4}{2(1 + \langle M \rangle)}\right).$ (5.17)

The formula explicitly shows how a self-reverse element consisting of only a scalar and 4-vector part 'nudges $(1 + M)$ back onto the motor manifold'. Treat $M = -1$ as $M = 1$ to avoid problems; the other noninvertible case cannot occur: since $R_4^2 \neq 1$, the quantity $1 + M$ is always invertible for motors with nonnegative scalar part.

Apart from a scalar normalization, (5.17) is, in essence,

$$\sqrt{M} \propto (1 + M)\left(1 + \langle M \rangle - \frac{1}{2}\langle M \rangle_4\right),$$

which may be more efficient in some applications.

- *Special Case Pure Rotations: The Square Root of a Quaternion.*

  A pure rotation **R** at the origin in 3D is a simple rotor (i.e., the exponential of a 2-blade). Therefore the motor has no grade-4 part, and the square root simplifies further to:

$$\text{square root of pure rotation } \mathbf{R}: \quad \sqrt{\mathbf{R}} = \frac{1+\mathbf{R}}{\sqrt{2(1+\langle\mathbf{R}\rangle)}}. \tag{5.18}$$

You can use this formula to interpolate unit quaternions (which are after all merely 3D rotors). In its explicit form in terms of the 2-blade **I** of the rotation plane and the rotation angle ϕ, it becomes a straightforward trigonometric equality,

$$\sqrt{\cos(\phi/2) - \mathbf{I}\sin(\phi/2)} = \frac{1+\cos(\phi/2) - \mathbf{I}\sin(\phi/2)}{\sqrt{2(1+\cos(\phi/2))}}$$
$$= \frac{2\cos^2(\phi/4) - 2\mathbf{I}\sin(\phi/4)\cos(\phi/4)}{2\sqrt{\cos^2(\phi/4)}}$$
$$= \pm\big(\cos(\phi/4) - \mathbf{I}\sin(\phi/4)\big)$$

(with the sign depending on that of $\cos(\phi/4)$, to make the resulting rotor have a non-negative scalar part). Using the correspondence between 3D rotor and unit quaternion (see e.g. [2]), this applies directly to quaternion interpolation.

- *Special Case Translations: Linear Interpolation.*

  A translation motor $T_t = 1 - \frac{1}{2}t n_\infty$ is also a simple rotor. Therefore there is no grade-4 part either. The square root formula now simplifies to:

$$\text{square root of pure translation } T_t: \quad \sqrt{T_t} = \frac{1}{2}(1 + T_t), \tag{5.19}$$

which is in agreement with the expected result $\sqrt{T_t} = T_{t/2}$.

5.3 Logarithms of Rotors in 3D CGA

A logarithm of a rotor R is a bivector L such that $R = e^L$. Such a bivector exists for every rotor. However, it may not be unique; a standard example is the 2π additive freedom in a rotation angle. It is therefore customary to compute a principal logarithm in a standard interval, and we will do so as well.

5.3.1 Logarithm of a Simple Rotor

Recall that a simple rotor R is the exponential of a 2-blade. Taking the 2-blade as $-B/2$ and writing the rotor R out in its grade components, we find

$$R = e^{-B/2} = \cosh(B/2) - \sinh(B/2).$$

Here the cosh() and sinh() functions on 2-blades are defined in terms of their Taylor series, as usual. Because a blade squares to a scalar, they can be mapped onto hyperbolic and trigonometric functions on scalars. For sinh(), this is explicitly for a 2-blade A:

$$\sinh(A) = A + \frac{1}{3!}A^3 + \cdots = \begin{cases} \frac{\sinh(\sqrt{A^2})}{\sqrt{A^2}} A & \text{if } A^2 > 0 \\ A & \text{if } A^2 = 0 \\ \frac{\sin(\sqrt{-A^2})}{\sqrt{-A^2}} A & \text{if } A^2 < 0. \end{cases} \tag{5.20}$$

One could define a function asinh() to invert this blade and apply it to the grade-2 part of the simple rotor R to retrieve A. Yet that does not work properly in the case $A^2 < 0$: an asinh() would have a range of $[-\pi/2, \pi/2]$, but we really need the full principal range $(-\pi, \pi]$. We can retrieve the proper quadrant by taking into account the value (or the sign) of $\cosh(A)$. So we define an atanh2() function on 2-blades to retrieve the correct value of A depending on a 2-blade argument s (the value of $\sinh(A)$) and a scalar argument c (the value of $\cosh(A)$):

$$\text{atanh2}(s, c) = \begin{cases} \frac{\text{asinh}(\sqrt{s^2})}{\sqrt{s^2}} s & \text{if } s^2 > 0 \\ s & \text{if } s^2 = 0 \\ \frac{\text{atan2}(\sqrt{-s^2}, c)}{\sqrt{-s^2}} s & \text{if } -1 \leq s^2 < 0. \end{cases} \tag{5.21}$$

In the $s^2 < 0$ part, the range of atan2 is $(-\pi, \pi]$ to provide the principal value. In the $s^2 \geq 0$ parts, we have used simpler functions, since then c satisfies $c \geq 1$ and is not required to establish the correct quadrant.

With this function defined, the principal logarithm of a simple rotor is

$$\text{Log}(R) = \text{atanh2}(\langle R \rangle_2, \langle R \rangle).$$

For general rotors, we will need to do quite some preliminary work before we can invoke the atanh2() to produce the logarithm.

5.3.2 The Split Structure of a Rotor

A general rotor is the exponential of a bivector, not a 2-blade. Following [3], we realize that a bivector can always be decomposed into a sum of commuting blades. In 3D CGA, which is 5D, two commuting 2-blades suffice, so we concentrate on that case.

Consider a rotor $R = \exp(-B/2)$ with B a bivector. Imagine the rotor bivector B split into two commuting blades $B = B_+ + B_-$ with $B_+ B_- = B_- B_+$. Since the blades commute, so do their exponentials: Then

$$R = e^{-B/2} = e^{-(B_+ + B_-)/2} = e^{-B_+/2} e^{-B_-/2}$$

$$= \cosh\left(\frac{1}{2}B_+\right) \cosh\left(\frac{1}{2}B_-\right)$$

$$- \left(\cosh\left(\frac{1}{2}B_-\right) \sinh\left(\frac{1}{2}B_+\right) + \cosh\left(\frac{1}{2}B_+\right) \sinh\left(\frac{1}{2}B_-\right)\right)$$

$$+ \sinh\left(\frac{1}{2}B_+\right) \sinh\left(\frac{1}{2}B_-\right). \tag{5.22}$$

The three terms are of grades 0, 2 and 4, respectively. No longer are the sinh() and cosh() values easily retrievable from R. Following [3], we study the curl of the rotor mapping to find simpler expressions.

5.3.3 Exterior Derivative of a Rotor Transformation in 3D CGA

The exterior derivative ('curl') of the orthogonal transformation $x \mapsto R x \tilde{R}$, characterized by the rotor R, is defined as $\partial_x \wedge (R x \tilde{R})$.

In the case of 3D CGA rotors (or rotors of any geometric algebra of a space with less than six dimensions), the exterior derivative can be expressed simply in terms of the grades R_i of R:

$$\partial_x \wedge (R x \tilde{R}) = \langle \partial_x (R x \tilde{R}) \rangle_2$$

$$= \langle (\partial_x (R_0 + R_2 + R_4) x) \tilde{R} \rangle_2$$

$$= \langle (5 R_0 + R_2 - 3 R_4) \tilde{R} \rangle_2$$

$$= \langle (R + 4(R_0 - R_4)) \tilde{R} \rangle_2$$

$$= \langle 1 + 4(R_0 - R_4) \tilde{R} \rangle_2$$

$$= 4(R_4 - R_0) R_2, \tag{5.23}$$

where we use the fact from Lemma 5.3 that $\langle R_4 R_2 \rangle_2 = R_4 R_2$.

We now define a bivector F from the rotor R as half the curl and compute:

$$F \equiv \frac{1}{2} \partial_x \wedge (R x \tilde{R}) = 2(R_4 - R_0) R_2$$

$$= 2\left(\cosh\left(\frac{1}{2}B_+\right) \cosh\left(\frac{1}{2}B_-\right) - \sinh\left(\frac{1}{2}B_+\right) \sinh\left(\frac{1}{2}B_-\right)\right)$$

$$\times \left(\cosh\left(\frac{1}{2}B_-\right) \sinh\left(\frac{1}{2}B_+\right) + \cosh\left(\frac{1}{2}B_+\right) \sinh\left(\frac{1}{2}B_-\right)\right)$$

$$= \cdots \text{ some straightforward rewriting } \cdots$$

$$= \sinh(B_+) + \sinh(B_-). \tag{5.24}$$

The curl-based quantity F therefore contains the information on $\sinh(B_\pm)$ much more cleanly than R itself. By (5.20), these components $\sinh(B_+)$ and $\sinh(B_-)$ are proportional to B_+ and B_-, respectively, and therefore also commuting 2-blades. Given R, we can determine F and then determine the separate values of $\sinh(B_+)$ and $\sinh(B_-)$ by performing a split of F into commuting 2-blades.

5.3.4 Split of a 3D CGA Bivector into Commuting 2-Blades

For 3D CGA, any bivector F splits into at most two commuting 2-blades.

Lemma 5.6 *Consider a non-null bivector F in a space of five dimensions or less. Define $\|F\| = \sqrt[4]{(2\langle F^2\rangle - F^2)F^2}$ and assume that $\|F\| \neq 0$. Then F can be split into a sum of commuting 2-blades $F = F_+ + F_-$ by*

$$F_\pm = \frac{1}{2}F\left(1 \pm \frac{\|F\|^2}{F^2}\right). \tag{5.25}$$

Proof Since F^2 is self-reverse, it is of the form 'scalar plus quadvector'; and because the quadvector squares to a scalar, $(F^2)^2$ should be expressible as a linear combination of F^2 and 1. Following the derivation of (5.8), this equation can be written as $(F^2)^2 - 2\langle F^2\rangle F^2 + \|F\|^4 = 0$, which defines $\|F\|^4$ analogously to (5.8) and which gives the same reasons for existence of its square root $\|F\|^2$ as (5.11). (Note that the lemma in fact involves $\|F\| = [[F^2]]$ with $[[\cdot]]$ defined as in (5.7).)

To prove the lemma, we need to show that $F_+ + F_- = F$ (which is trivial), that F_+ and F_- commute (and they do, for they only involve scalars and powers of F), and that they are 2-blades. According to [4], 'if the square of a bivector is real, then it is simple', i.e. a 2-blade. For invertible F, we find

$$F_\pm^2 = \frac{1}{4}F\left(1 \pm \frac{\|F\|^2}{F^2}\right)F\left(1 \pm \frac{\|F\|^2}{F^2}\right)$$
$$= \frac{1}{4}\left(F^2 \pm 2\|F\|^2 + \frac{\|F\|^4}{F^2}\right)$$
$$= \frac{1}{4}\left(F^2 \pm 2\|F\|^2 + (-F^2 + 2\langle F^2\rangle)\right)$$
$$= \frac{1}{2}(\langle F^2\rangle \pm \|F\|^2),$$

which is indeed scalar so that $F_\pm$ are blades. Incidentally, $F_- F_+ = \frac{1}{2}\langle F^2\rangle_4$. □

Equation (5.25) is identical to the formula for the bivector split in [3, p. 81], merely compactly reformulated in terms of our new 'norm' of F.

We have excluded $F^2 = 0$ or $\|F\|^2 = 0$ from the lemma, but they do occur in practice. As a consequence, there are several cases in producing a bivector split that lead to branches in an implementation.

- If $\|F\|^2 \neq 0$ and $F^2 \neq 0$, (5.25) gives the bivector split. This is the regular case. Notably, if $F^2 = \|F\|^2 \neq 0$, then F is a 2-blade, and we find the sensible outcome $F_+ = F$ and $F_- = 0$ as the bivector split.
- If $\|F\|^2 = 0$ and $F^2 = 0$, (5.25) cannot be applied since F is not invertible. But such null bivectors are all blades anyway, since they have a scalar square. Therefore there is no need to apply any split, and we then simply return $F_+ = F$ and $F_- = 0$ if $F^2 = 0$, to correspond with the case where F is a non-null 2-blade.
- If $\|F\|^2 = 0$ but $F^2 \neq 0$, F is noninvertible, and (5.25) would return the trivial split $F_\pm = F/2$. Although these bivectors are commuting and partition F, they are only blades if F was already—which it is not, or F^2 would have been equal to $\|F\|^2$.

    In this case the noninvertible bivector F actually does have a split into two 2-blades, but it is not unique. Two 2-blades in the split F_+ and F_- satisfy $F_+^2 = F_-^2$, and [3, p. 83], shows that there is a one-parameter family of splits, obtained from any member by application of a rotor that commutes with F. A rotor corresponding to this kind of F is called an 'isoclinic rotation' or a 'Clifford displacement'. An example in 3D CGA is $R = \exp((\mathbf{e}_{12} + \mathbf{e}_{3+})\phi/2)$.

    We have not found a source that produces any candidate from which the one-parameter family could be generated in this noninvertible case. But by a simple trick we can still use (5.25): add a small random bivector to F to make it invertible and then apply the formula (numerical stability may be an issue, though simulations show surprisingly robust behavior).

In summary, a bivector split always exists, (5.25) produces it for invertible F, and for noninvertible F, we have to take special measures.

5.3.5 The Principal Logarithm of a 3D CGA Rotor

The bivector split of half the curl F into $F = S_+ + S_-$ gives us $S_\pm = \sinh(B_\pm)$ (which is why we prefer to label the parts $S_\pm$ rather than $F_\pm$). But as in the simple rotor cases, to find $B_\pm$ properly, we also need the values $C_\pm = \cosh(B_\pm)$ (at least in the case of negative $B_\pm^2$). These can be retrieved from R^2 using $S_\mp$, as follows:

$$C_\pm = \cosh(B_\pm) = \begin{cases} -\langle\langle R^2\rangle_2/S_\mp\rangle & \text{if } S_\mp^2 \neq 0 \\ \langle R^2\rangle & \text{if } S_\mp^2 = 0. \end{cases} \tag{5.26}$$

Proof Define sinh() on a bivector B as the power series $\sinh(B) = B + B^3/3! + \cdots$, giving the usual hyperbolic and trigonometric relationships. Then

$$-\langle\langle R^2\rangle_2/S_-\rangle = \langle \sinh(B)/S_-\rangle$$
$$= \langle((\cosh(B_+)\sinh(B_-) + \cosh(B_-)\sinh(B_+))/\sinh(B_-)\rangle$$
$$= \cosh(B_+),$$

where the final step uses the commutativity of B_+ and B_- to observe that no mixed terms will contaminate the scalar part of the expression. The exceptional case of (5.26) relies on $\langle R^2 \rangle = \cosh(B_+)\cosh(B_-)$, with one of the coshs equal to 1. Isoclinic rotations are to be resolved as suggested above.   □

With the 2-blade and scalar values of $\sinh(B_\pm)$ and $\cosh(B_\pm)$ all available, we can retrieve B_+ and B_- by employing atanh2() defined for blades in (5.21). This yields:

$$\text{Log}(R) = -\frac{1}{2}\text{atanh2}(S_+, C_+) - \frac{1}{2}\text{atanh2}(S_-, C_-). \tag{5.27}$$

This explicitly solves the logarithm of a general conformal rotor in 3D CGA; the most general case known before was the scaled rigid body motion in [2].

Our logarithm method uses $F = \frac{1}{2}\partial_x \wedge (Rx\widetilde{R})$ and R^2, which are both clearly insensitive to a sign change in R. There is therefore an ambiguity in the logarithm, resulting in an ambiguity of sign in the logarithm-based rotor reconstruction $\exp(\log(R))$: we may reconstruct $-R$ rather than R.

In fact, we always have that

$$\langle \exp(\text{Log}(R)) \rangle = \cosh\left(\frac{1}{2}B_+\right)\cosh\left(\frac{1}{2}B_-\right) \geq 0.$$

Proof When $B_\pm^2 < 0$, the atan2() part of atanh2() will return bivector angles in the range $(-\pi, \pi]$ for the principal value of the logarithm. When exponentiating their half angles, this produces a positive cosine in the scalar part. When $B_\pm^2 \geq 0$, $\cosh(B_\pm)$ is always positive. Together, this implies that grade$(\exp(\text{Log}(R))) = \cosh(\frac{1}{2}B_+)\cosh(\frac{1}{2}B_-) \geq 0$.   □

Restriction to rotors with nonnegative scalar part eliminates this ambiguity of reconstruction.

5.4   Geometrical Interpretation of the Logarithm

5.4.1   The 2-Blade Generators in the Bivector Split

Geometrically, 2-blades of 3D CGA are mostly point pairs, or rather 1-spheres with a positive, negative or zero squared radius. The occurrence of the designated point at infinity (n_∞ in the case of Euclidean geometry) affects the geometrical interpretation, stretching the meaning of 'point pair' to include free vectors; see [2]. All cases are sketched in Fig. 5.2, and they can all occur for F_+ and F_- obtained from the exterior derivatives of rotor mappings.

More precisely,

$$B_-^2 \leq 0,$$

while B_+^2 can have any sign.

Proof Consider $F_\pm = \sinh(B_\pm)$, which have the same properties. Write the polar decomposition of F as $F = RS$ and note that $F^2 = -\widetilde{F}F = -S^2$. Observe that $\| -F^2 \|^2 = \|F^2\|^2$, so that (taking the positive root) $\| -F^2 \| = \|F^2\|$. Then $F_\pm^2 = \frac{1}{2}(-\langle -F^2 \rangle \pm \| -F^2 \|) = \frac{1}{2}(-\langle S^2 \rangle \pm \|S^2\|) = -\frac{1}{2}(\langle S^2 \rangle \mp \| -S^2 \|)$. We have shown in (5.12) that $\langle S^2 \rangle + \| -S^2 \| \leq 0$; therefore $F_-^2 \leq 0$, so that $B_-^2 \leq 0$. No such constraint can be given on F_+^2 and B_+^2. $\qquad\qquad\square$

In 2D and 3D, the split of the bivector into commuting 2-blades corresponds geometrically to writing the bivector as the product of two orthogonal point pairs. We have just seen that one of those is imaginary (or null), whereas the other may be real, null or imaginary. The orthogonality of such points pairs is achieved by placing them in a relative position, with an appropriate radius.

As an example, we compute the orthogonality condition for a real point pair P_1 and an imaginary point pair P_2. It is most easily derived in conveniently chosen coordinates (with $T_\mathbf{t}$ denoting the application of a translation versor over $\mathbf{t}$):

$$P_1 = T_{\alpha e_3/2}\left[\left(n_o - \frac{1}{2}\rho_1^2 n_\infty\right)\mathbf{e}_1\right],$$

$$P_2 = T_{-\alpha e_3/2}\left[\left(n_o + \frac{1}{2}\rho_2^2 n_\infty\right)\mathbf{e}_2\right].$$

Then computation yields:

$$P_1 P_2 = P_2 P_1 \quad\Longleftrightarrow\quad \alpha^2 = \frac{1}{2}(\rho_1^2 - \rho_2^2).$$

By changing the sign of ρ_1^2 and setting ρ_1^2 or ρ_2^2 to zero, one obtains the orthogonality conditions for other cases where P_1 and P_2 can be written in the above forms (such as the tangent vectors which are null). Not included are then the degenerate forms of dual lines, null directions and flat points; some of the combinations can never be orthogonal and will therefore not occur as bivector splits.

5.4.2 Orbits of Rotors

When a simple rotor (the exponential of a 2-blade $-B/2$) is applied repeatedly to an initial point x, it will generate points that are on the 3-blade $x \wedge B$ (this is easily verified by writing out $x \wedge B \wedge (e^{-B/2} x e^{B/2})$ using $e^{-B/2} = \cosh(B/2) - \sinh(B/2)$). That 3-blade is called the *R-orbit* of x. In 3D CGA, it is a circle (though it may degenerate to a line when either x or B contain the point at infinity n_∞).

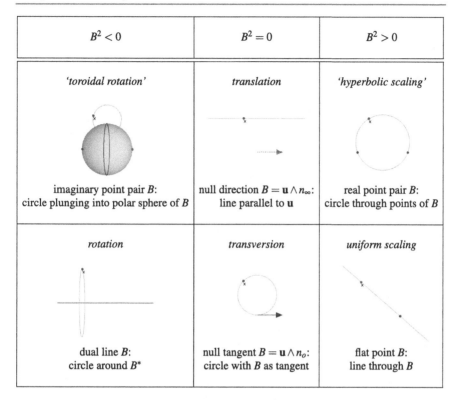

$B^2 < 0$	$B^2 = 0$	$B^2 > 0$
'toroidal rotation'	*translation*	*'hyperbolic scaling'*
imaginary point pair B: circle plunging into polar sphere of B	null direction $B = \mathbf{u} \wedge n_\infty$: line parallel to $\mathbf{u}$	real point pair B: circle through points of B
rotation	*transversion*	*uniform scaling*
dual line B: circle around B^*	null tangent $B = \mathbf{u} \wedge n_o$: circle with B as tangent	flat point B: line through B

Fig. 5.2 A geometrical classification of the orbits through a point x of simple rotors in CGA, characterized by properties of their 2-blade B. The name of the resulting transformation is indicated in *italics*. Color version of this figure online

We have shown how a general rotor in 3D CGA can be written as the product of two commuting simple rotors by means of (5.22) and (5.27). The geometry of those commuting rotors depends on the signature of their 2-blades $B_\pm$. A full classification of bivectors in 3D CGA and their orbits when used in a simple rotor is given in Fig. 5.2.

- When $B^2 < 0$, the rotor $e^{-B/2}$ is a 'toroidal rotation' around a circle that is the dual of B_+. If that dual happens to be a line, this is an ordinary rotation.
- When $B^2 = 0$, the rotor $e^{-B/2}$ is a translation or a (translated) transversion.
- When $B^2 > 0$, the rotor $e^{-B/2}$ is a 'hyperbolic scaling' with source and sink points corresponding to the two points in the point pair B.[2] In the special case that B is a *flat point* (when one of the points is n_∞), the rotor is a uniform scaling relative to the flat point location.

---

[2]It is the inversion of a uniform scaling with respect to an inversion sphere that has one of the points of B at its center and the other on its shell; dually it is $(B \pm \sqrt{B^2})(n_\infty \rfloor B) + 2B^2 n_\infty$, with n_∞ the point at infinity.

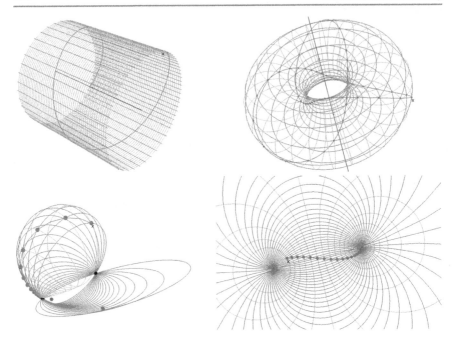

Fig. 5.3 Conformal coordinate grids induced by some rotors, with orbits for a point x indicated. See text for explanation

In (5.22), we have explicitly split the action of a general conformal rotor R into that of two simple rotors. Therefore we can analyze the incremental action of R by taking incremental steps along the circular (or line-shaped) orbits of these simple rotors.

The commuting nature of the 2-blades of the logarithm implies that the orbits of the corresponding simple rotors are orthogonal at each point (with a degenerate ambiguity in the special and rare case of isoclinic rotations). Drawing those orbits at successive steps of the incremental application of the rotor suggests a 'conformal coordinate grid' for the submanifold in which the orbit of a given point resides; see Fig. 5.3. The sketches in this figure clearly show that the orbits of general rotors are more involved than the circles that can be generated by simple rotors, though they can of course be composed from them incrementally.

As we will compute in Sect. 5.5.2, in the special case of motors, B_- is a dual line, B_+ is the generator of a translation, and Chasles' famous screw factorization of rigid body motions results. The grid is then a set of equidistantly placed circles on a cylinder scored by equidistant longitudinal lines; the composite orbits are screws (top left in Fig. 5.3). Other rotors may give grids that rule a torus or a Dupin cycloid (when we have two imaginary point pairs as generators; closed orbits are then typically knotted, top right); or that rule the inversion of a cone (for a real and imaginary point pair; the orbits are equiangular spirals on that cone, bottom left), with a plane as special case for coplanar point pairs (bottom right).

Table 5.1 Commutation of even versors in the conformal model. In the entries for VT and TV, A is the rotation/scaling formed by a rotation over $(1 + \mathbf{vt})/\|1 + \mathbf{vt}\|$ combined with a scaling by a factor $1/\|1 + \mathbf{vt}\|^2$, see [1]

	Rotation R	Scaling S	Translation T	Transversion V
R	$R_{\mathbf{I}\phi} R = RR_{\widetilde{R}\mathbf{I}\phi R}$	$RS_\sigma = S_\sigma R$	$RT_\mathbf{t} = T_{R\mathbf{t}\widetilde{R}} R$	$RV_\mathbf{v} = V_{R\mathbf{v}\widetilde{R}} R$
S	$S_\sigma R = RS_\sigma$	$S_\sigma S_\tau = S_\tau S_\sigma$	$S_\sigma T_\mathbf{t} = T_{\sigma\mathbf{t}} S_\sigma$	$S_\sigma V_\mathbf{v} = V_{\mathbf{v}/\sigma} S_\sigma$
T	$T_\mathbf{t} R = RT_{\widetilde{R}\mathbf{t} R}$	$T_\mathbf{t} S_\sigma = S_\sigma T_{\mathbf{t}/\sigma}$	$T_\mathbf{t} T_\mathbf{s} = T_\mathbf{s} T_\mathbf{t}$	$T_\mathbf{t} V_\mathbf{v} = A^{-1} V_{\mathbf{v}(1+\mathbf{tv})} T_{(1+\mathbf{tv})^{-1}\mathbf{t}}$
V	$V_\mathbf{v} R = RV_{\widetilde{R}\mathbf{v} R}$	$V_\mathbf{v} S_\sigma = S_\sigma V_{\sigma\mathbf{v}}$	$V_\mathbf{v} T_\mathbf{t} = A T_{\mathbf{t}(1+\mathbf{vt})} V_{(1+\mathbf{vt})^{-1}\mathbf{v}}$	$V_\mathbf{v} V_\mathbf{w} = V_\mathbf{w} V_\mathbf{v}$

5.5 Special Cases of the Rotor Logarithm

A general conformal transformation in $\mathbb{R}_{4,1}$ can be characterized as a sequence of a transversion $V_\mathbf{v} = \exp(n_o \mathbf{v})$, translation $T_\mathbf{t} = \exp(-\mathbf{t}n_\infty/2)$, a uniform scaling $S_{e^\gamma} = \exp(\gamma n_o \wedge n_\infty/2)$ and a rotation $R_{\mathbf{I}\phi} = \exp(-\mathbf{I}\phi/2)$, in any order. The following formula for half the exterior derivative of the particular sequence $RSTV$ is convenient for theoretical derivations. A more general sequence can be converted into this by commutation relationships, spelled out in [1] and summarized in Table 5.1.

$$
\begin{aligned}
F &= \frac{1}{2}\partial \wedge \left((R_{\mathbf{I}\phi} S_{e^\gamma} T_\mathbf{t} V_\mathbf{v})x(R_{\mathbf{I}\phi} S_{e^\gamma} T_\mathbf{t} V_\mathbf{v})^{-1}\right) \\
&= \sin(\phi)\mathbf{I} + \mathbf{v} \wedge (R\mathbf{t}/R) \\
&\quad - \frac{1}{2}\left(e^\gamma(1+\mathbf{tv})(1+\mathbf{vt}) - e^{-\gamma}\right)n_o \wedge n_\infty \\
&\quad + \left(e^{-\gamma}\mathbf{v} + R(\mathbf{v}+\mathbf{v}^2\mathbf{t})/R\right) \wedge n_o \\
&\quad + \frac{1}{2}\left(e^\gamma\mathbf{t} + R(\mathbf{t}+e^\gamma\mathbf{t}^2\mathbf{v})/R\right) \wedge n_\infty.
\end{aligned}
\tag{5.28}
$$

The general logarithm of this form is awkward to express in the characterizing parameters $\mathbf{v}$, $\mathbf{t}$, γ and $\mathbf{I}\phi$, though various special cases can be worked out in explicitly. We treat a few.

5.5.1 Log(TS)

As a special case of the method to illustrate the use of the atanh2(), let us combine a scaling with a translation to form the versor $T_\mathbf{a} S_{e^\gamma}$ (to conform to the order in [2]). We can use (5.28) above, setting $R = 1$ and $\mathbf{v} = 0$ to obtain an expression for $S_{e^\gamma} T_\mathbf{t} = T_{e^\gamma \mathbf{t}} S_{e^\gamma}$; apparently we should set $\mathbf{a} = e^\gamma \mathbf{t}$ to obtain:

$$
\begin{aligned}
F &= -\frac{1}{2}\left(e^\gamma - e^{-\gamma}\right)n_o \wedge n_\infty + \frac{1}{2}\left(1+e^\gamma\right)e^{-\gamma}\mathbf{a} \wedge n_\infty \\
&= \left(-\sinh(\gamma)n_o + \frac{1}{2}\left(1+e^{-\gamma}\right)\mathbf{a}\right) \wedge n_\infty.
\end{aligned}
\tag{5.29}
$$

We find that F^2 is scalar, so that there is only a single 2-blade required—obviously, since we have already shown F in factored form. When $F^2 = (\sinh(\gamma))^2 > 0$, we get:

$$
\begin{aligned}
\mathrm{Log}(T_{\mathbf{a}} S_{e^\gamma}) &= -\frac{1}{2}\,\mathrm{asinh}(F) \\
&= -\frac{1}{2}\frac{\mathrm{asinh}(|\sinh(\gamma)|)}{|\sinh(\gamma)|}\left(-\Big(\sinh(\gamma)n_o + \frac{1}{2}(1+e^{-\gamma})\mathbf{a}\Big)\wedge n_\infty\right) \\
&= -\frac{1}{2}\gamma\left(-n_o + \frac{1+e^{-\gamma}}{e^\gamma - e^{-\gamma}}\mathbf{a}\right)\wedge n_\infty \\
&= \frac{1}{2}\gamma\left(n_o + \frac{1}{1-e^\gamma}\mathbf{a}\right)\wedge n_\infty \\
&= \frac{1}{2}\gamma n_o \wedge n_\infty \exp\left(\frac{\mathbf{a}}{1-e^\gamma}n_\infty\right).
\end{aligned}
$$

This retrieves the result from [2], and the final form shows that it is a scaling by e^γ with center at $\mathbf{a}/(1-e^\gamma)$.

When $F^2 = (\sinh(\gamma))^2 = 0$, we find $\mathrm{Log}(T_{\mathbf{a}} S_{e^\gamma}) = -\frac{1}{2}\mathrm{atanh2}(F,1) = -\frac{1}{2}F = -\frac{1}{2}\mathbf{a}\wedge n_\infty$, and indeed the rotor is then a pure translation since $\gamma = 0$.

5.5.2   Log(RST): Generalized Chasles Theorem for Euclidean Similarities

We derive a generalized Chasles theorem for scaled rigid body motions (Euclidean similarities). Those are represented in CGA by rotors of the form $W = R_{\mathbf{I}\phi} S_{e^\gamma} T_{\mathbf{t}}$: a general combination of rotation, scaling and translation.

We find that any such motion can be represented as a rotation around a specific axis combined with a scaling relative to a specific point on that axis. Those two operations commute. When the scaling is trivial, Chasles' theorem results.

Specific steps in the computation are given. We define $E = n_o \wedge n_\infty$ for convenience of expression. Equation (5.28) provides the starting point:

$$
\begin{aligned}
F &= \sin(\phi)\mathbf{I} + \sinh(\gamma)E + \frac{1}{2}\big(e^\gamma \mathbf{t} + Rt\tilde{R}\big)n_\infty \\
&= \sin(\phi)\mathbf{I} + \sinh(\gamma)E + \mathbf{t}'n_\infty,
\end{aligned}
$$

defining $\mathbf{t}' = \frac{1}{2}(e^\gamma \mathbf{t} + Rt\tilde{R})$. Applying the bivector split, we obtain with $\|F\|^2 = \sin^2(\phi) + \sinh^2(\gamma)$:

$$
\begin{aligned}
F_- &= \sin(\phi)\mathbf{I}\left(1 + \frac{\sin(\phi) - \sinh(\gamma)\mathbf{I}}{\sin^2(\phi) + \sinh^2(\gamma)}(\mathbf{t}'\rfloor\mathbf{I})n_\infty\right) \\
&= \sin(\phi)\mathbf{I}\exp(\mathbf{t}_\| n_\infty),
\end{aligned}
$$

defining $\mathbf{t}_\| = \frac{\sin(\phi)-\sinh(\gamma)\mathbf{I}}{\sin^2(\phi)+\sinh^2(\gamma)}(\mathbf{t}'\rfloor\mathbf{I})$, and

$$F_+ = -\sinh(\gamma)E$$
$$\times\left(1 - \frac{\sin^2(\phi)/\sinh(\gamma)(\mathbf{t}'\wedge\mathbf{I})/\mathbf{I}-\sin(\phi)(\mathbf{t}'\rfloor\mathbf{I})+\sinh(\gamma)\mathbf{t}'}{\sin^2(\phi)+\sinh^2(\gamma)}n_\infty\right)$$
$$= -\sinh(\gamma)E\exp\big((\mathbf{t}_\|+\mathbf{t}_\perp)n_\infty\big),$$

defining $\mathbf{t}_\perp = -(\mathbf{t}'\wedge\mathbf{I})/\mathbf{I}/\sinh(\gamma)$. Then the atanh2() function on those parts gives the logarithm

$$\mathrm{Log}(W) = -\frac{1}{2}\phi\mathbf{I}\exp(\mathbf{t}_\| n_\infty) + \frac{1}{2}\gamma E\exp\big((\mathbf{t}_\|+\mathbf{t}_\perp)n_\infty\big)$$

This shows that W is indeed a rotation of ϕ around an axis with direction dual to $\mathbf{I}$ located at $\mathbf{t}_\|$ (with $\mathbf{t}_\|$ as defined above), combined with a scaling by e^γ with center $\mathbf{t}_\|+\mathbf{t}_\perp$. The two operations commute.

The expression above gives the logarithm of a rotor $W = RST$ in terms of its parameters γ, R, $\mathbf{I}\phi$ and $\mathbf{a}$. These can be obtained from W as $X = -n_o\rfloor(Wn_\infty)$, $\gamma = \log(X\tilde{X})$, $S = e^{\gamma E/2}$, $R = Xe^{-\gamma/2}$, $\mathbf{I}\phi = -2\mathrm{Log}(R)$ and $\mathbf{t} = -2n_o\rfloor(\tilde{S}\tilde{R}W)$ (see also [2]).

For motors (the rotors of rigid body motions), we have $\gamma = 0$, and this gives some simplification. Now $\mathbf{t}' = \frac{1}{2}(\mathbf{t} + Rt\tilde{R})$, and

$$F_- = \sin(\phi)\mathbf{I}\exp\left(\frac{\mathbf{t}'\rfloor\mathbf{I}}{\sin(\phi)}n_\infty\right),$$
$$F_+ = \big(\mathbf{t}'\wedge\mathbf{I}\big)/\mathbf{I}n_\infty,$$
$$\mathrm{Log}(M) = -\frac{1}{2}\phi\mathbf{I}\exp(\mathbf{t}_\| n_\infty) - \frac{1}{2}(\mathbf{t}'\wedge\mathbf{I})/\mathbf{I}\wedge n_\infty.$$

This is Chasles' theorem: the first term is a rotation around a translated axis, and the second term a translation along that axis. The theorem is more commonly given for a motor $M = T_\mathbf{a}R_{\mathbf{I}\phi} = R_{\mathbf{I}\phi}T_{\tilde{R}\mathbf{a}R}$, so that we should put $\mathbf{t} = \tilde{R}\mathbf{a}R$. Substituting this gives for the screw axis location $\mathbf{t}_\|$ and longitudinal motion vector $\mathbf{t}'_\perp$ (not to be confused with $\mathbf{t}_\perp$ above),

$$\mathbf{t}_\| = \frac{(\mathbf{a}\rfloor\mathbf{I})R}{2\sin(\frac{\phi}{2})} = (\mathbf{a}\rfloor\mathbf{I})/\mathbf{I}\frac{R}{2\langle R\rangle_2} \quad \text{and} \quad \mathbf{t}'_\perp = (\mathbf{a}\wedge\mathbf{I})/\mathbf{I},$$

after some rewriting. An alternative derivation is given in Chap. 6 of this volume.

5.5.3 Log(TV)

With the above techniques, it is possible to derive a closed-form expression for the logarithm of a versor of the type TV (a combination of transversion and translation), which was a case not tractable in [2]. We have done this. However, the resulting expressions in terms of the characterizing parameters are rather unwieldy, much more so than in the Chasles case above. In practice, it is probably better to apply the computational algorithm to the total rotor at hand, rather than to a fairly arbitrary parametric characterization of it.

5.6   Exercises

5.1 Verify that $X = US$ in (5.15).

5.2 Show that in $\mathbb{R}_{4,1}$, a nonzero null bivector cannot be an eigenbivector of multiplication by a quadvector with which it commutes. (Required in the proof of Lemma 5.4.)

5.3 Show that for a rotor R of $\mathbb{R}_{4,1}$, the condition $R^2 = 1$ is equivalent to either $\langle R \rangle^2 = 1$ or $\langle R \rangle_4{}^2 = 1$. A corollary is that $\langle R \rangle_2 = 0$.

5.4 A quadvector Q squaring to 1 is a rotor. It is easy enough to show that $\tilde{Q}Q = 1$, but can you write Q as the exponent of a bivector? Do this for $Q = (e_-)^*$ in $\mathbb{R}_{4,1}$.

5.5 Derive (5.29) using (5.23).

5.6 Work out the other cases in Sect. 5.4.1.

References

1. Dorst, L.: Conformal geometric algebra by extended Vahlen matrices. In: Skala, V., Hildenbrandt, D. (eds.) GraVisMa 2009 Workshop Proceedings, pp. 72–79 (2009)
2. Dorst, L., Fontijne, D., Mann, S.: Geometric Algebra for Computer Science: An Object Oriented Approach to Geometry, Morgan Kaufmann, San Mateo (2007/2009). See www.geometricalgebra.net
3. Hestenes, D., Sobczyk, G.: Clifford Algebra to Geometric Calculus. Reidel, Dordrecht (1984/1999)
4. Lounesto, P.: Clifford Algebras and Spinors, 2nd edn. Cambridge University Press, Cambridge (2001)
5. Valkenburg, R., Dorst, L.: Estimating motors from a variety of geometric data in 3D conformal geometric algebra. In: Dorst, L., Lasenby, J. (eds.) Guide to Geometric Algebra in Practice. Springer, London (2011). Chap. 2 in this book

Attitude and Position Tracking

6

Liam Candy and Joan Lasenby

Abstract

Several applications require the tracking of attitude and position of a body based on velocity data. It is tempting to use direction cosine matrices (DCM), for example, to track attitude based on angular velocity data, and to integrate the linear velocity data separately in a suitable frame. In this chapter we make the case for using bivectors as the attitude tracking method of choice since several features make their performance and flexibility superior to that of DCMs, Euler angles or even rotors. We also discuss potential advantages in using CGA to combine the integration of angular and linear velocities in one step, as the features that make bivectors attractive for tracking rotations extend to bivectors that represent general displacements.

6.1 Kinematics in Geometric Algebra

Several applications require attitude and position to be computed based on velocity data. This is a simple kinematic problem: integration of angular velocity data yields the total rotation, or attitude, of the body, and the integration of linear velocity data yields the current position of the body. The primary application under consideration in this chapter is that of inertial navigation. It is worth mentioning at the outset that there is a distinction to be drawn between strapdown inertial navigation (SDINS), where gyroscopes provide an output which is the integral of the angular velocity over each time interval, and kinematics applications where instantaneous angular

L. Candy (✉)
The Council for Scientific and Industrial Research (CSIR), Meiring Naude Rd, Pretoria, South Africa
e-mail: lcandy@csir.co.za

J. Lasenby
Engineering Department, Cambridge University, Trumpington Street, Cambridge, UK
e-mail: jl221@cam.ac.uk

L. Dorst, J. Lasenby (eds.), *Guide to Geometric Algebra in Practice*,
DOI 10.1007/978-0-85729-811-9_6, © Springer-Verlag London Limited 2011

velocity data is available at the interval boundaries. Here we will begin with the case where instantaneous angular velocity data is available; however, the essential nature of the problem in either case is the same, and the motivation for using bivectors as our representation of rotations and as a basis for our integration schemes is generally applicable.

This chapter deals initially with kinematic solutions in $\mathbb{R}^3$. The focus here will be on attitude computation in $\mathbb{R}^3$ as it is possible to show that the accuracy with which attitude is computed has the greatest impact on the overall kinematic solution. This is intuitively clear when considering that any error in attitude results in the linear velocity data being incorrectly transformed and the data being integrated in the 'wrong' direction.

Following the discussion of the $\mathbb{R}^3$ solution, we will move on to solving the attitude and position equations simultaneously in conformal geometric algebra (CGA).

6.2   Attitude Computation and Kinematics in 3D

Several methods exist for representing attitude in order to allow for the transformation of vectors between frames. The most commonly encountered in engineering, because they are covered in most undergraduate courses, are Euler angles and direction cosine matrices (DCM). Less common, but arguably superior, are quaternions (Euler–Rodrigues parameters) and rotation vectors. In geometric algebra, the counterparts to the quaternion and rotation vector are the rotor and rotation bivector. The natural consideration when selecting an attitude representation for a kinematics solution is computational efficiency. Other considerations are whether or not the solution contains singularities as well as the extent to which it can be optimised—via interpolation schemes for example. In all of these respects, not all attitude representations are created equal.

Table 6.1 provides a comparison of various attitude representations, most of which will be familiar to the reader. It is worth noting that Euler angles and DCMs, the most widely known attitude representations, are generally poor choices for handling kinematics. For Euler angles, awkward algorithmic work-arounds are required to ensure that singularities in the solution are avoided [17]. In the case of DCMs, the fact that the integration is performed on a manifold in $\mathbb{R}^9$ means that not only is the integration computationally expensive, but some re-projection is required to keep the solution on the manifold.

Quaternions and rotation vectors offer a far better alternative for handling attitude kinematics in $\mathbb{R}^3$. Both are computationally efficient, although quaternions do require a normalisation/reprojection step. Rotation vectors in particular are not only a minimal representation for rotations, but the kinematic equation lends itself well to efficient interpolation-based integration schemes such as Miller's algorithm [12]. Phillips and Haily [15] provide a comprehensive review of the attitude representations discussed up to this point.

Rotors and rotation bivectors, which will be introduced next, have all the advantages of quaternions and rotation vectors and are also more general in the sense that they are applicable to any dimension.

Table 6.1 Attitude representations

Representation	DOF (constraints)	Disadvantages	Advantages
Euler angles	3	Solution contains singularity	Outputs roll, pitch and yaw directly
DCM	9 (6)	Integration computationally expensive Requires normalisation/reprojection	Ubiquitous
Quaternions	4 (1)	Requires normalisation/reprojection	Integration computationally efficient
Rotation vectors	3		Integration very computationally efficient Minimal representation Excellent optimisation with interpolation
Rotors	4 (1)	Requires normalisation/reprojection	Integration computationally efficient Extends to nD CGA allows solution for general displacement
Bivectors	3		Integration very computationally efficient Minimal representation Excellent optimisation with interpolation Extends to nD CGA allows solution for general displacement

6.2.1 Rotation Bivectors

In geometric algebra the usual method of tracking the attitude of a rigid body with respect to some reference frame, would be to use the time dependent rotor $R(t)$. $R(t)$ can be computed using the kinematic equation for rotors [6],

$$\dot{R} = -\frac{1}{2}\Omega^r R \qquad (6.1)$$

where Ω^r is the angular velocity of the rigid body with respect to the reference frame, expressed as a bivector in the reference frame (the bivector defines the plane of rotation and has a magnitude equal to the rate of rotation). It is easy to show that if we express the angular velocity of the body with respect to the reference frame as a bivector in the body frame, Ω^b, then (6.1) becomes

$$\dot{R} = -\frac{1}{2}R\Omega^b \qquad (6.2)$$

Bivectors provide a natural way of representing rotations using a minimal set of parameters. It is possible to specify any finite rotation in a unit plane, defined by the unit bivector B, and through an angle α, using a rotation bivector,

$$\Phi = \alpha B \tag{6.3}$$

From this rotation bivector the corresponding rotation generator, or rotor, is easily obtained via

$$R = e^{-\frac{\Phi}{2}} = \cos\frac{\alpha}{2} - B\sin\frac{\alpha}{2} \tag{6.4}$$

No assumption is made that the plane defined by B must be limited to $\mathbb{R}^3$, and (6.3) and (6.4) generalise to any dimension. For use in a kinematic solution, a kinematic equation for rotation bivectors is required. This kinematic equation is given by

$$\dot{\Phi} = \Omega - \frac{\langle \Phi\Omega \rangle_2}{2} + \left(\frac{|\Phi|}{2}\cot\frac{|\Phi|}{2} - 1 \right)\left[\Omega + \frac{(\Omega \cdot \Phi)\Phi}{|\Phi|^2} \right] \tag{6.5}$$

The proof for this result is given below.

Proof Given (6.3), (6.2) and (6.4), find $\dot{\Phi}$ in terms of Φ and the measured body angular velocity Ω (where the superscript b has been dropped).

Differentiating (6.3) with respect to time gives

$$\dot{\Phi} = \alpha\dot{B} + \dot{\alpha}B \tag{6.6}$$

Expressions for $\dot{B}$ and $\dot{\alpha}$ in terms of the measured angular velocity Ω are now obtained via grade-wise comparison of (6.2) and the time derivative of (6.4):

$$\frac{d}{dt}R = \frac{d}{dt}\left[\cos\frac{\alpha}{2} - B\sin\frac{\alpha}{2} \right] \tag{6.7}$$

Finding an Expression for $\dot{\alpha}$ Equating the grade-0 components of (6.2) and (6.7) gives:

$$\langle R\Omega \rangle = \dot{\alpha}\sin\frac{\alpha}{2} \tag{6.8}$$

Noting that $\langle R\Omega \rangle = \Omega \cdot \langle R \rangle_2$ and substituting for the bivector component of R:

$$\dot{\alpha} = -\Omega \cdot B \tag{6.9}$$

Finding an Expression for $\dot{B}$ Equating the grade-2 components of (6.2) and (6.7) gives:

$$\langle R\Omega \rangle_2 = 2\dot{B} \sin \frac{\alpha}{2} + \dot{\alpha} B \cos \frac{\alpha}{2} \tag{6.10}$$

Using the fact that $\langle R\Omega \rangle_2 = \langle R \rangle \Omega + \langle\langle R \rangle_2 \Omega \rangle_2$ and substituting for $\langle R \rangle$ and $\langle R \rangle_2$ from (6.4) and for $\dot{\alpha}$ from (6.9), we have:

$$2\dot{B} \sin \frac{\alpha}{2} = \Omega \cos \frac{\alpha}{2} - \langle B\Omega \rangle_2 \sin \frac{\alpha}{2} - \dot{\alpha} B \cos \frac{\alpha}{2}$$
$$\dot{B} = \frac{1}{2}\cot\frac{\alpha}{2}\big[\Omega + (\Omega \cdot B)B\big] - \frac{\langle B\Omega \rangle_2}{2} \tag{6.11}$$

Completing the Derivation It is now possible to substitute the expressions for $\dot{B}$ and $\dot{\alpha}$ into the expression for $\dot{\Phi}$ in (6.6):

$$\dot{\Phi} = \Omega + \left(\frac{\alpha}{2}\cot\frac{\alpha}{2} - 1\right)\big[\Omega + (\Omega \cdot B)B\big] - \frac{\alpha}{2}\langle B\Omega \rangle_2 \tag{6.12}$$

Or, alternatively, by substituting for $B = \frac{\Phi}{|\Phi|}$ and $\alpha = |\Phi|$, it is possible to write (6.12) in terms of Φ only:

$$\dot{\Phi} = \Omega - \frac{\langle \Phi\Omega \rangle_2}{2} + \left(\frac{|\Phi|}{2}\cot\frac{|\Phi|}{2} - 1\right)\left[\Omega + \frac{(\Omega \cdot \Phi)\Phi}{|\Phi|^2}\right] \tag{6.13}$$

Equation (6.13) is the full kinematic equation for rotation bivectors, and its vector analog is known as the Bortz equation in the navigation community [3]. □

Equation (6.6) shows that $\dot{\Phi}$ consists of two components: one in the plane of rotation, $\dot{\alpha}B$, and a second component, $\alpha\dot{B}$, perpendicular to the plane of rotation. $\dot{B}$ is 'perpendicular' to B in the sense that $B \cdot \dot{B} = 0$; this can be seen from the fact that $|B^2| = 1$ implies $B\dot{B} + \dot{B}B = 0$, and therefore $B \cdot \dot{B} = 0$. While the final form of (6.5) does not keep these two perpendicular components separate, it is quite easy to show that in the event that $\dot{B} = 0$, which is to say there is no component of angular velocity perpendicular to the rotation bivector, $\dot{\Phi} = \Omega$. The final form of (6.5) is chosen because it can be shown [16] that

$$\dot{\Phi} \approx \Omega - \frac{\langle \Phi\Omega \rangle_2}{2} \tag{6.14}$$

since the term $(\frac{|\Phi|}{2} \cot \frac{|\Phi|}{2} - 1)[\Omega + \frac{(\Omega \cdot \Phi)\Phi}{|\Phi|^2}]$ is small and, in most practical algorithms, is neglected.

Equation (6.5) provides a more general bivector analog to the kinematic equation for rotation vectors since it generalises to n-D. The work of Bar-Itzhack [2] extends Euler's theorem to n-D, the Euler analogue in geometric algebra is given by (6.1),

Fig. 6.1 Sinusoidal body
frame oscillations with
magnitude a and angular
velocity ω that are out of
phase result in coning motion
where the third axis describes
a cone of half-angle a at
frequency ω

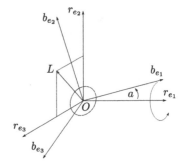

and is inherently general in n-D. This kinematic equation can be used directly in engineering problems such as SDINS algorithms implemented using geometric algebra. Alternatively, the traditional kinematic equation for rotation vectors can be obtained by substitution of $\Omega = I\omega$, $\Phi = I\phi$, multiplication of both sides by the pseudoscalar of $\mathscr{G}^3$, I, and simplification.

6.3   Practical Kinematics and Coning Motion

When performing experiments with gyroscopes on rate tables in the 1950s, researchers noticed that the drift of their algorithms appeared to increase near the resonant frequencies of the table supports [4]. What they had stumbled upon was the worst-case scenario for attitude computation, coning motion. Coning occurs when the b-frame experiences out of phase sinusoidal angular vibrations about any two body axes. The result is that the body frame angular velocity vector traces out a cone in space—hence the term *coning* motion. It is possible to show that the worst-case performance for numerical integration of angular velocity is observed under coning motion [5]. Since vibration is almost always present in mechanical systems, this effect is significant and consequently important when trying to evaluate the performance of attitude integration algorithms.

Coning motion is illustrated in Fig. 6.1. Here out of phase oscillations with magnitude a about the b_{e_2} and b_{e_3} axes result in the b_{e_1} axis tracing out a cone with half-angle a. The angular velocity vector (bivector dual shown as a dashed circle) lies in the $b_{e_{23}}$ plane and rotates about r_{e_1} at the coning frequency ω. The rotor and its corresponding bivector that specify the transformation from the reference to the body frame are given by

$$R = \cos\left(\frac{a}{2}\right) - \left[\sin\left(\omega t\right)e_{12} - \cos\left(\omega t\right)e_{13}\right]\sin\left(\frac{a}{2}\right) \tag{6.15}$$

$$\Phi = a\left[\sin(\omega t)e_{12} - \cos(\omega t)e_{13}\right] \tag{6.16}$$

and the body angular velocity by

$$\Omega = \omega \sin(a)\left[\cos(\omega t)e_{12} + \sin(\omega t)e_{13}\right] - 2\omega \sin^2\left(\frac{a}{2}\right)e_{23} \qquad (6.17)$$

The discussion that follows will be limited to (6.5), but analogous problems exist when considering the integration of any of the other kinematic equations: for quaternions, rotors, DCMs, etc.

The error of a numerical integration scheme will be dependent on some power of the step size, h, as well as some higher-order derivative of the integral function, both being dependent on the order of the integrator being used. Using the approximation to (6.5), $\dot{\Phi} \approx \Omega - \frac{\langle \Phi \Omega \rangle_2}{2}$, and a simple Euler integrator, we have:

$$\frac{d\Phi}{dt} = f\left(t, \Phi(t)\right) \qquad (6.18)$$

Then $\Phi(t_{i+1})$ is given by

$$\Phi(t_{i+1}) = \Phi(t_i) + hf\left(t_i, \Phi(t_i)\right) + \frac{h^2}{2}\ddot{\Phi}(\zeta_i) \qquad (6.19)$$

where the second-order error term is $\frac{h^2}{2}\ddot{\Phi}(\zeta_i)$, and ζ is some number in the interval (t_i, t_{i+1}).

Computing the derivative of $\dot{\Phi} \approx \Omega - \frac{\langle \Phi \Omega \rangle_2}{2}$ gives the expression for $\ddot{\Phi}$:

$$\ddot{\Phi} = \dot{\Omega} - \frac{1}{2}\left\langle \Phi\dot{\Omega} - \frac{1}{2}\langle \Phi\Omega \rangle_2 \Omega \right\rangle_2 \qquad (6.20)$$

In the case where there is no coning component in the motion of the b-frame, the axis of rotation is stationary—i.e. its direction is fixed in space—which implies that Φ and Ω lie in the same plane, and so, $\Phi\Omega = \langle \Phi\Omega \rangle$ and, as a result,

$$\ddot{\Phi} = \dot{\Omega} - \frac{1}{2}\langle \Phi\dot{\Omega} \rangle_2 \qquad (6.21)$$

since $\langle\langle \Phi\Omega \rangle\rangle_2 = 0$ by definition.

Furthermore, since the plane of the angular velocity Ω is fixed, when there is no coning motion, then not only do Φ and Ω lie in the same plane, but so does $\dot{\Omega}$. As a consequence, $\langle \Phi\dot{\Omega} \rangle_2 = 0$, leaving the error for the interval (t_i, t_{i+1}) as just $\frac{h^2}{2}\dot{\Omega}(\zeta_i)$.

On the other hand, for coning motion, from a consideration of (6.16) and (6.17) it is easy to show that the geometric product of Φ and Ω or any even-order derivative of Ω is a pure bivector, since Φ and Ω are perpendicular. Also, the geometric product of Φ and any odd-order derivative of Ω contain both scalar and bivector components. As a result, the error for the interval (t_i, t_{i+1}) for coning motion is $\frac{h^2}{2}[\dot{\Omega}(\zeta_i) - \frac{1}{2}\langle \Phi(\zeta_i)\dot{\Omega}(\zeta_i) - \frac{1}{2}\langle \Phi(\zeta_i)\Omega(\zeta_i) \rangle\rangle_2 \Omega(\zeta_i) \rangle_2]$.

In short, in the case of pure rotation our error is only in the magnitude of the rotation, whereas for any motion where the plane of rotation is changing, the numerical integration results in an error in both the magnitude and plane of rotation. The worst-case results are achieved in the case of coning motion where the orthogonality of Φ and the odd-order derivatives of Ω maximise the error term.

Historically the explanation in the literature for the coning error has been that it is due to the fact that finite rotations do not commute [1, 3, 4, 8–11, 13, 14, 17]. While the observation that finite rotations do not commute is legitimate, there is a clear absence in the literature explicitly linking this effect to the errors that are the result of the direct integration of (6.1). In Titterton and Weston, for example, the scenario of mixed order roll, pitch and yaw rotations is used to illustrate this effect, but it is left entirely to the reader to deduce how this might be applicable to tracking attitude. Here, however, we have shown that it is simply a numerical integration error.

6.3.1 Integration Schemes

The natural approach to solving a kinematics problem at this point would be to take one of our attitude representations and apply some standard integration scheme (Runge Kutta, for example). However, the ability to make sensible interpolations using bivectors allows for far more accurate and computationally cheap integration.

Assume that we can represent our kinematic data over two intervals (three boundary data points) by a quadratic polynomial, that is, $\Omega = At^2 + Bt + C$. This is always a reasonable assumption for smooth data as long as the sampling frequency is sufficiently high when compared to the highest frequency components of the input signal. We can then consider each interval pair independently, with the three samples Ω_0, Ω_1 and Ω_2 at $t = 0$, $t = T$ and $t = 2T$, respectively. This allows us to solve for A, B, and C to obtain

$$A = \frac{1}{4T^2}(2\Omega_2 - 4\Omega_1 + 2\Omega_0) \tag{6.22}$$

$$B = \frac{1}{2T}(-\Omega_2 + 4\Omega_1 - 3\Omega_0) \tag{6.23}$$

$$C = \Omega_0 \tag{6.24}$$

We can also easily show that at $t = 0$, we have $\Omega = C$, $\dot{\Omega} = B$, $\ddot{\Omega} = 2A$ and $\dddot{\Omega} = 0$. Using this result along with the kinematic equation for bivectors, (6.14), and the fact that for a given interval, we can set $\Phi(0) = 0$, we can show that

$$\dot{\Phi}(0) = C \tag{6.25}$$

$$\ddot{\Phi}(0) = B \tag{6.26}$$

$$\dddot{\Phi}(0) = 2A + \frac{1}{2}B \times C \tag{6.27}$$

noting that for bivectors $\langle AB \rangle_2 = A \times B$. Using a Taylor expansion, we can write the rotation bivector over two intervals as

$$\Phi(2T) = \Phi(0) + 2T\dot{\Phi}(0) + \frac{4T^2}{2}\ddot{\Phi}(0) + \frac{8T^3}{6}\dddot{\Phi}(0) + \cdots \tag{6.28}$$

Substituting into the above equation (6.25), (6.26) and (6.27), followed by (6.22), (6.23) and (6.24), gives the update equation

$$\Phi(2T) = T\left(\frac{1}{3}\Omega_0 + \frac{4}{3}\Omega_1 + \frac{1}{3}\Omega_2\right) + \frac{T^2}{3}(\Omega_0 \times \Omega_2 - 4\Omega_0 \times \Omega_1) \tag{6.29}$$

Equation (6.29) gives us a third-order approximation to the integral of the quadratic that fitted our data. The equation only needs to be evaluated on every second sample and has a computational complexity which is a fraction of that of an RK4 algorithm, for example. This update equation also performs exceptionally well as will be demonstrated in the next section.

It should be stressed that this equation, which is analogous to Miller's algorithm [12],[1] can only be obtained because of the favourable interpolation properties of bivectors.

6.3.2 Comparative Simulations

In order to illustrate how the various components of an attitude integration scheme and b-frame motion impact on the accuracy of the computation, several comparative simulations for coning motion and for mixed motion (coning motion with some fixed axis angular velocity component) are presented in Fig. 6.2 and Fig. 6.3. In all cases the coning angle is set at $1°$ and the input bandwidth is 50–100 Hz.[2] In all of the comparative simulations, identical integration schemes are used, except for the interpolation schemes specified. For the instantaneous angular velocity data, the integration scheme used is RK4, and for the gyroscope data, an Euler integrator is used. All of the plots consider the drift error rate against coning frequency. The angle through which the tracked body axes must be rotated to align them with the true body axes is considered the attitude error at any given time, and the drift error rate is the rate at which this error accumulates with time.

The first figure pair, Fig. 6.2, is for instantaneous velocity data. Figure 6.2(a) is presented to demonstrate numerically that the integration of (6.2), (6.5) and (6.14) produces *virtually* indistinguishable results. The qualifier here is that in our simulations the use of a naive normalisation required in the case of (6.2) can be a source of error, as we shall see in Fig. 6.3(b).

---

[1] Miller's algorithm is used for computing the update rotation Φ directly from gyroscope data. This algorithm is of order 5 and only has to be evaluated on every third sample.

[2] Modern gyroscope output bandwidths are typically 100–400 Hz.

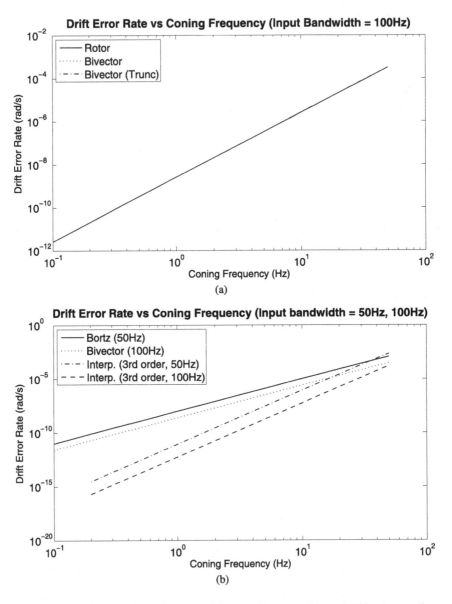

Fig. 6.2 Figure (**a**) shows the performance of the rotor, bivector and truncated bivector equations ((6.2), (6.5) and (6.14), respectively) for coning motion as a function of coning frequency—RK4 integrators are used in all cases. It is clear that under pure coning motion these equations exhibit almost identical performance. Figure (**b**) shows the impact that the sampling frequency has on performance. The far superior performance of the simple integration scheme presented in Sect. 6.3.1 is also illustrated

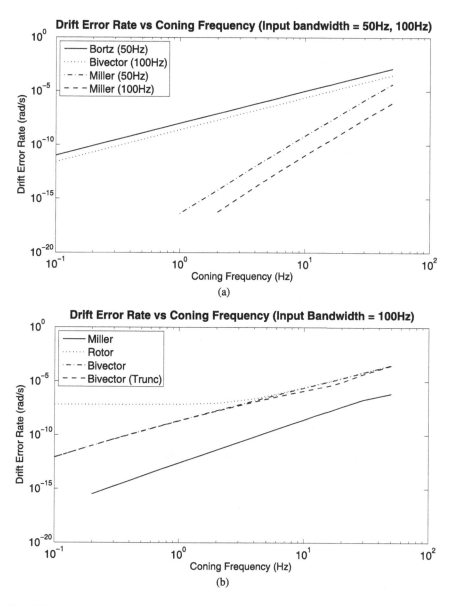

Fig. 6.3 The results in this figure are for simulations based upon gyroscope data and illustrate that the same relative performance of the different representations applies as in the case of instantaneous angular velocity data. Figure (**a**) is the counterpart to Fig. 6.2(b) for gyroscope data, the primary difference being that Miller's algorithm is a 5th-order interpolation scheme, as opposed to the 3rd-order scheme employed for Fig. 6.2(b). Figure (**b**) presents results for a blend of coning motion and fixed angular velocity. The most significant feature of this plot is the comparatively poor performance of the kinematic equation for rotors at low coning frequencies

Unsurprisingly, the attitude tracking performance drops in all cases in the presence of significant coning (Fig. 6.2(a) and Fig. 6.2(b)), and an increase in performance is seen at all coning frequencies when the input sampling frequency is increased (Fig. 6.2(b)). Figure 6.2(b) also demonstrates the dramatic performance increase to be gained from an interpolation-based integration scheme, which is also substantially computationally cheaper to implement than the RK4 integration against which it is being measured.

Figure 6.3 is for gyroscope data, i.e. the input data is the integral of the angular velocity bivector Ω^b. As before, for coning motion, there is a negligible difference in performance for the rotor, bivector or truncated bivector kinematic equation integration. Figure 6.3(a) is a plot equivalent to Fig. 6.2(b) and is meant to illustrate the dramatic performance gain of an interpolation-based integration scheme. In this case Miller's 5th-order interpolation scheme is used, which not only dramatically outperforms the Euler or RK4 integrator, but is far computationally cheaper than either (we need to evaluate an equivalent of a single commutator product per input sample).

Figure 6.3(b) plots the drift error rate against coning angle for mixed motion where the system is experiencing both a coning component and a component of rotation in a fixed plane. The influence of the coning motion is varied by changing the coning frequency; in other words, the right side of the plot shows a system where the dominant component of rotation is pure coning, while the left side shows a system where the motion is dominated by rotation in a fixed plane. Here it is clear that the performance of the bivector kinematic equations is superior to rotors when there is relatively little coning. Given that the bivector equations perform at least as well in the presence of coning, this factor alone should make them a compelling choice over rotors.

6.4   General Kinematics: Combining Rotation and Translation with CGA

In the previous sections spatial rotations were represented using a rotation bivector $\Phi = \alpha B$, where B is a unit bivector specifying the plane of rotation, and α is the angle of rotation. In conformal geometric algebra (CGA), spatial translations can also be represented by bivectors, where the bivector tn_∞ is used to represent a translation t. As is the case for pure rotation, the rotor is formed by the exponentiation of a bivector, which in this case is $-\frac{tn_\infty}{2}$:

$$R_t = e^{-\frac{tn_\infty}{2}} \tag{6.30}$$

which, employing Taylor series expansion gives

$$R_t = 1 - \frac{tn_\infty}{2} \tag{6.31}$$

Any general spatial displacement (rotation and translation) can now be represented by combining the rotation bivector and translation bivector: $\hat{\Phi} = \alpha B + t n_\infty$. As before, the rotor is written as an exponentiated bivector

$$R = e^{-\frac{\hat{\Phi}}{2}} \tag{6.32}$$

In order to evaluate this function, it is easiest to split the bivector into a pair of commuting blades, allowing $e^{-\frac{\hat{\Phi}}{2}}$ to be written as the product of two simpler exponential functions. Splitting $\hat{\Phi}$ into commuting blades F_1 and F_2 is always possible [7]. The rotor can then be written as

$$R = e^{-\frac{\hat{\Phi}}{2}} = e^{-\frac{F_1}{2}} e^{-\frac{F_2}{2}} \tag{6.33}$$

It turns out that $F_1 = \alpha B + t_\| n_\infty$ and $F_2 = t_\perp n_\infty$, where $t_\| = \frac{1}{2}(t + BtB)$ is the component of t that lies in the plane B, and $t_\perp = \frac{1}{2}(t - BtB)$ is the component of t that lies perpendicular to the plane B. Note that since $t_\|$ anticommutes with B and $t_\perp$ commutes with B, it can be shown that $F_1^2 = -\alpha^2$ and $F_2^2 = 0$. The exponentials in (6.33) can then be evaluated by using their Taylor expansions, giving

$$e^{-\frac{F_1}{2}} = \cos\frac{\alpha}{2} - \sin\frac{\alpha}{2}\left(B + \frac{t_\| n_\infty}{\alpha}\right) \tag{6.34}$$

and

$$e^{-\frac{F_2}{2}} = 1 - \frac{t_\perp n_\infty}{2} \tag{6.35}$$

By substituting (6.34) and (6.35) into (6.33) the expanded rotor equation is obtained:

$$\begin{aligned} R &= e^{-\frac{\hat{\Phi}}{2}} \\ &= \left[\cos\frac{\alpha}{2} - \sin\frac{\alpha}{2}\left(B + \frac{t_\| n_\infty}{\alpha}\right)\right]\left(1 - \frac{t_\perp n_\infty}{2}\right) \\ &= \left(1 - \frac{t'_\| n_\infty}{2}\right)\left(\cos\frac{\alpha}{2} - B\sin\frac{\alpha}{2}\right)\left(1 - \frac{t_\perp n_\infty}{2}\right) \\ &= \left(1 - \frac{t'_\| n_\infty}{2}\right)R_\alpha\left(1 - \frac{t_\perp n_\infty}{2}\right) \end{aligned} \tag{6.36}$$

where $t'_\| = t_\| \operatorname{sinc}\frac{\alpha}{2}\tilde{R}_\alpha$.

In other words, the rotor R formed by the exponentiated bivector $\hat{\Phi} = \alpha B + t n_\infty$ is a generator of a translation by $t_\perp$ followed by a rotation α in the plane B followed by a second translation of $t'_\parallel$.[3]

It would be useful to be able to form a bivector $\hat{\Phi}$ knowing that the total rotation will be defined by αB and the total translation by some vector s, in other words, $R = (1 - \frac{s n_\infty}{2}) R_\alpha$. To find a relationship between s and t, R is written as

$$R = \left(1 - \frac{t'_\parallel n_\infty}{2}\right)\left(1 - \frac{t_\perp n_\infty}{2}\right) R_\alpha$$

since $t_\perp$ is perpendicular to the plane of rotation B by definition, and therefore $R_\alpha(1 - \frac{t_\perp n_\infty}{2})\tilde{R}_\alpha = (1 - \frac{t_\perp n_\infty}{2})$. So;

$$R = \left(1 - \frac{s n_\infty}{2}\right) R_\alpha = \left(1 - \left(t_\perp + t'_\parallel\right)\frac{n_\infty}{2}\right) R_\alpha \qquad (6.37)$$

Therefore, $s = t_\perp + t'_\parallel$, or splitting s into its components perpendicular and parallel to the plane B,

$$t_\perp = s_\perp \qquad (6.38)$$

and

$$t'_\parallel = s_\parallel$$
$$t_\parallel \mathrm{sinc}\, \frac{\alpha}{2} \tilde{R}_\alpha = s_\parallel \qquad (6.39)$$
$$t_\parallel = \frac{s_\parallel R_\alpha}{\mathrm{sinc}\, \frac{\alpha}{2}}$$

where $s_\parallel = \frac{1}{2}(s + Bs B)$ and $s_\perp = \frac{1}{2}(s - Bs B)$.

6.4.1 Generalised Velocities

Consider a rotor R that takes vectors from some reference frame (r-frame) to some body frame (b-frame). This rotor can be written as a rotation, followed by a translation in the reference frame,

---

[3]Which is effectively Chasles' theorem; see also Chap. 5 in this volume.

$$R_r^b = R_t R_\alpha \tag{6.40}$$

where $R_\alpha = \cos\frac{\alpha}{2} - B\sin\frac{\alpha}{2}$ represents the rotational relationship between the frames, and $R_t = 1 + \frac{n_\infty s^r}{2}$ represents the translation between the frames in terms of the r-frame reference displacement vector s^r.

The kinematic equation that relates this rotor to a generalised velocity, $\hat{\Omega}$, is given by

$$\dot{R}_r^b = -\frac{1}{2}R_r^b \hat{\Omega}_{rb}^b \tag{6.41}$$

where $\hat{\Omega}_{rb}^b$ is the velocity of the b-frame, with respect to the r-frame, expressed as a bivector in the b-frame. Upon rearranging this equation becomes

$$\hat{\Omega}_{rb}^b = -2\tilde{R}_r^b \dot{R}_r^b \tag{6.42}$$

and from (6.40) and (6.42) we see that

$$
\begin{aligned}
\hat{\Omega}_{rb}^b &= -2\tilde{R}_r^b \dot{R}_r^b \\
&= -2\tilde{R}_\alpha \tilde{R}_t (\dot{R}_t R_\alpha + R_t \dot{R}_\alpha) \\
&= -2\tilde{R}_\alpha \tilde{R}_t \dot{R}_t R_\alpha - 2\tilde{R}_\alpha \dot{R}_\alpha
\end{aligned}
\tag{6.43}
$$

Now, since $\Omega_{rb}^b = -2\tilde{R}_\alpha \dot{R}_\alpha$ and $\tilde{R}_t \dot{R}_t = \frac{n_\infty \dot{s}^r}{2}$, it is possible to write, omitting the frame indicating subscripts and superscripts for legibility,

$$
\begin{aligned}
\hat{\Omega} &= \Omega - n_\infty \tilde{R}_\alpha \dot{R}_t R_\alpha \\
&= \Omega - n_\infty (\dot{s} + s \cdot \Omega)
\end{aligned}
\tag{6.44}
$$

In other words, the generalised velocity $\hat{\Omega}$ is simply the sum of the body frame referenced linear velocity bivector modified by a coriolis term and the rotational velocity bivector.

6.4.2 A Conformal Kinematic Equation for Bivectors

At this point, all of the conformal tools are available to produce an analogous derivation for a bivector kinematic equation.

Given

$$\hat{\Phi} = \alpha B + tn_\infty \tag{6.45}$$

$$R = \left[\cos\frac{\alpha}{2} - \sin\frac{\alpha}{2}\left(B + \frac{t_\| n_\infty}{\alpha}\right)\right]\left(1 - \frac{t_\perp n_\infty}{2}\right) \tag{6.46}$$

$$\dot{R} = -\frac{1}{2}R\hat{\Omega} \tag{6.47}$$

where $\hat{\Omega} = \Omega - n_\infty(\dot{s} + s \cdot \Omega)$, and the equations relating t to s,

$$t_\perp = s_\perp \tag{6.48}$$

$$t_\| = \frac{s_\| R_\alpha}{\operatorname{sinc} \frac{\alpha}{2}} \tag{6.49}$$

find $\dot{\hat{\Phi}}$ in terms of $\hat{\Phi}$ and the generalised velocity $\hat{\Omega}$.

Taking the derivative of (6.45) gives

$$\dot{\hat{\Phi}} = \dot{\alpha} B + \alpha \dot{B} + \dot{t} n_\infty \tag{6.50}$$

All that remains is to use (6.46) and (6.47) to find expressions for $\dot{\alpha}$ and $\dot{B}$, and (6.48) and (6.49) to find an expression for $\dot{t}$.

Finding Expressions for $\dot{\alpha}$ and $\dot{B}$ Taking the derivative of (6.46),

$$\begin{aligned}
\dot{R} &= \frac{d}{dt}\left[\left[\cos\frac{\alpha}{2} - \sin\frac{\alpha}{2}\left(B + \frac{t_\| n_\infty}{\alpha}\right)\right]\left(1 - \frac{t_\perp n_\infty}{2}\right)\right] \\
&= \left[-\frac{\dot{\alpha}}{2}\sin\frac{\alpha}{2} - \frac{\dot{\alpha}}{2}\cos\frac{\alpha}{2}\left(B + \frac{t_\| n_\infty}{\alpha}\right)\right. \\
&\quad \left. - \sin\frac{\alpha}{2}\left(\dot{B} + \frac{\alpha \dot{t}_\| n_\infty - \dot{\alpha} t_\| n_\infty}{\alpha^2}\right)\right]\left(1 - \frac{t_\perp n_\infty}{2}\right) \\
&\quad + \left[\cos\frac{\alpha}{2} - \sin\frac{\alpha}{2}\left(B + \frac{t_\| n_\infty}{\alpha}\right)\right]\frac{\dot{t}_\perp n_\infty}{2}
\end{aligned} \tag{6.51}$$

and substituting (6.46) into (6.47), we have

$$\dot{R} = -\frac{1}{2}\left[\cos\frac{\alpha}{2} - \sin\frac{\alpha}{2}\left(B + \frac{t_\| n_\infty}{\alpha}\right)\right]\left(1 - \frac{t_\perp n_\infty}{2}\right)(\Omega - n_\infty(\dot{s}_r + s \cdot \Omega)) \tag{6.52}$$

Comparing the scalar components of (6.51) and (6.52),

$$-\frac{\dot{\alpha}}{2}\sin\frac{\alpha}{2} = \frac{1}{2}\langle B\Omega\rangle \sin\frac{\alpha}{2}$$
$$\dot{\alpha} = -B \cdot \Omega \tag{6.53}$$

giving an expression for $\dot{\alpha}$. Comparing the bivector components of (6.51) and (6.52) that do not contain the special vector n_∞, we get

$$-\frac{\dot{\alpha}}{2}B\cos\frac{\alpha}{2} - \dot{B}\sin\frac{\alpha}{2} = -\frac{1}{2}\Omega\cos\frac{\alpha}{2} + \frac{1}{2}\langle B\Omega\rangle_2 \sin\frac{\alpha}{2}$$
$$\dot{B} = \frac{1}{2}\cot\frac{\alpha}{2}(\Omega - (\Omega \cdot B)B) - \frac{1}{2}\langle B\Omega\rangle_2 \tag{6.54}$$

giving an expression for $\dot{B}$.

The expressions for $\dot{\alpha}$ and $\dot{B}$ are the same as for the rotation only case. Comparison of the bivector components and the 4-vector components of (6.51) and (6.52) simply give a pair of identities that confirm that the equations are consistent.

Finding Expressions for $\dot{t}$ Taking the derivative of $t = t_\| + t_\perp$ where $t_\|$ and $t_\perp$ are given in (6.49) and (6.48), respectively,

$$\dot{t} = \dot{s}_\perp + \frac{R_\alpha}{2 \sin \frac{\alpha}{2}} \left(\dot{\alpha} s_\| - \frac{\alpha}{2} \cot \frac{\alpha}{2} \dot{\alpha} s_\| + \alpha \dot{s}_\| \right) + \frac{1}{2 \sin \frac{\alpha}{2}} (\alpha \dot{R}_\alpha s_\|) \tag{6.55}$$

Substituting for $\dot{\alpha}$, $\dot{B}$ and $\dot{R}_\alpha$ from (6.53), (6.54) and (6.47),

$$\dot{t} = \dot{s}_\perp + \left(\frac{\alpha}{2} \cot \frac{\alpha}{2} - 1 \right) \left[\frac{1}{\alpha} (B \cdot \Omega) \frac{\alpha R_\alpha s_\|}{2 \sin \frac{\alpha}{2}} \right] + \frac{\alpha R_\alpha}{2 \sin \frac{\alpha}{2}} \left(\dot{s}_\| - \frac{1}{2} \Omega s_\| \right) \tag{6.56}$$

which after some manipulation we show to be

$$\dot{t} = \dot{s} + s \cdot \Omega - \frac{1}{2} \langle \alpha B \dot{s} + \alpha B (s \cdot \Omega) + t \Omega \rangle_1$$

$$+ \left(\frac{\alpha}{2} \cot \frac{\alpha}{2} - 1 \right) \left[\dot{s} + s \cdot \Omega + \frac{1}{\alpha^2} (\alpha B \cdot \Omega) t \right]$$

$$- \frac{1}{2} s_\perp \cdot \Omega + \frac{\alpha}{2} B \cdot (\dot{s}_\perp + s_\perp \cdot \Omega)$$

$$- \left(\frac{\alpha}{2} \cot \frac{\alpha}{2} - 1 \right) \left[\dot{s}_\perp + s_\perp \cdot \Omega + \frac{1}{\alpha} (B \cdot \Omega) s_\perp \right] \tag{6.57}$$

Completing the Derivation   Substituting (6.53), (6.54) and (6.57) into (6.50), after some manipulation, this gives:

$$\dot{\hat{\Phi}} = \hat{\Omega} - \frac{1}{2} \langle \hat{\Phi} \hat{\Omega} \rangle_2 + \left(\frac{|\hat{\Phi}|}{2} \cot \frac{|\hat{\Phi}|}{2} - 1 \right) \left[\hat{\Omega} + \frac{(\hat{\Omega} \cdot \hat{\Phi}) \hat{\Phi}}{|\hat{\Phi}|^2} \right]$$

$$- \left(\frac{\alpha}{2} \cot \frac{\alpha}{2} - 1 \right) \left[\dot{s}_\perp + s_\perp \cdot \Omega + \frac{1}{\alpha} (B \cdot \Omega) s_\perp \right] n_\infty$$

$$- \frac{1}{2} (s_\perp \cdot \Omega) n_\infty + \frac{\alpha}{2} B \cdot (\dot{s}_\perp + s_\perp \cdot \Omega) n_\infty \tag{6.58}$$

This is the conformal kinematic equation for bivectors. Upon inspection, it is clear that this is equivalent to

$$\dot{\hat{\Phi}} = B(\hat{\Omega}, \hat{\Phi}) + \Theta(t) \tag{6.59}$$

where $B(\hat{\Omega}, \hat{\Phi})$ has the form of the ordinary 3D kinematic equation (6.13), and $\Theta(t)$ is a residual function of $\hat{\Phi}$ and $\hat{\Omega}$.

As before, we can show numerically that good results are achieved when dropping the majority of terms in (6.58). From the numerical simulations it is clear that (6.59) without the Θ term, which we shall call the *unmodified bivector equation*, i.e.

$$\dot{\hat{\Phi}} \approx \hat{\Omega} - \frac{1}{2}\langle\hat{\Phi}\hat{\Omega}\rangle_2 + \left(\frac{|\hat{\Phi}|}{2}\cot\frac{|\hat{\Phi}|}{2} - 1\right)\left[\hat{\Omega} + \frac{(\hat{\Omega}\cdot\hat{\Phi})\hat{\Phi}}{|\hat{\Phi}|^2}\right] \tag{6.60}$$

produces excellent results, as does the truncated version that we used in $\mathbb{R}^3$,

$$\dot{\hat{\Phi}} \approx \hat{\Omega} - \frac{1}{2}\langle\hat{\Phi}\hat{\Omega}\rangle_2 \tag{6.61}$$

6.4.3 Simulation Results

As was the case for attitude computation in $\mathbb{R}^3$, we would like to compare the various update equations for computing both attitude and velocity—for the case where the input is angular velocity and linear acceleration. For the purposes of the simulations, we consider a system which is coning, since this is the worst case, and is undergoing a linear acceleration along one axis. The coning angle is set at $1°$, and the input bandwidth is 100 Hz. The plots consider the drift error rate against coning frequency for both attitude and velocity.

The series indicated as '3D Rotor' in Fig. 6.4 is one of the conventional methods of performing this computation. In this case the attitude is tracked using the conventional kinematic equation for rotors. These rotors are then linearly interpolated to provide a midpoint rotor for each time interval. This midpoint rotor is used to transform the acceleration data, which is then integrated using an Euler integrator scheme to compute velocity.

As is quite apparent from the figure, the conformal bivector update equations (both the *unmodified bivector equation* and the truncated version of this) perform very favourably when compared to the equation for rotors and the 3D computation.

6.5 Conclusion

In this chapter we have provided compelling reasons to favour the linear space of bivectors over the rotor manifold for tracking attitude from velocity measurements. We have also extended this approach to CGA, where the advantage of the bivector update equations is also quite apparent. Aside from being computationally efficient, we have shown how bivectors lend themselves to integration schemes with dramatically improved performance and low computational load.

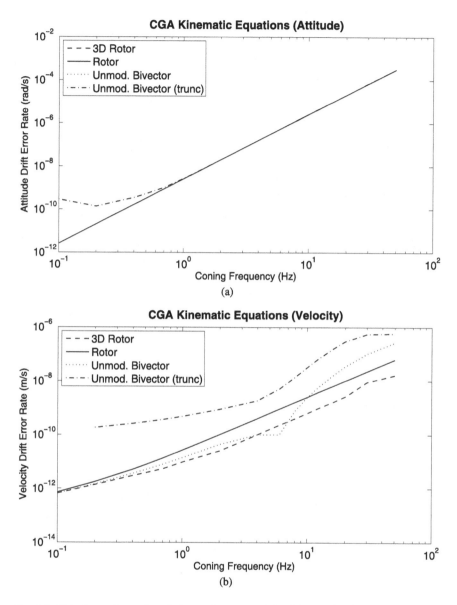

Fig. 6.4 This figure shows the relative performance of the CGA kinematic equations and the conventional method of computing attitude only and then transforming the acceleration data for the velocity integration. Figure (**a**) shows the essentially identical attitude performance of (6.47), (6.60), (6.61) and the 3D method. This is unsurprising, as we can show that the attitude component of the CGA equations is identical to the ordinary 3D case in Sect. 6.3.2. Also, it is worth noting that the truncated bivector can be used for tracking quite successfully

6.6   Exercises

6.1 For the kinematic equation for bivectors, demonstrate that the term

$$\left(\frac{|\Phi|}{2}\cot\frac{|\Phi|}{2} - 1\right)\left[\Omega + \frac{(\Omega \cdot \Phi)\Phi}{|\Phi|^2}\right]$$

is always small.

6.2 Show that $\alpha\dot{B} = \frac{1}{|\Phi|^2}[|\Phi|^2\dot{\Phi} + (\Phi \cdot \dot{\Phi})\Phi]$ and $\alpha^2 B\dot{B} = \langle\Phi\dot{\Phi}\rangle_2$.

6.3 Using the equation $\Omega = -2\tilde{R}\dot{R}$ as a starting point, find the inverse of the kinematic equation for bivectors. In other words, find the equation which, given a set of rotations Φ, will allow you to compute the angular velocities that cause the rotations, Ω.

   Hint: the relations in the previous question may come in handy.
   Answer: $\Omega = \dot{\Phi} + (\frac{1-\cos|\Phi|}{|\Phi|^2})\langle\Phi\dot{\Phi}\rangle_2 - \frac{1}{|\Phi|^2}(1 - \frac{\sin|\Phi|}{|\Phi|})[|\Phi|^2\dot{\Phi} + (\Phi \cdot \dot{\Phi})\Phi]$.

6.4 Use the method outlined in Sect. 6.3.1 to show that

$$\Phi = \Theta_s + 0.4125(\Theta_1 \wedge \Theta_3) + 0.7125\Theta_2 \wedge (\Theta_3 - \Theta_1) \tag{6.62}$$

where $\Theta_n = \int_{t_{n-1}}^{t_n} \Omega \, d\tau$ and $\Theta_s = \Theta_1 + \Theta_2 + \Theta_3$.
   This is Miller's algorithm in geometric algebra.

Acknowledgements   We would like to thank the Council for Scientific and Industrial Research in South Africa for sponsoring this research.

References

1. Bar-Itzhack, I.Y.: Navigation computation in terrestrial strapdown inertial navigation systems. IEEE Trans. Aerosp. Electron. Syst. **13**, 679–689 (1977)
2. Bar-Itzhack, I.Y.: Extension of Eulers theorem to n-dimensional spaces. IEEE Trans. Aerosp. Electron. Syst. **25**, 903–909 (1989)
3. Bortz, J.E.: A new mathematical formulation for strapdown inertial navigation. IEEE Trans. Aerosp. Electron. Syst. **7**, 61–66 (1971)
4. Goodman, L.E., Robinson, A.R.: Effect of finite rotations on gyroscopic sensing devices. J. Appl. Mech. 210–213 (June 1958)
5. Gusinsky, V.Z., Lesyuchevsky, V.M., Litmanovich, Yu.A.: New procedure for deriving optimized strapdown attitude algorithms. AIAA J. Guid. Control Dyn. **20**, 673–680 (1997)
6. Hestenes, D.: New Foundations for Classical Mechanics. Fundamental Theories of Physics. Springer, Berlin (1999)
7. Hestenes, D., Sobczyk, G.: Clifford Algebra to Geometric Calculus: A Unified Language for Mathematics and Physics. Reidel, Dordrecht (1984)
8. Ignani, M.B.: On the orientation vector differential equation in strapdown inertial navigation systems. IEEE Trans. Aerosp. Electron. Syst. **30**, 1076–1081 (1994)

9. Ignani, M.B.: New procedure for deriving optimized strapdown attitude algorithms. AIAA J. Guid. Control Dyn. **20**, 673–680 (1997)

10. Jiang, Y.F., Lin, Y.P.: On the rotation vector differential equation. IEEE Trans. Aerosp. Electron. Syst. **27**, 181–183 (1991)

11. Jiang, Y.F., Lin, Y.P.: Improved strapdown coning algorithms. IEEE Trans. Aerosp. Electron. Syst. **28**, 448–490 (1992)

12. Miller, R.: A new strapdown attitude algorithm. AIAA J. Guid. Control Dyn. **6**, 287–291 (1983)

13. Nazaroff, G.: The orientation vector differential equation. AIAA J. Guid. Control Dyn. **2**, 351–352 (1979)

14. Onunka, C., Bright, G.: A study on Direction Cosine Matrix (DCM) for autonomous navigation. In: CAD/CAM Robotics and Factories of the Future (2010)

15. Phillips, W.F., Haily, C.E.: Review of attitude representations used for aircraft kinematics. J. Aircr. **38**, 718–737 (2001)

16. Savage, P.: Strapdown inertial navigation integration algorithm design, part 1: Attitude algorithms. AIAA J. Guid. Control Dyn. **21**, 19–28 (1998)

17. Titterton, D., Weston, J.: Strapdown Inertial Navigation Technology. 2nd revised edn. IEEE Press, New York (2005)

Calibration of Target Positions Using Conformal Geometric Algebra

7

Robert Valkenburg and Nawar Alwesh

Abstract

This chapter describes an algorithm for calibrating the 3D positions of multiple stationary point targets which form part of an optical positioning system. A group of rigidly co-located calibrated cameras are moved to several positions and images of the targets acquired. The target pixel coordinates are extracted and transformed into 3D lines which are used as input data to the algorithm. A nonlinear solution is developed using geometric algebra and geometric calculus and expressed in the conformal model of Euclidean 3D space. A coordinate free approach to differentiating rotors is developed and used in the algorithm to differentiate motion rotors. Experiments are performed to evaluate the algorithm, and the results show that it performs well and is robust in the presence of noise.

7.1 Introduction

Measuring the position of many targets over a large area using a total station, for example, is a time-consuming task. When this must be performed frequently, it can be expensive.

This chapter describes a calibration algorithm to rapidly measure stationary point target positions. A group of rigidly co-located calibrated cameras are moved to several positions, and images of the targets are acquired. The target pixel coordinates are extracted and transformed to lines which are used as input data to the algorithm to estimate the 3D target locations. The targets and camera group comprise part of a 6 degree of freedom (DoF) optical positioning system. Every time the positioning

R. Valkenburg (✉) · N. Alwesh
Industrial Research Limited, Auckland, New Zealand
e-mail: r.valkenburg@irl.cri.nz

N. Alwesh
e-mail: n.alwesh@irl.cri.nz

L. Dorst, J. Lasenby (eds.), *Guide to Geometric Algebra in Practice*,
DOI 10.1007/978-0-85729-811-9_7, © Springer-Verlag London Limited 2011

Fig. 7.1 Camera group

system is deployed in a new site, the target positions must be determined, so it is important that this can be done rapidly.

The algorithm is developed using the conformal model of Euclidean 3D space and implemented with geometric algebra (GA). This chapter describes a re-expression of an algorithm implemented in the homogeneous model [7] to the conformal model with additional improvements.

The following geometric algebra conventions are used in this chapter. The geometric algebra over $\mathbb{R}$ with signature (p, q) (p positive and q negative basis elements) is denoted $\mathbb{R}_{p,q}$. When $q = 0$, we write $\mathbb{R}_p$, and when the signature and dimension do not matter, we simply write $\mathscr{G}$. $\mathscr{G}^r$ denotes the grade r elements of an algebra $\mathscr{G}$, and $\mathscr{G}^+$ denotes the even-grade elements of $\mathscr{G}$. If $A \in \mathscr{G}$ is a (possibly singular) blade with reciprocal blade $\bar{A}$ (i.e. $A \cdot \bar{A} = 1$), then $\mathscr{G}(A)$ denotes the geometric algebra generated by A and is defined as $\mathscr{G}(A) = \{X \in \mathscr{G} : X = (X \cdot \bar{A}) \cdot A\}$. We work with the CGA basis $\{n_o, e_1, e_2, e_3, n_\infty\}$ where $n_o = (-e_+ + e_-)/2$ and $n_\infty = e_+ + e_-$ are null vectors with $n_o \cdot n_\infty = -1$. The vector n_o represents the origin, and n_∞ represents the point at infinity. It is also convenient to define $I_4 = n_o I_3$, $\bar{I}_4 = I_3 n_\infty$ (note $I_4 \cdot \bar{I}_4 = 1$) and $E = n_o \wedge n_\infty = e_+ e_-$. The motors (*motion rotors*) are used to model rigid body motion, and the set of motors will be denoted $\mathscr{M}$. In addition, we will use the short-form notation $e_{ij} = e_i e_j$ for unit basis vectors e_i and $e_j, i \neq j$.

Fig. 7.2   Point targets implemented using ultra bright LEDs

7.2   Problem Statement

We have a set of C rigidly co-located calibrated cameras, referred to as the *camera group* (see Fig. 7.1, where $C = 6$). The cameras are organised to approximate an omnidirectional camera with their optical centres as coincident as physically possible, and the image planes providing maximum coverage with as little overlap as possible. The geometric relationship between each camera is known, so the camera group can be characterised by a single moving coordinate system denoted CSM. For example, CSM may be located at the centroid of the optical centres of all the cameras and aligned with one of the cameras. The cameras are synchronised so that a single frame capture event will grab C images. A world coordinate system, denoted CSW, is defined in terms of stationary point targets $q_k \in \mathbb{R}_{4,1}^1$, $k = 1, \ldots, K$ (e.g. one target is at the origin, another on the x-axis, and another in the xy-plane). Refer to Fig. 7.2, which shows targets implemented using ultra bright LEDs.

The camera group CSM is moved to N arbitrary positions in CSW. Let the motor $M_n \in \mathcal{M}$ represent the pose (orientation and position) of CSM in CSW for the nth position as in Fig. 7.3. At each position, a set of C images of the targets are captured (one from each camera). Not all the targets will necessarily be visible as the sensor field-of-views (FOV) do not give complete coverage, and some targets may also be occluded. Some targets may also be visible in more than one image because of small amounts of FOV overlap. In addition, false targets may appear

Fig. 7.3 Camera group
(CSM) moving to various
positions in CSW

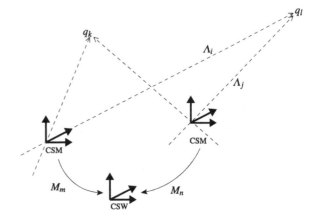

(in the same or different sensors) due to reflections off shiny surfaces. A target pixel
location is extracted using a subpixel estimator and then backprojected through the
camera model to produce a normalised (unit norm) line in the camera coordinate
system pointing out of the camera toward the target. Using the known relationships
between each camera, the line is rotated and translated to provide a normalised line
$\Lambda_j^o \in \mathbb{R}_{4,1}^3$ in CSM ready for subsequent processing. For each line Λ_j^o, the associated
target index $k = k(j)$ and camera group position index $n = n(j)$ are known. Each
line Λ_j^o can then be transformed to CSW using the associated motor M_n and is given
by

$$\Lambda_j = M_n \Lambda_j^o \tilde{M}_n, \quad n = n(j).$$

The problem can now be stated as: Given a set of measured normalised lines Λ_j^o,
$j \in J$, in CSM captured from N positions, we wish to recover the position of the
targets q_k, $k = 1, \dots, K$, in CSW.

7.3 Solution Using Geometric Algebra

The solution involves three steps: (1) Calculate initial estimates of M_n, $n =
1, \dots, N$, using a closed-form algorithm, (2) Calculate accurate estimates of M_n,
by nonlinear minimisation, (3) Reconstruct the target positions q_k, $k = 1, \dots, K$, in
CSW by triangulation using Λ_j^o and M_n.

Initially, all we have are the lines Λ_j^o expressed in CSM, we do not know the
location of the targets q_k and so cannot define CSW, and also we do not know the
positions of CSM, M_n, expressed in some common coordinate system. Therefore
we may temporarily identify CSW with one position of CSM to avoid the obvious ambiguity when both CSW and CSM are unknown. When the target positions
are reconstructed, we can move CSW to a more suitable location as described in
Sect. 7.3.3.

7.3.1 Initial Estimate of Poses

How the initial estimates are obtained depends greatly on the situation. Operational constraints can be imposed which eliminate, or significantly simplify, the process of obtaining an initial estimate. For example, by adding a compass and level to the camera group, all the data can be captured with rotational components of M_n almost constant.

One fairly general approach for obtaining an initial estimate is based on approximating the camera group by an idealised omnidirectional camera in which all the optical centres are coincident and positioned at the origin of CSM. This allows many standard results to be applied directly. These results can be expressed elegantly in terms of GA. A line in CSM going through the origin n_o can be represented $\Lambda = n_o \wedge v \wedge n_\infty = -Ev$ where $v \in \mathbb{R}^1_3$ is the direction vector.

Consider two positions of CSM, say CSM and CSM$'$. The direction vectors u, v' from the two positions, associated with a common target, satisfy the so called co-planarity constraint. This can be expressed $u \cdot \mathcal{E}v' = 0$ where $\mathcal{E} = \mathcal{A}\mathcal{R}$ is the well-known essential transformation [2]. Given sufficient vectors u and v', an estimate of $\mathcal{E}$ can be easily recovered. Orthogonal $\mathcal{R}$, defined by $v = \mathcal{R}v' = Rv'\widetilde{R}$, encodes the relative orientation R. Antisymmetric $\mathcal{A}$, defined by $\mathcal{A}v = (v \wedge t) \cdot I_3 = v \cdot (t \cdot I_3)$, encodes information about the relative position t. As $\mathcal{E}\bar{\mathcal{E}} = \mathcal{A}\bar{\mathcal{A}} = \bar{\mathcal{A}}\mathcal{A}$, it follows that $t \cdot \mathcal{E}\bar{\mathcal{E}}t = 0$. Therefore, given $\mathcal{E}$, an estimate of t (and hence $\mathcal{A}$) can be recovered up to a scale factor as the unit eigenvector associated with the smallest eigenvalue of $\mathcal{E}\bar{\mathcal{E}}$. Let $\{u_1, u_2, u_3\}$ be a basis of $\mathbb{R}^1_3$. As $\mathcal{R}\bar{\mathcal{E}} = \mathcal{A}$, the vectors $\bar{\mathcal{E}}u_k$ and $\mathcal{A}u_k$ are related by $\mathcal{R}$. It can be shown (see [6], which is Chap. 2 in this book) that an estimate of R is given by the eigenrotator of $\mathcal{L}$ associated with the largest eigenvalue, where $\mathcal{L} : \mathbb{R}^+_3 \rightarrow \mathbb{R}^+_3$ is defined by

$$\mathcal{L}X = \sum_{k=1}^{3}(\mathcal{A}u_k)X(\bar{\mathcal{E}}u_k).$$

By considering the lines at each position of CSM, estimates of the relative pose (with unit translation) between the positions of CSM can be obtained. These can be transformed into CSW to give initial estimates of R_n and the unit t_n. Any known distances (yardsticks) can be used to rescale t_n to consistent values. The motors are then given by $M_n = T_n R_n$ where $T_n = 1 - \frac{1}{2}t_n n_\infty$.

7.3.2 Accurate Estimate of Poses

In this section we define a suitable objective (error) function and form its gradient so that the problem can be formulated as a nonlinear optimisation. The lines associated with a given target k will nearly intersect. Image noise and quantisation, calibration errors, camera modelling errors, etc. will prevent them from intersecting exactly. The objective function is based on finding the motors M_n to minimise the dispersion of the lines Λ_j about their nominal intersection point. First, we present a number of useful general results about dual spheres and lines. It is worth noting that many results are best expressed in terms of spheres rather than points, even though we are only interested in a point location (given by the centre of the sphere). This is because a point can be regarded as a sphere with zero radius, so an additional constraint has to be imposed to enforce that the solution is a point. Considering spheres liberates us from the need to impose such constraints and makes the analysis simpler.

The squared distance between the centre of a normalised dual sphere $s \in \mathbb{R}^1_{4,1}$ and a line $\Lambda \in \mathbb{R}^3_{4,1}$ is given by

$$d^2(s, \Lambda) = \langle s\, S(s, \Lambda) \rangle, \tag{7.1}$$

where $S(X, Y) = X - (X \cdot Y) \cdot Y^{-1}$ for $X, Y \in \mathscr{G}$. To show this, let $s = q - \frac{1}{2}\rho^2 n_\infty$ be a dual sphere whose centre q is a distance d from Λ, so $d^2 = -(q \cdot \Lambda) \cdot (q \cdot \Lambda^{-1})$. Let $P_\Lambda(X) = (X \cdot \Lambda) \cdot \Lambda^{-1}$. Since $P_\Lambda(n_\infty) = n_\infty$, we have

$$\begin{aligned}
s \cdot S(s, \Lambda) &= s \cdot s - s \cdot P_\Lambda(s) \\
&= \rho^2 - \left(q \cdot P_\Lambda(q) + \rho^2\right) \\
&= \rho^2 - (q \cdot \Lambda) \cdot \left(q \cdot \Lambda^{-1}\right) - \rho^2 = d^2.
\end{aligned}$$

If Λ is normalised, then $\Lambda^{-1} = \Lambda$, and S can be simplified to $S(X, Y) = X - (X \cdot Y) \cdot Y$. We wish to triangulate a collection of lines which will make use of the following result. It is slightly more general than is usually presented because it allows us to express the inverse of a linear map without requiring an invertible pseudoscalar for the domain or range.

Lemma 7.1 *Let A be an r-blade, and B be an r-blade with reciprocal blade $\bar{B}$ (so $B \cdot \bar{B} = 1$). If $F : \mathscr{G}(A) \to \mathscr{G}(B)$ is an invertible linear operator, then*

$$F^{-1}(Y) = \frac{\bar{F}(Y \cdot \bar{B}) \cdot A}{\bar{F}(\bar{B}) \cdot A}.$$

Proof Let $Y = F(X)$. We have $F(A) = \alpha B$, so $\alpha = F(A) \cdot \bar{B} = \bar{F}(\bar{B}) \cdot A$, and we can define an associated reciprocal blade for A by $\bar{A} = \frac{\bar{F}(\bar{B})}{\bar{F}(\bar{B}) \cdot A}$. This gives

$$X = (X \cdot \bar{A}) \cdot A = \frac{(X \cdot \bar{F}(\bar{B})) \cdot A}{\bar{F}(\bar{B}) \cdot A} = \frac{\bar{F}(F(X) \cdot \bar{B}) \cdot A}{\bar{F}(\bar{B}) \cdot A}. \qquad \square$$

Lemma 7.2 *Let $\Lambda_j \in \mathbb{R}^3_{4,1}$, $j \in J$, be a set of lines, and $S(x) = \sum_{j \in J} S(x, \Lambda_j)$ for $x \in \mathbb{R}^1_{4,1}$. If $SI_3 \neq 0$, then the point $q \in \mathbb{R}^1_{4,1}$ closest to all the lines in the least squares sense is given by the centre of the normalised dual sphere*

$$s = -\frac{\langle S(I_3)I_4\rangle_1}{\langle S(I_3)I_3\rangle}. \tag{7.2}$$

The sum of squared distances d^2 from each line Λ_j to q is given by

$$d^2 = -\frac{\langle S(I_4)I_4\rangle}{\langle S(I_3)I_3\rangle}. \tag{7.3}$$

Proof The expression $s \cdot S(s) = \sum_{j\in J} s \cdot S(x, \Lambda_j) = \sum_{j\in J} d_j^2$ is the sum of squared distances to all lines Λ_j, $j \in J$. We wish to minimise $s \cdot S(s)$ subject to the constraint that s is normalised ($n_\infty \cdot s = -1$). The Lagrangian

$$L(s, \alpha) = \frac{1}{2} s \cdot S(s) + \alpha(n_\infty \cdot s + 1)$$

gives rise to the following first-order optimality conditions:

$$\partial_s L = S(s) + \alpha n_\infty = 0,$$
$$\partial_\alpha L = n_\infty \cdot s + 1 = 0.$$

Note that $S(n_\infty) = 0$, so S is not invertible on I, but we can restrict the domain of S to $I_4 = n_0 I_3$. Also note that S is symmetric, so for $x \in \mathbb{R}^1_{4,1}$, $0 = S(n_\infty) \cdot x = n_\infty \cdot S(x)$, and the range of S is in $\bar{I}_4 = I_3 n_\infty$. Using Lemma 7.1 with $A = \bar{B} = I_4$ gives

$$s = -\alpha \frac{S(n_\infty \cdot I_4) \cdot I_4}{S(I_4) \cdot I_4} = \alpha \frac{S(I_3) \cdot I_4}{S(I_4) \cdot I_4}.$$

We have $d^2 = s \cdot S(s) = \alpha$ and $-1 = n_\infty \cdot s = \alpha n_\infty \cdot (S(I_3) \cdot I_4)/(S(I_4) \cdot I_4) = \alpha(S(I_3) \cdot I_3))/(S(I_4) \cdot I_4)$, so $d^2 = -(S(I_4) \cdot I_4)/(S(I_3) \cdot I_3)$ and $s = -(S(I_3) \cdot I_4)/(S(I_3) \cdot I_3)$. $\qquad\square$

The situation associated with Lemma 7.2 is shown in Fig. 7.4. The condition $SI_3 = 0$ will only occur under rare circumstances such as all the lines are parallel. This result can be used directly to define a suitable objective function.

Each line Λ_j in CSW is obtained by rotating and translating the corresponding measured line in CSM and is given by

$$\Lambda_j = M_n \Lambda_j^o \tilde{M}_n, \quad n = n(j), \tag{7.4}$$

where M_n is the pose of CSM in CSW associated with the jth line. Let $d_k^2 = d_k^2(M_1, \ldots, M_N)$ be the sum of squared distances from the lines associated with the kth target (Λ_j, $j \in J_k$). In general, d_k^2 will involve lines from each position of

Fig. 7.4 The centre of the
dual sphere s gives the point
q closest to all the lines Λ_j

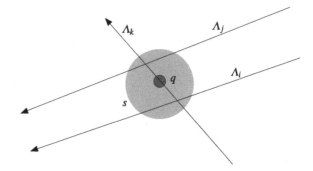

CSM, $n = 1, \ldots, N$, so d_k^2 depends on M_n, $n = 1, \ldots, N$. The objective function is
then defined by summing the errors for each target:

$$d^2 = d^2(M_1, \ldots, M_N) = \sum_{k=1}^{K} d_k^2.$$

Note that the target positions do not appear in this function, which only depends
on the motors M_n and the measured lines Λ_j^o. When the number of targets K is
large compared with the number of positions N, this has a significant impact on the
dimensionality of the optimisation problem.

Next we consider the gradient of d^2 which can be expressed in terms of the
multivector derivatives $\partial_{M_n} d^2$ for each $n = 1, \ldots, N$. Note that the variable $M_n \in
\mathcal{M}$ is constrained and satisfies $M_n \tilde{M}_n = 1$. The equation $M_n \tilde{M}_n = 1$ can be regarded
as a set of constraints. For example, if $X \in \mathbb{R}_{4,1}^{+}$, then $\tilde{X} X = \langle \tilde{X} X \rangle + \langle \tilde{X} X \rangle_4 =
1$ gives six constraints: $\langle \tilde{X} X \rangle = 1$ and $\langle e_J X \tilde{X} \rangle = 0$ where e_J is a basis for the
5D space of quadvectors. These constraints are incorporated into the derivative as
explained later. As

$$\partial_{M_n} d^2 = \sum_{k=1}^{K} \partial_{M_n} d_k^2, \qquad (7.5)$$

it suffices to initially consider the contribution of just the kth target. Let $S_k(p) =
\sum_{j \in J_k} S(p, \Lambda_j)$ denote the sum associated with the kth target.

If we let $N_k = \langle S_k(I_4) I_4 \rangle \in \mathbb{R}$ and $D_k = \langle S_k(I_3) I_3 \rangle \in \mathbb{R}$, then $d_k^2 = -N_k/D_k$. By
the product rule,

$$\partial_{M_n} d_k^2 = -\frac{\partial_{M_n} N_k D_k - N_k \partial_{M_n} D_k}{D_k^2}. \qquad (7.6)$$

For clarity, we will temporarily drop the subscripts n and k by identifying M
with M_n, d^2 with d_k^2, $J = \bigcup J_k$ with J_k, and $S(p)$ with $S_k(p)$. To calculate $\partial_M d^2$
it suffices to evaluate $\partial_M N$ and $\partial_M D$ which have similar structures. Recall that the

lines depend on $M \in \mathcal{M}$ as shown in (7.4). We could utilise the chain rule, but it is often easier to simply algebraically manipulate the expression into a form that is easy to differentiate, which is what we will do here.

We will make use of the following general result which applies immediately to the structure of N with $A_r = C_r = I_4$ and D with $A_r = C_r = I_3$.

Lemma 7.3 *Let $S : \mathcal{G}^1 \to \mathcal{G}^1$ be a linear operator which depends on a multivector X. Let $A_r = \bigwedge_{i \in I} a_i$ and C_r be r-blades independent of X, and define E_i so that $A_r = a_i \wedge E_i$. Then*

$$\partial_X \langle S(A_r)C_r \rangle = \sum_{i \in I} \dot{\partial}_X \langle \dot{S}(a_i)\big(S(E_i) \cdot C_r\big) \rangle.$$

Proof For each $k \in I$, $\langle S(A_r)C_r \rangle = \langle (S(a_k) \wedge S(E_k))C_r \rangle$. Hence,

$$\partial_X \langle S(A_r)C_r \rangle = \partial_X \Big\langle \Big(\bigwedge_{i \in I} S(a_i)\Big)C_r \Big\rangle$$

$$= \partial_X \Big\langle \prod_{i \in I} S(a_i)C_r \Big\rangle$$

$$= \sum_{i \in I} \dot{\partial}_X \langle (\dot{S}(a_i) \wedge S(E_i))C_r \rangle$$

$$= \sum_{i \in I} \dot{\partial}_X \langle \dot{S}(a_i)\big(S(E_i) \cdot C_r\big) \rangle. \qquad \square$$

Applying Lemma 7.3 with $S(x) = \sum_{j \in J} S(x, \Lambda_j)$ from Lemma 7.2 gives

$$\partial_M \langle S(A_r)C_r \rangle = \sum_{j \in J} \sum_{i \in I} \dot{\partial}_M \langle \dot{S}(a_i, \Lambda_j)\big(S(E_i) \cdot C_r\big) \rangle.$$

Observe that $v = S(E_i) \cdot C_r \in \mathbb{R}^1_{4,1}$. If $u, v \in \mathbb{R}^1_{4,1}$ and $\Lambda \in \mathbb{R}^3_{4,1}$, then $\langle S(u, \Lambda)v \rangle = u \cdot v - (u \cdot \Lambda) \cdot (v \cdot \Lambda)$, and we have

$$\dot{\partial}_M \langle \dot{S}(u, \Lambda)v \rangle = \dot{\partial}_M \langle \dot{\Lambda} B(u, v, \Lambda) \rangle, \tag{7.7}$$

where

$$B(u, v, \Lambda) = -u \wedge (v \cdot \Lambda) - v \wedge (u \cdot \Lambda) \in \mathbb{R}^3_{4,1}. \tag{7.8}$$

To evaluate $\dot{\partial}_M \langle \dot{\Lambda} B \rangle$ where $\Lambda = M \Lambda^o \tilde{M}$ we need to differentiate with respect to M while enforcing the constraint that M is in the set of motors, denoted $\mathcal{M}$. This can be done in a coordinate-free manner by observing that the tangent space of $\mathcal{M}$ at $M \in \mathcal{M}$, denoted $\mathcal{T}_M$, is simple to characterise in terms of M. The key concept is to define ∂_M so that $\langle A \partial_M \rangle = \langle P_{\mathcal{T}_M}(A) \partial_X \rangle$ for all multivectors A, where $P_{\mathcal{T}_M}$ is a projection on $\mathcal{T}_M$, and ∂_X is the derivative with respect to an unconstrained multivector $X \in \mathbf{X}$ where $\mathcal{T}_M \subset \mathbf{X}$. As $\langle P_{\mathcal{T}_M}(A) \partial_X \rangle = \langle A \bar{P}_{\mathcal{T}_M}(\partial_X) \rangle$ for all A, where $\bar{P}_{\mathcal{T}_M}$ is the adjoint of $P_{\mathcal{T}_M}$, we define

$$\partial_M = \bar{P}_{\mathcal{T}_M}(\partial_X). \tag{7.9}$$

In order to calculate $P_{\mathcal{T}_M}$ and $\bar{P}_{\mathcal{T}_M}$, we first establish the form of $\mathcal{T}_M$. It is convenient to define some additional linear spaces at the outset. Note that a motor lies in the 8D linear space $\mathbb{M} = \mathrm{sp}\{1, e_{12}, e_{13}, e_{23}, e_1 n_\infty, e_2 n_\infty, e_3 n_\infty, I_3 n_\infty\}$, with reciprocal space $\overline{\mathbb{M}} = \mathrm{sp}\{1, e_{21}, e_{31}, e_{32}, e_1 n_o, e_2 n_o, e_3 n_o, \tilde{I}_3 n_o\}$, and that if $X \in \mathbb{M}$ and $Y \in \mathbb{M}$, then $XY \in \mathbb{M}$. The bivector subspace of $\mathbb{M}$ will be denoted $\mathbb{B} = \mathrm{sp}\{e_{12}, e_{13}, e_{23}, e_1 n_\infty, e_2 n_\infty, e_3 n_\infty\}$.

Let $\psi(s)$ be a curve in $\mathcal{M}$ (a motor-valued function of a scalar), $M = \psi(0) \in \mathcal{M}$ and $\Delta = \psi'(0)$. Differentiating $\tilde{\psi} \psi = 1$ and evaluating at $s = 0$ gives $\tilde{M} \Delta = -\tilde{\Delta} M$. As $\Delta \in \mathbb{M}$, it follows that $\tilde{M} \Delta \in \mathbb{M}$ and the tangent space of $\mathcal{M}$ at $M \in \mathcal{M}$ is given by $\mathcal{T}_M = M\mathbb{B}$. It is natural to define the normal space of $\mathcal{M}$ at M (restricted to $\mathbb{M}$) by the orthogonal complement of $\mathcal{T}_M$ in $\mathbb{M}$, so $\mathcal{N}_M = \mathrm{sp}\{M, M I_3 n_\infty\}$ and $\mathbb{M} = \mathcal{T}_M \oplus \mathcal{N}_M$.

For any element $X \in \mathbb{M}$, $\tilde{M} X \in \mathbb{M}$, so X can be expressed $X = M\tilde{M} X = M\langle \tilde{M} X \rangle_2 + M(\langle \tilde{M} X \rangle + \langle \tilde{M} X \rangle_4)$. The first term is in $\mathcal{T}_M$, and the second is in $\mathcal{N}_M$. For $X \in \mathbb{M}$, we can define the projection on $\mathcal{T}_M$ along $\mathcal{N}_M$ by $P_{\mathcal{T}_M}(X) = M\langle \tilde{M} X \rangle_2$. It is clear that this is idempotent, onto $\mathcal{T}_M$, and has null space $\mathcal{N}_M$. This can be extended to $\mathbb{R}_{4,1}$ by first projecting into $\mathbb{M}$:

$$P_{\mathcal{T}_M}(X) = M\langle \tilde{M} P_{\mathbb{M}}(X) \rangle_2.$$

With $P_{\mathcal{T}_M}$ expressed in this form, we can easily find an expression for its adjoint $\bar{P}_{\mathcal{T}_M}$,

$$\langle P_{\mathcal{T}_M}(X)Y \rangle = \langle M\langle \tilde{M} P_{\mathbb{M}}(X) \rangle_2 Y \rangle$$
$$= \langle P_{\mathbb{M}}(X)\langle YM \rangle_2 \tilde{M} \rangle = \langle X\bar{P}_{\mathbb{M}}(\langle YM \rangle_2 \tilde{M}) \rangle,$$

giving

$$\bar{P}_{\mathcal{T}_M}(Y) = \bar{P}_{\mathbb{M}}(\langle YM \rangle_2 \tilde{M}).$$

This is expressed in terms of $\bar{P}_{\mathbb{M}}$ which can be efficiently calculated and, depending on exactly how the algebra is implemented, usually involves zeroing out some coordinates and a few additions.

Now consider applying ∂_M. Note that if $f(M) \in \mathbb{R}$ is a scalar valued function, then

$$\partial_M f = \bar{P}_{\mathcal{T}_M}(\partial_X) f = \bar{P}_{\mathcal{T}_M}(\partial_X f), \tag{7.10}$$

so we can differentiate as if the variable $X \in \mathbf{X}$ is unconstrained and then project on $\mathscr{T}_M$. As previously mentioned, $\mathscr{T}_M \subset \mathbf{X}$, so, for example, $\mathbf{X}$ could be $\mathbb{M}$, $\mathbb{R}^+_{4,1}$ or $\mathbb{R}_{4,1}$.

We wish to differentiate $\partial_M \langle M A \widetilde{M} B \rangle$, and (7.10) suggests we should first consider $\partial_X \langle X A \widetilde{X} B \rangle$. If $X \in \mathbf{X} = \mathrm{sp}\{e_I\}$ and $\{e^I\}$ is an associated reciprocal basis, then

$$\partial_X \langle X A \rangle = \sum_I e^I \langle e_I \partial_X \rangle \langle X A \rangle = \sum_I e^I \langle e_I A \rangle = \bar{\mathsf{P}}_{\mathbf{X}}(A), \tag{7.11}$$

from which we get

$$\partial_X \langle X A \widetilde{X} B \rangle = \bar{\mathsf{P}}_{\mathbf{X}}(A \widetilde{X} B + \widetilde{A} \widetilde{X} \widetilde{B}). \tag{7.12}$$

Normally $\mathsf{P}_{\mathbf{X}}$ and its adjoint $\bar{\mathsf{P}}_{\mathbf{X}}$ are the same, for example when $\mathbf{X} = \mathbb{R}^+_{4,1}$ or $\mathbb{R}_{4,1}$. However situations occur when this is not the case. This is because natural linear spaces $\mathbf{X}$ arise where we cannot find a reciprocal basis which also spans $\mathbf{X}$, so we must look outside of $\mathbf{X}$. For example, we have already seen that $\mathbb{M} \neq \overline{\mathbb{M}}$. Such linear spaces occur often with the introduction of null basis elements in CGA because important objects (e.g. lines and planes) and operations (e.g. motors) involve n_∞. Closely related, note also that ∂_X, as in (7.11), is expressed in terms of the directional derivatives $\langle e_I \partial_X \rangle$ (and not $\langle e^I \partial_X \rangle$) as the direction vectors e_I must be in $\mathbf{X}$. There is no need to make a distinction when e_I and e^I span the same space.

We can now evaluate (7.7). From (7.10) and (7.12), noting that $\bar{\mathsf{P}}_{\mathscr{T}_M} \bar{\mathsf{P}}_{\mathbf{X}} = \bar{\mathsf{P}}_{\mathscr{T}_M}$, we get

$$\begin{aligned} \partial_M \langle M A \widetilde{M} B \rangle &= \bar{\mathsf{P}}_{\mathscr{T}_M} \big(\partial_X \langle X A \widetilde{X} B \rangle \big) \big|_{X=M} \\ &= \bar{\mathsf{P}}_{\mathscr{T}_M} (A \widetilde{M} B + \widetilde{A} \widetilde{M} \widetilde{B}). \end{aligned} \tag{7.13}$$

If A and B are homogeneous (and of the same grade if the result is to be non-zero), which is by far the most common situation, we have $\widetilde{A} \widetilde{X} \widetilde{B} = A \widetilde{X} B$, and (7.13) reduces to

$$\partial_M \langle M A \widetilde{M} B \rangle = 2 \bar{\mathsf{P}}_{\mathscr{T}_M} (A \widetilde{M} B) = 2 \bar{\mathsf{P}}_{\mathscr{T}_M} \big(\widetilde{M} A' B \big), \tag{7.14}$$

where $A' = A'(M) = M A \widetilde{M}$ is the transformed A. This derivative turns up frequently, so it is convenient that it has such a simple form.

It is interesting to digress and examine how this relates to another way of differentiating $\langle X A \widetilde{X} B \rangle$ with respect to motors and rotators. Consider the behaviour of the function $G(X) = \langle X A X^{-1} B \rangle$ for $X \in \mathbb{M}$ as we move out from $M \in \mathscr{M}$ along the normal space $\mathscr{N}_M$. Let $\Psi = I_3 n_\infty$ and $N = M(\alpha + \beta \Psi) \in \mathscr{N}_M$. If A, B are homogeneous, we get

$$G(M + \tau N) = \langle M A \widetilde{M} B \rangle - \frac{\tau^2 \beta^2}{(1 + \tau \alpha)^2} \langle \Psi M A \widetilde{M} \Psi B \rangle.$$

As the linear term in τ vanishes, the directional derivative $\langle N \partial_X \rangle \langle XAX^{-1}B \rangle|_{X=M} = 0$. Now examine the behaviour of $G(X)$ on $\mathscr{T}_M$. Let $T \in \mathscr{T}_M$ and note that $(M + \tau T)^{-1} \approx \widetilde{M} + \tau \widetilde{T}$ to first order because $T\widetilde{M} = -M\widetilde{T}$. This gives

$$G(M + \tau T) = \langle MA\widetilde{M}B \rangle + \tau \langle TA\widetilde{M}B \rangle + \tau \langle MA\widetilde{T}B \rangle + O(\tau^2).$$

Here we have linear terms in τ, and they are the same linear terms that would arise if G were replaced by $\langle XA\widetilde{X}B \rangle$, so $\langle T\partial_X \rangle \langle XAX^{-1}B \rangle|_{X=M} = \langle T\partial_X \rangle \langle XA\widetilde{X}B \rangle|_{X=M}$. Thus, the directional derivatives of G vanish on $\mathscr{N}_M$ and coincide with the directional derivatives of $\langle XA\widetilde{X}B \rangle$ on $\mathscr{T}_M$. It follows that $\partial_X \langle XAX^{-1}B \rangle|_{X=M} = \partial_M \langle MA\widetilde{M}B \rangle$ for $X \in \mathbb{M}$ and A, B homogeneous. For the simpler case of rotators, denoted $\mathscr{R}$, we can even drop the restriction that A and B are homogeneous. As for motors, consider the behaviour of G for $X \in \mathscr{G}_3^+$ as we move out from $R \in \mathscr{R}$ along the normal space $\mathscr{N}_R$. If $N = R\alpha \in \mathscr{N}_R$ and A, B are general multivectors, we get

$$G(R + \tau N) = \langle R(1 + \tau\alpha)A(1 + \tau\alpha)^{-1}\widetilde{R}B \rangle = \langle RA\widetilde{R}B \rangle.$$

Now the quadratic term also vanishes, and G is constant on $\mathscr{N}_R$. Therefore, for general multivectors A, B, and $X \in \mathscr{G}_3^+$, we have $\partial_X \langle XAX^{-1}B \rangle|_{X=R} = \partial_R \langle RA\widetilde{R}B \rangle$. This has been exploited in [4] to provide a procedure for estimating rotators from pairs of points.

Returning to the original problem, we can use (7.14) to evaluate (7.7),

$$\dot{\partial}_M \langle \dot{A}B \rangle = 2\bar{\mathbf{P}}_{\mathscr{T}_M}(\widetilde{M} \wedge B), \tag{7.15}$$

which gives the following result:

Lemma 7.4 *Let S be defined as in Lemma 7.2, and A_r, C_r, a_i and E_i be defined as in Lemma 7.3. If Λ_j, $j \in J_M$, is the set of lines which depend on M, then*

$$\dot{\partial}_M \langle \dot{S}(A_r)C_r \rangle = 2 \sum_{j \in J_M} \bar{\mathbf{P}}_{\mathscr{T}_M}(\widetilde{M} \wedge \Lambda_j B_j),$$

where $B_j = \sum_{i \in I} B(a_i, S(E_i) \cdot C_r, \Lambda_j) \in \mathbb{R}_{4,1}^3$, and B is defined in (7.8).

Let J_{kn} denote the set of lines associated with the kth target and nth position. For each target, only a few lines will be associated with a specific motor M_n. Normally there will be just one. If the target was occluded or located where there was no image coverage, there will be none. If it is located in a region of overlap, there might be two or more. Hence J_{kn} will only have a few elements. Reintroducing the subscripts n and k and applying Lemma 7.4 to $N_k = \langle S_k(I_4)I_4 \rangle$ with $C_r = I_4$ and $A_r = I_4 = e_i \wedge E_i$ (i.e. $E_0 = I_3$, $E_1 = -n_o \wedge e_2 \wedge e_3$, $E_2 = n_o \wedge e_1 \wedge e_3$, $E_3 = -n_o \wedge e_1 \wedge e_2$) gives

$$\partial_{M_n} N_k = 2 \sum_{j \in J_{kn}} \bar{\mathsf{P}}_{\mathscr{T}_{M_n}} (\tilde{M}_n \Lambda_j B_{kj}),$$

$$B_{kj} = \sum_{i=0}^{3} B\left(e_i, S_k(E_i) \cdot I_4, \Lambda_j\right) \in \mathbb{R}_{4,1}^3. \tag{7.16}$$

Calculating the derivatives $\partial_{M_n} N_k$ is implemented efficiently by iterating over all lines $j \in J = \bigcup J_k$ and updating the derivatives $\partial_{M_n} N_k$ where $n = n(j)$. Similarly applying Lemma 7.4 to $D_k = \langle S_k(I_3) I_3 \rangle$ with $C_r = I_3$ and $A_r = I_3 = e_i \wedge E_i$ gives

$$\partial_{M_n} D_k = 2 \sum_{j \in J_{kn}} \bar{\mathsf{P}}_{\mathscr{T}_{M_n}} (\tilde{M}_n \Lambda_j B_{kj}),$$

$$B_{kj} = \sum_{i=1}^{3} B\left(e_i, S_k(E_i) \cdot I_3, \Lambda_j\right) \in \mathbb{R}_{4,1}^3. \tag{7.17}$$

Using (7.5), (7.6), (7.16), and (7.17), we can calculate $\partial_{M_n} d^2$, $n = 1, \ldots, N$.

7.3.2.1 Constraints

Because the optical centres of the cameras are very close, the distances between them do not provide effective yardsticks to scale the problem. This can be remedied by adding additional larger yardsticks such as the distance between two targets or the distance between two poses.

Distance Between Two Targets   Let d_{kl}^o be the measured distance between the kth and lth targets. The normalised dual sphere s with centre q in Lemma 7.2, representing the closest point to all the lines, has radius $|\mathbf{q}|$. It follows that the distance between the kth and lth targets is given by $d_{kl} = |t_{kl}|$ where $t_{kl} = s_k - s_l$, and we can impose the constraint

$$\varepsilon_{kl} = d_{kl} - d_{kl}^o = 0.$$

To evaluate the derivative $\partial_{M_n} \varepsilon_{kl}^2$, we make use of the following result:

Lemma 7.5 *Let $F(X)$ be a multivector-valued function of a multivector X such that $|F|^2 = \langle F \tilde{F} \rangle$. Then*

$$\partial_X |F| = \dot{\partial}_X \langle \dot{F} \tilde{U} \rangle \quad \text{where } U = \frac{F}{|F|}.$$

Proof $\partial_X |F|^2 = 2|F| \partial_X |F|$ and $\partial_X |F|^2 = \partial_X \langle F \tilde{F} \rangle = 2 \dot{\partial}_X \langle \dot{F} \tilde{F} \rangle.$ □

The condition $|F|^2 = \langle F \tilde{F} \rangle$ will clearly be met if F is Euclidean. As $t_{kl} \in \mathbb{R}_3^1$ is Euclidean, Lemma 7.5 gives

$$\partial_{M_n}\varepsilon_{kl}^2 = 2\varepsilon_{kl}\partial_{M_n}|t_{kl}|$$
$$= 2\varepsilon_{kl}\partial_{M_n}\langle\hat{t}_{kl}u\rangle$$
$$= 2\varepsilon_{kl}\left(\partial_{M_n}\langle\hat{s}_k u\rangle - \partial_{M_n}\langle\hat{s}_l u\rangle\right),$$

where $u = \frac{t_{kl}}{|t_{kl}|}$. Note that

$$\langle s_k u\rangle = \frac{\langle S_k(I_3)(u \cdot I_4)\rangle}{\langle S_k(I_3)I_3\rangle} = \frac{N_k}{D_k}.$$

Using the product rule, we can express $\partial_{M_n}\langle\hat{s}_k u\rangle$ in terms of $\partial_{M_n}N_k$ and $\partial_{M_n}D_k$. Using Lemma 7.4 with $A_r = I_3 = e_i \wedge E_i$ and $C_r = u \cdot I_4$ or $C_r = I_3$ gives

$$\partial_{M_n}N_k = 2\sum_{j\in J_{kn}}\bar{\mathrm{P}}_{\mathscr{T}_{Mn}}(\tilde{M}_n\Lambda_j B_{kj}), \quad B_{kj} = \sum_{i=1}^{3} B\left(e_i, S_k(E_i)\cdot(u\cdot I_4), \Lambda_j\right),$$

$$\partial_{M_n}D_k = 2\sum_{j\in J_{kn}}\bar{\mathrm{P}}_{\mathscr{T}_{Mn}}(\tilde{M}_n\Lambda_j B_{kj}), \quad B_{kj} = \sum_{i=1}^{3} B\left(e_i, S_k(E_i)\cdot I_3, \Lambda_j\right).$$

Similarly $\partial_{M_n}\langle\hat{s}_l u\rangle$ can be computed by replacing S_k with S_l in the above equations.

Distance Between Two Poses Similarly, a measured distance d_{kl}^o between two poses $M_k = T_k R_k$ and $M_l = T_l R_l$ can be used to scale the problem. The relative pose is given by $M = \tilde{M}_k M_l = \tilde{R}_k\tilde{T}_k T_l R_l$. We can factor this as $M = TR$ where $T = \tilde{R}_k\tilde{T}_k T_l R_k$ and $R = \tilde{R}_k R_l$. The distance between the two poses can be written as $d_{kl} = 2|n_o \cdot M| = |t_k - t_l|$, and we can impose the constraint

$$\varepsilon_{kl} = d_{kl} - d_{kl}^o = 0.$$

As $|Q|^2 = \langle Q\tilde{Q}\rangle$ where $Q = n_o \cdot M$, Lemma 7.5 gives

$$\partial_{M_n}\varepsilon_{kl}^2 = 4\varepsilon_{kl}\partial_{M_n}|n_o \cdot M|$$
$$= 4\varepsilon_{kl}\partial_{M_n}\langle(n_o \cdot \dot{M})\tilde{U}\rangle$$

with $U = \frac{Q}{|Q|}$. As $\langle(n_o \cdot M)\tilde{U}\rangle = \langle n_o M\tilde{U}\rangle$, we get $\langle(n_o \cdot M)\tilde{U}\rangle = \langle M_l\tilde{U}n_o\tilde{M}_k\rangle = \langle M_k n_o U\tilde{M}_l\rangle$. It follows that

$$\partial_{M_k}\varepsilon_{kl}^2 = 4\varepsilon_{kl}\bar{\mathrm{P}}_{\mathscr{T}_{Mk}}(n_o U\tilde{M}_l),$$
$$\partial_{M_l}\varepsilon_{kl}^2 = 4\varepsilon_{kl}\bar{\mathrm{P}}_{\mathscr{T}_{Ml}}(\tilde{U}n_o\tilde{M}_k).$$

7.3.2.2 Parameterisation

These multivector derivatives are not suitable for using directly with standard optimisation tools such as the NAG library which make use of information such as gradient vectors and Jacobian matrices. However we can easily interface with such tools using the chain rule. Let $X(\alpha) \in \mathscr{G}$, where $\alpha \in \mathbb{R}^p$, be a parameterisation of a multivector, $s(X) \in \mathbb{R}$ be a scalar-valued function of a multivector, and define $g : \mathbb{R}^p \to \mathbb{R}$ by $g(\alpha) = s(X(\alpha))$. The gradient $\nabla_\alpha g$ is given by

$$[\nabla_\alpha g]_i = \partial_{\alpha_i} g = \langle \partial_{\alpha_i} X(\alpha) \partial_X s(X) \rangle.$$

A parameterisation of a motor $M : \mathbb{R}^6 \to \mathscr{M} \subset \mathbb{R}_{4,1}$ is given by $M(\alpha) = \exp(-\frac{1}{2}B(\alpha))$ where $B(\alpha) = \sum_I \alpha_I B_I$ and $B_I \in \{e_{12}, e_{13}, e_{23}, e_1 n_\infty, e_2 n_\infty, e_3 n_\infty\}$. Making use of Chasles's theorem allows $M = \exp(-\frac{1}{2}B)$ and $B = -2 \ln M$ to be efficiently calculated [1]. We have

$$\partial_{\alpha_I} M(\alpha) = -\frac{1}{2} \partial_{\alpha_I} B(\alpha) * \partial_{B'} \exp(B') \big|_{B' = -\frac{1}{2} B(\alpha)}$$

$$= -\frac{1}{2} B_I * \partial_{B'} \exp(B') \big|_{B' = -\frac{1}{2} B(\alpha)}.$$

To evaluate this, we make use of the following result:

Lemma 7.6 *Let A and B be bivectors. Then*

$$A * \partial_B e^B = e^B F(A, B), \tag{7.18}$$

where $F(A, B)$ is bivector-valued function given by

$$F(A, B) = \sum_k 2^k T_k / (k+1)!$$

and $T_k = \langle T_{k-1} B \rangle_2 = \langle \langle \cdots \langle \langle AB \rangle_2 B \rangle_2 \cdots \rangle_2 B \rangle_2$ with $T_0 = A$.

This can be easily proved using (2.31) from [3] on the left-hand side of (7.18), expanding the right-hand side, and comparing terms. By observing that $AB^2 = B^2 A$ for $A, B \in \mathbb{B}$ the function $F(A, B)$ can be expressed as

$$F(A, B) = A + \frac{1}{2}(1 - 2B - e^{-2B})B^{-2} \langle BA \rangle_2,$$

which makes Lemma 7.6 of practical interest. Thus we have $\partial_{\alpha_I} M(\alpha) = M F(-\frac{1}{2} B_I, -\frac{1}{2} B)$. If $\alpha \in \mathbb{R}^6$ are the parameters associated with $M_n = M_n(\alpha)$, then the gradient component $\nabla_\alpha d^2$ associated with M_n is given by

$$[\nabla_\alpha d^2]_i = \langle \partial_{\alpha_i} M \partial_{X_n} d^2 \rangle = \langle \partial_{\alpha_i} M \partial_{M_n} d^2 \rangle.$$

The last equality holds because, by definition, $\partial_{\alpha_i} M(\alpha) \in \mathcal{T}_M$ and for any $\Delta \in \mathcal{T}_M$, we have $\langle \Delta \partial_X \rangle = \langle P_{\mathcal{T}_M}(\Delta) \partial_X \rangle = \langle \Delta \bar{P}_{\mathcal{T}_M}(\partial_X) \rangle = \langle \Delta \partial_M \rangle$. In this way the multivector derivatives $\partial_{M_n} d^2$, $n = 1, \ldots, N$, can be used to form a $6N$ gradient vector. An important point is that the derivative of the parameterisation is completely decoupled from the multivector derivative and the exact choice of parameterisation can be delayed and easily changed.

It is apparent that, in the current situation, d^2 could simply have been differentiated with respect to unconstrained $M \in \mathbb{M}$, rather than constrained $M \in \mathcal{M}$ (by constraining the derivative to $\mathcal{T}_M$) as this would have still given rise to $\nabla_\alpha d^2$ when the scalar products are formed with $\partial_{\alpha_i} M(\alpha)$. Indeed we could have differentiated d^2 with respect to any variable $X \in \mathbf{X} \supset \mathcal{T}_M$ because $\langle \Delta \partial_X \rangle = \langle \Delta \partial_M \rangle$. However, we ultimately wish to optimise directly using elements of the algebra, and in this case we will want to step in the tangent space.

Given the objective function, constraints, their gradients, and an initial estimate, the problem can be formulated in a standard way as a constrained optimisation, or an optimisation with barrier functions.

7.3.3 Reconstructing the Target Positions

Given the lines Λ_j^o in CSM and poses M_n, $n = 1, \ldots, N$, of CSM in CSW, it is a simple matter to reconstruct target positions in CSW. First the lines are mapped into CSW using (7.4):

$$\Lambda_j = M_n \Lambda_j^o \tilde{M}_n, \quad n = n(j).$$

The lines associated with the kth target, Λ_j, $j \in J_k$, are then triangulated using (7.2) in Lemma 7.2:

$$q_k = -\frac{1}{2} s_k n_\infty s_k = s_k + \frac{1}{2} s_k^2 n_\infty, \quad s_k = -\frac{\langle S_k(I_3) I_4 \rangle_1}{\langle S_k(I_3) I_3 \rangle}.$$

Several target calibrations of the same target field, made at different times for example, can be merged into a single optimal (as defined below) set of targets. Each calibration $\gamma = 1, \ldots, \Gamma$ will give rise to a set of target points $q_{k,\gamma}^o$, $k = 1, \ldots, K$, each in a different coordinate system. As with the line data (compare (7.4)), we can transform the points to bring them into a common coordinate system

$$q_{k,\gamma} = M_\gamma q_{k,\gamma}^o \tilde{M}_\gamma, \quad k = 1, \ldots, K.$$

Ideally, when M_γ are chosen correctly, $q_{k,\gamma}$, $\gamma = 1, \ldots, \Gamma$, will coincide, but measurement noise on $q_{k,\gamma}$ will prevent this from happening exactly.

In general, given several measurements of a point q_γ, $\gamma = 1, \ldots, \Gamma$, the mean $\bar{q} = \frac{1}{\Gamma} \sum_{\gamma=1}^{\Gamma} q_\gamma$ is an imaginary dual sphere

$$\bar{q} = \bar{\mathbf{q}} + n_o + \frac{1}{2}\overline{\mathbf{q}^2}n_\infty$$

$$= \bar{\mathbf{q}} + n_o + \frac{1}{2}\overline{\mathbf{q}}^2 n_\infty - \frac{1}{2}(\overline{\mathbf{q}^2} - \overline{\mathbf{q}}^2)n_\infty.$$

The squared radius of this mean sphere is the variance of the points about their centroid

$$d^2 = -\bar{q}^2 = \overline{\mathbf{q}^2} - \overline{\mathbf{q}}^2.$$

We can use this to merge the points by minimising the sum of these errors for all targets

$$d^2 = d^2(M_1, \ldots, M_\Gamma) = -\sum_k \bar{q}_k^2.$$

The multivector derivative $\partial_{M_\gamma} d^2$ can also be calculated:

$$\partial_{M_\gamma} d^2 = -\sum_k \partial_{M_\gamma} \bar{q}_k^2$$

$$= -2\sum_k \dot{\partial}_{M_\gamma} \langle \dot{\bar{q}}_k \bar{q}_k \rangle$$

$$= -\frac{2}{\Gamma} \sum_k \dot{\partial}_{M_\gamma} \langle M_\gamma q_{k,\gamma}^o \tilde{M}_\gamma \bar{q}_k \rangle$$

$$= -\frac{4}{\Gamma} \dot{\mathbf{P}}_{\mathscr{I}_{M_\gamma}} \left(\tilde{M}_\gamma \sum_k q_{k,\gamma} \bar{q}_k \right).$$

With the derivatives $\partial_{M_\gamma} d^2$, we can minimise d^2 over M_γ, $\gamma = 1, \ldots, \Gamma$, to obtain a merged data set. The new targets are then taken as $q_k = -\frac{1}{2}\bar{q}_k n_\infty \bar{q}_k$.

As previously mentioned, CSW was temporarily identified with one of the positions of CSM. The motor M representing the pose of the temporary CSW in the actual CSW can now be calculated, and the target positions mapped into the true CSW,

$$q_k' = M q_k \tilde{M}, \quad k \in K.$$

Exactly how M is obtained is application specific and depends on how the true CSW is defined. For example, three or more targets, say q_k^o, $k \in K'$, may have been surveyed in the true CSW. M can then be easily estimated from $\{q_k, q_k^o\}$, $k \in K'$, as a small 4D eigenrotator problem (see [6], which is Chap. 2 in this book). Alternatively, one target q_o may be assigned to the origin, another on the x-axis q_x, and another in the xy-plane q_{xy}. Again the motor can be easily calculated. Recall that the targets and camera group comprise a 6-DoF optical positioning system. Another approach

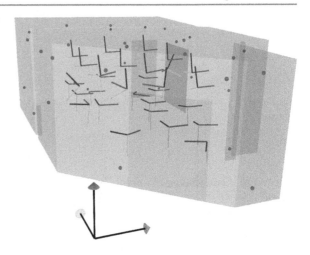

Fig. 7.5 Model of laboratory showing targets as small spheres and positions of camera group as coordinate systems (axis length is 1 m)

is to move the camera group to a position in space that we wish to define as CSW. Some line data can be gathered and the motor M estimated by recovering the pose CSM in the temporary CSW.

7.4   Results

The algorithm has been used in practice for a number of years to calibrate targets for an optical positioning system when it is deployed in new locations. In this section we test the algorithm on some measured line data in a laboratory. The camera group comprised $C = 6$ calibrated cameras mounted on a mechanical rig so that the image planes were parallel with the faces of a cube as shown in Fig. 7.1. Each camera had an image size of 640×480 pixels with a field of view of approximately $90°$ across the larger image dimension. The intrinsic parameters of each camera were accurately calibrated. As we used wide angle lenses, a lens distortion model which includes three radial distortion terms and two decentering terms was used. In addition, the pose of each camera in CSM was accurately calibrated using conformal geometric algebra as described in [5]. The procedure is quite similar to the current algorithm for target calibration. The $K = 32$ targets were located on the walls and ceiling of a laboratory with dimensions 8.7 meters long by 5.1 meters wide by 2.9 meters high. Figure 7.5 shows a model of the laboratory with the targets shown as small spheres.

The camera group was moved to $N = 28$ positions in the laboratory (an example is shown as coordinate systems in Fig. 7.5) and line data Λ_j^o, $j \in J = \bigcup_{k=1}^{K} J_k$, gathered as described in Sect. 7.2. A constraint was introduced by measuring the distance between a pair of poses. The acquisition for a single position takes about 0.8 seconds, and the total acquisition for all positions takes a few minutes. This procedure was repeated four times to give four independent sets of line data. The algorithm was then applied to each of these four sets of data.

Table 7.1 Error between the measured lines and reconstructed target positions (m)

	avg	rms	max
lines and targets (m)	0.001797	0.002123	0.007908

Table 7.2 Standard deviation of target estimates about their centroids

	avg	rms	max
target std. dev. (m)	0.000668	0.000813	0.003121

Table 7.3 Error between measured and model positions of CSM

	avg	rms	max
distance error (m)	0.00012	0.00014	0.00034

The results presented in Table 7.1 are consolidated for all four data sets. It gives an indication of model fit adequacy by summarising how close the lines came to the reconstructed target positions. To get a feel for repeatability, the four sets of reconstructed target positions were merged together into a single set of target positions by minimising the dispersion associated with each target, as described in Sect. 7.3.3. Table 7.2 shows the dispersion of the target sets about the optimal merged target set. It is difficult to obtain a reference measurement of the target positions with, for example, a total station because the target LEDs are not ideal for measuring (the light penetrates into them). Flat retro-reflective targets would have been better suited for this task. We can however estimate target location accuracy indirectly by using the system as a 6-DoF positioning sensor. We translated the system to 12 equally spaced positions along a linear stage and calculated the motors M_n at each position with respect to the target locations. Ideally the origin of CSM given by the data points $p_n^o = \widetilde{M}_n n_o M_n$ will fall exactly on a line and be equally separated by the known amount. A perfect model of 12 equally spaced points p_n on a line was fitted to the data p_n^o in the total least squares sense, and the results are presented in Table 7.3. The results shown in Table 7.1 to Table 7.3 are typical for multiple runs with the same camera group and approximate target configuration. Ideally we would like the measured line data $\Lambda_j \in J_k$ to pass exactly through their associated targets q_k. The quality of the measured line data is affected by several factors which can be broadly separated into noise errors and modelling errors. The subpixel estimates of the target locations will not be perfect, and camera sensor noise, quantisation, imperfect (wide angle) optics, etc. will contribute to the error. These estimates are backprojected through the camera model to provide lines in the camera coordinate system. The accuracy of this step is affected by the quality of the camera intrinsic calibration. The lines are then rotated and translated into CSM to give the data Λ_j, and this step is affected by the quality of pose calibration for each camera in CSM. The intrinsic camera parameters are easy to calibrate to very high accuracy. Based on the quantity of data used and the residuals obtained during camera intrinsic calibration, we can eliminate this as a significant source of error. The camera poses are

Fig. 7.6 Regions of interest from an image showing a target: (*left*) from centre image; (*right*) from bottom right of image showing significant distortion

more difficult to determine, but if we gather enough data (by moving to 200 or more positions), we can again eliminate this as a significant source of error. However, the camera lenses are very wide angle and exhibit some large aberrations in the outer field. This results in significant deformation of the target image shape with small circular targets becoming comet shaped (see Fig. 7.6). We are investigating better lenses and expect improvements to the results when they have been incorporated. The repeatability results are influenced by three targets which are responsible for the six largest errors. Removing these three targets brings the maximum error down to just over a millimetre (and the average and root mean square values drop accordingly). The three targets are in geometrically poor positions. For example, the worse errors came from the target in the bottom right in Fig. 7.5. Because of the shape of the laboratory, this target can only be viewed from a narrow range of angles.

Recall that CSW is temporarily identified with a position of CSM as described in Sect. 7.3. This avoided over-parameterisation in the system model by fixing one of the motors at a constant value ($M_{n'} = 1$). If this was not done, we could multiply all the motors M_n, $n = 1, \ldots, N$, by an arbitrary motor M and get the same error. However, with many optimisers this step of fixing a motor is unnecessary and actually counter-productive. We observed the optimiser converged approximately 30% faster if we did not fix a motor, and the error was the same as expected. This can be explained as follows. By fixing $M_{n'} = 1$, all other motors M_n must adjust a little further to find the solution. When $M_{n'}$ is also allowed to adjust, a solution in the "middle" is found which is closer to all motors on average. Even though the problem is overdetermined, the optimiser has no difficultly terminating.

7.5 Conclusions

In this chapter an algorithm for calibrating the 3D position of multiple stationary point targets has been presented. The targets and camera group form part of an optical 6-DoF positioning system. The algorithm is developed in the conformal model of Euclidean space, expressed in terms of the Geometric Algebra $\mathbb{R}_{4,1}$.

The conformal model proved very useful for the theoretical and practical development of the calibration algorithm. In general, expressions were simpler than those for the homogeneous model [7]. The ability to differentiate with respect to a motor (rather than a rotator and a translation vector) significantly simplified the implementation of the gradient. The choice of parameterisation can be delayed and readily changed without affecting $\partial_{M_n} d^2$. The ability to represent many objects and operators (e.g. vectors, points, lines, dual spheres, motors) as elements of a single algebra simplified the software implementation of the algorithm. The extension of linear operators to multivectors via outermorphism allowed a compact representation of results (e.g. (7.2) and (7.3)). Expressing inversion in terms of duality simplified expressions involving inverses, and in particular allowed them to be more easily differentiated. For example, (7.2) for the closest point to a number of lines is expressed in terms of duality. This formulation is more elegant than equivalent formulation in terms of matrix theory, where it is most natural to involve a matrix inverse.

The error function used in the algorithm is simple and elegant but does not accurately reflect the sources of noise in the measured data such as the subpixel image locations of the targets. It would be useful to develop new error functions which better model noise and compare the results.

7.6   Exercises

7.1 Show that $\partial_R \langle RA\widetilde{R}B \rangle = \partial_X \langle XAX^{-1}B \rangle|_{X=R}$ where $X \in \mathbb{R}_3^+$, A, B are general multivectors, and $R \in \mathscr{R} \subset \mathbb{R}_3^+$ is a rotator, without resorting to directional derivatives as in the text. Consider only the derivative $\partial_Y \langle YAY^{-1}B \rangle$ and note that the relevant projections satisfy $P_{\mathbb{R}_3^+}(Y) = \bar{P}_{\mathscr{T}_R}(Y) + \langle YR \rangle \widetilde{R}$ for any multivector Y.

7.2 In the text the necessary derivatives are developed largely by algebraically manipulating the expressions into a form which is easily to differentiate. Compare this approach to direct application of the multivector chain rule.

7.3 In Lemma 7.2 we find the optimal dual sphere s using constrained optimisation. Show that you can arrive at the first-order optimality condition $S(s) + \alpha n_\infty = 0$ directly from geometric considerations. If we project a dual sphere on all the lines and take the average, we get a new dual sphere. Consider finding the dual sphere whose centre is at the same position as its associated average sphere. This gives $\frac{1}{N}\sum_{j \in J} P_{A_j}(s) = s + \frac{1}{N}d^2 n_\infty$. From this we also recover the value of the Lagrange multiplier as d^2.

Acknowledgements   This work was supported by the New Zealand Foundation for Research, Science and Technology.

References

1. Dorst, L., Fontijne, D., Mann, S.: Geometric Algebra for Computer Science, An Object-Oriented Approach to Geometry. The Morgan Kaufman Series in Computer Graphics. Elsevier, Amsterdam (2007)
2. Faugeras, O.: Three-Dimensional Computer Vision: A Geometric Viewpoint. MIT Press, Cambridge (1993)
3. Hestenes, D., Sobczyk, G.: Clifford Algebra to Geometric Calculus. Reidel, Dordrecht (1984)
4. Lasenby, J., Lasenby, A.N., Doran, C.J.L., Fitzgerald, W.J.: New geometric methods for computer vision: an application to structure and motion estimation. Int. J. Comput. Vis. **36**(6), 191–213 (1998)
5. Valkenburg, R.J., Alwesh, N.: Calibration of the relative poses of multiple cameras. In: Image and Vision Computing New Zealand 2009. Wellington, New Zealand, 2009
6. Valkenburg, R., Dorst, L.: Estimating motors from a variety of geometric data in 3D conformal geometric algebra. In: Dorst, L., Lasenby, J. (eds.) Guide to Geometric Algebra in Practice. Springer, London (2011), Chap. 2 in this book
7. Valkenburg, R., Lin, X., Klette, R.: Self-calibration of target positions using geometric algebra. In: Proceedings of Image and Vision Computing New Zealand, pp. 221–226 (2004)

Part III
Image Processing

Apart from the obviously geometrical applications in tracking and 3D reconstruction, geometric algebra finds inroads in image processing at the signal description level. It can provide more symmetrical ways to encode the geometrical properties of the 2D or 3D domain of such signals.

Quaternion Atomic Function for Image Processing

8

Eduardo Bayro-Corrochano and Eduardo Ulises Moya-Sánchez

Abstract

In this work we introduce a new kernel for image processing called the atomic function. This kernel is compact in the spatial domain, and it can be adapted to the behavior of the input signal by broadening or narrowing its band ensuring a maximum signal-to-noise ratio. It can be used for smooth differentiation of images in the quaternion algebra framework. We discuss the role of the quaternion atomic function with respect to monogenic signals. We then propose a steerable quaternion wavelet scheme for image structure and contour detection. Making use of the generalized Radon transform and images processed with the quaternion wavelet atomic function transform, we detect shape contours in color images. We believe that the atomic function is a promising kernel for image processing and scene analysis.

8.1 Introduction

This work introduces the atomic function for grey scale and color image processing using the quaternion algebra framework. The goal of this book is to present applications of geometric algebra techniques; for example, conformal geometric algebra has been recently used in neural computing and robot vision. The quaternion algebra is a sub-algebra of the 3D Euclidean geometric algebra, 4D motor algebra and the 3D conformal geometric algebra. This chapter shows how quaternion atomic filters can be used as smoothing and differentiator filters to carry out differential geometry on the visual manifold. These techniques can be further extended to compute, for

E. Bayro-Corrochano (✉) · E.U. Moya-Sánchez
Campus Guadalajara, CINVESTAV, Jalisco, Mexico
e-mail: edb@gdl.cinvestav.mx

E.U. Moya-Sánchez
e-mail: emoya@gdl.cinvestav.mx

L. Dorst, J. Lasenby (eds.), *Guide to Geometric Algebra in Practice*,
DOI 10.1007/978-0-85729-811-9_8, © Springer-Verlag London Limited 2011

example, Riemannian differential geometry. In this regard the quaternion atomic filters appear to be useful for differential geometry. A practical example is their use for optical flow detection which, together with epipolar geometry, makes possible the development of very useful algorithms for 3D robot vision. Thus our chapter contributes to this book by presenting a novel filter for image processing which brings geometric computing into play.

The atomic function AF was introduced in the 1970s by V.L. and V.A. Rvachev. Since then, it has given rise to new research areas in mathematical analysis, signal processing, numerical methods, and so forth [7, 18, 20]. This work presents the theory and some applications of the quaternion atomic function Qup in image processing as a novel quaternionic wavelet. We use the AF because it is versatile, easy to differentiate (simply shift), compact in the spatial domain, and can represent any polynomial by means of its translations [9, 12, 18, 20]. We developed the $Qup(x)$ using the quaternion algebra $\mathbb{H}$. The quaternion atomic function permits us to extract the phase information from the image [1, 3, 11]. In this work we also discuss the role of the quaternion atomic function with respect to monogenic signals. Furthermore, we propose a steerable quaternion wavelet scheme for image structure detection. Using the generalized Radon transform and images processed with the quaternion wavelet atomic function, we look for contours in color images, showing that the atomic function is a promising kernel for image processing and scene analysis.

We structure this work as follows: Sect. 8.2 is devoted to presenting the AF and the main characteristics; the subject of Sect. 8.3 is the quaternion algebra; Sect. 8.4 introduces the quaternion atomic function. In Sect. 8.5, we briefly comment on monogenic signals and atomic functions. Section 8.6 describes the quaternion wavelet atomic function transform. In Sect. 8.7, we outline the generalized Radon transform, explaining how we detect shape contours using images processed with the quaternion wavelet atomic function transform. In Sect. 8.8, we present the applications; finally, in Sect. 8.9, we present the conclusions.

8.2   Atomic Functions

Atomic functions are compactly supported, infinitely differentiable solutions of differential equations with a shifted argument [12, 18], namely,

$$Lf(x) = \lambda \sum_{k=1}^{M} c(k) f\big(ax - b(k)\big), \quad |a| < 1, \tag{8.1}$$

where $L = \frac{d^n}{dx^n} + a_1 \frac{d^{n-1}}{dx^{n-1}} + \cdots + a_n$ is a linear differential operator with constant coefficients. In the AF class, the function $up(x)$ is the simplest and, at the same time, most useful primitive function to generate other kinds of atomic functions [12]. It satisfies the equation

$$f(x)' = 2\big(f(2x + 1) - f(2x - 1)\big) \tag{8.2}$$

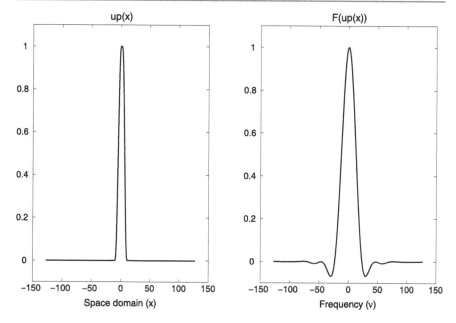

Fig. 8.1 Atomic function $up(x)$ and the Fourier transform of $up(x)$

with compact support. The main characteristics of this function are well described in [7, 12, 20]. The function $up(x)$ is infinitely differentiable, $up(0) = 1$, $up(-x) = up(x)$. Other types of AF satisfying (8.1) are $fup_n(x)$, $\Xi_n(x)$, $h_a(x)$ [7]. In this work we only use the functions $up(x)$ and $dup(x)$, see (8.6).

8.2.1   The Atomic Function $up(x)$

In general, the atomic function $up(x)$ is generated by infinite convolutions of rectangular impulses. The function $up(x)$ has the following representation in terms of the Fourier transform [7, 18, 20]:

$$up(x) = \frac{1}{2\pi} \int_{\infty}^{\infty} e^{iux} \prod_{k=1}^{\infty} \frac{\sin(u2^{-k})}{u2^{-k}} \, du. \tag{8.3}$$

Figure 8.1 shows the $up(x)$ and its Fourier transform $F(up(x))$. Atomic windows were compared with classic ones [7, 12] by means of a system of parameters such as the equivalent noise bandwidth, the 50% overlapping region correlation, the parasitic modulation amplitude, the maximum conversion losses (in decibels), the maximum side lobe level (in decibels), the asymptotic decay rate of the side lobes (in decibels per octave), the window width at the six-decibel level, and the coherent gain. The properties of all atomic windows exceed those of the classic ones showing higher asymptotic decay rate [7, 12].

The main properties of the $up(x)$ are as follows: $up(x)$ is even, its maximum value is $up(0) = 1$, $\int_{-1}^{1} up(x) = 1$, the support domain or supp $up(x) = [-1, 1]$, and the first derivative has a simple expression.

8.2.2   The Differentiator Atomic Function $dup(x)$

Some mask operators, such as the Sobel, Prewitt, and Kirsch, are used to process images. A common drawback to these approaches is that it is impossible to ensure the required characteristics over a wide range of the working band of the processed signals and difficult to retune to adapt to the signal parameters [9]. This means that adaptation of the differential operator to the behavior of the input signal by broadening or narrowing its band is desirable, in order to ensure a maximum signal-to-noise ratio [9].

This problem reduces to the synthesis of infinitely differentiable finite functions with small-diameter carriers that are used for constructing the weighting windows [7, 9, 18]. One of the most effective solutions is obtained with the help of the atomic functions [9]. The AF can be used in two ways: for the construction of a window in a certain frequency region to improve the properties of the impulse response or the direct synthesis based in (8.1) [9]. Therefore, the function $up(x)$ satisfies the functional (8.2), and if we compute the Fourier transform of (8.2), we obtain

$$iu F_{up}(2u) = \left(e^{iu} - e^{-iu}\right) F_{up}(2u),\tag{8.4}$$
$$F\left(dup(x)\right) = 2i\,\sin(u) F\left(up(2x)\right),\tag{8.5}$$

with $dup(\cdot)$ introduced below.

Therefore, the function $up(x)$ satisfies (8.2) and also satisfies (8.1), as mentioned. Upon differentiating (8.2) term by term, we obtain the derivatives of $up(x)$ called $d^{(n)}up(x)$ [18]:

$$d^{(n)}up(x) = 2^{n(n+1)/2} \sum_{k=1}^{2^n} \delta_k up\left(2^n x + 2^n + 1 - 2k\right),\tag{8.6}$$

where $\delta_{2k} = -\delta_k$, $\delta_{2k-1} = \delta_k$, $\delta_{2k} = 1$. We abbreviate this family of derivative functions as $dup(x)$. They provide a good window in the spatial frequency regions since, in this case (frequency), the side lobes have been completely eliminated [9]. Figure 8.2 illustrates the first derivative, dup (8.2) in 1D and 2D, and also shows the 2D convolution with a test image. This differentiator, dup, can also be used as an oriented line detector via simple rotation.

8.3   Quaternion Algebra

The quaternion algebra $\mathbb{H}$ was invented by W.R. Hamilton in 1843 [8]. It is an associative, non-commutative, four-dimensional algebra

Fig. 8.2 Convolution of *dup*(*x*, *y*) with the test image: (*upper row*) *dup*(*x*); (*middle row*) *dup*(*x*, *y*); (*lower row*) (**a**) test image; results of convolutions with *dup*(*x*, *y*) oriented at (**b**) 0°; (**c**) 45°; (**d**) 135°

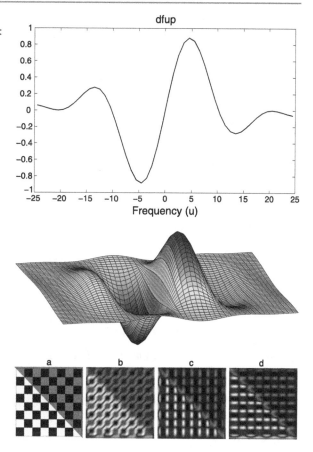

$$\mathbb{H} = \{ \mathbf{q} = s + xi + yj + zk \mid s, x, y, z \in \mathbb{R} \}, \tag{8.7}$$

where the orthogonal imaginary numbers i, j, and k obey the following multiplicative rules:

$$i^2 = j^2 = -1, \qquad k = ij = -ji \rightarrow k^2 = -1. \tag{8.8}$$

The conjugate of a quaternion is given by

$$\bar{\mathbf{q}} = s - xi - yj - zk. \tag{8.9}$$

For the quaternion $\mathbf{q}$, we can compute its partial angles as

$$\arg_i(q) = \arctan2(x, s), \qquad \arg_j(q) = \arctan2(y, s),$$
$$\arg_k(q) = \arctan2(z, s), \tag{8.10}$$

and its partial moduli and its projections on its imaginary axes as

$$\text{mod}_i(q) = \sqrt{s^2 + x^2}, \qquad \text{mod}_j(q) = \sqrt{s^2 + y^2},$$

$$\text{mod}_k(q) = \sqrt{s^2 + z^2}, \qquad \text{mod}_i(q)\exp\bigl(i\,\arg_i(q)\bigr) = s + xi, \tag{8.11}$$

$$\text{mod}_j(q)\exp\bigl(j\,\arg_j(q)\bigr) = s + yj,$$

$$\text{mod}_k(q)\exp\bigl(k\,\arg_k(q)\bigr) = s + zk.$$

The concept of a quaternionic Hermitian function is very useful for the computation of any kind of inverse quaternionic transforms using the quaternionic analytic signal. As an extension of the Hermitian function $f : \mathbb{R} \to \mathbb{C}$ with $f(x) = f^*(-x)$ for every $x \in \mathbb{R}$, we regard $f : \mathbb{R}^2 \to \mathbb{H}$ as a quaternionic Hermitian function if it fulfills the following nontrivial involution rules:

$$\begin{aligned}
f(-x, y) &= -jf(x, y)j := \mathcal{T}_j\bigl(f(x, y)\bigr), \\
f(x, -y) &= -if(x, y)i := \mathcal{T}_i\bigl(f(x, y)\bigr), \\
f(-x, -y) &= -if(-x, y)i = -i\bigl(-jf(x, y)j\bigr)i \\
&= (-i - j)f(x, y)(ji) \\
&= -kf(x, y)k := \mathcal{T}_k\bigl(f(x, y)\bigr).
\end{aligned} \tag{8.12}$$

Similar to the complex numbers, which can be expressed in a polar representation, we can also represent a quaternion $\mathbf{q} = r + xi + yj + zk$ in a polar form:

$$\mathbf{q} = |\mathbf{q}|e^{i\phi}e^{k\psi}e^{j\theta}, \tag{8.13}$$

where the phase ranges are delimited as follows:

$$(\phi, \theta, \psi) \in [-\pi, \pi[\times \left[-\frac{\pi}{2}, \frac{\pi}{2}\right[\times \left[-\frac{\pi}{4}, \frac{\pi}{4}\right].$$

For a unit quaternion $\mathbf{q} = q_0 + q_x i + q_y j + q_z k$, $|\mathbf{q}| = 1$, its phase can be evaluated first by computing

$$\psi = -\frac{\arcsin(2(q_x q_y - q_0 q_z))}{2} \tag{8.14}$$

and then by checking that it adheres to the following rules:

- If $\psi \in]-\frac{\pi}{4}, \frac{\pi}{4}[$, then $\phi = \frac{\arg_i(\mathbf{q}\mathcal{T}_j(\bar{\mathbf{q}}))}{2}$ and $\theta = \frac{\arg_j(\mathcal{T}_i(\bar{\mathbf{q}})\mathbf{q})}{2}$.
- If $\psi = \pm\frac{\pi}{4}$, then select either $\phi = 0$ and $\theta = \frac{\arg_j(\mathcal{T}_k(\bar{\mathbf{q}})\mathbf{q})}{2}$ or $\theta = 0$ and $\phi = \frac{\arg_i(\mathbf{q}\mathcal{T}_k(\bar{\mathbf{q}}))}{2}$.
- If $e^{i\phi}e^{k\psi}e^{j\theta} = -\mathbf{q}$ and $\phi \geq 0$, then $\phi \to \phi - \pi$.
- If $e^{i\phi}e^{k\psi}e^{j\theta} = -\mathbf{q}$ and $\phi < 0$, then $\phi \to \phi + \pi$.

8.4 Quaternion Atomic Function $Qup(x)$

The $up(x)$ function is easily extendible to two dimensions. Since a 2D signal can be split into even (e) and odd (o) parts [3],

$$f(x, y) = f_{ee}(x, y) + f_{oe}(x, y) + f_{eo}(x, y) + f_{oo}(x, y), \tag{8.15}$$

one can then separate the four components of (8.3) and represent them as a quaternion as follows:

$$
\begin{aligned}
Qup(x, y) &= up(x, y)\big[\cos(w_x)\cos(w_y) + i\sin(w_x)\cos(w_y) \\
&\quad + j\cos(w_x)\sin(w_y) + k\sin(w_x)\sin(w_y)\big] \\
&= Qup_{ee}(x, y) + iQup_{oe}(x, y) + jQup_{eo}(x, y) + kQup_{oo}(x, y) \\
&= \Phi^q(x, y) + i\Psi_i^q(x, y) + j\Psi_j^q(x, y) + k\Psi_k^q(x, y).
\end{aligned}
\tag{8.16}
$$

Note that in (8.16) we rename Qup as Φ and Ψ, because we want to use a similar notation to that used for multiresolution wavelet pyramids. Figure 8.3 shows a quaternion atomic function Qup in the space and frequency domains with its four components: the real part Qup_{ee} and the imaginary parts Qup_{eo}, Qup_{oe}, and Qup_{oo}. We can clearly see the differences in each part of our filter.

Next we show the performance of Qup to detect edge changes using the phase concept. In Fig. 8.4, we see the phase changes along a transversal cut of an image with straight edges. Figure 8.5 shows the quaternionic phases by three transverse lines along a circle, a square, and a group of squares. We can see that the phases yield very useful information about shape contour changes.

8.5 Monogenic Signal and the Atomic Function

First we will outline the concept of a monogenic signal introduced by Felsberg, Bülow, and Sommer [5]. If we embed $\mathbb{R}^3$ into a subspace of $\mathbb{H}$ spanned by just $\{1, i, j\}$ according to

$$q = (i, j, 1)\mathbf{x} = x_3 + x_1 i + x_2 j, \tag{8.17}$$

and further embed the vector field $\mathbf{g}$ as follows:

$$g_\mathbb{H} = (-i, -j, 1)\mathbf{g} = g_3 - g_1 i - g_2 j, \tag{8.18}$$

then $\nabla \times g(\mathbf{x}) = 0$ and $\nabla \cdot g(\mathbf{x}) = 0$ are equivalent to the generalized Cauchy–Riemann equations from Clifford analysis [2, 16]. All functions that fulfill these equations are known as *monogenic* functions. Using the same embedding, the monogenic signal can be defined in the frequency domain as follows:

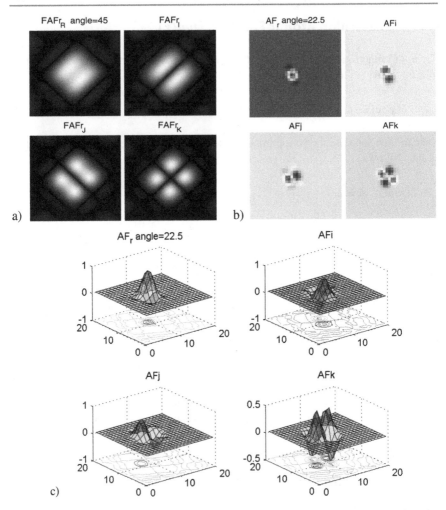

Fig. 8.3 Quaternion atomic function, elongated by $s_x = 0.3$, $s_y = 0.25$ (see (8.28)), and oriented at 45°. (**a**) In the frequency domain, the magnitude and the three imaginary quaternionic components; (**b**) in the spatial domain, the magnitude and the three imaginary quaternionic components; (**c**) 3D shapes of the quaternionic components in the spatial domain

$$F_M(u) = G_3(u_1, u_2, 0) - iG_1(u_1, u_2, 0) - jG_2(u_1, u_2, 0)$$

$$= F(\mathbf{u}) - (i, j)\mathbf{F}_R(\mathbf{u}) = \frac{|\mathbf{u}| + (1, k)\mathbf{u}}{|\mathbf{u}|} F(\mathbf{u}), \qquad (8.19)$$

where the inverse Fourier transform of $F_M(u)$ is given by

$$f_M(\mathbf{x}) = f(\mathbf{x}) - (i, j)\mathbf{f}_R(\mathbf{x}) = f(\mathbf{x}) + h_R \star f(\mathbf{x}), \qquad (8.20)$$

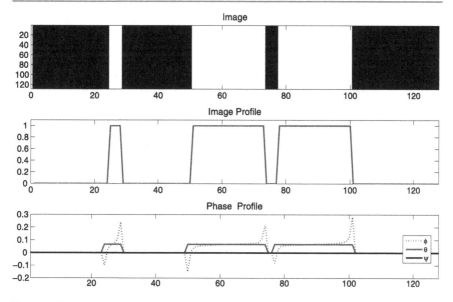

Fig. 8.4 (From *up* to *bottom*) Image; image of a transverse cut; quaternionic phases with respect to the transverse cut

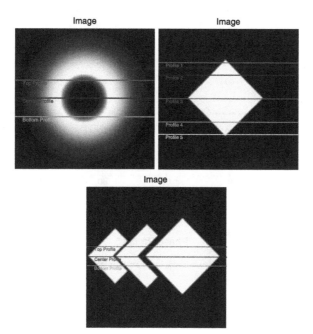

Fig. 8.5 Quaternionic phases by three transversal lines along (**a**) a circle; (**b**) a square; and (**c**) a group of squares

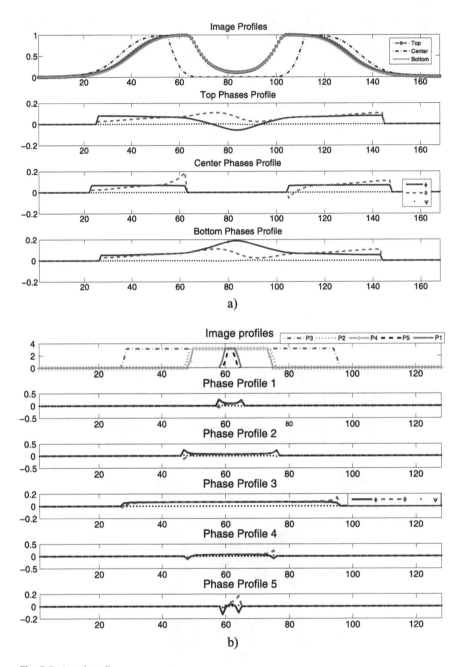

Fig. 8.5 (continued)

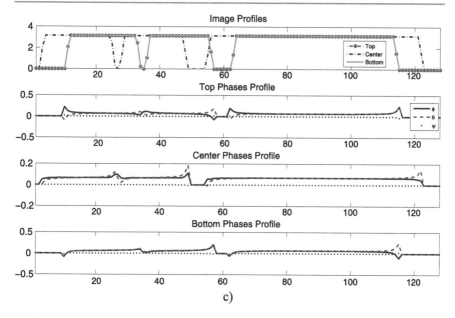

Fig. 8.5 (continued)

where $\mathbf{f}_R(\mathbf{x})$ stands for the Riesz transform [4] obtained by taking the inverse transform of $\mathbf{F}_R(\mathbf{u})$ as follows:

$$\mathbf{F}_R(\mathbf{u}) = \frac{i\mathbf{u}}{|\mathbf{u}|}F(\mathbf{u}) = \mathbf{H}_R(\mathbf{u})F(\mathbf{u}) \longleftrightarrow \mathbf{f}_R(\mathbf{x}) = -\frac{\mathbf{x}}{2\pi|\mathbf{x}|^2} \star f(\mathbf{x})$$
$$= h_R(\mathbf{x}) \star f(\mathbf{x}), \qquad (8.21)$$

where $\star$ stands for the convolution operation. Note that here the Riesz transform is the generalization of the 1D Hilbert transform. Using the fundamental solution of the 3D Laplace equation restricted to the open half-space $z > 0$ with boundary condition, the solution is defined as

$$f_M(x, y, z) = h_P \star f(x, y, z) + h_P \star h_R \star f(x, y, z)$$
$$= h_P \star (1 + h_R\star) f(x, y, z), \qquad (8.22)$$

where h_P stands for the 2D Poisson kernel. Setting in $f_M(x, y, z)$ the variable z equal to zero, we obtain the so-called monogenic signal. Some authors have used the Gauss kernel instead of the Poisson kernel, because the Poisson kernel establishes a linear scale space similar to the Gaussian scale space.

The atomic function is also an LSI operator; therefore, it appears that its use ensures a computation in a linear scale space as well. In a paper to appear elsewhere, we will give the theoretical foundations to prove that the *AF* satisfies a set of necessary axioms.

The monogenic functions are the solutions of generalized Cauchy–Riemann equations or Laplace-type equations. The atomic function can be used to compute compactly supported solutions of functional differential equations, for example, (8.1). Conditions under which the type of equations (8.1) have solutions with compact support and explicit form were obtained by Kravchenko et al. [12]. Compactly supported solutions of equations of the type (8.1) are called atomic functions.

Now, for the case of 2D signal processing, we can apply the wavelet steerability and the Riesz transform. In this regard we will utilize the quaternion wavelet atomic function, which will be discussed in more depth in the next section.

8.6   Quaternion Wavelet Atomic Function Transform

The word *wavelet* was used for the first time in the thesis of Alfred Haar in 1909. Surprisingly, the wavelet transform (WT) has become a useful signal processing tool only in the last few decades, mainly due to the contributions in the areas of applied mathematics and signal processing [10, 13–15]. Generally speaking, the WT is an approach that overcomes the shortcomings of the windowed Fourier transform as the WT detects changes in both in the spatial and in the frequency domains. Thanks to the development of the quaternion Fourier transform (QFT) [3], the generalization of the real and complex wavelet transform to the Quaternion Wavelet Transform (QWT) was straightforward.

The QWT is a natural extension of the real and complex wavelet transform, taking into account the axioms of the quaternion algebra, the quaternionic analytic signal [3], and its separability property. The QWT can be applied to signals of 2D or higher dimensions.

Multi-resolution analysis can also be straightforwardly extended to the quaternionic case; we can therefore improve the power of the phase concept, which is not possible in real wavelets and, in the case of complex wavelets, is limited to only one phase. Thus, in contrast to the similarity distance used in the complex wavelet pyramid [19], we favor the quaternionic phase concept for top–down parameter estimation.

For the quaternionic versions of the wavelet scale function h and the wavelet function g, in general we choose two quaternionic modulated filters in quadrature as follows:

$$
\begin{aligned}
h^q &= h(x, y, \sigma_1, \varepsilon) \exp\left(i \frac{c_1 \omega_1 x}{\sigma_1}\right) \exp\left(j \frac{c_2 \varepsilon \omega_2 y}{\sigma_1}\right) \\
&= h_{ee}^q + h_{oe}^q i + h_{eo}^q j + h_{oo}^q k, \\
g^q &= g(x, y, \sigma_2, \varepsilon) \exp\left(i \frac{\tilde{c}_1 \tilde{\omega}_1 x}{\sigma_2}\right) \exp\left(j \frac{\tilde{c}_2 \varepsilon \tilde{\omega}_2 y}{\sigma_2}\right) \\
&= g_{ee}^q + g_{oe}^q i + g_{eo}^q j + g_{oo}^q k,
\end{aligned}
\tag{8.23}
$$

where the parameters σ_1, σ_2, c_1, c_2, $\tilde{c}_1$, $\tilde{c}_2$, w_1, w_2, $\tilde{w}_1$, $\tilde{w}_2$ are selected for the fulfillment of the requirements of bandwidth and cut frequency. Note that the horizontal axis x is related to i and the vertical axis y is related to j; both imaginary numbers of the quaternion algebra fulfill the equation $k = ji$.

The right-hand sides of (8.23) and (8.23) obey a natural decomposition of a quaternionic analytic function: the subindex ee is an even–even symmetric filter, eo even–odd or oe odd–even are both unsymmetrical filters, and oo odd–odd is also an unsymmetrical filter. Thus, we can clearly see that h^q and g^q of (8.23) and (8.23) are powerful filters for disentanglement of the symmetries of 2D signals.

In the Fourier transform of a 2D signal, the phase component carries the main part of image information. We use this phase information in the quaternionic wavelet multi-resolution analysis. This technique can be easily formulated in terms of the quaternion AF mother wavelet; for a more detailed explanation, see [1]. For the 2D image function $f(x, y)$, a quaternionic wavelet can be written as

$$f(x, y) = A_n^q f + \sum_{\alpha=1}^{n} \left[D_{\alpha,1}^q f + D_{\alpha,2}^q f + D_{\alpha,3}^q f \right]. \tag{8.24}$$

The upper index q indicates a *quaternion* 2D signal. We can characterize each approximation function $A_\alpha^q f(x, y)$ and the difference components $D_{\alpha,p}^q f(x, y)$ for $p = 1, 2, 3$ via a 2D scaling function $\Phi^q(x, y)$ and its associated wavelet functions $\Psi_p^q(x, y)$ as follows:

$$A_\alpha^q f(x, y) = \sum_{\beta=-\infty}^{+\infty} \sum_{\gamma=-\infty}^{+\infty} a_{\alpha,\beta,\gamma} \Phi_{\alpha,\beta,\gamma}^q(x, y),$$
$$D_{\alpha,p}^q f(x, y) = \sum_{\beta=-\infty}^{+\infty} \sum_{\gamma=-\infty}^{+\infty} d_{\alpha,p,\beta,\gamma} \Psi_{\alpha,p,\beta,\gamma}^q(x, y), \tag{8.25}$$

where

$$\Phi_{\alpha,\beta,\gamma}^q(x, y) = \frac{1}{2^\alpha} \Phi^q\left(\frac{x - \beta}{2^\alpha}, \frac{y - \gamma}{2^\alpha} \right), \quad (\alpha, \beta, \gamma) \in Z^3,$$
$$\Psi_{\alpha,p,\beta,\gamma}^q(x, y) = \frac{1}{2^\alpha} \Psi_p^q\left(\frac{x - \beta}{2^\alpha}, \frac{y - \gamma}{2^\alpha} \right), \tag{8.26}$$

and

$$a_{\alpha,\beta,\gamma}(x, y) = \langle f(x, y), \Phi_{\alpha,\beta,\gamma}^q(x, y) \rangle,$$
$$d_{\alpha,p,\beta,\gamma} = \langle f(x, y), \Psi_{\alpha,p,\beta,\gamma}^q(x, y) \rangle. \tag{8.27}$$

In order to carry out a separable quaternionic multi-resolution analysis, using (8.16) and considering the separability of the atomic function, we decompose the scaling

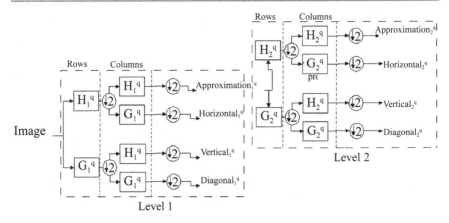

Fig. 8.6 Abstraction of two levels of the quaternionic wavelet pyramid

function $\boldsymbol{\Phi}^q(x, y)_\alpha$ and the wavelet functions $\boldsymbol{\Psi}_p^q(x, y)_\alpha$ for each level α as follows:

$$
\begin{aligned}
\boldsymbol{\Phi}^q(x, y)_\alpha &= \phi^i(x)_\alpha \phi^j(y)_\alpha = s_x s_y up(x, y)_\alpha \cos(w_x)_\alpha \cos(w_y)_\alpha, \\
\boldsymbol{\Psi}_1^q(x, y)_\alpha &= \phi^i(x)_j \psi^j(y)_\alpha = s_x s_y up(x, y)_\alpha \cos(w_x)_\alpha \sin(w_y)_\alpha, \\
\boldsymbol{\Psi}_2^q(x, y)_\alpha &= \psi^i(x)_j \phi^j(y)_\alpha = s_x s_y up(x, y)_\alpha \sin(w_x)_\alpha \cos(w_y)_\alpha, \\
\boldsymbol{\Psi}_3^q(x, y)_\alpha &= \psi^i(x)_j \psi^j(y)_\alpha = s_x s_y up(x, y)_\alpha \sin(w_x)_\alpha \sin(w_y)_\alpha,
\end{aligned}
\tag{8.28}
$$

where $\phi_\alpha^i(x) = s_x up(x)_\alpha \cos(x)_\alpha$ and $\psi(x)_\alpha^i = s_x up(x)_\alpha \sin(x)_\alpha$ are the 1D complex filters applied along the rows and columns, respectively. Note that in ϕ and ψ, we use the imaginary numbers i, j of quaternions.

By using these formulas, we can build quaternionic atomic wavelet function pyramids. Figure 8.6 shows the two primary levels of the pyramid (fine to coarse). According to (8.28), the approximation after the first level $A_1^q f(x, y)$ is the output of $\boldsymbol{\Phi}^q(x, y)_1$, and the differences $D_{1,1}^q f, D_{1,2}^q f, D_{1,3}^q f$ are the outputs of $\boldsymbol{\Psi}_{1,1}^q(x, y)$, $\boldsymbol{\Psi}_{1,2}^q(x, y)$, and $\boldsymbol{\Psi}_{1,3}^q(x, y)$. The procedure continues through the j levels, decimating the image at the outputs of the levels (indicated in Fig. 8.6 within the circle).

The quaternionic atomic function wavelet analysis from level $\alpha - 1$ to level α corresponds to the transformation of one quaternionic approximation to a new quaternionic approximation and three quaternionic differences, i.e.,

$$
\{A_{\alpha-1}^q\} \rightarrow \{A_\alpha^q, D_{\alpha,p}^q, p = 1, 2, 3\}.
\tag{8.29}
$$

Note that we do not use the idea of a mirror tree [1]. As a result, the quaternionic wavelet tree is a compact and economic processing structure that can be used for n-dimensional multi-resolution analysis.

The procedure for quaternionic wavelet multi-resolution analysis depicted partially in Fig. 8.6 is as follows:

- Convolve the 2D real signal at level n and convolve it with the scale and wavelet filters H_α^q and G_α^q along the rows of the 2D signal. The latter filters are the discrete versions of those filters given in (8.23).
- H_α^q and G_α^q are convolved with the columns of the previous responses of the filters H_α^q and G_α^q.
- Subsample the responses of these filters by a factor of two ($\downarrow 2$).
- The real part of the approximation at level j is taken as input at the next level $\alpha\alpha$. This process continues through all the levels $\alpha = 1, \ldots, n$, repeating steps 1→4.

8.7   Radon Transform of Functionals

The aim of this section is to put in context of the Radon transform (RT) theory the use of the QWT for the detection of contours of arbitrary shapes using either grey scale or color images. Some essential concepts are outlined briefly, however we give enough references of the literature for any reader interested to go into much more detail about these theoretic issues.

The Radon transform (RT), introduced by J. Radon in 1917, describes a function in terms of its (integral) projections [17]. The RT can be seen as the mapping from the function onto the projections of the Radon transform. The original formulation of the RT was given by

$$\mathscr{R}\{I\}(d,\phi) = \int_{\mathbb{R}} I(d\cos\phi - s\sin\phi, d\sin\phi + s\cos\phi)\,ds, \tag{8.30}$$

which represents the projections of the function I along the lines $c_l(d,\phi)$. The inverse of the RT corresponds to the back-reconstruction of the function from the projections. Furthermore, one can use the RT to detect the shape; for this purpose, one reformulates (8.30) to detect lines

$$\mathscr{R}\{I\}(d,\phi) = \int_{\mathbf{x}\text{ on }c_l(d,\phi)} I(x,y)\,dx\,dy$$
$$= \int_{\mathbb{R}} \delta(x\cos\phi + y\sin\phi - d)\,dx\,dy. \tag{8.31}$$

The generalization of the RT to detect arbitrary shape's contours $c_l(d,\phi)$ appears now to be straightforward, so let us consider the following formulas:

$$\mathscr{R}_{c(p)}\{I\}(\mathbf{p}) = \int_{\mathbf{x}\text{ on }c(p)} I(\mathbf{x})\,d\mathbf{x}$$
$$= \int_{\mathbb{R}^N} I(c(\mathbf{s};\mathbf{p}))\left\|\frac{\partial c(\mathbf{s})}{\partial \mathbf{s}}\right\|\,d\mathbf{s}$$
$$= \int_{\mathbb{R}^D} I(\mathbf{x})\delta(C(\mathbf{x};\mathbf{p}))\,d\mathbf{x}. \tag{8.32}$$

The first formulation is aimed at assigning votes at the point $\mathbf{p}$ of the Radon parameter space based on the integral of $I(\mathbf{x})$ for points lying on the shape's contour $c(\mathbf{p})$. The second formulation is the same following the absolute value of the gradient $\frac{\partial c(\mathbf{s})}{\partial \mathbf{s}}$ along the segment of contour $c(\mathbf{s})$, and the third votes for points along a contour described by the null constraint formulated in terms of a Dirac delta function; for more details, see [6]. The RT builds in the Radon parameter space a function $P(\mathbf{p})$ having peaks for those parameter vectors $\mathbf{p}$ for which the corresponding shape $c(\mathbf{p})$ is present in the image. As a result, the problem of shape detection has been simplified to a task of peak detection. The third formulation of the RT stresses the importance of the use of generalized functions. In fact, we can recognize the form of a linear integral (Fredholm) operator $\mathscr{L}_C$ with kernel C:

$$(\mathscr{L}_C I)(\mathbf{p}) = \int_{\mathbb{R}^D} C(\mathbf{p}, \mathbf{x}) I(\mathbf{x}) \, d\mathbf{x}. \tag{8.33}$$

In (8.32), the kernel C is of the form $C(\mathbf{p}, \mathbf{x}) = \delta(C(\mathbf{x}; \mathbf{p}))$. With respect to shape detection, the role of the operator $\mathscr{L}_C$ is to compute via the inner product the match between the image shape and a template C for a given parameter set $\mathbf{p}$. Note the close connection between the Radon transform and template matching. Since the matching criterion is the inner product between the template $T = \delta(C(\mathbf{x}; \mathbf{p}))$ and the image I, (8.32) can be rewritten with respect to a parameter vector $\mathbf{p}$ as a parameter response function $P(\mathbf{p})$ as follows:

$$P(\mathbf{p}) = \int_{\mathbb{R}^D} T(\mathbf{x}, \mathbf{p}) I(\mathbf{x}) \, d\mathbf{x}. \tag{8.34}$$

Although this technique can be used to detect gray-value blobs in I, usually this equation is applied to an edge/line map $E(\mathbf{x})$, which contains the contours of the shapes instead of being applied directly to the image $I(x, y)$. In this regard, (8.34) reads

$$P(\mathbf{p}) = \int_{(x,y,\varphi) \text{ on } c^{[\varphi]}(\mathbf{x};\mathbf{p})} I^{[\varphi]}(x, y, \varphi) \, dx \, dy \, d\varphi, \tag{8.35}$$

where φ indicates the edge orientation. Finally, if we use the quaternionic wavelet atomic function for curves parameterized in terms of the quaternionic parameters $c^{[\varphi(\phi,\theta)]}(s; \mathbf{p}) = (x(s; \mathbf{p}), y(s; \mathbf{p}), \varphi(\phi(s; \mathbf{p}), \theta(s; \mathbf{p})))$, the standard RD equation can be formulated using the 2D quaternionic phase concept as follows:

$$P(\mathbf{p}) = \int_{(x,y,\phi) \text{ on } c^{[\varphi(\phi,\theta)]}(\mathbf{x};\mathbf{p})} I^{[\varphi(\phi,\theta)]}(x, y, \varphi) \, dx \, dy \, d\varphi. \tag{8.36}$$

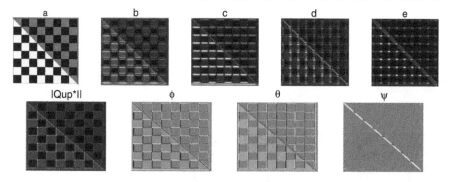

Fig. 8.7 Convolution of the test image with the *qup*. (*Upper row*, from *left* to *right*) Original image, after convolution follow: real-part; i-part; j-part; k-part; (*lower row*) magnitude; and the phases ϕ; θ; φ

In fact, (8.34) is equivalent to (8.30) of the RT of the $\mathbb{H}$-embedded Riesz transform $f_R(\mathbf{x}) = (i, j) f_r(\mathbf{x})$ of a 2D signal $f(\mathbf{x})$ given by the Hilbert transform $(h_1(t))$ of the RT of $f(\mathbf{x})$ according to

$$\mathscr{R}\{I_R\}(d, \theta) = (i, j)\mathbf{n}_\theta h_1(t) * \mathscr{R}\{f\}(t, \theta), \tag{8.37}$$

where the I_R is the Riesz transform of the image. This equation is the RT of the $\mathbb{H}$-embedded Riesz transform and is computed with respect to a line with orientation θ and Hesse distance d.

8.8   Applications of the Quaternion Atomic Function $Qup(x)$

This section presents applications of the $Qup(x)$ as a filter and for multi-resolution analysis. In certain levels of the multiresolution pyramid, one can compute the RT for extracting shapes.

8.8.1   Convolution with an Image

Figure 8.7 shows the original image and four resulting images after convolution with components of the filter. This figure illustrates how the Qup filter works in different directions, such as the vertical direction in Fig. 8.7.c, the horizontal in Fig. 8.7.d, and a combination of both for diagonal detection in Fig. 8.7.e. The direct convolution with the image is sensitive to the contrast of the image. Figure 8.7 also shows the amplitude (f) (real part) and the three phases (ϕ, θ, φ). The phase information is immune to changes of the contrast. This experiment clearly shows how the phase information can be used to localize and extract more information independently of the contrast of the image. Additionally, we show in Fig. 8.8 the application of a Qup using its quaternionic phases to detect the circle structure. Note how each phase detects edge phase changes around the circles.

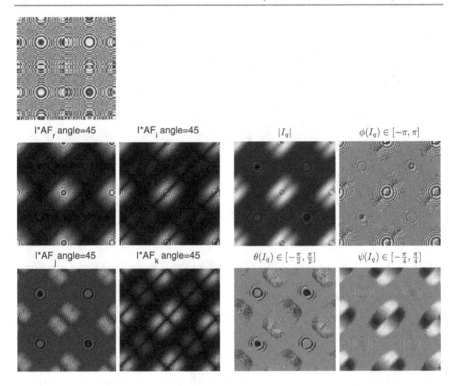

Fig. 8.8 (From the *left*) Test image filtered with a quaternionic wavelet atomic function elongated by $s_x = 0.3$, $s_y = 0.25$ and oriented at 45°: (*first group*) in space domain: magnitude and quaternionic components; (*second group*) in frequency domain: magnitude and quaternionic phases of the filtered image (ϕ, θ, φ)

8.8.2 Multi-resolution Analysis Using the Quaternion Wavelet Atomic Function

The *Qup* kernel was used as the mother wavelet in the multi-resolution pyramid. Figure 8.9 presents the three quaternionic phases at two scale levels of the pyramid. The bottom row shows the phases after thresholding to enhance the phase structure. You can see how vertical lines and crossing points are highlighted.

The *Qup* mother wavelet kernel was steered: elongation $s_x = 0.3$ and $s_y = 0.25$ and angles $\{0°, 22.5°, 45°, 77.25°, 90°\}$ through the multi-resolution pyramid. Figure 8.10 shows the detected structure.

8.8.3 Radon Transform for Circle Detection in Color Images Using the Quaternion Atomic Functions

First the circles image of Fig. 8.11 was filtered to round the sharp contours, then Gaussian noise was added to distort the circle contours, and salt and pepper noise was added to each of the r, g, and b color image components. The image was then

Scale=1

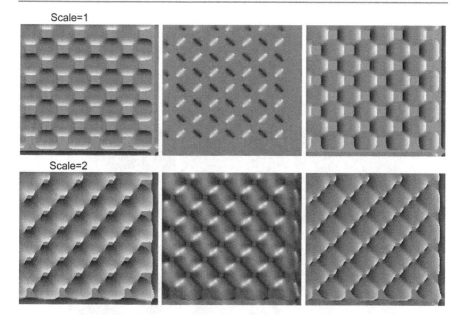

Scale=2

Fig. 8.9 (*Upper row*) Thresholded quaternionic phases (ϕ, θ, φ) at the first scale of resolution. (*Second row*) Thresholded quaternionic phases (ϕ, θ, φ) at the next scale

filtered by steering a quaternion wavelet atomic function, and at a certain level of the multi-resolution pyramid we applied the RD. Figure 8.11 shows how the cyan circle was extracted. For the extraction and the RT transform, we used as parameters the color and a combination of phase values.

Finally, one can improve performance of the multi-resolution approach by applying the RT from coarse to fine levels to identify possible shape contours in the upper level. Then the shapes are oversampled and overlapped with the next-lower level to improve the definition of the shapes. At the finest level, one gets the parameters of all well-supported contours.

8.9   Conclusion

This paper introduces the theory and some applications of the quaternion atomic function wavelet in image processing. We comment on the relevance of the atomic function with respect to the theory of monogenic signals. Since the information from the three phases is independent of illumination changes, algorithms using the quaternion atomic function wavelet can be less sensitive than those using metric tensors, which are affected by illumination changes. We also present the use of the quaternionic atomic function wavelet for multi-resolution analysis. Making use of the generalized Radon transform and images processed with the quaternion wavelet atomic function transform, we also detect shape contours in color images. We hope that this work will encourage computer scientists and practitioners to use quaternion

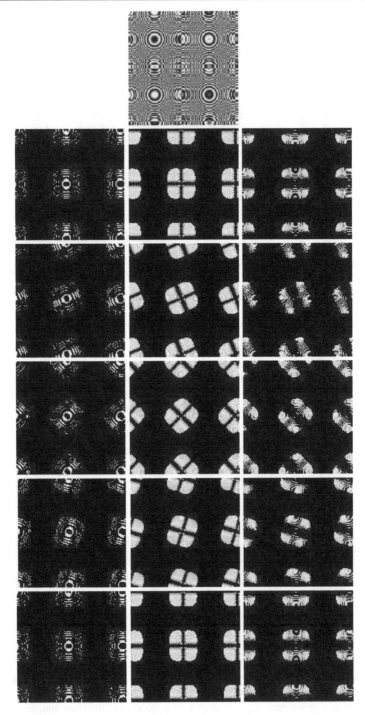

Fig. 8.10 (*Three columns*) Thresholded quaternionic phases (ϕ, θ, φ), filter elongation $s_x = 0.3$, $s_y = 0.25$, steering angles for each row are: $0°, 22.5°, 45°, 77.25°, 90°$

Fig. 8.11 (*Upper row*) Image of two circles in color and gray scales; (*the next four images*) the magnitude $|I_q|$ and the quaternionic phases of the filtered image (ϕ, θ, φ) and (*the next four images*) the quaternion components after the convolution $I \star AF_r$, $I \star AF_i$, $I \star AF_j$ and $I \star AF_k$; (*lower row*) gray scale image, its log FFT and the filtered contour of the cyan circle (larger circle). Color version of this figure online

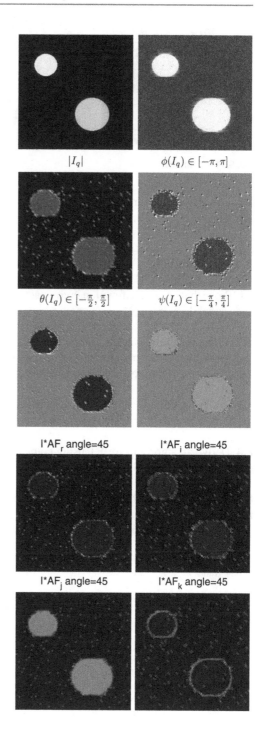

QWAT   with   angle=45 : Image (grey value), LogFFT, extracted color circle

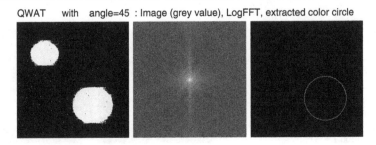

Fig. 8.11 (continued)

wavelet atomic functions to tackle various problems in image processing and scene analysis.

8.10   Exercises

8.1 Prove that an atomic function is compact. Discuss its properties in the frequency and spatial domains.

8.2 Compare the quaternion atomic function with the monogenic signal (Poisson kernel and Riesz transform). Can the quaternion atomic function be used for smoothing the scale space and steered wavelet image processing? Justify.

8.3 Derive a sort of quaternion Harris detector for color image processing using the quaternion atomic filter which only detects corners of a selected color.

8.4 Derive an algorithm for RT using color images and quaternion atomic filter to detect lines, circles and ellipsoids of one particular color.

8.5 Apply a conformal transformation to map the color image plane to the sphere, then using a quaternion atomic filter derive a line and circle detector on the sphere which detects with respect to only a particular color.

Acknowledgements   We want to thank the financial support of the SEP/CONACYT - 2007-1 82084 grant.

References

1. Bayro-Corrochano, E.: Geometric Computing for Wavelet Transforms, Robot Vision, Learning, Control and Action. Springer, Berlin (2010)
2. Brackx, F., Delanghe, R., Sommen, F.: Clifford Analysis. Pitman, Boston (1982)
3. Bülow, T.: Hypercomplex spectral signal representations for the processing and analysis of images. PhD thesis, Christian-Albert, Kiel University (1999)
4. Delanghe, R.: Clifford analysis: history and perspective. In: Computational Methods and Function Theory, vol. 1, pp. 107–153 (2001)

5. Felsberg, M., Sommer, G.: The monogenic signal. IEEE Trans. Signal Process. **49**(2), 3136–3144 (2007)
6. Gelf'and, M.I., Graev, M.I., Vilenkin, N.Ya.: Generalized Functions, vol. 5, Integral Geometry and Representation Theory. Academic Press, San Diego (1966)
7. Guyaev, Yu.V., Kravchenko, V.F.: A new class of WA-systems of Kravchenko–Rvachev functions. Moscow Dokl. Math. **75**(2), 325–332 (2007)
8. Hamilton, W.R.: Elements of Quaternions. Chelsea, New York (1969). Longmans Green, London (1866)
9. Gorshkov, A., Kravchenko, V.F., Rvachev, V.A.: Estimation of the discrete derivative of a signal on the basis of atomic functions. Izmer. Tekh. **1**(8), 10 (1992)
10. Kingsbury, N.: Image processing with complex wavelets. Philos. Trans. R. Soc. Lond. A **357**, 2543–2560 (1999)
11. Kovesi, P.: Invariant measures of images features from phase information. PhD thesis, University of Western Australia (1996)
12. Kravchenko, V.F., Perez-Meana, H.M., Ponomaryov, V.I.: Adaptive Digital Processing of Multidimensional Signals with Applications. Fizmatlit, Moscow (2009)
13. Magarey, J.F.A., Kingsbury, N.G.: Motion estimation using a complex-valued wavelet transform. IEEE Trans. Image Process. **6**, 549–565 (1998)
14. Mallat, S.: A theory for multiresolution signal decomposition: the wavelet representation. IEEE Trans. Pattern Anal. Mach. Intell. **11**(7), 674–693 (1989)
15. Mitrea, M.: Clifford Waveletes, Singular Integrals and Hardy Spaces. Lecture Notes in Mathematics, vol. 1575. Springer, Berlin (1994)
16. Nabighian, M.N.: Toward a three-dimensional automatic interpretation of potential field data via generalized Hilbert transforms: fundamental relations. Geophysics **49**(6), 780–786 (1984)
17. Radon, J.: Über die Bestimmung von Funktionen durch ihre Integralwerte längs gewisser Mannigfaltigkeiten. Ber. Sächs. Akad. Wiss., Leipzig. Math.-Phys. Kl. **69**, 262–277 (1917)
18. Rvachev, V.A.: Compactly supported solution of functional–differential equations and their applications. Russ. Math. Surv. **45**(1), 87–120 (1990)
19. Pan, H.-P.: Uniform full information image matching complex conjugate wavelet pyramids. In: XVIII ISPRS Congress, vol. XXXI, Vienna, July 1996
20. Kolodyazhnya, V.M., Rvachev, V.A.: Atomic functions: generalization to the multivariable case and promising applications. Cybern. Syst. Anal. **43**(6), 893–911 (2007)

Color Object Recognition Based on a Clifford Fourier Transform

9

Jose Mennesson, Christophe Saint-Jean, and Laurent Mascarilla

Abstract

The aim of this chapter is to propose two different approaches for color object recognition, both using the recently defined color Clifford Fourier transform. The first one deals with so-called Generalized Fourier Descriptors, the definition of which relies on plane motion group actions. The proposed color extension leads to more compact descriptors, with lower complexity and better recognition rates, than the already existing descriptors based on the processing of the r, g and b channels separately. The second approach concerns color phase correlation for color images. The idea here is to generalize in the Clifford framework the usual means of measuring correlation from the well-known shift theorem. Both methods necessitate to choose a 2-blade B of $\mathbb{R}_4$ which corresponds to an analysis plane in the color space. The relevance of proposed methods for classification purposes is discussed on several color image databases. In particular, the influence of parameter B is studied regarding the type of images.

9.1 Introduction

Most of already existing works on image recognition make use of discriminative and invariant descriptors. Among them, moment-based descriptors [7] such as Hu invariants, Legendre moments or Zernike moments are well known. Beside these approaches, SIFT (Scale-Invariant Feature Transform) descriptors are a popular choice

J. Mennesson (✉) · C. Saint-Jean · L. Mascarilla
Laboratory of Mathematics, Images and Applications, University of La Rochelle,
La Rochelle, France
e-mail: jose.mennesson@univ-lr.fr

C. Saint-Jean
e-mail: christophe.saint-jean@univ-lr.fr

L. Mascarilla
e-mail: laurent.mascarilla@univ-lr.fr

L. Dorst, J. Lasenby (eds.), *Guide to Geometric Algebra in Practice*,
DOI 10.1007/978-0-85729-811-9_9, © Springer-Verlag London Limited 2011

giving very good results [11]. An alternative to these methods is to define descriptors in the frequency domain. In this framework, our chapter concerns two extensions of existing methods based on a Fourier transform. Clearly, Fourier coefficients do not respect the classical invariances (translation, rotation and scale) and must be processed to obtain invariant descriptors. This chapter proposes an extension of a recent advance concerning invariant Generalized Fourier Descriptors (*GFD*) defined by F. Smach et al. [15]. The extension of these descriptors to color images is generally based on a marginal processing of the three channels (r, g, b). Then, descriptors extracted from each channel are concatenated to form the description vector. In order to avoid this marginal processing, our proposal is to extract descriptors from a color Clifford Fourier transform as defined by Batard et al. [1]. A second proposal is the extension of the classical color phase correlation by means of the same Fourier transform.

9.2   A Clifford Fourier Transform for Color Image Processing

As relating to color image processing, the usual Fourier transform corresponds in fact to three two-dimensional Fourier transforms applied on each color channel, that is a marginal processing. To emphasize the role of color, several authors have proposed to embed the color space in a more pertinent and meaningful geometric space such as the space of quaternions. For instance, Ell and Sangwine [6] propose a luminance/chrominance Fourier analysis replacing the imaginary complex i by the quaternion $\mu = \frac{i+j+k}{\sqrt{3}}$ corresponding to the gray-level axis. It already appears in this work that one has to focus on an analysis direction (here given by μ).

Recently, Batard et al. [1] have defined a Fourier transform for $L^2(\mathbb{R}^m; \mathbb{R}^n)$ functions. This one is mathematically rigorous and clarifies relations between the Fourier transform and the action of the translation group through an action of a spinor group. They show that the previously proposed generalizations for color images (i.e. $n = 3$, the number of color channels) are particular cases of their definition. In this chapter, only the particular case $m = 2$ and $n = 3$ is considered and described briefly in the following. Firstly, in the equation of the classical 2D Fourier transform,

$$\widehat{f}(u_1, u_2) = \int_{\mathbb{R}^2} f(x_1, x_2) e^{-i(u_1 x_1 + u_2 x_2)} \, dx_1 \, dx_2 \qquad (9.1)$$

the term $e^{-i(u_1 x_1 + u_2 x_2)}$ $(= e^{-i\langle \mathbf{u}, \mathbf{x} \rangle}$ with $\mathbf{u} = (u_1, u_2)$ and $\mathbf{x} = (x_1, x_2))$ rotates $f(x_1, x_2)$ in the complex plane $\mathbb{C}$. From a mathematical point of view, it corresponds to the action of the group S^1 on $\mathbb{C}$ which can be identified as the group action of $SO(2)$ on $\mathbb{R}^2$. In order to generalize this principle to color images, one has to consider the action of the matrix group $SO(3)$ on $\mathbb{R}^3$. As described in [1], a general and elegant expression may be written if the function corresponding to the color image is embedded in the Clifford algebra $\mathbb{R}_4$:

$$f(x, y) = r(x, y)\mathbf{e_1} + g(x, y)\mathbf{e_2} + b(x, y)\mathbf{e_3} + 0\mathbf{e_4}$$

Within this framework, the rotation of a vector $\mathbf{v}$ by the angle $-\theta$, in the plane generated by a unitary 2-blade B is given by the action $s^{-1}\mathbf{v}s$ of a spinor s and can be written as

$$\mathbf{v} \to s^{-1}\mathbf{v}s = e^{\frac{\theta}{2}B}\mathbf{v}e^{-\frac{\theta}{2}B}$$

For this type of functions, the following Fourier transform is considered:

$$\widehat{f_B}(\mathbf{u}) = \int_{\mathbb{R}^2} e^{\frac{\langle \mathbf{u},\mathbf{x}\rangle}{2}I_4 B}e^{\frac{\langle \mathbf{u},\mathbf{x}\rangle}{2}B}f(\mathbf{x})e^{-\frac{\langle \mathbf{u},\mathbf{x}\rangle}{2}B}e^{-\frac{\langle \mathbf{u},\mathbf{x}\rangle}{2}I_4 B}\,d\mathbf{x} \tag{9.2}$$

where I_4 is the pseudo-scalar of $\mathbb{R}_4$, $I_4 B$ is a unitary 2-blade orthogonal to B, $\mathbf{u} = (u_1, u_2)$ and $\mathbf{x} = (x_1, x_2)$. From the geometric point of view, two independent rotations in orthogonal planes are acting on $f(\mathbf{x})$. As these rotations are of the same angle, the chosen Fourier transform involves isoclinic rotations of f in $\mathbb{R}^4$ [10]. Let us emphasize that considering more general rotations in $\mathbb{R}^4$ leads to other definitions of the Fourier transform and yields additional parameters which are hard to set in practice.

9.2.1 The Shift Theorem for the Clifford Fourier Transform

The color phase correlation subsequently proposed relies on the Fourier shift theorem which states that a translation in the spatial domain induces a phase shift in the frequency domain. By construction, this property still holds for our transform and takes the following form.

Theorem 9.1 *Let $f, g \in L^2(\mathbb{R}^2, \mathbb{R}^3)$, B be a unit 2-blade in $\mathbb{R}_4$, and $\mathbf{\Delta} = (\Delta_1, \Delta_2)$ the vector containing the translation parameters.*
If $g(\mathbf{x}) = f(\mathbf{x} + \mathbf{\Delta})$, then

$$\widehat{g_B}(\mathbf{u}) = e^{-\frac{\langle \mathbf{u},\mathbf{\Delta}\rangle}{2}I_4 B}e^{-\frac{\langle \mathbf{u},\mathbf{\Delta}\rangle}{2}B}\widehat{f_B}(\mathbf{u})e^{\frac{\langle \mathbf{u},\mathbf{\Delta}\rangle}{2}B}e^{\frac{\langle \mathbf{u},\mathbf{\Delta}\rangle}{2}I_4 B} \tag{9.3}$$

Proof The proof is essentially based on the fact that rotations in orthogonal planes can commute. Let $\mathbf{x_\Delta} = \mathbf{x} + \mathbf{\Delta}$.

$$\widehat{g_B}(\mathbf{u})$$
$$= \int_{\mathbb{R}^2} e^{\frac{\langle \mathbf{u},\mathbf{x}\rangle}{2}(B+I_4 B)}f(\mathbf{x}+\mathbf{\Delta})e^{-\frac{\langle \mathbf{u},\mathbf{x}\rangle}{2}(B+I_4 B)}\,d\mathbf{x}$$
$$= \int_{\mathbb{R}^2} e^{\frac{\langle \mathbf{u},(\mathbf{x_\Delta}-\mathbf{\Delta})\rangle}{2}(B+I_4 B)}f(\mathbf{x_\Delta})e^{-\frac{\langle \mathbf{u},(\mathbf{x_\Delta}-\mathbf{\Delta})\rangle}{2}(B+I_4 B)}\,d\mathbf{x_\Delta}$$
$$= \int_{\mathbb{R}^2} e^{-\frac{\langle \mathbf{u},\mathbf{\Delta}\rangle}{2}(B+I_4 B)}e^{\frac{\langle \mathbf{u},\mathbf{x_\Delta}\rangle}{2}(B+I_4 B)}f(\mathbf{x_\Delta})e^{-\frac{\langle \mathbf{u},\mathbf{x_\Delta}\rangle}{2}(B+I_4 B)}e^{\frac{\langle \mathbf{u},\mathbf{\Delta}\rangle}{2}(B+I_4 B)}\,d\mathbf{x_\Delta}$$

$$= e^{-\frac{\langle \mathbf{u}, \Delta \rangle}{2}(B+I_4B)} \left(\int_{\mathbb{R}^2} e^{\frac{\langle \mathbf{u}, \mathbf{x}_\Delta \rangle}{2}(B+I_4B)} f(\mathbf{x}_\Delta) \right.$$

$$\left. \times e^{-\frac{\langle \mathbf{u}, \mathbf{x}_\Delta \rangle}{2}(B+I_4B)} d\mathbf{x}_\Delta \right) e^{\frac{\langle \mathbf{u}, \Delta \rangle}{2}(B+I_4B)}$$

$$= e^{-\frac{\langle \mathbf{u}, \Delta \rangle}{2}(B+I_4B)} \widehat{f_B}(\mathbf{u}) e^{\frac{\langle \mathbf{u}, \Delta \rangle}{2}(B+I_4B)} \tag{9.4}$$

$\square$

Even if (9.4) is more compact than (9.3), this formulation emphasizes that two independent rotations apply. Simpler equations can be obtained using the decomposition of f as the sum of its parallel projection $f_{\|B}$ on the plane defined from B and its perpendicular projection $f_{\perp B}$ on the plane defined from $I_4 B$ (see later for details on the decomposition). Skipping some technical details, (9.2) can be rewritten following this decomposition as

$$\widehat{f_B}(\mathbf{u}) = \widehat{f_{\|B}}(\mathbf{u}) + \widehat{f_{\perp B}}(\mathbf{u}) \tag{9.5}$$

where

$$\widehat{f_{\|B}}(\mathbf{u}) = \int_{\mathbb{R}^2} e^{\frac{\langle \mathbf{u}, \mathbf{x} \rangle}{2}B} f_{\|B}(\mathbf{x}) e^{-\frac{\langle \mathbf{u}, \mathbf{x} \rangle}{2}B} d\mathbf{x} = \int_{\mathbb{R}^2} f_{\|B}(\mathbf{x}) e^{-\langle \mathbf{u}, \mathbf{x} \rangle B} d\mathbf{x} \tag{9.6}$$

$$\widehat{f_{\perp B}}(\mathbf{u}) = \int_{\mathbb{R}^2} e^{\frac{\langle \mathbf{u}, \mathbf{x} \rangle}{2}I_4B} f_{\perp B}(\mathbf{x}) e^{-\frac{\langle \mathbf{u}, \mathbf{x} \rangle}{2}I_4B} d\mathbf{x} = \int_{\mathbb{R}^2} f_{\perp B}(\mathbf{x}) e^{-\langle \mathbf{u}, \mathbf{x} \rangle I_4B} d\mathbf{x} \tag{9.7}$$

Later on, (9.6) and (9.7) will provide a practical and efficient way to implement our transform. According the same decomposition, Lemma 9.1 becomes:

Theorem 9.2 *Let $f, g \in L^2(\mathbb{R}^2, \mathbb{R}^3)$, and B be a unit 2-blade in $\mathbb{R}_4$.*
If $g(\mathbf{x}) = f(\mathbf{x} + \Delta)$, then

$$\widehat{g_{\|B}}(\mathbf{u}) = e^{-\frac{\langle \mathbf{u}, \Delta \rangle}{2}B} \widehat{f_{\|B}}(\mathbf{u}) e^{\frac{\langle \mathbf{u}, \Delta \rangle}{2}B} = \widehat{f_{\|B}}(\mathbf{u}) e^{\langle \mathbf{u}, \Delta \rangle B} \tag{9.8}$$

$$\widehat{g_{\perp B}}(\mathbf{u}) = e^{-\frac{\langle \mathbf{u}, \Delta \rangle}{2}I_4B} \widehat{f_{\perp B}}(\mathbf{u}) e^{\frac{\langle \mathbf{u}, \Delta \rangle}{2}I_4B} = \widehat{f_{\perp B}}(\mathbf{u}) e^{\langle \mathbf{u}, \Delta \rangle I_4B} \tag{9.9}$$

A unit 2-blade B can be obtained from the geometric product of two unit orthogonal vectors as $B = \mathbf{c} \wedge \mathbf{e}_4$ or $B = \mathbf{c}_1 \wedge \mathbf{c}_2$ (where $\mathbf{c}$, $\mathbf{c}_1$ and $\mathbf{c}_2$ are RGB colors). These two settings appear to be analogous up to a sign change since the dualization $I_4 B$ of B gives also a 2-blade of the other form. In the following, only the choice $B = \mathbf{c} \wedge \mathbf{e}_4$ will be considered in the experiments.

9.2.2 Computation of the Clifford Fourier Transform

The Clifford Fourier transform can be efficiently computed using two complex FFTs. Whereas $\{\mathbf{c}, \mathbf{e}_4\}$ is a trivial basis for the plane given by B, an orthonormal basis $\{\mathbf{v}, \mathbf{w}\}$ for the plane generated by $I_4 B$ has to be constructed. A possible solu-

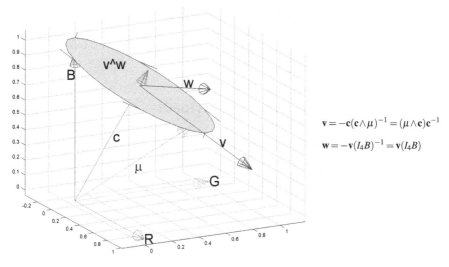

$$\mathbf{v} = -\mathbf{c}(\mathbf{c} \wedge \mu)^{-1} = (\mu \wedge \mathbf{c})\mathbf{c}^{-1}$$

$$\mathbf{w} = -\mathbf{v}(I_4 B)^{-1} = \mathbf{v}(I_4 B)$$

Fig. 9.1 Illustration of the basis $\{\mathbf{c}, \mathbf{e}_4, \mathbf{v}, \mathbf{w}\}$ of I_4 using GABLE [4]

tion is to choose a unit vector μ with no $\mathbf{e}_4$ component and different from $\mathbf{c}$.[1] Vector $\mathbf{v}$ is taken as the rejection of μ on $\mathbf{c}$ and $\mathbf{w}$ as an orthogonal vector to $\mathbf{v}$ in subspace represented by blade $I_4 B$ (see Fig. 9.1). Then, the function f can decomposed as

$$
\begin{aligned}
f(\mathbf{x}) &= f_{\|B}(\mathbf{x}) + f_{\perp B}(\mathbf{x}) \\
&= \mathbf{c}\big[\big(f(\mathbf{x}) \cdot \mathbf{c} \big) + \big(f(\mathbf{x}) \cdot \mathbf{c}B \big)B \big] + \mathbf{v}\big[\big(f(\mathbf{x}) \cdot \mathbf{v} \big) \\
&\quad + \big(f(\mathbf{x}) \cdot \mathbf{v}I_4 B \big)I_4 B \big]
\end{aligned}
\tag{9.10}
$$

Each of the two square brackets can identified as a complex number since $BB = I_4 B I_4 B = -1$. The computation of $\widehat{f_B}$ is now reduced to the computation of two usual Fourier transforms of a real function and of a complex function.

Depending on the intended application, it is not always necessary to reconstruct $\widehat{f_B}$ from $\widehat{f_{\|B}}$ and $\widehat{f_{\perp B}}$.[2] If so, the following properties of the vectorial function $\widehat{f_B}$

$$
\big(\widehat{f_{\|B}(\mathbf{x})} \big)_B = \mathbf{c}\big[\big(\widehat{f_B}(\mathbf{x}) \cdot \mathbf{c} \big) + \big(\widehat{f_B}(\mathbf{x}) \cdot \mathbf{c}B \big)B \big]
\tag{9.11}
$$

$$
\big(\widehat{f_{\perp B}(\mathbf{x})} \big)_B = \mathbf{v}\big[\big(\widehat{f_B}(\mathbf{x}) \cdot \mathbf{v} \big) + \big(\widehat{f_B}(\mathbf{x}) \cdot \mathbf{v}I_4 B \big)I_4 B \big]
\tag{9.12}
$$

give a set of four linear equations with four unknowns.

---

[1] A typical setting for μ is $(\mathbf{e}_1 + \mathbf{e}_2 + \mathbf{e}_3)/\sqrt{3}$, which corresponds to select the achromatic axis.

[2] More precisely, the two functions in the square brackets in (9.10).

9.3   Generalized Color Fourier Descriptors

In this section, we propose an extension of the Generalized Fourier Descriptors of Smach et al. (initially dedicated to grayscale images) to color images.

9.3.1   Generalized Fourier Descriptors (GFD)

Generalized Fourier descriptors introduced by [15] are defined from the group action of M_2. This group is composed of translations and rotations on the plane. Two kinds of descriptors have been defined:
- "Spectral densities"-type invariants:

$$I_1^r(f) = \int_0^{2\pi} \left| \widehat{f}(r, \theta) \right|^2 d\theta$$

- "Shift of phases"-type invariants:

$$I^{\xi_1, \xi_2}(f) = \int_0^{2\pi} \widehat{f}\left(R_\theta(\xi_1 + \xi_2)\right) \overline{\widehat{f}\left(R_\theta(\xi_1)\right) \widehat{f}\left(R_\theta(\xi_2)\right)} \, d\theta$$

where f is the image, $\widehat{f}(r, \theta)$ is the Fourier transform expressed in polar coordinates in the frequency plane, ξ_1 and ξ_2 are variables of the frequency plane, and R_θ is a rotation of angle θ. It must be emphasized that, by construction, I_1^r and I^{ξ_1, ξ_2} are strictly invariant in $\mathbb{R}^2$ with respect to the action of M_2.

Then, the descriptor vector for the first family of invariants, namely I_1^r, is defined as follows:

$$GFD1(f) = \left\{ I_1^0(f) = |\widehat{f}(0,0)|^2, \frac{I_1^1(f)}{I_1^0(f)}, \ldots, \frac{I_1^m(f)}{I_1^0(f)} \right\}$$

where m is the number of computed descriptors. In the same way, we define the GFD2 descriptor vector from the second family of invariants I^{ξ_1, ξ_2}.

9.3.2   Generalized Color Fourier Descriptors (GCFD)

In order to deal with color images, a commonly used approach consists in computing descriptors on each color channel separately. Then, they are concatenated into a unique vector (e.g. [15]). This method implies three FFTs and three sets of descriptors. However, this marginal processing induces a loss of colorimetric information that can be avoided by using the color Clifford Fourier transform.

Equation (9.5) shows that the Clifford Fourier transform can be decomposed into two parts. So, two descriptor vectors are defined: $GCFD1_{\|B}$ and $GCFD1_{\perp B}$, each of them implying two complex FFTs. According to the definition of $f_{\|B}$:

$$GCFD1_{\|B}(f) = \left\{ I_{\|B}^{0}(f) = |\widehat{f_{\|B}}(0,0)|^2, \frac{I_{\|B}^{1}(f)}{I_{\|B}^{0}(f)}, \dots, \frac{I_{\|B}^{m}(f)}{I_{\|B}^{0}(f)} \right\}$$

where $I_{\|B}^{r}(f) = \int_{0}^{2\pi} |\widehat{f_{\|B}}(r,\theta)|^2 \, d\theta$, and m is the number of computed descriptors. Similarly, $GCFD1_{\perp B}$ is defined thanks to $\widehat{f_{\perp B}}$. Finally, the descriptor vector length is $2 \times m$:

$$GCFD1_B(f) = \{GCFD1_{\|B}(f), GCFD1_{\perp B}(f)\}$$

The same construction based on I^{ξ_1,ξ_2} leads to $GCFD2_B$.

9.4 Color Phase Correlation

In the literature, phase correlation [14] is a well-established method that is used for a lot of applications such as image recognition, video stabilization, motion estimation, stereo disparity analysis, vector flow analysis [5]. As it is based on a direct application of the Fourier shift theorem, its definition depends on the chosen Fourier transform. Before presenting what can be a phase correlation method for color images, we recall now the principle of this method for grayscale images.

9.4.1 Phase Correlation for Grayscale Images

Let f and g be two grayscale images which are spatial shifted version of each one another. According to the Fourier shift theorem,

$$\widehat{g}(\mathbf{u}) = \widehat{f}(\mathbf{u}) \, e^{i\langle \mathbf{u}, \Delta \rangle} \tag{9.13}$$

Ideally, it is possible to extract the phase shift between $\widehat{f}$ and $\widehat{g}$ through the computation of their cross-power spectrum

$$\begin{aligned} R(\mathbf{u}) &= \frac{\widehat{f}(\mathbf{u}) \, \overline{\widehat{g}(\mathbf{u})}}{|\widehat{f}(\mathbf{u})\widehat{g}(\mathbf{u})|} = \frac{\widehat{f}(\mathbf{u}) \overline{\widehat{f}(\mathbf{u})} e^{-i\langle \mathbf{u}, \Delta \rangle}}{|\widehat{f}(\mathbf{u}) \overline{\widehat{f}(\mathbf{u})} \, e^{-i\langle \mathbf{u}, \Delta \rangle}|} \\ &= \frac{|\widehat{f}(\mathbf{u})|^2 e^{-i\langle \mathbf{u}, \Delta \rangle}}{||\widehat{f}(\mathbf{u})|^2 e^{-i\langle \mathbf{u}, \Delta \rangle}|} = \frac{e^{-i\langle \mathbf{u}, \Delta \rangle}}{|e^{-i\langle \mathbf{u}, \Delta \rangle}|} = e^{-i\langle \mathbf{u}, \Delta \rangle} \end{aligned} \tag{9.14}$$

where the operator $^-$ is the usual complex conjugate. Ideally again, the exact translation $\Delta = (\Delta_1, \Delta_2)$ can be obtained by taking the inverse Fourier transform of $R(\mathbf{u})$:

$$r(\mathbf{x}) = \check{R}(\mathbf{u}) = \delta_{-\Delta} \tag{9.15}$$

where δ is the Dirac distribution.

In practice, the best estimated translation and correlation score ρ are given by

$$\mathbf{\Delta} = -\operatorname*{argmax}_{\mathbf{x}}\left(\left|r(\mathbf{x})\right|\right), \qquad \rho = \max_{\mathbf{x}}\left(\left|r(\mathbf{x})\right|\right)$$

The coefficient ρ should be equal to 1 when g is a translated version of f, and it could be used as a similarity index between images in a recognition process. Note that the phase correlation method is invariant under translations but not under rotations. The invariance under rotations can be achieved by converting images in log-polar domain, but it will not be discussed here.

9.4.2 Phase Correlation for Color Images

Phase correlation for color images is much more difficult than for grayscale images. A first tentative approach is to work directly on the relation between $\widehat{f_B}$ and $\widehat{g_B}$. According to (9.3) and for any unit 2-blade B, $\widehat{g_B}$ is ideally obtained by an isoclinic rotation of $\widehat{f_B}$. After some calculations similar to those of (9.14), it should be possible to obtained this rotation as a spinor represented by a multivector containing non vectorial terms. Unfortunately, definitions of the Fourier transform and Fourier inverse transform are not yet available for general multivectorial functions. So, an easier approach is to use the decompositions of $\widehat{f_B}$ and $\widehat{g_B}$ with respect to B.

According to (9.8) and (9.9), the phase correlation now relies on the detection of *simultaneous Dirac at the same location*:

$$R_{\parallel B}(\mathbf{u}) = \frac{\widehat{f_{\parallel B}}(\mathbf{u})\,\overline{\widehat{g_{\parallel B}}(\mathbf{u})}}{|\widehat{f_{\parallel B}}(\mathbf{u})\widehat{g_{\parallel B}}(\mathbf{u})|} = e^{-\langle \mathbf{u},\mathbf{\Delta}\rangle B} \quad\Rightarrow\quad r_{\parallel B}(\mathbf{x}) = \delta_{-\mathbf{\Delta}} \qquad (9.16)$$

$$R_{\perp B}(\mathbf{u}) = \frac{\widehat{f_{\perp B}}(\mathbf{u})\overline{\widehat{g_{\perp B}}(\mathbf{u})}}{|\widehat{f_{\perp B}}(\mathbf{u})\widehat{g_{\perp B}}(\mathbf{u})|} = e^{-\langle \mathbf{u},\mathbf{\Delta}\rangle I_4 B} \quad\Rightarrow\quad r_{\perp B}(\mathbf{x}) = \delta_{-\mathbf{\Delta}} \qquad (9.17)$$

In practice, one has to cope with the aggregation of $r_{\parallel B}$ and $r_{\perp B}$. The experimental section, Sect. 9.5, gives some results for different aggregation criteria. The whole process is illustrated in Fig. 9.2.

There are many rotations in $\mathbb{R}^4$ which map $\widehat{f_B}(\mathbf{u})$ to $\widehat{g_B}(\mathbf{u})$. Among these, one can choose the unique one that leaves invariant the plane generated by $\widehat{f_B}(\mathbf{u})$ and $\widehat{g_B}(\mathbf{u})$. This leads to the spinor $\tau(\mathbf{u})$ such that $g_B{}^\sharp(\mathbf{u}) = \tau(\mathbf{u}) f_B{}^\sharp(\mathbf{u})\tau^{-1}(\mathbf{u})$:

$$\tau_B(\mathbf{u}) = \exp\left[\frac{\theta(\mathbf{u})}{2} \frac{g_B{}^\sharp(\mathbf{u}) \wedge f_B{}^\sharp(\mathbf{u})}{|g_B{}^\sharp(\mathbf{u}) \wedge f_B{}^\sharp(\mathbf{u})|}\right] = \frac{1 + g_B{}^\sharp(\mathbf{u}) f_B{}^\sharp(\mathbf{u})}{\sqrt{2(1 + g_B{}^\sharp(\mathbf{u}) \cdot f_B{}^\sharp(\mathbf{u}))}} \qquad (9.18)$$

where $f_B{}^\sharp(\mathbf{u}) = \widehat{f_B}(\mathbf{u})/|\widehat{f_B}(\mathbf{u})|$ and $g_B{}^\sharp(\mathbf{u}) = \widehat{g_B}(\mathbf{u})/|\widehat{g_B}(\mathbf{u})|$. This rotation is classified as a simple rotation by Lounesto [10]. Such an approach deviates from the conditions of the shift theorem by relaxing the constraint on the type of rotation between vectors $\widehat{f_B}$ and $\widehat{g_B}$. Here again, the inverse Fourier transform of $\tau(\mathbf{u})$ is

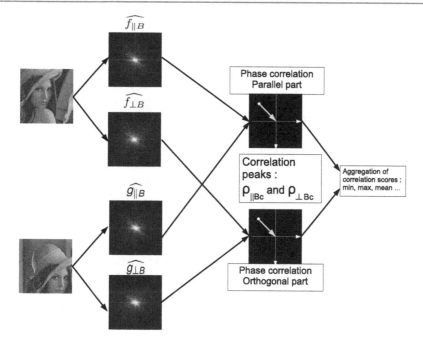

Fig. 9.2 Image similarity as a score aggregation

not accessible. However, it is possible to neglect the bivectorial part of the spinor by transforming it to a constant bivector identifiable with complex imaginary i. A correlation score ρ_B then can be built on $\theta(\mathbf{u})$:

$$R_B(\mathbf{u}) = e^{i\theta(\mathbf{u})} \quad \Rightarrow \quad \rho_B = \max_{\mathbf{x}}\left(|r_B(\mathbf{x})|\right) \tag{9.19}$$

where $r_B(\mathbf{x}) = \check{R}_B(\mathbf{u})$. An alternative formulation of such a criterion is given by the cosine between $\widehat{f_B}$ and $\widehat{g_B}$:

$$\Re\left(R_B(\mathbf{u})\right) = \cos\left(\widehat{f_B}(\mathbf{u}), \widehat{g_B}(\mathbf{u})\right) = \frac{\widehat{f_B}(\mathbf{u}) * \widetilde{\widehat{g_B}}(\mathbf{u})}{|\widehat{f_B}(\mathbf{u})||\widehat{g_B}(\mathbf{u})|} \tag{9.20}$$

where $\widetilde{}$ and $*$ denote the reverse operator and the Hestenes scalar product [8].

9.5   Experiments

Our goal is to evaluate if the proposed descriptors are good at classifying images. More precisely, the unique label of the request image is predicted from an entire set of labeled images. Both synthetic and standard image databases are considered, and the choice of the 2-blade is discussed.

9.5.1   Image Database

The databases used in this section are COIL-100, color FERET and, to check robustness again noise, a noisy version of this last dataset.

- *COIL-100 (Columbia Object Image Library) database* [12] is composed of 7200 color images of size 128×128 of 100 different objects. Each picture has been taken with a black background and 72 different angles of view. This database, used in similar works [15], can be qualified as "easy" from an image classification context as every image background is removed.
- *Color FERET database* [13] is composed of face images of 1408 different persons, taken from different angles of view. In our tests, a set of 2992 images containing 272 persons equally represented by 11 pictures is selected, and the size of images is reduced to 128×128. This database is more difficult than the first one due to background illumination changes.
- *Noisy color FERET database* is derived from the color FERET database, but Gaussian noise is added to each color plane of the images. The parameter θ is fixed to 0.23, which is the maximum noise level used in [15].

9.5.2   Descriptors Extraction

Regarding the *GFD*, 64 descriptors are extracted for each color channel. For *GFD*1, it consists of 64 values of radius r in I_1^r, and *GFD*2 is built from equally spaced values of ξ_2 in its polar domain $[0, 2\pi] \times [1, 8]$ and ξ_1 set to $(0, 1)$. As it is argued in [15], such ξ_1 and ξ_2 choices allow us to take into account low frequencies of the image, i.e. the shape. For *GCFD*1 and *GCFD*2, the length of the descriptor vectors are 64×2 (parallel and orthogonal part of the Clifford Fourier transform). Regarding the phase correlation, one score corresponding to the correlation peak is extracted for each image pair. Two cases are considered: either phase correlation is computed from parallel and orthogonal parts of the Clifford Fourier transform, and the two correlation scores, $\rho_{\parallel B}$ and $\rho_{\perp B}$, are aggregated to obtain a single score, or the correlation ρ_B is computed from reconstructed Fourier transform.

9.5.3   Classification

The classification step is performed using a standard SVM (Support Vector Machine) [2]. As *GFD* and *GCFD* are vector descriptors, they can directly be used as input for such a kernel based classifier. In this chapter, an RBF kernel is selected, and the value of the two parameters θ and C are empirically determined to maximize the recognition rates for each database. The phase correlation methods directly result in a real-valued score assessing the matching quality, and this similarity measure cannot be used as input in a standard SVM. Various solutions to address such cases have been proposed in the literature (see [3] for a recent review). Fortunately, in the phase correlation case, one can slightly modify the SVM algorithm by replacing the inner product values of the Gram matrix by a symmetric similarity measure ensuring its semi-positive definiteness. Such a property is guaranteed by taking as a measure the mean value of the correlation scores between $\widehat{f_B}$ and $\widehat{g_B}$ and between $\widehat{g_B}$ and $\widehat{f_B}$. Validation of the classification results is done by a 10-fold cross-validation.

9.5.4 Evaluation of the *GCFD*

In this section, the Generalized Clifford based Fourier Descriptors, *GCFD*1 and *GCFD*2, are evaluated to assess their classification performance relatively to the usual Generalized Fourier Descriptors, *GFD*1 and *GFD*2. The classification performance of *GFD* and *GCFD* is tested on the COIL-100, the color FERET and the noisy color FERET database. Each database (see Table 9.1, Table 9.2, Table 9.3) is processed using the same choices of 2-blades, one 2-blade per row of the tables. The first one, denoted by B_r, is obtained using a red vector (i.e. $B_r = \mathbf{r} \wedge \mathbf{e_4}$), the three next 2-blades B_g, B_b, B_μ respectively refer to green, blue colors and the achromatic axis. The next row gives the best classification rate, the mean and standard deviation of classification rates obtained by randomly choosing 100 2-blades. These five first rows provide results for single 2-blades, while the last two concern triple-size 2-blades. The first one, denoted by $B_r + B_g + B_b$, uses as descriptor the concatenation of descriptors computed from B_r, B_g and B_b (descriptors of size 384). The last row, denoted by $B_1 + B_2 + B_3$, provides results obtained by using an automatic selection algorithm to select the three 2-blades maximizing the classification rate. This so-called "SFFS" (Sequential Floating Forward Selection) algorithm is a suboptimal selection procedure that avoids exhaustive search in the feature space, here defined by the space of the **c** normalized color vectors. An interested reader can refer to [9] for comparison to similar optimization techniques.

COIL-100 being an easy dataset from a classification point of view, any descriptor provides excellent results, very close to 100%. In Table 9.1, the descriptors' size is recalled for each method, and the best results for each method are bold-faced. The bold-faced values in the upper part (resp. the lower part) of Table 9.1 show the best results for one 2-blade (resp. for three 2-blades [obtained by concatenation or by the SFFS procedure]). It must be noted that such a classification rate validates the choice of Fourier descriptors on this kind of image where color background is homogeneous and similar for all images considered. The standard deviations obtained for B_{rand} are small and suggest little influence of the 2-blade choice on this database, and consequently experiments were not carried further in that direction.

Regarding the *color FERET* dataset, it can be checked that the *GCFD*1 outperforms both *GFD*1 and *GFD*2 in terms of classification rate for any 2-blade choice. *GCFD*2 while providing better results than *GFD*1 and *GFD*2 is not better than *GCFD*1. This may be due to the choice of the ξ_1 and ξ_2 parameters, but none of our experiments led to an improvement in that respect. Anyway, these results clearly show the added benefit of the Clifford Fourier transform with regard to classification. Concatenation ($B_r + B_g + B_b$) of 2-blades improves the results for *GFD* but

Table 9.1 COIL-100: Recognition rates in % with $GFD1$, $GCFD1$, $GFD2$ and $GCFD2$

Bivectors	COIL-100			
	$GFD1$ (64 desc.)	$GCFD1$ (128 desc.)	$GFD2$ (64 desc.)	$GCFD2$ (128 desc.)
B_r	98.04	99.83	98.69	99.81
B_g	98.06	99.56	99.39	**99.85**
B_b	96.90	**99.86**	94.03	**99.85**
B_μ	**98.49**	99.25	**99.40**	99.47
$B_{rand}(\times 100)$	98.42 ± 0.3	99.54 ± 0.3	98.88 ± 0.82	99.82 ± 0.1
max.	98.87	99.89	99.47	99.96
	192 desc.	384 desc.	192 desc.	384 desc.
$B_r + B_g + B_b$	**99.9**	99.92	**99.89**	99.87
$B_1 + B_2 + B_3$ (SFFS)	99.86	**99.96**	99.89	**99.96**

Table 9.2 Color FERET: Recognition rates in % with $GFD1$, $GCFD1$, $GFD2$ and $GCFD2$

Bivectors	Color FERET			
	$GFD1$ (64 desc.)	$GCFD1$ (128 desc.)	$GFD2$ (64 desc.)	$GCFD2$ (128 desc.)
B_r	**76.70**	**87.90**	77.31	**84.42**
B_g	73.66	79.65	**77.37**	80.01
B_b	70.49	84.49	75.87	82.31
B_μ	73.03	78.10	77.30	82.12
$B_{rand}(\times 100)$	73.72 ± 1	85.34 ± 2.92	77.54 ± 0.69	84.50 ± 2.06
max.	76.14	90.37	78.91	89.57
	192 desc.	384 desc.	192 desc.	384 desc.
$B_r + B_g + B_b$	**88.03**	85.53	**84.26**	82.55
$B_1 + B_2 + B_3$ (SFFS)	85.46	**93.15**	82.89	**90.07**

not for *GCFD*. This is probably due to the better information encoding done by the Clifford Fourier transform, and *GCFD* descriptors obtained for various 2-blades are probably more redundant than the marginal *GFD*. This is confirmed by the SFFS method: selection of three "optimal" 2-blades pushes *GCFD* to the best results. In the random 2-blade choice experiment, *GCFD1* and *GCFD2* standard deviations are quite important compared to the ones of *GFD1* and *GFD2*. This clearly reveals that the *GCFD* depends on the choice of the 2-blade. To inspect its influence, the random experiment results are detailed in Fig. 9.3. Color of each dot denotes the **c** color chosen to define the 2-blade, and best ones are mostly blue. Visual inspection of the database confirms that this color corresponds to the background color of most images. As the 2-blade is unique for a given database, it must be chosen either ac-

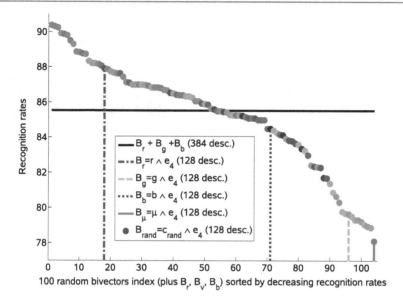

Fig. 9.3 Color FERET: Recognition rates with *GCFD*1 for 100 random 2-blades

Table 9.3 Noisy color FERET: Recognition rates in % with *GFD*1, *GCFD*1, *GFD*2 and *GCFD*2

Bivectors	Noisy color FERET			
	*GFD*1 (64 desc.)	*GCFD*1 (128 desc.)	*GFD*2 (64 desc.)	*GCFD*2 (128 desc.)
B_r	45.32	71.05	73.46	**83.49**
B_g	46.83	61.99	75.26	78.64
B_b	48.49	**73.46**	74.77	81.78
B_μ	**55.28**	62.03	**77.34**	80.98
$B_{rand} (\times 100)$	54.23 ± 1.75	69.64 ± 3.21	76.59 ± 0.74	82.56 ± 1.80
max.	57.55	77.27	78.41	87.00
	192 desc.	384 desc.	192 desc.	384 desc.
$B_r + B_g + B_b$	**73.16**	72.16	83.25	81.12
$B_1 + B_2 + B_3$ (SFFS)	71.52	**80.62**	**83.36**	**88.24**

cording to some prior knowledge or according to an empirical search method like SFFS.

Noisy color FERET: Influence of noise on classification rate is given in Table 9.3. *GFD*1 is the most sensitive to noise, and, as expected (see [15]), *GFD*2 is much more robust.

Fig. 9.4 Synthetic data: *rectangles* of color c_1 and c_2

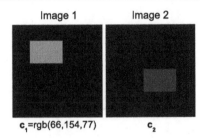

Image 1   Image 2

c_1=rgb(66,154,77) c_2

Table 9.4 Correlation scores between image 1 and 2 for various choices of c_2 and B. From image 1 to 2, the rectangle has color changed from $c_1 = $ rgb(66, 154, 77) to c_2, and a translation is applied

c_2	B														
	B_μ			B_r			B_g			B_b			B_{c_1}		
	$\rho_{\|B}$	$\rho_{\perp B}$	ρB	$\rho_{\|B}$	$\rho_{\perp B}$	ρB	$\rho_{\|B}$	$\rho_{\perp B}$	ρB	$\rho_{\|B}$	$\rho_{\perp B}$	ρB	$\rho_{\|B}$	$\rho_{\perp B}$	ρB
c_1	1	1	1	1	1	1	1	1	1	1	1	1	1	–	1
rgb(66, 0, 0)	1	1	0.43	1	–	0.36	–	1	0.55	–	1	0.91	1	–	0.36
rgb(0, 154, 0)	1	1	0.84	–	1	0.93	1	–	0.83	–	1	0.91	1	–	0.84
rgb(0, 0, 77)	1	1	0.50	–	1	0.93	–	1	0.55	1	–	0.42	1	–	0.42
μ	1	–	0.93	1	1	0.96	1	1	0.93	1	1	0.97	1	–	0.93

9.5.5 Evaluation of the Phase Correlation

In this chapter, two methods have been proposed to compute phase correlation for color images, both depending on a 2-blade B_c, where c is the chosen color. The first one depends on two correlation scores, $\rho_{\|B_c}$ and $\rho_{\perp B_c}$, given by the parallel and orthogonal part of the Clifford Fourier transform and requires an aggregation step to give final score. The second one, denoted ρ_{B_c}, does not require such a processing.

Synthetic data: Two simple images (see Fig. 9.4) are considered; they contain the same shape (a rectangle) on a black background, but the second is translated, and its color c_2 changed from experiments to experiments. Bivectors B_μ, B_r and B_{c_1}, where c_1 is the color of the first rectangle, are used to compute the correlation scores. In Table 9.4, the '–' symbol denotes a value that cannot be computed because the parallel (resp. orthogonal) part is null. Taking the example in which $c_2 = $ rgb(66, 0, 0), one can see that $\rho_{\|B_{c_2}} = \rho_{\perp B_{c_2}} = 1$. This is not conclusive as two rectangles which have different colors must be considered as different. However, the correlation score $\rho_{B_{c_2}}$ computed from the reconstructed Clifford Fourier transform gives scores lower than one and depends on the amount of color that the two rectangles have in common. The same remarks apply to c_2 taken equal to rgb(0, 154, 0) or rgb(0, 0, 77). The behavior for the gray axis level, μ, is different as ρ_{B_μ} is always high: unsurprisingly, it mostly depends on the shape without taking into account color information.

Table 9.5 COIL-100: Recognition rates in % with the phase correlation for color images

ρ		$\rho_{\|B}$	$\rho_{\perp B}$	$\rho_{\|,\perp B}^{min}$	$\rho_{\|,\perp B}^{max}$	$\rho_{\|,\perp B}^{mean}$	ρ_B
94.96	B_r	95.38	96.33	96.46	96.14	96.58	**97.50**
	B_g	95.29	96.79	96.68	95.90	96.68	**97.49**
	B_b	95.08	96.58	96.60	95.89	96.51	**97.49**
	B_{c_i}	95.33	83.92	82.60	96.08	95.50	**97.53**

Table 9.6 Color FERET: Recognition rates in % with the phase correlation for color images

ρ		$\rho_{\|B}$	$\rho_{\perp B}$	$\rho_{\|,\perp B}^{min}$	$\rho_{\|,\perp B}^{max}$	$\rho_{\|,\perp B}^{mean}$	ρ_B
66.37	B_r	66.51	66.00	67.64	68.05	**69.15**	66.74
	B_g	66.00	**67.91**	67.31	66.95	67.51	66.41
	B_b	65.57	67.41	**68.01**	66.95	67.71	66.81
	B_{c_i}	66.34	74.50	76.84	71.52	**78.38**	66.94

COIL-100: Results with B_r, B_g, B_b and B_{c_i} 2-blades are given. Notice that c_i corresponds to the choice of one 2-blade per request image, this color being the dominant color of the image. Table 9.5 clearly shows that the correlation computed from the reconstructed Clifford Fourier transform is the best method for this database. It has more discriminative power for color objects than other methods, and more importantly, it seems to be quite insensitive to the 2-blade choice. One can also see that most of color phase correlation methods give better recognition rates than the classical phase correlation for grayscale images. The different choices of 2-blades or aggregation functions do not give really improve the results; nevertheless recognition rates are high. One can see on Table 9.6 that the results for the *color FERET* database are quite different. Indeed, ρ_{B_r} is not the best method anymore, but still very stable. This relatively low performance is due to confusions induced by the different colors constituting the background. So, if the reconstructed Clifford Fourier transform is considered, all the color information is aggregated. Hence the choice of one 2-blade per request image, corresponding to the dominant color, separates the background and the foreground. This choice is the best among B_r, B_g and B_b and gives the best recognition rate using the mean as an aggregation function.

9.6   Conclusion

Two descriptors for color object recognition based on Clifford Fourier transform and with the viewpoint of group actions are proposed. Better classification results than those of analogous marginal methods are provided. Specially, Clifford Fourier descriptors enhance Generalized Fourier Descriptors with lower computation cost and size (only two FFT instead of three). Mathematically sound phase correlation computation for color images would imply an inverse Clifford Fourier transform of a spinor which is not available for now. Although some workarounds are proposed in this chapter, future work will give more efficient and consistent solutions.

9.7   Exercises

9.1 Verify (9.5) and develop the construction of elements B, I_4B, $\mathbf{v}$, $\mathbf{w}$ for the choice $\mathbf{c} = (\mathbf{e_1} + \mathbf{e_2})/\sqrt{2}$ (see Sect. 9.2.2).

9.2 Verify that for a bivector $B = \mathbf{c_1} \wedge \mathbf{e_4}$, I_4B is proportional to $\mathbf{c_2} \wedge \mathbf{c_3}$, where $\mathbf{c_1}$, $\mathbf{c_2}$ and $\mathbf{c_3}$ are RGB colors. Conclude that the choice of the form $\mathbf{c_1} \wedge \mathbf{e_4}$ or $\mathbf{c_2} \wedge \mathbf{c_3}$ for B are equivalent up to a sign change.

9.3 Give the MATLAB® code for computing the parallel and orthogonal parts of the color Clifford Fourier transform using the basis $\{\mathbf{e_1}, \mathbf{e_4}\}$ for the plane generated by the bivector $B = \mathbf{e_1} \wedge \mathbf{e_4}$ and the basis $\{\mathbf{e_3}, \mathbf{e_2}\}$ for the plane generated by $I_4B = \mathbf{e_3} \wedge \mathbf{e_2}$. (Hint: in this particular case, the projection on the bases is obvious.)

9.4 Give the MATLAB® code for computing the phase correlation on the parallel and orthogonal part of the Clifford Fourier transform using the previous code and taking the maximum operator as aggregation criterion.

9.5 Using the trigonometric rule $\cos(\theta) = 2\cos^2(\theta/2) - 1$, verify (9.18).

Acknowledgements   This work is partially supported by the ONR Grant N00014-09-1-0493 and "La Région Poitou-Charentes".

References

1. Batard, T., Berthier, M., Saint-Jean, C.: Clifford Fourier transform for color image processing. In: Bayro-Corrochano, E., Scheuermann, G. (eds.) Geometric Algebra Computing in Engineering and Computer Science, pp. 135–161. Springer, Berlin (2010), Chap. 8
2. Chang, C.C., Lin, C.J.: Libsvm: a library for support vector machines (2001). Software available at http://www.csie.ntu.edu.tw/cjlin/libsvm
3. Chen, Y., Garcia, E.K., Gupta, M.R., Rahimi, A., Cazzanti, L.: Similarity-based classification: concepts and algorithms. J. Mach. Learn. Res. **10**, 747–776 (2009)
4. Dorst, L., Mann, S., Bouma, T.: GABLE: A Matlab tutorial for geometric algebra (2002)
5. Ebling, J., Scheuermann, G.: Clifford Fourier transform on vector fields. IEEE Trans. Vis. Comput. Graph. **11**, 469–479 (2005)
6. Ell, T.A., Sangwine, S.J.: Hypercomplex Fourier transforms of color images. IEEE Trans. Image Process. **16**(1), 22–35 (2007)
7. Flusser, J., Suk, T., Zitova, B.: Moments and Moment Invariants in Pattern Recognition. Wiley, New York (2009)
8. Hestenes, D., Sobczyk, G.: Clifford Algebra to Geometric Calculus. Reidel, Dordrecht (1984)
9. Jain, A., Zongker, D.: Feature-selection: evaluation, application, and small sample performance. IEEE Trans. Pattern Anal. Mach. Intell. **19**(2), 153–158 (1997)
10. Lounesto, P.: Clifford Algebras and Spinors, 2nd edn. Cambridge University Press, Cambridge (2001)

11. Lowe, D.G.: Object recognition from local scale-invariant features. In: International Conference on Computer Vision 99, pp. 1150–1157 (1999)
12. Nene, S.A., Nayar, S.K., Murase, H.: Columbia object image library (coil-100). Technical Report CUCS-006-96 (1996)
13. Phillips, P.J., Wechsler, H., Huang, J., Rauss, P.: The FERET database and evaluation procedure for face recognition algorithms. Image Vis. Comput. **16**(5), 295–306 (1998)
14. Reddy, B., Chatterji, B.: An FFT-based technique for translation, rotation, and scale-invariant image registration. IEEE Trans. Image Process. **5**(8), 1266–1271 (1996)
15. Smach, F., Lemaître, C., Gauthier, J.P., Miteran, J., Atri, M.: Generalized Fourier descriptors with applications to objects recognition in SVM context. J. Math. Imaging Vis. **30**(1), 43–71 (2008)

Part IV
Theorem Proving and Combinatorics

A recurrent theme in this book is how the right representation can improve encoding and solving geometrical problems. This also affects traditionally combinatorial fields like theorem proving, constraint satisfaction and even cycle enumeration. The null elements of algebras turn out to be essential!

On Geometric Theorem Proving with Null Geometric Algebra

10

Hongbo Li and Yuanhao Cao

Abstract

The bottleneck in symbolic geometric computation is middle expression swell. Another embarrassing problem is geometric explanation of algebraic results, which is often impossible because the results are not invariant under coordinate transformations. In classical invariant-theoretical methods, the two difficulties are more or less alleviated but stay, while new difficulties arise.

In this chapter, we introduce a new framework for symbolic geometric computing based on conformal geometric algebra: the algebra for describing geometric configuration is null Grassmann–Cayley algebra, the algebra for advanced invariant manipulation is null bracket algebra, and the algebra underlying both algebras is null geometric algebra. When used in geometric computing, the new approach not only brings about amazing simplifications in algebraic manipulation, but can be used to extend and generalize existing theorems by removing some geometric constraints from the hypotheses.

10.1 Introduction

In algebraic approaches to geometric computing, the general procedure is as follows [19]: first, the geometric configuration, including both the hypotheses and the conclusion, is translated into an algebraic formulation in a prerequisite algebraic language; second, algebraic computations are carried out to the conclusion by utilizing the computational rules of the algebra and the given hypotheses; third, the result of the computations is translated back to geometry or, in other words, is given

H. Li (✉) · Y. Cao
Key Laboratory of Mathematics Mechanization, Academy of Mathematics and Systems Science, Chinese Academy of Sciences, Beijing 100190, P.R. China
e-mail: hli@mmrc.iss.ac.cn

Y. Cao
e-mail: ppxhappy@126.com

L. Dorst, J. Lasenby (eds.), *Guide to Geometric Algebra in Practice*,
DOI 10.1007/978-0-85729-811-9_10, © Springer-Verlag London Limited 2011

a geometric interpretation. In geometric reasoning and theorem proving, the input of a geometric problem is formulated by a set of symbols and their algebraic relations, and the algebraic computing, if geometrically meaningful, is called "symbolic geometric computation" [18].

The most commonly used algebraic formulation is Cartesian coordinates and its variations. In this setting, geometric relations are represented by polynomial equalities of coordinates. When coordinates are used in geometric computation, two typical difficulties occur:

1. Middle expression swell [1]: It is quite often that both the input and output are small but the polynomials in middle steps are huge. Some computations are thus possible only theoretically, at least for the current publicly available PCs and computer algebra systems.
2. Geometric inexplicability [19]: The result of algebraic computation is usually difficult to explain geometrically. In fact, most results produced do not have any geometric meaning—they are not invariant under coordinate transformations and thus are geometrically meaningless.

In the second half of the 19th century, several algebras of geometric covariants and invariants were proposed. When used in geometric computing, such algebras may help alleviating the difficulties, because they keep more geometry within their algebraic structures [16].

Classical invariant theory deals with invariance under the transformation group $GL_n(\mathcal{K})$. The corresponding geometry is projective geometry. The corresponding algebra of covariants for describing projective incidence relations is called "Grassmann–Cayley algebra." This is an algebra equipped with two products that are dual to each other: the outer product as in exterior algebra represents the extension of geometric entities, and the meet product represents the intersection of the entities.

In classical invariant theory, the algebra of invariants is the algebra of determinants of homogeneous coordinates, called "bracket algebra" [18]. For example, let $\mathbf{1}, \mathbf{2}, \mathbf{3}$ be three points in the 2D projective plane, let their homogeneous coordinates be (x_i, y_i, z_i) for $i = 1, 2, 3$, respectively. Then

$$[\mathbf{123}] = \begin{vmatrix} x_1 & y_1 & z_1 \\ x_2 & y_2 & z_2 \\ x_3 & y_3 & z_3 \end{vmatrix}. \tag{10.1}$$

Obviously, the bracket is multilinear and antisymmetric with respect to its components $\mathbf{1}, \mathbf{2}, \mathbf{3}$. The two properties, however, do not suffice to define bracket algebra completely.

In bracket algebra, people do not resort to Laplace expansions of the brackets; instead they use the brackets as basic indeterminates and take the algebraic relations among the brackets as "syzygies" [16]. Again take as example the 2D projective geometry. Let there be five points $\mathbf{1}, \mathbf{2}, \mathbf{3}, \mathbf{4}, \mathbf{5}$ in the projective plane. The 3D bracket algebra generated by the five points is the bracket polynomials with $C_5^3 = 10$ indeterminates

$[123], [124], [125], \ldots, [345]$.

The ten brackets are not algebraically independent. They satisfy five algebraic relations which generate all other relations:

$$[123][145] - [124][135] + [125][134] = 0,$$
$$[123][245] - [124][235] + [125][234] = 0,$$
$$[123][345] - [134][235] + [135][234] = 0, \qquad (10.2)$$
$$[124][345] - [134][245] + [145][234] = 0,$$
$$[125][345] - [135][245] + [145][235] = 0.$$

These generating relations are the syzygies defining the bracket algebra, called the "Grassmann–Plücker syzygies" of the five coplanar points [16]. These syzygies are still not algebraically independent from each other. For example, among the five syzygies in (10.2), only three are algebraically independent, e.g., the first three. They form a "bracket basis" of the syzygies.

Brackets have obvious representational advantage over coordinates. For five points **1** to **5** in the projective plane, the corresponding bracket algebra contains monomials like $[123][145]$ and binomials like $[123][145] + [124][135]$. In coordinates, however, their expanded forms are much longer:

$[123][145]$

$$
\begin{aligned}
= \; & x_2 z_1 y_3 x_1 y_4 z_5 - x_2 z_1 y_3 x_1 z_4 y_5 - x_2 z_1 y_3 x_4 y_1 z_5 + x_2 z_1^2 y_3 x_4 y_5 \\
& - x_1 y_2 z_3 x_4 y_1 z_5 + x_1 y_2 z_3 x_4 z_1 y_5 + x_1 y_2 z_3 x_5 y_1 z_4 - x_1 y_2 z_3 x_5 z_1 y_4 \\
& - x_3 z_1 y_2 x_1 y_4 z_5 + x_3 z_1 y_2 x_1 z_4 y_5 + x_3 z_1 y_2 x_4 y_1 z_5 - x_3 z_1^2 y_2 x_4 y_5 \\
& + x_3 z_1^2 y_2 x_5 y_4 + x_1^2 y_2 z_3 y_4 z_5 - x_1^2 y_2 z_3 z_4 y_5 - x_1 z_2 y_3 x_5 y_1 z_4 \\
& + x_1 z_2 y_3 x_5 z_1 y_4 - x_2 y_1 z_3 x_1 y_4 z_5 + x_2 y_1 z_3 x_1 z_4 y_5 + x_2 y_1^2 z_3 x_4 z_5 \\
& - x_2 y_1 z_3 x_4 z_1 y_5 - x_2 y_1^2 z_3 x_5 z_4 + x_2 y_1 z_3 x_5 z_1 y_4 + x_2 z_1 y_3 x_5 y_1 z_4 \\
& - x_2 z_1^2 y_3 x_5 y_4 + x_3 y_1 z_2 x_1 y_4 z_5 - x_3 y_1 z_2 x_1 z_4 y_5 - x_3 y_1^2 z_2 x_4 z_5 \\
& + x_3 y_1 z_2 x_4 z_1 y_5 + x_3 y_1^2 z_2 x_5 z_4 - x_3 y_1 z_2 x_5 z_1 y_4 - x_1^2 z_2 y_3 y_4 z_5 \\
& + x_1^2 z_2 y_3 z_4 y_5 + x_1 z_2 y_3 x_4 y_1 z_5 - x_1 z_2 y_3 x_4 z_1 y_5 - x_3 z_1 y_2 x_5 y_1 z_4,
\end{aligned}
$$

$[123][145] + [124][135]$

$$
\begin{aligned}
= \; & -x_1 y_2 z_4 x_3 y_1 z_5 - x_1 y_2 z_4 x_5 z_1 y_3 + x_1^2 z_2 y_4 z_3 y_5 \\
& - x_1 z_2 y_4 x_5 y_1 z_3 - x_2 y_1 z_4 x_1 y_3 z_5 + x_2 y_1^2 z_4 x_3 z_5 - x_2 y_1 z_4 x_3 z_1 y_5 \\
& + 2 x_2 z_1 y_3 x_1 y_4 z_5 - x_2 z_1 y_3 x_1 z_4 y_5 - x_2 z_1 y_3 x_4 y_1 z_5 + x_2 z_1^2 y_3 x_4 y_5
\end{aligned}
$$

$$- x_1 y_2 z_3 x_4 y_1 z_5 + 2 x_1 y_2 z_3 x_4 z_1 y_5 + 2 x_1 y_2 z_3 x_5 y_1 z_4 - x_1 y_2 z_3 x_5 z_1 y_4$$

$$- x_3 z_1 y_2 x_1 y_4 z_5 + 2 x_3 z_1 y_2 x_1 z_4 y_5 + 2 x_3 z_1 y_2 x_4 y_1 z_5 - 2 x_3 z_1^2 y_2 x_4 y_5$$

$$+ x_3 z_1^2 y_2 x_5 y_4 + x_1^2 y_2 z_3 y_4 z_5 - 2 x_1^2 y_2 z_3 z_4 y_5 - x_1 z_2 y_3 x_5 y_1 z_4$$

$$+ 2 x_1 z_2 y_3 x_5 z_1 y_4 - x_2 y_1 z_3 x_1 y_4 z_5 + 2 x_2 y_1 z_3 x_1 z_4 y_5 + x_2 y_1^2 z_3 x_4 z_5$$

$$- x_2 y_1 z_3 x_4 z_1 y_5 - 2 x_2 y_1^2 z_3 x_5 z_4 + 2 x_2 y_1 z_3 x_5 z_1 y_4 + 2 x_2 z_1 y_3 x_5 y_1 z_4$$

$$- 2 x_2 z_1^2 y_3 x_5 y_4 + 2 x_3 y_1 z_2 x_1 y_4 z_5 - x_3 y_1 z_2 x_1 z_4 y_5 - 2 x_3 y_1^2 z_2 x_4 z_5$$

$$+ 2 x_3 y_1 z_2 x_4 z_1 y_5 + x_3 y_1^2 z_2 x_5 z_4 - x_3 y_1 z_2 x_5 z_1 y_4 - x_4 z_1 y_2 x_1 y_3 z_5$$

$$- x_4 z_1 y_2 x_5 y_1 z_3 + x_4 z_1^2 y_2 x_5 y_3 - 2 x_1^2 z_2 y_3 y_4 z_5 + x_1^2 z_2 y_3 z_4 y_5$$

$$+ 2 x_1 z_2 y_3 x_4 y_1 z_5 - x_1 z_2 y_3 x_4 z_1 y_5 - x_3 z_1 y_2 x_5 y_1 z_4 + x_1^2 y_2 z_4 y_3 z_5$$

$$- x_4 y_1 z_2 x_1 z_3 y_5 + x_4 y_1^2 z_2 x_5 z_3 - x_4 y_1 z_2 x_5 z_1 y_3 - x_2 z_1 y_4 x_1 z_3 y_5$$

$$- x_1 z_2 y_4 x_3 z_1 y_5 - x_2 z_1 y_4 x_3 y_1 z_5 + x_2 z_1^2 y_4 x_3 y_5. \tag{10.3}$$

The representational advantage of brackets does not necessarily lead to any manipulational advantage. Since the brackets are not algebraically independent, one may consider using only a minimum set of algebraically independent brackets and representing all other brackets by elements in the minimum set. If brackets are used in this way, then they are equivalent to coordinates. For example, in (10.2) the first three syzygies form a bracket basis. If only such syzygies are used, then the following algebraically independent brackets can represent the other brackets via the syzygies and are such a minimum set:

$$[\mathbf{123}], [\mathbf{124}], [\mathbf{125}], [\mathbf{234}], [\mathbf{235}], [\mathbf{134}], [\mathbf{135}]. \tag{10.4}$$

Then essentially points $\mathbf{1, 2, 3}$ are taken as a basis of the 3D vector space realizing the 2D projective plane, and (10.4) is composed of the volume $[\mathbf{123}]$ of the basis and the homogeneous coordinates of points $\mathbf{i} = \mathbf{4, 5}$ with respect to the basis

$$x_i = \frac{[\mathbf{12i}]}{[\mathbf{123}]}, \qquad y_i = \frac{[\mathbf{13i}]}{[\mathbf{123}]}, \qquad z_i = \frac{[\mathbf{23i}]}{[\mathbf{123}]}.$$

Let us see how invariants are manipulated in classical invariant theory. Classical invariant-theoretical method employs a Gröbner basis of the ideal generated by the Grassmann–Plücker syzygies in bracket algebra, called "straightening syzygies" [16]. All bracket polynomials form a $\mathscr{Z}$-module with elements in the straightening syzygies as a basis. Any bracket polynomial can be written in a unique manner as a linear combination of the basis elements with integer coefficients. The latter is the "normal form" of the bracket polynomial. In the procedure of normalization, a non-straightened bracket monomial is "exploded" into many terms many times. This procedure does not have any control to the middle expression swell.

Geometric interpretation is also a problem for bracket algebra. Although each bracket, as a determinant of homogeneous coordinates of the constituent points, can be interpreted in affine geometry as the signed volumes of the simplex spanned by the points as vertices, a bracket polynomial is by no means easily interpretable with geometric terms. If the polynomial can be written as a rational monomial in a suitable covariant algebra in which the basic elements and their products are geometrically meaningful, then the polynomial finds its geometric interpretation. According to a theorem by Sturmfels [17], theoretically this procedure is always successful, called "Cayley factorization." However, there is no algorithm to produce this factorization, except for the simplest case where every point in the bracket polynomial occurs only once [18].

So in the setting of classical invariant theory, the two major problems faced by the coordinate approach are still alive, although in some special cases the algebraic manipulations can be simplified because of the simplicity in algebraic representation. Due to the algebraic dependencies among brackets, new difficulties arise, which are by no means easy to handle. In invariant-theoretical methods, people do not get rid of algebraic dependencies; otherwise it becomes a traditional coordinate method. The following are some newly invoked problems [4, 11]:

- *Representation*: A geometric entity or relation often has many representations in invariant algebra. How to choose a suitable one in computation? Can the computing be made robust against the choice?

   This problem has never been studied before. A typical example is a conic formed by five points in the projective plane. It has fifteen equal but different forms when represented as a degree-four binomial of brackets. Different choice of the representation can lead to drastic difference in complexity in subsequent algebraic manipulations.

- *Contraction*: Reduce the number of terms of a bracket polynomial.

   This problem does not exist in polynomials of coordinates. In bracket algebra this problem is wide open: people do not know how to judge and how to find a minimum-sized form for a bracket polynomial.

- *Expansion*: The reverse procedure of Cayley factorization is called "Cayley expansion" [11]. It is to translate a scalar-valued expression of the algebra of covariants into a polynomial in the algebra of invariants.

   This problem turns out to be rather complicated. A simple example is the bracket $[\mathbf{a}\mathbf{a}'\mathbf{a}'']$ formed by three intersections of pairs of lines in the projective plane:

$$\mathbf{a} = 12 \cap 34, \qquad \mathbf{a}' = 1'2' \cap 3'4', \qquad \mathbf{a}'' = 1''2'' \cap 3''4''.$$

To compute the bracket, by substituting the expressions of the $\mathbf{a}$'s in Grassmann–Cayley algebra into it, we get

$$[\{(\mathbf{1} \wedge \mathbf{2}) \vee (\mathbf{3} \wedge \mathbf{4})\}\{(\mathbf{1}' \wedge \mathbf{2}') \vee (\mathbf{3}' \wedge \mathbf{4}')\}\{(\mathbf{1}' \wedge \mathbf{2}') \vee (\mathbf{3}' \wedge \mathbf{4}')\}].$$

It has 16847 different expansion results.

It is an appalling fact that classical invariant-theoretical method is far from being well developed for symbolic computation. Basic computing tasks like choosing optimal representations in the procedure of computing, different expansions of covariant expressions, contraction of invariant expressions, and factorization in both invariant and covariant algebras, are either open or overlooked. The bottleneck in symbolic computing, i.e., middle expression swell, is not taken care of. Because of this, although invariant algebra can provide simplification in algebraic description, its cost is significantly high complexity in algebraic manipulation.

When it comes to Euclidean geometry, the algebra of basic invariants is "inner-product bracket algebra" [3]. This algebra contains, besides brackets, the inner products of vectors as basic elements. In its defining syzygies there is a polynomial of the form

$$[\mathbf{i}_1 \mathbf{i}_2 \cdots \mathbf{i}_n][\mathbf{j}_1 \mathbf{j}_2 \cdots \mathbf{j}_n] - \det(\mathbf{i}_k \cdot \mathbf{j}_l)_{k,l=1,\ldots,n},$$

which equates the product of two determinants to the determinant of the inner products of the constituent column vectors of the two determinants. This syzygy contains as many as $n! + 1$ terms. So the task to control the limit of middle expression swell is much heavier. Further to people's dismay, the invariants and covariants in geometric computing are often complicated rational polynomials of basic ones. This suggests that basic invariants are too low-level. As a result, symbolic computation in Euclidean geometry with inner-product bracket algebra is much more difficult than in projective geometry.

This is the background of our research in recent years on invariant symbolic computation in classical geometry. In the course of eight years, we have proposed a new invariant framework, called "null geometric algebra," and a new guideline for computation, called "BREEFS" [6–10, 12, 13, 15]. The former is a system of monomial representations of Euclidean incidence geometric constructions and an associated hierarchy of infinitely graded advanced invariants grown out of Clifford multiplication, and the latter is a stepwise size control strategy based on syzygies of the invariant algebra. The new framework and guideline can help achieving significant simplification in invariant algebraic manipulation and thus lead to much better computation both in efficiency (size control) and in quality (geometric interpretation), as follows:

- In projective geometry and Euclidean geometry, generally the size of an expression being computed is controlled to within two terms.
- Some geometric computing tasks, which have proved to be very hard if using only coordinates or basic invariants, can be finished with our new system.

This chapter intends to provide an introduction to the new system. This is done in Sect. 10.2. Section 10.3 is a practical application of the new system to the problem of "geometric factorization, decomposition, and theorem completion," that is, to explore the quantitative and geometric relationship between the hypotheses and the conclusion of a geometric theorem, and to explore and discover geometrically meaningful new conclusions by reducing the number of hypotheses in the theorem.

10.2   Null Grassmann–Cayley Algebra and Null Bracket Algebra

10.2.1 Grassmann–Cayley Algebra, Bracket Algebra and Inner-Product Bracket Algebra

First recall the definition of *Grassmann–Cayley algebra* [18]. Let $\mathscr{V}^n$ be an nD linear space over a field of characteristic not 2. Let $\Lambda(\mathscr{V}^n)$ be the Grassmann space over the base space $\mathscr{V}^n$. Define in $\Lambda(\mathscr{V}^n)$ the following meet product, which is dual to the outer product: for any $\mathbf{A}, \mathbf{B} \in \Lambda(\mathscr{V}^n)$, their meet product $\mathbf{A} \vee \mathbf{B}$ is defined by [9]

$$(\mathbf{A} \vee \mathbf{B})^* := \mathbf{B}^* \wedge \mathbf{A}^*, \tag{10.5}$$

where "*" is the dual operator in Grassmann algebra, and "$\wedge$" is the outer product.

Grassmann–Cayley algebra is a language for describing projective incidence constructions. Any vector of $\mathscr{V}^n$ represents a point of $(n - 1)$D projective space, and the representation is *homogeneous* in that it is unique up to scale: any two vectors represent the same projective point if and only if they differ by scale only. The line extended by two points is represented by their outer product, and the plane extended by three points is represented by the outer product of any three vectors representing the three points. In projective space, the intersection of a line and a plane is represented by their meet product.

Let there be m projective points. An nD *bracket algebra* generated by a sequence $\mathscr{S}$ of m symbols $\mathbf{1}, \mathbf{2}, \ldots, \mathbf{m}$ representing the points, where $m > n + 1$, is the polynomial ring generated by all subsequences of $\mathscr{S}$ of length n, denoted by square brackets, modulo the ideal generated by the left side of the following identity, called "Grassmann–Plücker syzygies":

$$\sum_{k=1}^{n+1}(-1)^{k+1}[\mathbf{i}_1\mathbf{i}_2 \cdots \mathbf{i}_{n-1}\mathbf{j}_k][\mathbf{j}_1\mathbf{j}_2 \cdots \check{\mathbf{j}}_k \cdots \mathbf{j}_{n+1}] = 0, \tag{10.6}$$

where the $\mathbf{i}$'s and $\mathbf{j}$'s are symbols in $\mathscr{S}$, and $\check{\mathbf{j}}_k$ denotes that $\mathbf{j}_k$ does not occur in the subsequence. By requiring antisymmetry among them, the elements in each bracket do not need to follow their original order in the sequence $\mathbf{1}, \mathbf{2}, \ldots, \mathbf{m}$.

The proof of (10.6) is trivial: expand

$$(\mathbf{i}_1 \wedge \mathbf{i}_2 \wedge \cdots \wedge \mathbf{i}_{n-1}) \vee (\mathbf{j}_1 \wedge \mathbf{j}_2 \wedge \cdots \wedge \mathbf{j}_{n+1}) \tag{10.7}$$

using the "shuffle formula" in Grassmann–Cayley algebra [16] to distribute the $\mathbf{j}$'s to the sequence $\mathbf{i}_1\mathbf{i}_2 \cdots \mathbf{i}_{n-1}$, once for each $\mathbf{j}$ but with alternating signs. That (10.7) equals zero follows from the fact that any $n + 1$ vectors in an nD vector space are linearly dependent, so their outer product equals zero. Thus,

$$\mathbf{j}_1 \wedge \mathbf{j}_2 \wedge \cdots \wedge \mathbf{j}_{n+1} = 0.$$

The meet product of zero with any element is zero. This proves (10.6).

Bracket algebra is established for projective geometry and, after some revision, for affine geometry. For Euclidean geometry, a new structure called inner product is needed. A bracket algebra supplemented by an inner product is an *inner-product bracket algebra* [3]. Formally, an nD inner-product bracket algebra generated by a sequence $\mathscr{S}$ of m symbols of vectors $\mathbf{1}, \mathbf{2}, \ldots, \mathbf{m}$, where $m \geq n$, is the quotient of the polynomial ring generated by two classes of subsequences of $\mathscr{S}$ of length 2 and n, denoted by dot and square bracket, respectively, modulo the ideal generated by the following syzygies:

• GP1:

$$\sum_{k=1}^{n+1}(-1)^{k+1}\mathbf{i}\cdot\mathbf{j}_k[\mathbf{j}_1\mathbf{j}_2\cdots\check{\mathbf{j}}_k\cdots\mathbf{j}_{n+1}]. \tag{10.8}$$

• GP2:

$$[\mathbf{i}_1\mathbf{i}_2\cdots\mathbf{i}_n][\mathbf{j}_1\mathbf{j}_2\cdots\mathbf{j}_n]-\det(\mathbf{i}_k\cdot\mathbf{j}_l)_{k,l=1,\ldots,n}. \tag{10.9}$$

The order of elements in the subsequences can be violated by requiring that the dot structure is symmetric while the bracket structure is antisymmetric.

The proof of the syzygies is also trivial. (10.8) is the expansion of the inner product

$$\mathbf{i}\cdot(\mathbf{j}_1\wedge\mathbf{j}_2\wedge\cdots\wedge\mathbf{j}_{n+1}), \tag{10.10}$$

which equals zero because the outer product of the $n+1$ $\mathbf{j}$'s is zero. (10.9) follows from

$$\begin{aligned}[\mathbf{i}_1\mathbf{i}_2\cdots\mathbf{i}_n]&[\mathbf{j}_1\mathbf{j}_2\cdots\mathbf{j}_n]\\&=(\mathbf{i}_1\wedge\mathbf{i}_2\wedge\cdots\wedge\mathbf{i}_n)^*(\mathbf{j}_1\wedge\mathbf{j}_2\wedge\cdots\wedge\mathbf{j}_n)^*\\&=\big((\mathbf{i}_1\wedge\mathbf{i}_2\wedge\cdots\wedge\mathbf{i}_n)\cdot(\mathbf{j}_n\wedge\mathbf{j}_{n-1}\wedge\cdots\wedge\mathbf{j}_1)\big)\\&=\det(\mathbf{i}_k\cdot\mathbf{j}_l)_{k,l=1,\ldots,n},\end{aligned}$$

where the $\mathbf{e}$'s are a basis of the nD vector space defining the dual operator, and the dot denotes the inner product in geometric algebra. The Grassmann–Plücker syzygies (10.6) can be obtained from (10.9) directly.

To obtain geometrically explicitly meaningful results, one needs to resort to advanced algebraic invariants. An *algebraic invariant* is a polynomial function of finitely many geometric entities, each represented by coordinates, such that under all kinds of coordinate transformations specified by the defining group of the geometry, the function remains either invariant or rescaled by a power of the determinant of the transformation. It is a classical theorem that all projective algebraic invariants are generated by brackets and that all Euclidean algebraic invariants are generated by brackets and inner products of vector pairs. Conversely, it is clear that

Fig. 10.1 Basic invariants in
2D Euclidean geometry

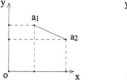

all brackets are projective invariants and all inner products are Euclidean invariants.

Brackets and inner products of vector pairs are *basic Euclidean invariants*. For example, in 2D Euclidean geometry the two basic Euclidean invariants are shown in Fig. 10.1: the distance between two points $\mathbf{a}_1 = (x_1, y_1)$, $\mathbf{a}_2 = (x_2, y_2)$:

$$|\mathbf{a}_1 - \mathbf{a}_2|^2 = (x_1 - x_2)^2 + (y_1 - y_2)^2$$
$$= x_1^2 + x_2^2 - 2x_1 x_2 + y_1^2 + y_2^2 - 2y_1 y_2, \tag{10.11}$$

and twice the signed volume of a simplex spanned by vertexes $\mathbf{a}_j = (x_j, y_j)$ for $1 \le j \le 3$:

$$\left[(\mathbf{a}_1 - \mathbf{a}_2)(\mathbf{a}_1 - \mathbf{a}_3)\right] = \begin{vmatrix} x_1 - x_2 & x_1 - x_3 \\ y_1 - y_2 & y_1 - y_3 \end{vmatrix}$$
$$= x_1 y_2 - x_2 y_1 - x_1 y_3 + x_3 y_1 + x_2 y_3 - x_3 y_2. \tag{10.12}$$

All other algebraic invariants are polynomial functions of the basic ones.

An invariant has three different forms of appearance: the *coordinate form* such as the right side of (10.12); the *expanded form* such as the right side of

$$\left[(\mathbf{a}_1 - \mathbf{a}_2)(\mathbf{a}_1 - \mathbf{a}_3)\right] = [\mathbf{a}_1 \mathbf{a}_2] - [\mathbf{a}_1 \mathbf{a}_3] + [\mathbf{a}_2 \mathbf{a}_3]; \tag{10.13}$$

and the *compact form* such as $[(\mathbf{a}_1 - \mathbf{a}_2)(\mathbf{a}_1 - \mathbf{a}_3)]$. Obviously, the compact form is the most convenient in reading out geometric interpretation. On the other hand, without expanding the parentheses in a compact form, algebraic manipulations are much more complicated.

An *advanced invariant* is an algebraic invariant having a compact form that is a monomial in a geometric algebra. The geometric interpretation of an advanced invariant is immediate from the compact form. Advanced invariant theory studies the geometric and algebraic properties of advanced invariants. CGA (*conformal geometric algebra*) provides a natural tool to construct advanced Euclidean invariants.

10.2.2   From Conformal Geometric Algebra to Null Grassmann–Cayley Algebra and Null Bracket Algebra

To start with, consider the expansion of (10.11), i.e., the squared length of line segment $\mathbf{a}_1 \mathbf{a}_2$:

$$d_{ab}^2 = |\mathbf{a}_1 - \mathbf{a}_2|^2 = (\mathbf{a}_1 - \mathbf{a}_2) \cdot (\mathbf{a}_1 - \mathbf{a}_2)$$

$$= \mathbf{a}_1 \cdot \mathbf{a}_1 + \mathbf{a}_2 \cdot \mathbf{a}_2 - 2\mathbf{a}_1 \cdot \mathbf{a}_2. \tag{10.14}$$

A geometric point can be represented by the vector drawn from the origin of the coordinate frame to the point; so $\mathbf{a}_1 \cdot \mathbf{a}_1$ represents the squared distance between the origin and the point. When the coordinate frame changes, so does the squared distance. Hence, $\mathbf{a}_1 \cdot \mathbf{a}_1$ is geometrically meaningless, and so is each term on the right side of (10.14). We are confronted with a bunch of geometrically meaningless terms when expanding a squared distance.

To avoid such expansions, a natural idea is to introduce a new inner product such that if vectors $\mathbf{a}_1, \mathbf{a}_2$ represent two geometric points, then $\mathbf{a}_1 \cdot \mathbf{a}_2$ is a geometric quantity relying on the two points only. Any such a candidate must be a function of the distance between the two points.

Then $\mathbf{a}_1 \cdot \mathbf{a}_1$ has to be independent of $\mathbf{a}_1$, because it has to be a function of the distance zero between a point and itself. The simplest choice is to set $\mathbf{a}_1 \cdot \mathbf{a}_1 = 0$, i.e., vector $\mathbf{a}_1$ is *null*. Then from (10.14) we get

$$\mathbf{a} \cdot \mathbf{b} = -\frac{d_{ab}^2}{2}. \tag{10.15}$$

To realize nD Euclidean geometry in an inner-product space having property (10.15), the smallest dimension is $n + 2$, and the space is Minkowski. Such a realization, called the *conformal model*, has its root in the work of Wachter (1830s) and later occurred in S. Lie's dissertation (1870s). For $n = 3$, the null-vector representation of a point $(x, y, z) \in \mathbb{R}^3$ in Minkowski space $\mathbb{R}^{4,1}$ is the following: let $(\mathbf{e}_1, \mathbf{e}_2, \mathbf{e}_3)$ be a basis of $\mathbb{R}^3$, and let $(\mathbf{e}_1, \mathbf{e}_2, \mathbf{e}_3, \mathbf{n}_\infty, \mathbf{n}_0)$ be a basis of $\mathbb{R}^{4,1}$ with inner-product matrix

$$\begin{pmatrix} 1 & & & & \\ & 1 & & & \\ & & 1 & & \\ & & & 0 & -1 \\ & & & -1 & 0 \end{pmatrix}; \tag{10.16}$$

then

$$(x, y, z) \in \mathbb{R}^3 \mapsto \left(x, y, z, 1, -\frac{x^2 + y^2 + z^2}{2} \right) \in \mathbb{R}^{4,1} \tag{10.17}$$

is an isometry. When $x^2 + y^2 + z^2 \to \infty$, the right side of (10.17) tends to the direction of vector $\mathbf{n}_\infty$. So $\mathbf{n}_\infty$ represents a unique point at infinity compactifying a Euclidean space.

The classical *conformal model* of nD Euclidean space is the following set:

$$\mathcal{N}_{\mathbf{n}_\infty} = \left\{ \mathbf{x} \in \mathscr{R}^{n+1,1} \mid \mathbf{x} \cdot \mathbf{x} = 0, \ \mathbf{x} \cdot \mathbf{n}_\infty = -1 \right\}. \tag{10.18}$$

Here $\mathbf{n}_\infty$ is a null vector in the $(n+2)$D Minkowski space $\mathscr{R}^{n+1,1}$. Elements in $\mathscr{N}_{\mathbf{n}_\infty}$ are in one-to-one correspondence with points in nD Euclidean space. Let null vector $\mathbf{n}_0 \in \mathscr{N}_e$ be the origin. In the conformal model, a point $\mathbf{x}$ in $\mathscr{R}^n$ is represented by the null vector

$$\vec{\mathbf{x}} = \mathbf{n}_0 + \mathbf{x} + \frac{\mathbf{x}^2}{2}\mathbf{n}_\infty. \tag{10.19}$$

Because they never occur simultaneously, later on, both a Euclidean vector and its null vector representation are denoted by the same letter without arrow top.

From the definition, it is clear that the conformal model depends on the choice of the origin $\mathbf{n}_0$. The *homogeneous model* [3] is a more general formulation of the classical conformal model. The model is composed of the set of null vectors

$$\mathscr{N} = \left\{ \mathbf{x} \in \mathscr{R}^{n+1,1} \mid \mathbf{x} \cdot \mathbf{x} = 0 \right\} \tag{10.20}$$

and a null vector $\mathbf{n}_\infty \in \mathscr{N}$. An element $\mathbf{x} \in \mathscr{N}$ represents a finite point if and only if $\mathbf{x} \cdot \mathbf{n}_\infty \neq 0$. Two elements in $\mathscr{N}$ represents the same point if and only if they differ by scale. This representation is homogeneous, and the model is conformal instead of isometric. Because of this, it can represent classical geometries of different metrics, where the "point at infinity" $\mathbf{n}_\infty$ remains a nonzero vector, but not necessarily a null vector. To unclutter the formulas, we will denote it by $\mathbf{e}$ in the remainder of this chapter.

The geometric algebra established upon the homogeneous model is called *conformal geometric algebra* (CGA).[1] It is the covariant algebra for Euclidean incidence relations, including collinearity, cocircularity, parallelism, perpendicularity, and tangency. CGA provides the following algebraic representations for incidence geometric constructions in Euclidean geometry:

(1) The line passing through two points $\mathbf{a}, \mathbf{b}$ is represented by $\mathbf{e} \wedge \mathbf{a} \wedge \mathbf{b}$, where the vectors representing points are null. A circle passing through three points $\mathbf{a}, \mathbf{b}, \mathbf{c}$ is represented by $\mathbf{a} \wedge \mathbf{b} \wedge \mathbf{c}$.

(2) The above constructions are "extension constructions" based on points. Duality provides intersection constructions. The intersection of two circles/lines $A_1 = \mathbf{a}_1 \wedge \mathbf{b}_1 \wedge \mathbf{c}_1$ and $A_2 = \mathbf{a}_2 \wedge \mathbf{b}_2 \wedge \mathbf{c}_2$ is a 0D circle represented by the meet product $A_1 \vee A_2$.

Besides geometric constructions based on points, CGA also provides representations of geometric constructions based on symmetry generators. There are three kinds of nonzero vectors in a Minkowski space: *positive*, *null*, and *negative*. They are vectors whose inner product with itself is positive, zero, and negative, respectively. We have seen that a null vector represents in Euclidean geometry a point or the point at infinity. Below we introduce the other two kinds of vectors as symmetry generators.

---

[1] *Editorial note*: This characterization of CGA is more restrictive than elsewhere in this book, where (10.19) is employed.

Fig. 10.2 Reflection, inversion, and antipodal inversion. *Left*: $x \mapsto x'$, reflection. *Right*: $x \mapsto x'$, inversion; $x \mapsto x''$, antipodal inversion

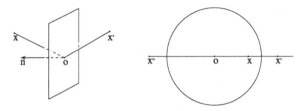

Any positive vector in Minkowski space $\mathbb{R}^{n+1,1}$ can be written up to scale as either $\mathbf{n} + \delta \mathbf{e}$ or $\mathbf{c} - \rho^2 \mathbf{e}/2$, where $\mathbf{n} \in \mathbb{R}^n$ is a unit vector, $\mathbf{c}$ is a null vector representing a point, and $\delta, \rho \in \mathbb{R}$.

Let $\mathbf{x}$ be a null vector representing a point. Let $\mathbf{s}$ be a positive vector. Then $\mathbf{x} \cdot \mathbf{s} = 0$ iff point $\mathbf{x}$ is on the hyperplane or sphere represented by $\mathbf{s}^*$. When $\mathbf{s} = \mathbf{n} + \delta \mathbf{e}$, where $\delta = -(\mathbf{a} \cdot \mathbf{n})/(\mathbf{a} \cdot \mathbf{e})$ for a vector $\mathbf{a} \in \mathcal{N}$ representing a point, then $\mathbf{s}^*$ is the hyperplane normal to $\mathbf{n}$ and passing through point $\mathbf{a}$; when $\mathbf{s} = \mathbf{c} - \rho^2 \mathbf{e}/2$, then $\mathbf{s}^*$ is the sphere centering at $\mathbf{c}$ with radius $|\rho|$.

When $\mathbf{x} \cdot \mathbf{s} \neq 0$, then $\mathbf{x} \wedge \mathbf{s}$ represents a pair of points: $\mathbf{x}$ and the reflection/inversion of $\mathbf{x}$ with respect to hyperplane/sphere $\mathbf{s}^*$, as shown in Fig. 10.2.

Any negative vector $\mathbf{t} \in \mathbb{R}^{n+1,1}$ can be written up to scale as $\mathbf{c} + \rho^2 \mathbf{e}/2$, where $\mathbf{c}$ is a null vector representing a point, and $\rho \in \mathbb{R}^+$. Let $\mathbf{x}$ be a null vector representing a point. Let $\mathbf{t}$ be a negative vector. Then $\mathbf{x} \wedge \mathbf{t}$ represents a pair of points, $\mathbf{x}$ and the *antipodal inversion* of $\mathbf{x}$ with respect to sphere $(\mathbf{c}, \rho)$. As shown in Fig. 10.2 (right): $\mathbf{x}$ is mapped to $\mathbf{x}''$ such that $\overrightarrow{\mathbf{ox}}'' = -\rho \overrightarrow{\mathbf{ox}}^{-1}$.

Hence in CGA, the extension product is generalized to include not only generating objects (e.g., points), but also symmetries (e.g., reflection, inversion, and antipodal inversion) of the resulting object. Dually, the intersection of two geometric objects can be their common symmetry. CGA extends Grassmann's original extension of linear (flat) objects to include not only round objects, but also symmetries.

Now consider the simplest 2D case and the representations of points of intersection. Let $\mathbf{ab}_1 \mathbf{c}_1$ and $\mathbf{ab}_2 \mathbf{c}_2$ be two circles/lines. Their intersection is a pair of points, one of which may be the point at infinity:

$$(\mathbf{a} \wedge \mathbf{b}_1 \wedge \mathbf{c}_1) \vee (\mathbf{a} \wedge \mathbf{b}_2 \wedge \mathbf{c}_2) = \mathbf{a} \wedge \left([\mathbf{ab}_1\mathbf{c}_1\mathbf{c}_2]\mathbf{b}_2 - [\mathbf{ab}_1\mathbf{c}_1\mathbf{b}_2]\mathbf{c}_2\right)$$
$$= \mathbf{a} \wedge \left([\mathbf{ab}_1\mathbf{b}_2\mathbf{c}_2]\mathbf{c}_1 - [\mathbf{ac}_1\mathbf{b}_2\mathbf{c}_2]\mathbf{b}_1\right). \quad (10.21)$$

The point other than $\mathbf{a}$ at the intersection is called the *second point of intersection*.

Suppose that we are already given one point of intersection $\mathbf{a}$, and we want to have an expression for the second point of intersection. First of all, the expression is neither vector $[\mathbf{ab}_1\mathbf{c}_1\mathbf{c}_2]\mathbf{b}_2 - [\mathbf{ab}_1\mathbf{c}_1\mathbf{b}_2]\mathbf{c}_2$ nor vector $[\mathbf{ab}_1\mathbf{b}_2\mathbf{c}_2]\mathbf{c}_1 - [\mathbf{ac}_1\mathbf{b}_2\mathbf{c}_2]\mathbf{b}_1$, because both vectors are not null. Second, the second point of intersection must be of the form $[\mathbf{ab}_1\mathbf{c}_1\mathbf{c}_2]\mathbf{b}_2 - [\mathbf{ab}_1\mathbf{c}_1\mathbf{b}_2]\mathbf{c}_2 + \lambda \mathbf{a}$ or $[\mathbf{ab}_1\mathbf{b}_2\mathbf{c}_2]\mathbf{c}_1 - [\mathbf{ac}_1\mathbf{b}_2\mathbf{c}_2]\mathbf{b}_1 + \mu \mathbf{a}$, where λ, μ are scalars to make the whole expression into a null vector. In order to obtain a monomial and symmetric representation with respect to $\mathbf{ab}_1\mathbf{c}_1$ and $\mathbf{ab}_2\mathbf{c}_2$, we need to introduce a new meet product to unify the two different forms, and another product to convert a non-null vector into a null one.

Hence, to represent the second intersection point by a null vector multiplicatively in a monomial manner, we need to introduce two more products into CGA, the *reduced meet product* and the *nullification product*. The reduced meet product of $\mathbf{b}_1 \wedge \mathbf{c}_1$ and $\mathbf{b}_2 \wedge \mathbf{c}_2$ with base $\mathbf{a}$ is denoted by $(\mathbf{b}_1 \wedge \mathbf{c}_1) \vee_\mathbf{a} (\mathbf{b}_2 \wedge \mathbf{c}_2)$ and defined by

$$\mathbf{a} \wedge \left\{ (\mathbf{b}_1 \wedge \mathbf{c}_1) \vee_\mathbf{a} (\mathbf{b}_2 \wedge \mathbf{c}_2) \right\} = (\mathbf{a} \wedge \mathbf{b}_1 \wedge \mathbf{c}_1) \vee (\mathbf{a} \wedge \mathbf{b}_2 \wedge \mathbf{c}_2). \tag{10.22}$$

The above identity indicates that the reduced meet product defined in the CGA over $\mathbb{R}^{3,1}$ with base $\mathbf{a}$ is unique only *modulo* $\mathbf{a}$, i.e., if both $\mathbf{u}, \mathbf{v} \in \mathbb{R}^{3,1}$ satisfy

$$\mathbf{a} \wedge \mathbf{u} = \mathbf{a} \wedge \mathbf{v} = (\mathbf{a} \wedge \mathbf{b}_1 \wedge \mathbf{c}_1) \vee (\mathbf{a} \wedge \mathbf{b}_2 \wedge \mathbf{c}_2), \tag{10.23}$$

then $\mathbf{u} = \mathbf{v} + \lambda \mathbf{a}$ for some scale λ. Despite the uncertainty, $\mathbf{a}\{(\mathbf{b}_1 \wedge \mathbf{c}_1) \vee_\mathbf{a} (\mathbf{b}_2 \wedge \mathbf{c}_2)\}$ and $\{(\mathbf{b}_1 \wedge \mathbf{c}_1) \vee_\mathbf{a} (\mathbf{b}_2 \wedge \mathbf{c}_2)\}\mathbf{a}$ are both fixed.

An important property of the Minkowski plane is that it has two and only two null directions, and the two directions can be interchanged by any reflection in the plane. Since in Geometric Algebra a reflection is generated by the graded adjoint action of an invertible vector [2], the second intersection of circles/lines $\mathbf{ab}_1\mathbf{c}_1$ and $\mathbf{ab}_2\mathbf{c}_2$ can be represented by reflecting vector $\mathbf{a}$ with respect to vector $(\mathbf{b}_1 \wedge \mathbf{c}_1) \vee_\mathbf{a} (\mathbf{b}_2 \wedge \mathbf{c}_2)$:

$$\frac{1}{2}\left\{ (\mathbf{b}_1 \wedge \mathbf{c}_1) \vee_\mathbf{a} (\mathbf{b}_2 \wedge \mathbf{c}_2) \right\}\mathbf{a}\left\{ (\mathbf{b}_1 \wedge \mathbf{c}_1) \vee_\mathbf{a} (\mathbf{b}_2 \wedge \mathbf{c}_2) \right\}. \tag{10.24}$$

In CGA, the *nullification product* of $\mathbf{a}$ by $\mathbf{b}$ is defined by

$$N_\mathbf{b}(\mathbf{a}) := \frac{1}{2}\mathbf{aba}. \tag{10.25}$$

An nD *null Grassmann–Cayley algebra* refers to a Grassmann–Cayley algebra whose generating vectors are null and whose algebraic operators include not only the outer product, meet product, dual and bracket operators, but also i-grading operators where i takes values in $\{0, 1, n-1, n\}$, the reduced meet product, and the nullification product.

Null Grassmann–Cayley algebra is a language for describing Euclidean incidence constructions. For example, in the plane there are three points $\mathbf{1}, \mathbf{2}, \mathbf{3}$. The line passing through point $\mathbf{1}$ and parallel to line $\mathbf{23}$ has the following monomial representation:

$$\mathbf{e} \wedge \mathbf{1} \wedge \langle \mathbf{e23} \rangle_1, \tag{10.26}$$

where $\langle \mathbf{e23} \rangle_1$ is a vector describing the direction of line $\mathbf{23}$. As a second example, the line passing through point $\mathbf{1}$ and perpendicular to line $\mathbf{23}$ has the following monomial representation:

$$\mathbf{e} \wedge \mathbf{1} \wedge \langle \mathbf{e23} \rangle_3^*, \tag{10.27}$$

where $\langle \mathbf{e23} \rangle_3^*$ is a vector describing the normal direction of line $\mathbf{23}$.

Fig. 10.3 Geometric interpretations of $\langle \mathbf{a}_1\mathbf{a}_2\mathbf{a}_3\mathbf{a}_4 \rangle$ and $[\mathbf{a}_1\mathbf{a}_2\mathbf{a}_3\mathbf{a}_4]$

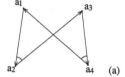
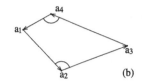

CGA also provides a hierarchy of advanced invariants for geometric computing. In the previous subsection, we have shown that there are two basic invariants, $\langle \mathbf{a}_1\mathbf{a}_2 \rangle = \mathbf{a}_1 \cdot \mathbf{a}_2 = \langle \mathbf{a}_1\mathbf{a}_2 \rangle_0$, and $[\mathbf{a}_1\mathbf{a}_2 \cdots \mathbf{a}_n] = (\mathbf{a}_1 \wedge \cdots \wedge \mathbf{a}_n)^* = \langle \mathbf{a}_1\mathbf{a}_2 \cdots \mathbf{a}_n \rangle_n^*$, where $\langle \ \rangle_0$ and $\langle \ \rangle_n$ are grading operators of grade 0 and n, respectively. The geometric product prolongs the two basic invariants to the following two sequences of advanced invariants: for any $k, l \geq 0$,

$$\langle \mathbf{a}_1\mathbf{a}_2 \cdots \mathbf{a}_{2k} \rangle := \langle \mathbf{a}_1\mathbf{a}_2 \cdots \mathbf{a}_{2k} \rangle_0,$$
$$[\mathbf{a}_1\mathbf{a}_2 \cdots \mathbf{a}_{n+2l}] := \langle \mathbf{a}_1\mathbf{a}_2 \cdots \mathbf{a}_{n+2l} \rangle_n^*. \tag{10.28}$$

The two kinds of advanced invariants have nice geometric interpretations and algebraic properties. For example, if the $\mathbf{a}_i$ are null vectors representing points such that $\mathbf{a}_i \cdot \mathbf{e} = -1$, then

$$\langle \mathbf{a}_1\mathbf{a}_2 \cdots \mathbf{a}_{2k} \rangle = \frac{1}{2} \langle \overrightarrow{\mathbf{a}_1\mathbf{a}_2}\overrightarrow{\mathbf{a}_2\mathbf{a}_3} \cdots \overrightarrow{\mathbf{a}_{2k-1}\mathbf{a}_{2k}}\overrightarrow{\mathbf{a}_{2k}\mathbf{a}_1} \rangle,$$
$$[\mathbf{a}_1\mathbf{a}_2 \cdots \mathbf{a}_{n+2l}] = (-1)^n \frac{1}{2} [\overrightarrow{\mathbf{a}_1\mathbf{a}_2}\overrightarrow{\mathbf{a}_2\mathbf{a}_3} \cdots \overrightarrow{\mathbf{a}_{n+2l}\mathbf{a}_1}]. \tag{10.29}$$

Here $\overrightarrow{\mathbf{a}_i\mathbf{a}_j}$ denotes the displacement vector from point $\mathbf{a}_i$ to point $\mathbf{a}_j$ in Euclidean geometry. In particular, when $n = 2$, $k = 2$, and $l = 1$, then

$$\langle \mathbf{a}_1\mathbf{a}_2\mathbf{a}_3\mathbf{a}_4 \rangle = -\frac{d_{\mathbf{a}_1\mathbf{a}_2}d_{\mathbf{a}_2\mathbf{a}_3}d_{\mathbf{a}_3\mathbf{a}_4}d_{\mathbf{a}_4\mathbf{a}_1}}{2} \cos(\angle \mathbf{a}_1\mathbf{a}_2\mathbf{a}_3 + \angle \mathbf{a}_3\mathbf{a}_4\mathbf{a}_1),$$
$$[\mathbf{a}_1\mathbf{a}_2\mathbf{a}_3\mathbf{a}_4] = -\frac{d_{\mathbf{a}_1\mathbf{a}_2}d_{\mathbf{a}_2\mathbf{a}_3}d_{\mathbf{a}_3\mathbf{a}_4}d_{\mathbf{a}_4\mathbf{a}_1}}{2} \sin(\angle \mathbf{a}_1\mathbf{a}_2\mathbf{a}_3 + \angle \mathbf{a}_3\mathbf{a}_4\mathbf{a}_1), \tag{10.30}$$

where $d_{\mathbf{ab}}$ denotes the Euclidean distance between points $\mathbf{a}, \mathbf{b}$.

In the plane, if $\mathbf{a}_1, \mathbf{a}_2, \mathbf{a}_4, \mathbf{a}_3$ are the sequence of vertexes of a quadrilateral (Fig. 10.3(a)), then $\angle \mathbf{a}_1\mathbf{a}_2\mathbf{a}_3, \angle \mathbf{a}_3\mathbf{a}_4\mathbf{a}_1$ have opposite signs; if $\mathbf{a}_1, \mathbf{a}_2, \mathbf{a}_3, \mathbf{a}_4$ are the sequence of vertexes of a quadrilateral (Fig. 10.3(b)), then $\angle \mathbf{a}_1\mathbf{a}_2\mathbf{a}_3, \angle \mathbf{a}_3\mathbf{a}_4\mathbf{a}_1$ have the same sign. By (10.30), $[\mathbf{a}_1\mathbf{a}_2\mathbf{a}_3\mathbf{a}_4] = 0$ iff $\angle \mathbf{a}_1\mathbf{a}_2\mathbf{a}_3 + \angle \mathbf{a}_3\mathbf{a}_4\mathbf{a}_1 = 0 \bmod \pi$ or, equivalently, iff points $\mathbf{a}_1, \mathbf{a}_2, \mathbf{a}_3, \mathbf{a}_4$ are on the same circle or line.

The two advanced invariants have the following reversion symmetries and shift symmetries:

$$\langle \mathbf{a}_1\mathbf{a}_2 \cdots \mathbf{a}_{2k} \rangle = \langle \mathbf{a}_{2k}\mathbf{a}_{2k-1} \cdots \mathbf{a}_1 \rangle = \langle \mathbf{a}_{2k}\mathbf{a}_1\mathbf{a}_2 \cdots \mathbf{a}_{2k-1} \rangle,$$
$$[\mathbf{a}_1\mathbf{a}_2 \cdots \mathbf{a}_{n+2l}] = (-1)^{\frac{n(n-1)}{2}} [\mathbf{a}_{n+2l} \cdots \mathbf{a}_2\mathbf{a}_1] = (-1)^{n-1} [\mathbf{a}_{n+2l}\mathbf{a}_1 \cdots \mathbf{a}_{n+2l-1}]. \tag{10.31}$$

Clifford bracket algebra [3] is the algebra of advanced invariants generated by the above two kinds of brackets. The elements are naturally graded by their lengths. Formally, an nD Clifford bracket algebra generated by a sequence $\mathscr{S}$ of m symbols of vectors $\mathbf{1}, \mathbf{2}, \ldots, \mathbf{m}$, where $m \geq n$, is the quotient of the polynomial ring generated by (1) subsequences of length n, denoted by square brackets, (2) symmetric pairs of vectors, denoted by angular brackets, (3) repeatable permutations of vectors of length $n + 2k$ for $k > 0$, denoted by square brackets, (4) another group of repeatable permutations of vectors of length $2l + 2$ for $l > 0$, denoted by angular brackets, modulo the ideal generated by GP1 in (10.8), GP2 in (10.9), where the dot products are replaced by angular brackets of length two, and the following SB and AB:

- SB:

$$[\mathbf{i}_1 \mathbf{i}_2 \cdots \mathbf{i}_{n+2k}] - \sum_{1 \leq \sigma \leq n+2k} \text{sign}(\sigma, \breve{\sigma}) \langle \mathbf{i}_{\sigma(1)} \mathbf{i}_{\sigma(2)} \cdots \mathbf{i}_{\sigma(2k)} \rangle$$
$$\times [\mathbf{i}_{\breve{\sigma}(1)} \mathbf{i}_{\breve{\sigma}(2)} \cdots \mathbf{i}_{\breve{\sigma}(n)}], \tag{10.32}$$

where $\sigma, \breve{\sigma}$ run over all permutations of $1, 2, \ldots, n + 2k$ such that $\sigma(1) < \sigma(2) < \cdots < \sigma(2k)$ and $\breve{\sigma}(1) < \breve{\sigma}(2) < \cdots < \breve{\sigma}(n)$.

- AB:

$$\langle \mathbf{i}_1 \mathbf{i}_2 \cdots \mathbf{i}_{2l} \rangle - \sum_{k=2}^{2l} (-1)^k \langle \mathbf{i}_1 \mathbf{i}_k \rangle \, \langle \mathbf{i}_2 \cdots \breve{\mathbf{i}}_k \cdots \mathbf{i}_{2l} \rangle. \tag{10.33}$$

In fact, $\langle \mathbf{i}_1 \mathbf{i}_2 \cdots \mathbf{i}_{2l} \rangle$ is just the Pfaffian of the $\mathbf{i}$'s, and (10.33) is the recursive relation of Pfaffians. (10.32) is the nth-grade *Caianiello expansion* [5] of the geometric product of the $\mathbf{i}$'s. The term Pfaffian was introduced by A. Cayley, who used the term in 1852 to honor of the German mathematician J. Pfaff [14].

The Clifford bracket algebra generated by points and the point at infinity $\mathbf{e}$ in conformal geometric algebra is *null bracket algebra* [7]. It is an $(n + 2)$D Clifford bracket algebra with the special requirement that all symbolic vectors are null, i.e., the following *null syzygies*:

$$\langle \mathbf{ii} \rangle = 0. \tag{10.34}$$

10.3 Applications: Geometric Factorization, Decomposition, and Theorem Completion

In this section, we present an example of automated discovering of new geometric theorems to show the essential role played by null geometric algebra. The scenario is as follows [7]: for a geometric theorem, if one of its hypotheses is removed, then the conclusion is no longer true. However, the conclusion should contain the removed hypothesis as a factor. If the other factors of the conclusion are all geometrically meaningful, the factorization of the conclusion is called *geometric*.

If more than one hypothesis is removed, by Hilbert's Nullstellensatz, some power of the conclusion can be written as a linear combination of the removed hypotheses. If the coefficients in the combination are geometrically meaningful, then this decomposition is called *geometric*. However, a geometric decomposition does not provide any clear geometric interpretation to the conclusion other than the quantitative contribution of each hypothesis to the conclusion in geometrical terms. If instead the conclusion can be written in some suitable covariant algebra into the form $f = 0$ where f is a monomial, then it has clear geometric interpretation, and a new theorem is created (or discovered), called the geometric *completion* of the original theorem. It generalizes an existing theorem by reducing its hypotheses.

In the following example, first the geometric factorization is carried out by removing one hypothesis, then a geometric completion is reached by removing one more hypothesis, and finally a geometric decomposition is obtained by expanding the completion. The original theorem is very easy:

Example 10.1 In the plane two circles intersect at points $\mathbf{1}, \mathbf{1}'$, respectively. Draw two secant lines through them, which intersect the two circles at points $\mathbf{2}, \mathbf{3}$ and $\mathbf{2}', \mathbf{3}'$, respectively, then $\mathbf{22}'//\mathbf{33}'$. See Fig. 10.4 (left).

The first question is this: if one constraint is absent, say $\mathbf{1}, \mathbf{1}', \mathbf{2}, \mathbf{2}'$ are no longer cocircular (see Fig. 10.4 (right)), then how far are lines $\mathbf{22}'$ and $\mathbf{33}'$ away from being parallel?

A beautiful formula is obtained with null bracket algebra:

$$\frac{1}{2}\frac{[e22'e33']}{(e \cdot 2)(e \cdot 2')(3 \cdot 3')} = \frac{[e13]}{[e12]}\frac{[e131']}{[e31'2']}\frac{[11'22']}{(1 \cdot 1')(1 \cdot 3)}. \tag{10.35}$$

Its geometric interpretation follows the list below:

(1) $e \cdot 2 = e \cdot 2' = -1$;

(2) $3 \cdot 3' = \dfrac{d_{33'}}{2}, \quad 1 \cdot 1' = \dfrac{d_{11'}}{2}, \quad 1 \cdot 3 = \dfrac{d_{13}}{2}$;

(3) $[e13'1'] = 2 S_{131'}, \quad [e31'2'] = 2 S_{31'2'}$;

(4) $[e22'e33'] = -2(\mathbf{23} \times \mathbf{2'3'})$;

(5) $\dfrac{[e13]}{[e12]} = \dfrac{d_{13}}{d_{12}}\varepsilon_{123}$.

Here (a) $S_{131'}$ denotes the signed area of triangle $\mathbf{131}'$ with respect to the orientation of the plane; (b) $\overrightarrow{\mathbf{22}'} \times \overrightarrow{\mathbf{33}'}$ denotes the signed length of the vector product of vectors $\overrightarrow{\mathbf{22}'}, \overrightarrow{\mathbf{33}'}$ with respect to the unit normal of the plane; (c) $\varepsilon_{123} = 1$ if point $\mathbf{1}$ is inside line segment $\mathbf{23}$ and -1 otherwise.

Fig. 10.4 *Left*: the original
theorem; *right*: one
hypothesis removed

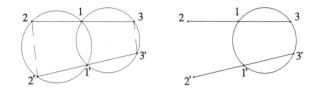

The computing of (10.35) is executed by a general algorithm as follows:

Input   (1) Geometric objects constructed sequentially, with free objects first.
(2) Target $conc = [\mathbf{e}22'\mathbf{e}33']$, which is a Clifford algebraic expression.

The construction sequence of the configuration is as follows:
Free points **1, 2, 1', 2'**.
Semifree point **3** on line **12**.
Intersection $\mathbf{3}' = \mathbf{1}'\mathbf{e}2' \cap \mathbf{1}'\mathbf{13}$.
This means that **1', 3'** are the intersection of circle **1'13** and line **1'2'**.

Output $conc/conc'$ after canceling their common factors, where $conc'$ is an expression to homogenize $conc$, and for which we choose $\mathbf{3} \cdot \mathbf{3}'$. Below we explain the term "homogenization."

In the homogeneous model, any geometric relation occurs as a homogeneous equality. Unfortunately, this is no longer true in algebraic representations of geometric entities. For example, let **a** be the intersection of lines **12** and **1'2'**. In Grassmann–Cayley algebra,

$$\mathbf{a} = (\mathbf{1} \wedge \mathbf{2}) \vee \left(\mathbf{1}' \wedge \mathbf{2}'\right). \tag{10.36}$$

Obviously, this is not a homogeneous relation. The five vectors can be scaled arbitrarily and independently, so the equality can only be understood as an equality up to an arbitrary scale. When we compute the quantitative relations among geometric objects, we certainly do not want a result with arbitrary scale.

There is a remedy for this. For example, in a rational expression in which the degree of point **a** in the numerator equals that in the denominator, the substitution of (10.36) into the expression does not cause any arbitrary scaling. If we compute like this:

$$\frac{[\mathbf{abc}]}{[\mathbf{ab'c'}]} = \frac{(\mathbf{1} \wedge \mathbf{2}) \vee (\mathbf{1}' \wedge \mathbf{2}') \vee (\mathbf{b} \wedge \mathbf{c})}{(\mathbf{1} \wedge \mathbf{2}) \vee (\mathbf{1}' \wedge \mathbf{2}') \vee (\mathbf{b}' \wedge \mathbf{c}')},$$

then we get an equality invariant under the scaling of **a**.

Homogenization is to change a nonhomogeneous equality into a homogeneous one. To achieve this, we need to compute a second expression $conc'$ containing the same constrained vector variables with their degrees inclusive as those in $conc$. Then we compute $conc/conc'$.

Part 1. Elimination   (1) Eliminate the last entity from *conc*. Expand and simplify the result. (2) Go to the beginning of Step 1 if *conc* contains any constrained entity.

In this example there are two constructions, the second point of intersection

$$3' = \frac{1}{2} \{ (e \wedge 2') \vee_{1'} (1 \wedge 3) \} 1' \{ (e \wedge 2') \vee_{1'} (1 \wedge 3) \} \tag{10.37}$$

and a free point **3** on line **12**; from $e \wedge 1 \wedge 2 \wedge 3 = 0$ we get the following Cramer's rule:

$$[e12]3 = [123]e - [e23]1 + [e13]2, \tag{10.38}$$

where the brackets are based on the 3D space spanned by $e, 1, 2, 3$.

The eliminations are made by substituting the expressions of the constructions into the conclusion expression and then making simplification:

$$[e22'e33'] \stackrel{\text{eliminate } 3'}{=} \frac{1}{2} [e22'e3\{(e \wedge 2') \vee_{1'} (1 \wedge 3)\} 1'\{(e \wedge 2') \vee_{1'} (1 \wedge 3)\}]$$

$$\stackrel{\text{expand}}{=} \frac{1}{2} \underbrace{[e31'2'][e131']}[e22'e311'2']$$

$$\stackrel{\text{simplify}}{=} \underbrace{e \cdot 2'}[2'2e311']$$

$$\stackrel{\text{eliminate } 3}{=} \frac{[e13]}{[e12]}[2'2e211']$$

$$\stackrel{\text{simplify}}{=} 2\underbrace{(e \cdot 2)}[121'2']. \tag{10.39}$$

Explanation:
(1) Step 1 substitutes the expression of $3'$ into *conc*.
(2) Step 2 expands the reduced meet products in a monomial manner: since

$$(e \wedge 2') \vee_{1'} (1 \wedge 3) = [1'e2'3]1 - [1'e2'1]3$$
$$= [1'e13]2' - [1'2'13]e \quad \mathrm{mod}\ 1',$$

if $3'$ occurs in an expression where $3'$ is neighbor to either null vector **1** or null vector **3**, by selecting the first expansion one can control the conclusion expression to be 1-termed. Alternatively, if $3'$ is neighbor to either **e** or $2'$ in the conclusion expression, then choosing the last expansion leads to a monomial result. This is called *monomial expansion* in null Grassmann–Cayley algebra.

In the above computing, the first meet product in the expression is neighbor to **3**, while the second is neighbor to **e** by shift symmetry, so up to scale, the two meet products are replaced by vectors **1** and $2'$, respectively. The under braced factors in the result do not participate in further eliminations and simplifications, and are removed from succeeding computing procedure.

Fig. 10.5 Two hypotheses
removed

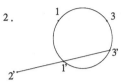

(3) Step 3 is simplification after expansions and is based on the symmetries

$$\left[e22'e311'2'\right] = -\left[2'e22'e311'\right] = \left[2'e2'2e311'\right]$$

and monomial expansion $2'e2' = 2(e \cdot 2')2'$.

(4) Step 4 substitutes the expression of **3** from Cramer's rule into *conc*.

(5) The last step is based on $2e2 = 2(e \cdot 2)2$ and antisymmetry within a bracket of length $n + 2 = 4$.

Part 2. Homogenization   Use the algorithm in Part 1 to compute $conc' = 3 \cdot 3'$.

$$3 \cdot 3' \overset{\text{eliminate 3'}}{=} \frac{1}{2}\langle 3\{(e \wedge 2') \vee_{1'} (1 \wedge 3)\}1'\{(e \wedge 2') \vee_{1'} (1 \wedge 3)\}\rangle$$

$$\overset{\text{expand}}{=} \frac{1}{2}\underbrace{\left[e31'2'\right]^2}\langle 311'1\rangle$$

$$\overset{\text{simplify}}{=} (1 \cdot 1')(1 \cdot 3).$$

The ratio *conc/conc'* gives the desired identity (10.35), which is an extended theorem. It provides a quantization of the dependency of the conclusion upon the cocircularity of points $1, 2, 1', 2'$.

In the above computing procedure, the conclusion expression remains 1-termed. The result is naturally in factored form containing the desired factor $[121'2']$. Furthermore, (10.35) is a quantitative description of the conclusion.

Next, we remove one more hypothesis. We remove straight line **123**. The new configuration, as shown in Fig. 10.5, has only two constraints: cocircularity $[131'3'] = 0$ and collinearity $[e1'2'3'] = 0$ in the homogeneous model.

The new configuration can be constructed as follows: points $1, 2, 3, 1', 2'$ are free in the plane, and points $1', 3'$ are at the intersection of line $1'2'$ and circle $131'$. In (10.39), we have already obtained the result of eliminating $3'$ after the third step. Again, choose $conc' = 3 \cdot 3'$. We have

$$\frac{conc}{conc'} = \frac{[e22'e33']}{3 \cdot 3'} = \frac{e \cdot 2'[e131'][e311'2'2]}{(1 \cdot 1')(1 \cdot 3)[e31'2']}. \tag{10.40}$$

(10.40) is the *geometric completion* of the original theorem under the constraints that $1, 3, 1', 3'$ are cocircular and $1', 2', 3'$ are collinear. Its geometric meaning is immediate from

$$\left[\mathbf{e}311'2'2\right] = \frac{1}{2} d_{31} d_{11'} d_{1'2'} d_{2'2} \sin(\angle(\overrightarrow{3\mathbf{1}}, \overrightarrow{1\mathbf{1}'}) + \angle(\overrightarrow{1'2'}, \overrightarrow{2'2})). \qquad (10.41)$$

The *geometric decomposition* of $[\mathbf{e}22'\mathbf{e}33']$ with respect to the two removed hypotheses $[121'2'] = 0$ and $[\mathbf{e}123]$ is obtained by the following *rational binomial expansion* [9] of $[\mathbf{e}311'2'2]$:

$$\begin{aligned}
\left[\mathbf{e}311'2'2\right] &= -\left[2\mathbf{e}311'2'\right] \\
&= -\frac{1}{2(1 \cdot 2)} \left[2\mathbf{e}31\underline{21}1'2'\right] \\
&= -\frac{1}{2(1 \cdot 2)} \left(\langle 2\mathbf{e}31 \rangle \left[211'2'\right] + \langle 211'2' \rangle [2\mathbf{e}31]\right).
\end{aligned} \qquad (10.42)$$

An algebraic proof is said to be a *monomial* (or *binomial*) one if throughout the proving procedure, the expressions in process are monomials (or binomials at most). By now we have tested over 100 theorems in Euclidean geometry involving circles and angles. About two thirds are given binomial proofs, and about one third are given monomial proofs.

10.4 Conclusion

In symbolic geometric computation, the bottleneck is middle expression swell, which makes many computations possible only theoretically. Another problem is geometric explanation of algebraic results. Often this is impossible if using coordinates, especially when the results are not invariant under coordinate transforms. In classical invariant-theoretical methods the two problems remain, and new difficulties arise.

In this chapter, we introduce a new invariant framework based on Clifford algebra and the homogeneous model of classical geometry. In geometric computing, the advanced invariants introduced in this framework bring about amazing simplifications in algebraic manipulations. The proofs generated by such advanced invariants have the features that the symbolic manipulations are easy and succinct, the input and output are both geometrically meaningful, and the proofs provide *quantitative descriptions* of the relations among the conclusion and the hypotheses.

Still this is just the beginning. A variety of open problems, old and new, are waiting there for us to solve. Their solving may ultimately lead to a revolution in symbolic geometric computing, which is a revitalization of synthetic covariant approach to classical geometry.

10.5 Exercises

10.1 Using the nullification product show that $N_{\mathbf{e}}(\mathbf{a})$ for a non-null vector $\mathbf{a}$ representing a sphere is proportional to the point at the center of the sphere (remembering that $\mathbf{e}$ represents the point at infinity).

10.2 Verify the geometric interpretation of (10.30).

10.3 Verify the geometric interpretation of all factors in (10.35).

10.4 Verify the geometric interpretation of all factors in (10.40) using hint (10.42).

Acknowledgements   Both authors are supported partially by NSFC 10871195, NSFC 60821002/ F02 and NKBRSF 2011CB302404.

References

1. Chou, S.-C., Gao, X.-S., Zhang, J.-Z.: Machine Proofs in Geometry. World Scientific, Singapore (1994)
2. Hestenes, D., Sobczyk, G.: Clifford Algebra to Geometric Calculus. Reidel, Dordrecht (1984)
3. Li, H.: Automated theorem proving in the homogeneous model with Clifford bracket algebra. In: Dorst, L., et al. (eds.) Applications of Geometric Algebra in Computer Science and Engineering, pp. 69–78. Birkhäuser, Boston (2002)
4. Li, H.: Algebraic representation and elimination and expansion in automated geometric theorem proving. In: Winkler, F. (ed.) Automated Deduction in Geometry, pp. 106–123. Springer, Heidelberg (2004)
5. Li, H.: Automated geometric theorem proving, Clifford bracket algebra and Clifford expansions. In: Qian, T., et al. (eds.) Trends in Mathematics: Advances in Analysis and Geometry, pp. 345–363. Birkhäuser, Basel (2004)
6. Li, H.: Clifford algebras and geometric computation. In: Chen, F., Wang, D. (eds.) Geometric Computation, pp. 221–247. World Scientific, Singapore (2004)
7. Li, H.: Symbolic computation in the homogeneous geometric model with Clifford algebra. In: Gutierrez, J. (ed.) Proceedings of International Symposium on Symbolic and Algebraic Computation 2004, pp. 221–228. ACM Press, New York (2004)
8. Li, H.: A recipe for symbolic geometric computing: long geometric product, BREEFS and Clifford factorization. In: Brown, C.W. (ed.) Proc. ISSAC 2007, pp. 261–268. ACM Press, New York (2007)
9. Li, H.: Invariant Algebras and Geometric Reasoning. World Scientific, Singapore (2008)
10. Li, H., Huang, L.: Complex brackets balanced complex differences, and applications in symbolic geometric computing. In: Proc. ISSAC 2008, pp. 181–188. ACM Press, New York (2008)
11. Li, H., Wu, Y.: Automated short proof generation in projective geometry with Cayley and Bracket algebras I. Incidence geometry. J. Symb. Comput. **36**(5), 717–762 (2003)
12. Li, H., Wu, Y.: Automated short proof generation in projective geometry with Cayley and Bracket algebras II. Conic geometry. J. Symb. Comput. **36**(5), 763–809 (2003)
13. Li, H., Hestenes, D., Rockwood, A.: Generalized homogeneous coordinates for computational geometry. In: Sommer, G. (ed.) Geometric Computing with Clifford Algebras, pp. 27–60. Springer, Heidelberg (2001)
14. Muir, T.: A Treatise on the Theory of Determinants. Macmillan & Co., London (1882)
15. Sommer, G. (ed.): Geometric Computing with Clifford Algebras. Springer, Heidelberg (2001)
16. Sturmfels, B.: Algorithms in Invariant Theory. Springer, Wien (1993)
17. Sturmfels, B., White, N. (eds.): Invariant-Theoretic Algorithms in Geometry. Special Issue. J. Symb. Comput. **11** (5/6) (2002)
18. White, N. (ed.): Invariant Methods in Discrete and Computational Geometry. Kluwer, Dordrecht (1994)
19. Wu, W.-T.: Mathematics Mechanization. Kluwer and Science Press, Beijing (2000)

On the Use of Conformal Geometric Algebra in Geometric Constraint Solving

11

Philippe Serré, Nabil Anwer, and JianXin Yang

Abstract

To model a geometrical part in Computer Aided Design systems, declarative modeling is a well-adapted solution to declare and specify geometric objects and constraints. In this chapter, we are interested in the representation of geometric objects and constraints using a new language of description and representation, Geometric Algebra (GA). GA is used here in association with the conformal model of Euclidean geometry (CGA) which requires two extra dimensions comparing to the usual vector space model. Topologically and Technologically Related Surfaces (TTRS) Theory is introduced here as a unified framework for geometric objects representation and geometric constraints solving. Based on TTRS, this chapter shows the capability of the CGA to represent geometric objects and geometric constraints through symbolic geometric constraints solving and algebraic classification.

P. Serré (✉)
LISMMA, Institut Supérieur de Mécanique de Paris, Paris, France
e-mail: philippe.serre@supmeca.fr

N. Anwer
LURPA, École Normale Supérieure de Cachan, Cachan, France
e-mail: anwer@lurpa.ens-cachan.fr

J. Yang
Robotics and Machine Dynamics Laboratory, Beijing University of Technology, Beijing, P.R. China
e-mail: yangjx@tsinghua.org.cn

L. Dorst, J. Lasenby (eds.), *Guide to Geometric Algebra in Practice*,
DOI 10.1007/978-0-85729-811-9_11, © Springer-Verlag London Limited 2011

11.1   Declarative Modeling of Geometric Systems

The aim of declarative modeling in mechanical design is to provide high-level tools
and methods to assist the designer to solve technical problems and find suitable
solutions. The associated geometric problem can be described through a set of variables on geometrical objects, and design requirements as a set of constraints related
to those variables. In contrast with procedural and rule-based modeling, declarative
modeling explicitly describes properties of designed objects but not how to construct
them.

From a formal point of view, declarative modeling consists of a set of variables
and constraints, which state properties and requirements by defining relations between variables. The model does not specify how to satisfy those constraints, but
rather a constraint solver is used to determine values for the variables such that all
constraints are satisfied [20]. In general, there can be no, one or many solutions.

Declarative modeling can be divided in three phases [8, 21].

- The *description* step defines the interaction language that allows the designer to
  provide properties and relations between the design objects. Global (total description of the model) and incremental (partial description based on trial and error
  strategy) approaches are used during this phase.
- The *generation* step performs an exploration of the solution space (totally or partially) to find one, several or all the solutions that match the properties given by
  the designer. Different constraint solving methods have been investigated such as
  Constraint Satisfaction Problem (CSP) techniques. If no solution can be found
  that satisfies all constraints, the generation step should provide more intelligence
  on problem solving thanks to automated deduction, training or user preferences.
- The *lookup* step allows the designer to visualize the results. Most declarative
  modelers choose to present only one solution to the designer. When no solution
  can be provided, it is necessary to detect and explain the causes of the failure
  so that the designer can make some modifications. When several solutions are
  generated, specially adapted tools are required in order to help the designer to
  determine the most appropriate solution.

The three phases described above are suited to the geometric design of mechanical
systems by declarative modeling when following specific processes largely embedded in parametric Computer-Aided Design (CAD) systems. The geometric modeling of mechanical systems relies on the description of a set of primitives related to
shapes (geometric objects) and geometric constraints that define size, position and
orientation of geometric objects.

11.2   Geometric Constraint Solving

Geometric Constraint Solving is considered as a critical function of modern CAD
systems. Most of CAD systems provide parametric and feature-based modeling
methods to enable the designer semantic-based and intuitive interactivity for digital models and to support frequent model changes. The geometric objects and con-

straints at the description step mentioned in the previous section have a significant impact on how easily a design can be constructed and modified.

A typical geometric constraint problem requires finding a configuration of points, lines and planes with prescribed pair-wise constraints between these geometric objects. A pair-wise constraint may be the distance or the angle between them.

Betig [2] characterized Geometric Constraint Solving as follows:

> *Given a set of geometric objects, such as points, lines and circles; given a set of constraints, such as distance and angle; and given an embedding space, such as the Euclidean plane or the Euclidean 3D space; assign coordinates to the geometric objects such that the constraints are satisfied, or report that no such assignment has been found.*

Geometric constraint solving methods can be classified in three broad categories: graph-based, logic-based and algebraic-based [1, 12, 13]. Solving a geometric constraint problem can be also considered a problem of automated proving of geometric theorems.

The graph-based geometric constraint solving methods works in two steps. First the geometric problem is translated into a graph whose vertices represent the set of geometric elements and whose edges are the constraints. Then the constraint problem is solved by decomposing the graph into a collection of sub-graphs each representing a standard problem [13]. The typical problems solved by these methods are ruler and compass constructive problems.

The Rule-based geometric constraint solving methods rely on the predicates formulation considering a set of geometric assertions and axioms and an inference engine is used to derive the solution by exhaustively applying rules. Although they provide a qualitative study of geometric constraints, their exhaustive searching and matching computations make them inappropriate for real-world CAD applications.

The Algebraic-based geometric constraint solving methods translate the constraint problem into a set of equations. The equations are in general nonlinear. The problem can then be solved using any of the known methods for solving nonlinear equations. The main advantages of algebraic solvers are their generality, dimension independence and the ability to deal with symbolic constraints in a natural way. Algebraic methods can be further classified according to the specific technique used to solve the system of equations, namely into numerical and symbolic methods. Numerical methods have the potential to solve large nonlinear systems, while symbolic methods can be used only for small systems.

The geometric objects considered in our work, based on TTRS model, are embedded in 2D or 3D Euclidean space. They are defined by a list of elements chosen among the point, the line and the plane (Minimum Geometrical Reference Element), and 13 pair-wise constraints are then defined among them [7]. An initial digital representation is needed to initialize the topology of the object to be created. Obviously, this initial digital representation does not respect all the constraints, and there is no request to find an exact solution of the given problem, but the processing time should be shortened as is currently the case with the existing solvers in CAD.

11.3   Topologically and Technologically Related Surfaces (TTRS)

Rigid body motions are fundamental for the design of mechanical systems. The classification of three-dimensional surfaces according to their invariant properties under the action of rigid motions was investigated in mechanisms theory by Hervé [9] and, in the context of Geometric Dimensioning and Tolerancing, by Clément and Srinivasan [6, 19]. The Geometrical Product Specification (GPS) program under ISO TC213 [4] also adopted the classification of three-dimensional surfaces based on their invariance under the action of rigid motions.

In Topologically and Technologically Related Surfaces (TTRS) Theory [7], three-dimensional surfaces or features are classified according to their respective degree of invariance under the action of rigid motions. Basically, seven main features equivalent to kinematic lower pairs are identified: planar feature, cylindrical feature, revolution feature, spherical feature, prismatic feature, helicoidal feature and complex feature. Each main feature is then described by a unique minimum geometrical reference element (MGRE) allowing the positioning in the Euclidean space. An MGRE is set as a combination of elementary geometrical objects: Point, Line and Plane. TTRS Theory has been adopted by ISO TC213 and successfully implemented in the CATIA v5 CAD system to manage assembly constraints and tolerance annotations.

The classification described above can be also applied to a collection of two surfaces, resulting in a composed surface again classifiable according to the classification of the two previous surfaces. All the possible combinations of elementary surfaces or features are then reduced to only 28 situations. 13 reclassifying cases between Points, Lines and Planes have been defined.

11.3.1 Lie Algebra of the Group of Rigid Motions

From a theoretical point of view, the above classification derives from the properties of the twelve connected Lie subgroups of $T(3) \times SO(3)$ (the group of translations and rotations in the three-dimensional Euclidean space). A comprehensive system of Lie subalgebras and corresponding Lie subgroups for rigid body motion was presented by Hervé [10]. The most general formulation of the rigid body motion is the screw motion. According to Lie's theory of continuous groups, an infinitesimal motion or displacement is represented by an operator acting on the points of the three-dimensional Euclidean space. This operator includes a field of moments called screws which are a six-dimensional vector that combines rotational and translational motions and velocities of a rigid body. Screws are elements of the Lie algebra of the group of rigid motions [16].

The classification of three-dimensional surfaces based on their invariance under the action of the Lie group of rigid motions is extremely powerful but needs more mathematical formalization and computability [5].

11.3.2 Geometric Algebra Considerations

Geometrical representations of Points, Lines and Planes, the group of rigid body motions, the Lie algebra of this group and screw theory have already been investigated using Geometric and Clifford Algebras [16, 17]. According to Hestenes, the conformal model is becoming a standard for applications of Euclidean Geometry and will promote a rejuvenation of screw theory [11]. Just like the resurrection of screw theory, the prosperity of Geometric Algebra in kinematics needs some time [15].

11.4   CGA for Geometric Constraint Solving

In the following sections, we study the capability of the conformal model to represent the geometric objects (here Point, Line and Plane) and geometric constraints between them. Two examples are presented to demonstrate practical applications of CGA in the field of declarative modeling of geometric system for CAD applications.

11.4.1 Example of Symbolic Solving

In this section, we will illustrate the usage of CGA to solve symbolically geometric problems defined by constraints. The aim of the development presented below is to highlight the three steps of the algebraic treatment. The first step is the representation of geometric elements. The second step is to write equations that define the geometric constraints. The third step is the symbolic solving itself.

The development presented is the case study given in Fig. 11.1. It is a problem defined by four geometric elements and five constraints between them. Geometric elements are: two points M_1 and M_2, line L and plane P. Geometric constraints are: three incidence relations between M_1 and L, between M_2 and L, between M_1 and P, one distance constraint between M_2 and P and one angular constraint between L and P. The distance is equal to d. The angle is equal to α.

Blades of CGA are used to represent geometric elements. Geometric constraints are expressed by algebraic relations between vectors. The geometric problem is symbolically solved step by step by algebraic operations. The complete sequence is described in the following.

11.4.1.1 Representation of Geometric Elements
The method begins with the declaration of geometric objects. Each of them is represented by CGA elements.
1. The four points M_1, M_2, M_3 and M_4 are represented by null vectors called m_1, m_2, m_3 and m_4;
2. The line L passes through M_3 with unit direction vector $\mathbf{t}$. L is represented by the CGA 3-blade L,

$$L = m_3 \wedge \mathbf{t} \wedge n_\infty. \tag{11.1}$$

Let a point X be represented by the CGA vector x; then $\forall x \in L$, $x \wedge L = 0$;

Fig. 11.1 Case study

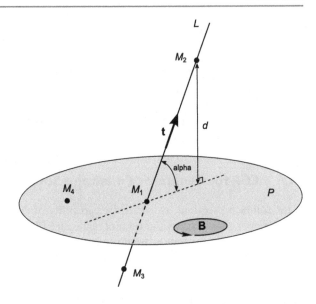

3. The plane P passes through M_4 with unit direction bivector $\mathbf{B}$. P is represented by the CGA 4-blade P,

$$P = m_4 \wedge \mathbf{B} \wedge n_\infty. \tag{11.2}$$

Let a point X be represented by the CGA vector x; then $\forall x \in P,\ x \wedge P = 0$.

11.4.1.2 Representation of Geometric Constraints
The second step is to define the algebraic relations between the CGA elements. Some of them are given in [18].

1. C_1 constraint: M_1 and L are incident. This translates into the following algebraic relation:

$$m_1 \wedge L = 0. \tag{11.3}$$

2. C_2 constraint: M_2 and L are incident. This translates into the following algebraic relation:

$$m_2 \wedge L = 0. \tag{11.4}$$

3. C_3 constraint: M_1 and P are incident. This translates into the following algebraic relation:

$$m_1 \wedge P = 0. \tag{11.5}$$

4. C_4 constraint: The distance between M_2 and P is equal to d. This translates into the following algebraic relation:

$$m_2 \wedge P = \varepsilon d I_5 \tag{11.6}$$

with $\varepsilon = \pm 1$ and $I_5 = n_o \wedge e_1 \wedge e_2 \wedge e_3 \wedge n_\infty$ the pseudoscalar of CGA space.

This relation indicates that $m_2 \wedge P$ is proportional to I_5. Because P is a unit 4-blade, the norm of $m_2 \wedge P$ is equal to the distance between M_2 and P. The sign of ε defines whether M_2 is above or below P.

5. C_5 constraint: The angle between P and L is equal to α. This translates into the following algebraic relation:

$$\mathbf{t} \wedge \mathbf{B} = \sin \alpha \mathbf{I_3} \tag{11.7}$$

with $\alpha \in [-\frac{\pi}{2}, \frac{\pi}{2}]$.

This relation indicates that $\mathbf{t} \wedge \mathbf{B}$ is proportional to $\mathbf{I_3} = e_1 \wedge e_2 \wedge e_3$, the pseudoscalar of Euclidean space. Because $\mathbf{t}$ is a unit vector and $\mathbf{B}$ a unit bivector, the norm of $\mathbf{t} \wedge \mathbf{B}$ is the sinus of the angle between L and P.

The relations are now known, and the last step is to solve them symbolically.

11.4.1.3 Symbolic Solving

In this section, we detail the symbolic resolution of the five previous relations. Initially, the 3-blade L is expressed in terms of m_3 and $\mathbf{t}$. The first operation is to express L in terms of m_1 and $\mathbf{t}$. Because the constraint C_1 implies Eq. (11.3), we have:

$$m_1 \wedge L = 0 \quad \Rightarrow \quad \begin{cases} m_1 = m_3 + \mathbf{x} + \gamma_1 n_\infty \\ \mathbf{x} \wedge \mathbf{t} = 0 \end{cases}$$

$$\Rightarrow \quad \begin{cases} m_3 = m_1 - \mathbf{x} - \gamma_1 n_\infty \\ \mathbf{x} \wedge \mathbf{t} = 0 \end{cases} \tag{11.8}$$

with a Euclidean vector $\mathbf{x}$ collinear with $\mathbf{t}$ and a real number γ_1 such that $\gamma_1 = \frac{1}{2}\mathbf{m_1}^2$.

By replacing m_3 in Eq. (11.1),

$$L = m_3 \wedge \mathbf{t} \wedge n_\infty$$
$$= (m_1 - \mathbf{x} - \gamma_1 n_\infty) \wedge \mathbf{t} \wedge n_\infty$$
$$= m_1 \wedge \mathbf{t} \wedge n_\infty \tag{11.9}$$

because $\mathbf{x} \wedge \mathbf{t}$ vanishes.

Initially, the 4-blade P is expressed in terms of m_4 and $\mathbf{B}$. The second operation is to express P in terms of m_1 and $\mathbf{B}$. Because C_3 constraint implies Eq. (11.5), we have:

$$m_1 \wedge P = 0 \quad \Rightarrow \quad \begin{cases} m_1 = m_4 + \mathbf{y} + \gamma_2 n_\infty \\ \mathbf{y} \wedge \mathbf{B} = 0 \end{cases}$$

$$\Rightarrow \quad \begin{cases} m_4 = m_1 - \mathbf{y} - \gamma_2 n_\infty \\ \mathbf{y} \wedge \mathbf{B} = 0 \end{cases} \tag{11.10}$$

with a Euclidean vector $\mathbf{y}$ coplanar with $\mathbf{B}$ and a real number γ_2 such that $\gamma_2 = \frac{1}{2}\mathbf{m_1}^2$.

By replacing m_4 in Eq. (11.2),

$$
\begin{aligned}
P &= m_4 \wedge \mathbf{B} \wedge n_\infty \\
&= (m_1 - \mathbf{y} - \gamma_2 n_\infty) \wedge \mathbf{B} \wedge n_\infty \\
&= m_1 \wedge \mathbf{B} \wedge n_\infty
\end{aligned}
\tag{11.11}
$$

because $\mathbf{y} \wedge \mathbf{B}$ vanishes.

The third operation is to establish relation between m_1 and m_2, because they belong to the same line. Because C_3 constraint implies Eq. (11.4) and thanks to Eq. (11.9), we have:

$$
m_2 \wedge L = 0 \quad \Rightarrow \quad \begin{cases} m_2 = m_1 + \mathbf{z} + \gamma_3 n_\infty \\ \mathbf{z} \wedge \mathbf{t} = 0 \end{cases}
\tag{11.12}
$$

with a Euclidean vector $\mathbf{z}$ collinear with $\mathbf{t}$ and a real number γ_3 such that $\gamma_3 = \frac{1}{2}\mathbf{m_2}^2$.

The fourth operation is to expand the term $m_2 \wedge P$ using expression (11.12) for m_2 and expression (11.11) for P:

$$
\begin{aligned}
m_2 \wedge P &= (m_1 + \mathbf{z} + \gamma_3 n_\infty) \wedge P \\
&= (m_1 + \mathbf{z} + \gamma_3 n_\infty) \wedge (m_1 \wedge \mathbf{B} \wedge n_\infty) \\
&= \mathbf{z} \wedge m_1 \wedge \mathbf{B} \wedge n_\infty \\
&= -m_1 \wedge \mathbf{z} \wedge \mathbf{B} \wedge n_\infty.
\end{aligned}
\tag{11.13}
$$

Because $\mathbf{z}$ and $\mathbf{t}$ are vectors and $\mathbf{z} \wedge \mathbf{t} = 0$, so $\mathbf{z} = \lambda \mathbf{t}$ with λ a real number. Therefore Eq. (11.13) becomes

$$
m_2 \wedge P = -\lambda(m_1 \wedge \mathbf{t} \wedge \mathbf{B} \wedge n_\infty).
\tag{11.14}
$$

Development of expression (11.14) continues, replacing $\mathbf{t} \wedge \mathbf{B}$. Because C_5 constraint implies Eq. (11.7), we have:

$$
\begin{aligned}
m_2 \wedge P &= -\lambda \sin\alpha (m_1 \wedge \mathbf{I_3} \wedge n_\infty) \\
&= -\lambda \sin\alpha (n_o \wedge \mathbf{I_3} \wedge n_\infty) \\
&= -\lambda \sin\alpha I_5.
\end{aligned}
\tag{11.15}
$$

The last operation is to eliminate $m_2 \wedge P$ in Eqs. (11.15) and (11.6). To do that, C_4 constraint is used.

$$
\begin{cases} m_2 \wedge P = -\lambda \sin\alpha I_5 \\ m_2 \wedge P = \varepsilon d I_5 \end{cases} \quad \Rightarrow \quad \varepsilon d I_5 = -\lambda \sin\alpha I_5.
\tag{11.16}
$$

Fig. 11.2 Association between L_i and L_j

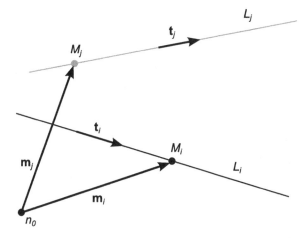

Finally, the following scalar relation encapsulates the geometric problem

$$\lambda = \varepsilon \frac{d}{\sin \alpha}. \tag{11.17}$$

Solving the preceding relation permits to construct in the 3D affine space, a geometric object consistent with the constraints. If the reference frame of the 3D affine space is defined by the point O and the Euclidean coordinate frame (e_1, e_2, e_3), a possible geometric construction is: point M_1 is the origin O, P is the (e_1, e_2) plane, and L is coplanar with the (e_2, e_3) plane. Vector $\mathbf{t}$ is defined by the coordinates $(0, \cos \alpha, \sin \alpha)$. Point M_2 is defined by the coordinates $\varepsilon \frac{d}{\sin \alpha} (0, \cos \alpha, \sin \alpha)$.

Fortunately, the outcome of the computation for this example is as expected. The authors do not claim to have given a general solution to the problem of the symbolic resolution of geometric problems defined by constraints, because the model chosen and the proposed sequence are not unique. A detailed study of this domain lies in the research works of Shang-Ching Chou on automated geometry theorem proving [3] and Hongbo Li on automatic symbolic simplification [14] (see also Chap. 10 of this volume).

11.4.2 Example of Geometric Classification

The objective of this section is to present the role of CGA to classify the association of two elementary geometric objects. One result of TTRS theory is the classification of two lines in three different cases: coincident, parallel or general. We will describe a solution using CGA to obtain algebraically the classification of two lines L_i and L_j.

11.4.2.1 Representation of Geometric Elements

Figure 11.2 indicates the geometric elements used in the development. Lines are drawn in arbitrary position.

Point M_i (resp. M_j) is represented by the null vector called m_i (resp. m_j). Line L_i (resp. L_j) of direction $\mathbf{t}_i$ (resp. $\mathbf{t}_j$) passing through a point m_i (resp. m_j) is represented by the 3-blade: $L_i = m_i \wedge \mathbf{t}_i \wedge n_\infty$ (resp. $L_j = m_j \wedge \mathbf{t}_j \wedge n_\infty$).

Association between two lines is represented by a new CGA element called $TTRS_5$. This new element is the geometric product of L_i and L_j:

$$TTRS_5 = L_i L_j. \tag{11.18}$$

11.4.2.2 Geometric Interpretation of $TTRS_5$

Equation (11.18) is easily expandable by remarking that geometric product of $(m_i \wedge \mathbf{t}_i)n_\infty$ and $n_\infty(m_j \wedge \mathbf{t}_j)$ vanishes because n_∞ is a null vector. Therefore,

$$(m_i \wedge \mathbf{t}_i)n_\infty n_\infty(m_j \wedge \mathbf{t}_j) = 0 \quad \Rightarrow \quad (L_i + \mathbf{t}_i)(L_j - \mathbf{t}_j) = 0. \tag{11.19}$$

By replacing $L_i L_j$ in Eq. (11.19), $TTRS_5$ is represented by the following expression:

$$TTRS_5 = \mathbf{t}_i \mathbf{t}_j + L_i \mathbf{t}_j - \mathbf{t}_i L_j. \tag{11.20}$$

It reveals a rotor $\mathbf{t}_i \mathbf{t}_j$ that carries out the angular specification between the two lines. The expression $L_i \mathbf{t}_j - \mathbf{t}_i L_j$ is more difficult to interpret. An algebraic rewriting enables to compute the Euclidean vectors $\mathbf{p}$ and $\mathbf{d}_{ij}$. $\mathbf{p}$ is the position of the common perpendicular L of the two lines. $\mathbf{d}_{ij}$ is the direction of L. The norm of $\mathbf{d}_{ij}$ is the distance between the two lines. These elements are drawn in Fig. 11.3.

$$L_i \mathbf{t}_j - \mathbf{t}_i L_j = \mathbf{t}_j \cdot L_i - \mathbf{t}_i \cdot L_j - \mathbf{t}_j \wedge L_i - \mathbf{t}_i \wedge L_j$$
$$= \mathbf{t}_j \cdot L_i - \mathbf{t}_i \cdot L_j + \big((m_j - m_i) \wedge \mathbf{t}_i \wedge \mathbf{t}_j \wedge n_\infty\big). \tag{11.21}$$

This can be written in the form

$$L_i \mathbf{t}_j - \mathbf{t}_i L_j = \big((\mathbf{t}_i \cdot \mathbf{t}_j)\mathbf{d}_{ij} + (\mathbf{t}_i \wedge \mathbf{t}_j)\mathbf{p} + (\mathbf{d}_{ij} \wedge \mathbf{t}_i \wedge \mathbf{t}_j)\big) \wedge n_\infty, \tag{11.22}$$

where $\mathbf{p}$ and $\mathbf{d}_{ij}$ are Euclidean vectors. We will show that this specifies them completely and that their semantics is as in Fig. 11.3.

From here and for practical reasons, $TTRS_5$ is rewritten as

$$TTRS_5 = \mathbf{X} + \mathbf{Y} \wedge n_\infty \tag{11.23}$$

with $\mathbf{X} = TTRS_5 - (n_o \cdot TTRS_5) \wedge n_\infty$ and $\mathbf{Y} = n_o \cdot TTRS_5$.

Because $\mathbf{X}$ is a sum of a scalar and 2-blade, and $\mathbf{Y}$ is a sum of 1-blade and 3-blade, Eq. (11.23) can be expanded as

$$TTRS_5 = \langle \mathbf{X} \rangle_0 + \langle \mathbf{X} \rangle_2 + \langle \mathbf{Y} \rangle_1 \wedge n_\infty + \langle \mathbf{Y} \rangle_3 \wedge n_\infty. \tag{11.24}$$

Therefore, expressions of $\mathbf{d}_{ij}$ and $\mathbf{p}$ are given in terms of $\mathbf{X}$ and $\mathbf{Y}$:

Fig. 11.3 Geometric interpretation of α, $\mathbf{p}$ and $\mathbf{d}_{ij}$

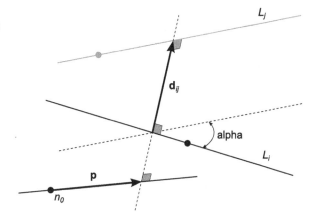

$$\mathbf{d}_{ij} = \frac{\langle \mathbf{Y} \rangle_3}{\langle \mathbf{X} \rangle_2},$$

$$\mathbf{p} = \frac{\langle \mathbf{X} \rangle_0 \langle \mathbf{Y} \rangle_3 - \langle \mathbf{Y} \rangle_1 \langle \mathbf{X} \rangle_2}{\langle \mathbf{X} \rangle_2{}^2}.$$

(11.25)

This study shows that the CGA multivector $TTRS_5$ embeds the geometrical constraints of distance and angle between the two lines. It can be used as a conceptual model of the geometrical constraint between two lines.

11.4.2.3 Classification of $TTRS_5$

Now is the time to find algebraically the classification proposed by the TTRS theory. Recall that $TTRS_5 = \mathbf{X} + \mathbf{Y} \wedge n_\infty$.

First, assume that the lines are coincident. Then L_i is proportional to L_j, and the geometric product between them is a scalar. Therefore, expression of $TTRS_5$ is:

$$TTRS_5 = \langle \mathbf{X} \rangle_0 + \cancel{\langle \mathbf{X} \rangle_2} + \cancel{\langle \mathbf{Y} \rangle_1 \wedge e_\infty} + \cancel{\langle \mathbf{Y} \rangle_3 \wedge n_\infty}$$

$$= \langle \mathbf{X} \rangle_0. \tag{11.26}$$

Second, assume that the lines are parallel. In this case, the terms $\langle \mathbf{X} \rangle_2$ and $\langle \mathbf{Y} \rangle_3$ vanish because $\mathbf{t}_i$ and $\mathbf{t}_j$ are collinear:

$$TTRS_5 = \langle \mathbf{X} \rangle_0 + \cancel{\langle \mathbf{X} \rangle_2} + \langle \mathbf{Y} \rangle_1 \wedge e_\infty + \cancel{\langle \mathbf{Y} \rangle_3 \wedge n_\infty}$$

$$= \langle \mathbf{X} \rangle_0 + \langle \mathbf{Y} \rangle_1 \wedge n_\infty. \tag{11.27}$$

Third, assume that the lines are coplanar. In this case, the term $\langle \mathbf{Y} \rangle_3$ vanishes because $\mathbf{d}_{ij}$, $\mathbf{t}_i$ and $\mathbf{t}_j$ are coplanar.

$$TTRS_5 = \langle \mathbf{X} \rangle_0 + \langle \mathbf{X} \rangle_2 + \langle \mathbf{Y} \rangle_1 \wedge n_\infty + \cancel{\langle \mathbf{Y} \rangle_3 \wedge n_\infty}$$

$$= \langle \mathbf{X} \rangle_0 + \langle \mathbf{X} \rangle_2 + \langle \mathbf{Y} \rangle_1 \wedge n_\infty. \tag{11.28}$$

In the general case, no terms vanish.

Fig. 11.4 Opposite chirality
for same geometric constraint

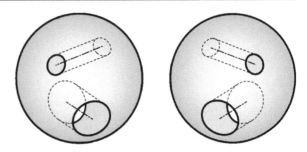

We see from this example that CGA gives an algebraic solution to classify the relationship of geometric objects. This is possible because $TTRS_5$ contains the metric and the orientation that characterize the association. In further work, we must study the generality of this representation for other TTRS.

11.5 Open Problems

To improve characteristics of existing solvers, it is possible to act in several directions. We have chosen to present two improvements that concern the declaration phase. If they were made, they would be of benefit for the designers of mechanical systems. The first is related to the specification of the chirality. The second is related to the specification of the mobility.

11.5.1 Chirality Specification

Today, in case of iso-constrained geometric problem, existing solvers generate only one solution even if the geometric problem has several. Generally, this is not troublesome for CAD applications except when the solver generates a symmetric solution to that expected by the designer. For instance, the designer defines the geometric object drawn in Fig. 11.4 with distance and angle constraints between the holes. The solver generates the left solution while the designer wants the right one.

Given this situation, the designer cannot change the generated solution interactively as he did with an under-constrained problem. The practical solution is to remove one or more constraints and modify interactively the generated solution that is now under-constrained. The new form is closer to that desired. Finally, he must reactivate suppressed constraints and solve again, hoping that this time the solver generates the expected solution.

This method is not suitable, because the same problem can occur several times during the design process. A better solution would be that the designer specifies what solution he wants. It could be done using the chirality constraint. We believe that CGA could represent this new kind of constraint, as we have seen previously with $TTRS_5$. In this case, the signs of $\langle \mathbf{Y}_{\text{left}} \rangle_3$ and $\langle \mathbf{Y}_{\text{right}} \rangle_3$ are opposite.

Fig. 11.5 Bennett linkage

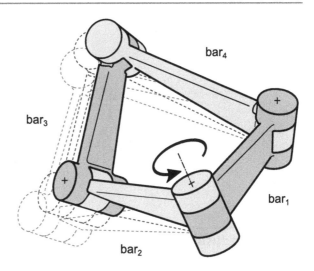

11.5.2 Mobility Specification

Geometric constraint solvers used today do not allow designers to specify the mobility of desired solutions.

Consider the over-constrained mechanism shown in Fig. 11.5, called a Bennett linkage. It is a spatial single degree of freedom mechanism consisting of four bars connected by revolute joints. Each of the bars is specified by two geometric parameters, the common perpendicular length between the revolute joints axes and the angle between these axes. If a designer uses a GCS to construct this system, the generated solution will be mostly the rigid structure presented in Fig. 11.6 and very rarely the expected mechanism of Fig. 11.5. Distances, angles of bars and geometric constraints between them are respected, but not the mobility because the designer can not specify it.

It is a weakness of GCS modules because it is a common request in the field of mechanical design. Today, the solution adopted is to use specific applications of kinematic synthesis as Lincages-2000[1] or Synthetica.[2] However, there is a lack of the interoperability between these systems and CAD software.

A better solution would be that the designer specifies the number of degree of freedom that the geometric system should have. It could be done using the mobility constraint. We believe that CGA could represent this new kind of constraint, because rotor and geometric constraints can be described in the same space.

---

[1] http://www.me.umn.edu/labs/lincages/new.html.

[2] http://www.umbc.edu/engineering/me/vrml/research/software/synthetica/index.html.

Fig. 11.6 Example of
solution generated by GCS
solver

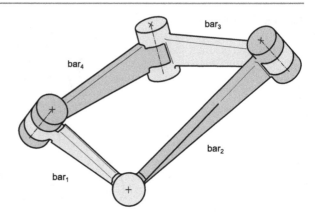

11.6   Conclusion

This chapter has shown that Geometric Algebra language and Conformal Geometric
Algebra (CGA) enable to model geometric objects and constraints by a system of
algebraic relations. This result is central for the development of a declarative modeling approach for mechanical systems in the context of Computer-Aided Design of
mechanical parts.

CGA has been successfully implemented for symbolic geometric constraints
solving and algebraic classification. Open problems were also discussed through
CGA perspectives. The main interest of CGA is to support a coordinate free formulation and to develop a new geometric framework for TTRS models and their
underlying Lie algebras and screw-based representations. Geometric reasoning and
automatic symbolic simplification based on CGA can be also considered as a perspective direction of the work presented here.

Finally, the scope of geometric algebra is its universality to model all the physical
sciences as was clearly demonstrated by D. Hestenes. This chapter has shown that,
with this mathematical tool, we could model CAD geometric artifacts. Thus, we
pave the way for the creation of general declarative models for Mechatronics. These
models will represent physical phenomena that interact with geometry (mechanical,
electrical, electromagnetism, . . .).

11.7   Exercises

11.1 Interpreting Eq. (11.25), show that $\mathbf{d}_{ij}$ and $\mathbf{p}$ are Euclidean vectors and have
the semantics of Fig. 11.1.

11.2 The association between two planes is represented by a CGA element called
$TTRS_9$. This element is the geometric product of P_i and P_j,

$$TTRS_9 = P_i P_j.$$

We know that $P_i = m_i \wedge \mathbf{B}_i \wedge n_\infty$ and $P_j = m_j \wedge \mathbf{B}_j \wedge n_\infty$. Give the expression for $TTRS_9$ in terms of P_i, P_j, $\mathbf{B}_i$ and $\mathbf{B}_j$.

11.3 The association between a line and a plane is represented by two CGA elements called $TTRS_6$ and $TTRS_8$, depending on the order. These two elements are the geometric product of P_i and L_j:

$$TTRS_6 = L_i P_j,$$

$$TTRS_8 = P_i L_j.$$

We know that $P_i = m_i \wedge \mathbf{B}_i \wedge n_\infty$ and $L_i = m_i \wedge \mathbf{t}_i \wedge n_\infty$. Give the expressions for $TTRS_6$ and $TTRS_8$ in terms of L_i, L_j, $\mathbf{t}_i$, $\mathbf{t}_i$, P_i, P_j, $\mathbf{B}_i$ and $\mathbf{B}_j$.

11.4 Consider the example of Fig. 11.1. Add a point N and a line L_2, and specify four new geometric constraints: C_6 is an incidence between L_2 and M_1, C_7 is an incidence between L_2 and N, C_8 is a distance constraint between P and N called d_2, and finally C_9 is a distance constraint between L and N called d_3.

If the unit direction vector of L_2 is $\mathbf{t}_2$, show that:

1. $n \wedge L = -\lambda_2(n \wedge \mathbf{t}_2 \wedge \mathbf{t} \wedge n_\infty)$ with λ_2 a real number.
2. $n \wedge P = -\lambda_2(n \wedge \mathbf{t}_2 \wedge \mathbf{B} \wedge n_\infty)$ with λ_2 the same real number.
3. Let the angle between P and L_2 be denoted as θ, and the angle between L and L_2 be denoted as ω. Give the expression for θ in terms of ω, d_2 and d_3.
4. Give the semantics of the Euclidean bivector $\mathbf{t} \wedge \mathbf{t}_2$.

References

1. Ait-Aoudia, S., Bahriz, M., Salhi, L.: 2D geometric constraint solving: an overview. In: Proceedings of 2nd International Conference in Visualisation (VIZ), Barcelona (Spain), July 15–17, 2009, pp. 201–206. IEEE Comput. Soc., Los Alamitos (2009)
2. Bettig, B., Hoffmann, C.M.: Geometric constraint solving in parametric computer-aided design. doi:10.1115/1.3593408
3. Chou, S.-C.: Mechanical Geometry Theorem Proving. Springer, Berlin (1988)
4. Chiabert, P., Orlando, M.: About a cat model consistent with iso/tc 213 last issues. Achievements in Mechanical and Materials Engineering Conference. J. Mater. Process. Technol. **157–158**, 61–66 (2004)
5. Chiabert, P., Lombardi, F., Vaccarino, F.: Analysis of kinematic methods for invariants based classification in the ISO/TC213 framework. In: Proceedings of the 10th CIRP International Seminar on Computer-Aided Tolerancing, Erlangen (Germany), March 21–23, 2007
6. Clément, A., Rivière, A., Temmerman, M.: Cotation tridimensionnelle des systèmes mécaniques – Théorie et pratique. PYC, Ivry-sur-Seine (1994)
7. Clément, A., Rivière, A., Serré, P., Valade, C.: The TTRS: 13 constraints for dimensioning and tolerancing. In: Proceedings of the 5th CIRP International Seminar on Computer-Aided Tolerancing, pp. 28–29 (1997)
8. Gaildrat, V.: Declarative modelling of virtual environments, overview of issues and applications. In: Plemenos, D., Miaoulis, G. (eds.) Proceedings of International Conference on Computer Graphics and Artificial Intelligence (3IA), Athens (Greece), May 30–31, 2007
9. Hervé, J.-M.: The mathematical group structure of the set of displacements. Mech. Mach. Theory **29**(1), 73–81 (1994)

10. Hervé, J.-M.: The Lie group of rigid body displacements, a fundamental tool for mechanism design. Mech. Mach. Theory **34**(5), 719–730 (1999)
11. Hestenes, D.: New tools for computational geometry and rejuvenation of screw theory. In: Bayro-Corrochano, E., Scheuermann, G. (eds.) Geometric Algebra Computing, pp. 3–33. Springer, London (2010)
12. Hoffmann, C.M., Joan-Arinyo, R.: A brief on constraint solving. Comput-Aided Des. Appl. **2**(5), 655–663 (2005)
13. Joan-Arinyo, R.: Basics on geometric constraint solving. In: Proceedings of 13th Encuentros de Geometrfa Computacional (EGC09), Zaragoza (Spain), June 29–July 1, 2009
14. Li, H.: Invariant Algebras and Geometric Reasoning. World Scientific, Singapore (2008)
15. Luo, Z., Dai, J.S.: Mathematical methodologies in computational kinematics. In: 14th Biennial Mechanisms Conference, Chong Qing (China), 2004
16. Selig, J.M., Bayro-Corrochano, E.: Rigid body dynamics using Clifford algebra. Adv. Appl. Clifford Algebras **20**, 141–154 (2010)
17. Selig, J.M.: Clifford algebra of points, lines and planes. Robotica **18**(5), 545–556 (2000)
18. Serré, P., Moinet, M., Clément, A.: Declaration and specification of a geometrical part in the language of geometric algebra. In: Advanced Mathematical and Computational Tools in Metrology and Testing VIII. Series on Advances in Mathematical for Applied Sciences, vol. 78, pp. 298–308 (2009)
19. Srinivasan, V.: A geometrical product specification language based on a classification of symmetry groups. Comput. Aided Des. **31**(11), 659–668 (1999)
20. van der Meiden, H.A., Bronsvoort, W.F.: A constructive approach to calculate parameter ranges for systems of geometric constraints. Comput. Aided Des. **38**(4), 275–283 (2006)
21. Zaragoza, J., Ramos, F., Orozco, H.R., Gaildrat, V.: Creation of virtual environments through knowledge-aid declarative modeling. In: LAPTEC, pp. 114–132 (2007)

On the Complexity of Cycle Enumeration for Simple Graphs

12

René Schott and G. Stacey Staples

Abstract

We show how a number of combinatorial problems, such as determining the number of cycles in graphs, can be recast using a graded commutative algebra constructed within a real Grassmann exterior algebra. The computational complexity of this approach is then measured by considering operations at the basis blade level of the algebra. In particular, the worst-case time complexity of counting arbitrary length cycles in simple n-vertex graphs via nilpotent adjacency matrix methods is shown to be $\mathscr{O}(n^{\alpha+1}2^n)$, where $\alpha \leq 3$ is the exponent representing the complexity of matrix multiplication. The storage complexity of the nilpotent adjacency matrix approach is $\mathscr{O}(n^2 2^n)$. A probabilistic model is used to describe a class of graphs in which the average-case time complexity of cycle enumeration is $\mathscr{O}(n^3(1+q)^n)$ for fixed $0 < q < 1$. For reference, experimental results detailing computation times (in seconds) are compared with algorithms based on the approaches of Bax and Tarjan.

12.1 Introduction

The complexity of a number of NP-class combinatorial problems can be solved using only a polynomial number of multivector operations in a 2^n-dimensional algebra generated by n mutually commuting null-squares, as we have shown in [5]. In particular, by defining a "nilpotent adjacency matrix" associated with a finite graph

R. Schott
IECN and LORIA, Nancy Université, Université Henri Poincaré, BP 239, 54506
Vandoeuvre-lès-Nancy, France
e-mail: schott@loria.fr

G.S. Staples (✉)
Department of Mathematics and Statistics, Southern Illinois University Edwardsville,
Edwardsville, IL 62026-1653, USA
e-mail: sstaple@siue.edu

L. Dorst, J. Lasenby (eds.), *Guide to Geometric Algebra in Practice*,
DOI 10.1007/978-0-85729-811-9_12, © Springer-Verlag London Limited 2011

on n vertices, the problem of enumerating cycles of length k requires $\mathcal{O}(n^\alpha \log k)$ multivector operations in the algebra, where $\alpha \leq 3$ denotes the exponent associated with matrix multiplication. While $\alpha < 3$ for ordinary matrix multiplication (see [2]), such algorithmic speedups do not necessarily hold for matrices whose elements are multivectors.

In this chapter, the computational complexity of enumerating cycles of arbitrary length in graphs is studied in greater detail by counting algebraic operations at the basis blade level of the algebra as opposed to the multivector level. For practical reference, the theoretical complexity of this approach is compared to that of the algorithms of Bax and Tarjan. Moreover, experimental comparisons using MATHE-MATICA illustrate practical advantages of the nilpotent adjacency matrix approach, particularly in the case of sparse graphs.

The algebra used in the construction can be regarded as a commutative subalgebra of the exterior algebra $\bigwedge \mathbb{R}^{2n}$. Its generators commute and square to zero so that linear combinations of the generators are nilpotent. This nilpotent nature is the key to combinatorial applications.

All MATHEMATICA examples were computed on a 2.4 GHz MacBook Pro with 4 GB of 667 MHz DDR2 SDRAM running MATHEMATICA 6 for MAC OS X with the *Combinatorica* package. Cycle enumeration is accomplished using the nilpotent adjacency matrix approach, Bax's approach, and the HamiltonianCycle procedure found in the MATHEMATICA package *Combinatorica*. Time plots comparing the three approaches are included. MATHEMATICA code used to generate examples can be found online through the second-named author's web page, http://www.siue.edu/~sstaple.

12.2 Essential Background

A *graph* $G = (V, E)$ is a collection of vertices V and a set E of unordered pairs of vertices called *edges*. Two vertices $v_i, v_j \in V$ are said to be *adjacent* if there exists an edge $e_{ij} = \{v_i, v_j\} \in E$. In this case, the vertices v_i and v_j are said to be *incident* with e_{ij}.

A *k-walk* $\{v_0, \dots, v_k\}$ in a graph G is a sequence of vertices in G with *initial vertex* v_0 and *terminal vertex* v_k such that there exists an edge $(v_j, v_{j+1}) \in E$ for each $0 \leq j \leq k - 1$. A k-walk contains k edges. A *k-path* is a k-walk in which no vertex appears more than once. A *closed k-walk* is a k-walk whose initial vertex is also its terminal vertex. A *k-cycle* is a closed k-path with $v_0 = v_k$.

The *diameter* of a graph G is defined to be the length of the longest path in G. The *girth* and *circumference* of a graph are defined to be the lengths of the graph's shortest and longest cycles, respectively. If the graph has no cycles, its girth and circumference are defined to be ∞.

If $G = (V, E)$ is a graph, A is its adjacency matrix, and $S \subseteq V$, define the modified adjacency matrix A_S by

$$[A_S]_{ij} = \begin{cases} A_{ij} & \text{if } i \in S \text{ and } j \in S, \\ 0 & \text{otherwise.} \end{cases}$$

Theorem 12.1 (Bax) *Each main diagonal element of*

$$\sum_{S\subseteq V}(-1)^{|V|-|S|}(A_S)^{|V|}$$

contains the number of Hamiltonian cycles in G if $|V| > 0$.

Bax's approach to cycle enumeration uses powers of a graph's adjacency matrix with the principle of inclusion–exclusion to count all Hamiltonian cycles in $\mathscr{O}(2^n \text{poly}(n))$ time and $\text{poly}(n)$ storage [1]. Enumerating only those cycles of length k is accomplished by applying Bax's algorithm to all k-vertex subgraphs. For fixed k, this is $\mathscr{O}(\text{poly}(n))$ since $\binom{n}{k} \leq n^k$ for all $n \geq k$. For k increasing with n, the complexity remains $\mathscr{O}(2^n \text{poly}(n))$, which can be verified by a rearrangement of Bax's formula (cf. Exercise 12.1).

Tarjan's algorithm (based on pruning with look-ahead) enumerates all cycles in a graph on n vertices with time complexity $\mathscr{O}((n + |E|)(C + 1))$ when applied to a graph with C cycles [7]. The storage complexity is $\mathscr{O}(n + |E| + S)$, where S is the sum of the lengths of all cycles. Note that the number of cycles on a k-vertex subgraph is potentially of order $k!$ while the number of subgraphs supporting such cycles is of order $\binom{n}{k}$.

A convenient and practical Tarjan-type implementation is the HamiltonianCycle procedure found in the MATHEMATICA package *Combinatorica*. The algorithm uses backtracking and look-ahead to enumerate all Hamiltonian cycles in a graph on n vertices. The implementation utilized for the examples in this paper enumerates cycles of length k in an n-vertex graph G by applying HamiltonianCycle to all k-vertex subgraphs of G. Implementations of this Tarjan-like approach are referred to henceforth as "CombiTarjan." Tarjan's algorithm actually lists cycles, which can result in $\mathscr{O}(n!)$ space complexity.

12.3   Technical Considerations

Fix positive integer n and let $\gamma = \{e_i : 1 \leq i \leq 2n\}$ be an orthonormal basis for $\mathbb{R}^{2n}$. Note that the pair $(\gamma, \wedge)$ generates a nonabelian semigroup A of order 2^{2n}. Let

$$Z = \{e_{2i-1} \wedge e_{2i} : 1 \leq i \leq n\} \subset A, \tag{12.1}$$

and note that the pair $(Z, \wedge)$ generates an abelian subsemigroup $\mathbb{R}\Psi_n$ of A. It should be clear that all elements of Ψ_n square to zero.

Definition 12.1   Let $\mathbb{R}\Psi_n$ denote the real abelian semigroup algebra of Ψ_n. For convenience, the generators of Ψ_n are rewritten as $\zeta_i = e_{2i-1} \wedge e_{2i}$, and henceforth the wedge operator is implicit.

Remark 12.1   The algebra $\mathbb{R}\Psi_n$ is isomorphic to the *n-particle zeon algebra* $\mathscr{C}\ell_n{}^{\text{nil}}$ appearing in the earlier work [5].

It is evident that the dimension of $\mathbb{R}\Psi_n$ is 2^n and that an arbitrary element $u \in \mathbb{R}\Psi_n$ can be expanded as

$$u = \sum_{I \in 2^{[n]}} u_I \zeta_I, \tag{12.2}$$

where $I \in 2^{[n]}$ is a subset of $[n] = \{1, 2, \ldots, n\}$ used as a multiindex, $u_I \in \mathbb{R}$, and $\zeta_I = \prod_{\iota \in I} \zeta_\iota$.

A canonical basis element ζ_I is referred to as a *blade*. The number of elements in the multiindex I is referred to as the *grade* of the blade ζ_I.

The *scalar sum* evaluation of an element $u \in \mathbb{R}\Psi_n$, denoted $\langle\langle u \rangle\rangle$, is defined by

$$\langle\langle u \rangle\rangle = \left\langle\left\langle \sum_{I \in 2^{[n]}} u_I \zeta_I \right\rangle\right\rangle = \sum_{I \in 2^{[n]}} u_I. \tag{12.3}$$

Definition 12.2 Let G be a graph on n vertices, either simple or directed with no multiple edges, and let $\{\zeta_i\}$, $1 \le i \le n$, denote the nilpotent generators of $\mathbb{R}\Psi_n$. Define the *nilpotent adjacency matrix* associated with G by

$$\mathscr{A}_{ij} = \begin{cases} \zeta_j & \text{if } (v_i, v_j) \in E(G), \\ 0 & \text{otherwise.} \end{cases} \tag{12.4}$$

Recalling Dirac notation, the ith row of $\mathscr{A}$ will be conveniently denoted by $\langle v_i | \mathscr{A}$, while the jth column will be denoted by $\mathscr{A} | v_j \rangle$.

A graph-theoretic interpretation of the nilpotent adjacency matrix can be stated thusly: $\langle v_i | \mathscr{A} | v_j \rangle = \zeta_j$ if and only if one can reach v_j from v_i in one step. Moreover, that "step" algebraically corresponds to multiplication by the null-square generator ζ_j. Extending by induction, nonzero terms of $\langle v_i | \mathscr{A}^k | v_j \rangle$ correspond to k-step walks from v_i to v_j in which each walk is "accomplished" in the algebra by computing a product of null-square generators. The null-square property then naturally "sieves out" walks on distinct generators, i.e., *self-avoiding* walks. This is all made precise in the next theorem.

Theorem 12.2 *Let $\mathscr{A}$ be the nilpotent adjacency matrix of an n-vertex graph G. For any $k > 1$ and $1 \le i, j \le n$,*

$$\langle v_i | \mathscr{A}^k | v_j \rangle = \sum_{\substack{(w_1,\ldots,w_k) \in V^k \\ (w_k = v_j) \wedge (m \ne \ell \Rightarrow w_m \ne w_\ell)}} \zeta_{\{w_1,\ldots,w_k\}} = \sum_{\substack{I \subseteq V \\ |I| = k}} \omega_I \zeta_I, \tag{12.5}$$

where ω_I denotes the number of k-step walks from v_i to v_j visiting each vertex in I exactly once when initial vertex $v_i \notin I$, and revisiting v_i exactly once when $v_i \in I$. In particular, for any $k \ge 3$ and $1 \le i \le n$,

$$\langle v_i | \mathscr{A}^k | v_i \rangle = \sum_{\substack{I \subseteq V \\ |I|=k}} \omega_I \zeta_I, \tag{12.6}$$

where ω_I denotes the number of k-cycles on vertex set I based at $v_i \in I$.

Proof Because the generators of $\mathbb{R}\Psi_n$ square to zero, a straightforward inductive argument shows that the nonzero terms of $\langle v_i | \mathscr{A}^k | v_j \rangle$ are multivectors corresponding to two types of k-walks from v_i to v_j: self-avoiding walks (i.e., walks with no repeated vertices) and walks in which v_i is repeated exactly once at some step but are otherwise self-avoiding. Walks of the second type are zeroed in the kth step when the walk is closed. Hence, terms of $\langle v_i | \mathscr{A}^k | v_i \rangle$ represent the collection of k-cycles based at v_i.   □

In light of this theorem, the name "nilpotent adjacency matrix" is justified by the following corollary.

Corollary 12.1 *Let $\mathscr{A}$ be the nilpotent adjacency matrix of a simple graph on n vertices. For any positive integer $k \leq n$, the entries of $\mathscr{A}^k$ are homogeneous elements of grade k in $\mathbb{R}\Psi_n$. Moreover, $\mathscr{A}^k = \mathbf{0}$ for all $k > n$.*

Another immediate corollary is that

$$\langle\langle \mathrm{tr}(\mathscr{A}^k) \rangle\rangle = k \big| \{k\text{-cycles in } G\} \big|, \tag{12.7}$$

since each k-cycle appears with k choices of base point along the main diagonal of $\mathscr{A}^k$.

In earlier work (see [5, 6]), the authors defined complexity in terms of the number of multivector operations in $\mathscr{C}\ell_n^{\mathrm{nil}}$, or "$\mathscr{C}\ell$ops", required. In contrast, this chapter work considers complexity at the level of the basis blades of $\mathbb{R}\Psi_n \cong \mathscr{C}\ell_n^{\mathrm{nil}}$.

Definition 12.3 A *blade operation* in $\mathbb{R}\Psi_n$ is defined as computing the sum or product of two basis blades. In particular, for multiindices I and J, each of the following computations is regarded as a blade operation:

$$(a\zeta_I)(b\zeta_J) = \begin{cases} 0 & \text{if } I \cap J \neq \emptyset, \\ (ab)\zeta_{I \cup J} & \text{otherwise;} \end{cases} \tag{12.8}$$

$$a\zeta_I + b\zeta_J = \begin{cases} (a+b)\zeta_I & \text{if } I = J, \\ a\zeta_I + b\zeta_J & \text{otherwise.} \end{cases} \tag{12.9}$$

Recalling the correlation between subsets of $[n]$ and bit strings of length n, each basis blade ζ_I is uniquely associated with a binary string $\underline{I}$. The cost of a basis blade multiplication in $\mathbb{R}\Psi_n$ is then equal to that of computing first the bitwise AND and then the bitwise OR of two n-bit words, which is known to be $\mathcal{O}(n)$. Summing a pair of basis blades is similarly $\mathcal{O}(n)$.

Given arbitrary elements $u, v \in \mathbb{R}\Psi_n$ and letting ν_u and ν_v denote the respective numbers of nonzero coefficients in the canonical expansions of u and v, the number of blade multiplications required to compute uv is then $\nu_u \nu_b$. The number of blade additions is similarly $\mathscr{O}(\nu_u \nu_v)$. Taking the costs of the blade operations themselves into consideration, the complexity of expanding the product uv is thus seen to be $\mathscr{O}(n \nu_u \nu_v)$.

The MATHEMATICA implementation of $\mathbb{R}\Psi_n$ used in the examples contained herein is based on subset operations rather than binary representations of subsets and bit operations. The additional overhead is offset by the relatively low dimensions of the examples.

12.4 Theoretical Complexity

Lemma 12.1 *Enumerating cycles in a simple graph on n vertices using nilpotent adjacency matrix methods has storage complexity $\mathscr{O}(n^2 2^n)$.*

Proof The nilpotent matrix method requires construction of $n \times n$ matrices whose entries are elements of a 2^n-dimensional algebra; i.e., in the worst case, $\mathscr{O}(2^n)$ coefficients must be associated with each matrix entry. Consequently, the space complexity is $\mathscr{O}(n^2 2^n)$.   □

Theorem 12.3 *The worst-case time complexity of enumerating cycles of arbitrary length in a graph on n vertices using the nilpotent adjacency matrix method is $\mathscr{O}(n^{\alpha+1} 2^n)$.*

Proof In light of Theorem 12.2, for any $k \le n$, computing $\mathscr{A}^k = \mathscr{A}^{k-1} A$ requires computing

$$\langle v_i | \mathscr{A}^k | v_j \rangle = \sum_{\ell=1}^{n} \langle v_i | \mathscr{A}^{k-1} | v_\ell \rangle \langle v_\ell | \mathscr{A} | v_j \rangle \tag{12.10}$$

for all $1 \le i, j \le n$. Entries of $\mathscr{A}^{k-1}$ are homogeneous grade-$(k-1)$ elements of $\mathscr{L}_n$. Moreover, terms in the canonical expansion of $\langle v_i | \mathscr{A}^{k-1} | v_\ell \rangle$ must be indexed by subsets containing v_ℓ, while in all cases, $\langle v_\ell | \mathscr{A} | v_j \rangle$ is either 0 or ζ_{v_j}.

Thus, the maximum number of blade multiplications performed in computing the product $\langle v_i | \mathscr{A}^{k-1} | v_\ell \rangle \langle v_\ell | \mathscr{A} | v_j \rangle$ is $\binom{n-1}{k-2}$ for each $1 \le \ell \le n$.

Computing the product $\mathscr{A}^{k-1} \mathscr{A}$ then requires at most $n^\alpha \binom{n-1}{k-2}$ blade multiplications. Applying this result recursively, computing $\mathscr{A}^k$ requires

$$n^\alpha \sum_{\ell=2}^{k} \binom{n-1}{\ell-2} < n^\alpha 2^{n-1} \tag{12.11}$$

blade multiplications. Since each blade multiplication is of complexity $\mathcal{O}(n)$, the result follows. $\square$

Note that an immediate consequence of the theorem is that the worst-case complexity of computing the girth and circumference of a graph on n vertices is also $\mathcal{O}(n^{\alpha+1}2^n)$.

Recalling that the diameter of a graph is defined as the length of the graph's longest *path*, another corollary is obtained.

Corollary 12.2 *The worst-case time complexity of computing the diameter of a graph on n vertices using the nilpotent adjacency matrix method is $\mathcal{O}(n^{\alpha+1}2^n)$.*

Proof Letting Δ be the $n \times n$ diagonal matrix whose main diagonal entries are $\Delta_{ii} = \zeta_i$, it can be shown that the off-diagonal entries of $\Delta \mathscr{A}^k$ are homogeneous grade-$(k+1)$ multivectors corresponding to k-paths in the graph associated with $\mathscr{A}$. The effect of left multiplication by Δ is to account for the initial vertex of any walk. Hence, the diameter of the graph is given by the largest positive integer k for which $\Delta \mathscr{A}^k$ is *not* the zero matrix. $\square$

Note that the complexity of computing $\mathscr{A}^k$ may vary depending on various methods of computing powers. The *iterated method* requires $k-1$ matrix products to compute

$$\mathscr{A}^k := \begin{cases} \mathscr{A} & \text{if } k=1, \\ \mathscr{A}^{k-1}\mathscr{A} & \text{otherwise.} \end{cases} \tag{12.12}$$

Given the binary representation of positive integer k, the *successive squares method* requires $\lfloor \log_2 k \rfloor$ matrix products and matrix sums to compute. In particular, letting $\underline{k}$ be a set of nonnegative integers such that $k = \sum_{\ell \in \underline{k}} 2^\ell$, then

$$\mathscr{A}^k = \sum_{\ell \in \underline{k}} \mathscr{A}^{2^\ell}. \tag{12.13}$$

While the successive squares method is generally more efficient than the iterated method, the application to nilpotent adjacency matrices is not straightforward. All discussion is henceforth restricted to the iterated method.

Example 12.1 Computation times in seconds are given for enumerating $\lfloor n/2 \rfloor$-cycles in randomly generated n-vertex graphs in Fig. 12.1. The experimental results illustrate practical advantages of the nilpotent adjacency matrix approach.

n	p	Zeon Time	Graph	Bax Time	CombiTarjan Time	cycle size	#{k-cycles}
8	0.25	0.003565	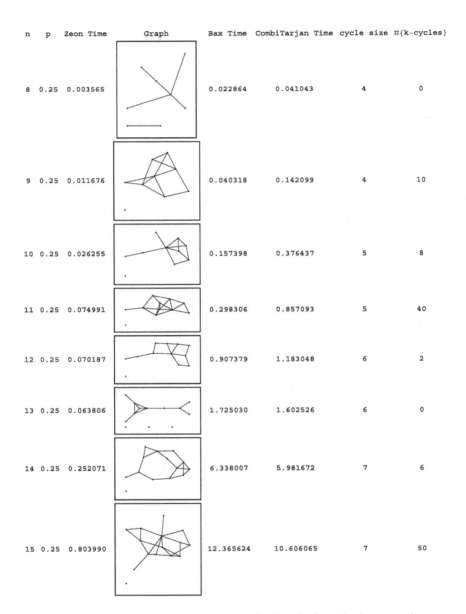	0.022864	0.041043	4	0
9	0.25	0.011676		0.040318	0.142099	4	10
10	0.25	0.026255		0.157398	0.376437	5	8
11	0.25	0.074991		0.298306	0.857093	5	40
12	0.25	0.070187		0.907379	1.183048	6	2
13	0.25	0.063806		1.725030	1.602526	6	0
14	0.25	0.252071		6.338007	5.981672	7	6
15	0.25	0.803990		12.365624	10.606065	7	50

Fig. 12.1 Times (in seconds) required to enumerate $\lfloor n/2 \rfloor$-cycles in randomly generated n-vertex graphs having equiprobable edges ($p = 0.25$)

12.4.1 Average-Case Complexity in "Suitably Sparse" Graphs

Discussion now turns to the average-case complexity of cycle enumeration in homogeneous random graphs. A *homogeneous random graph* $G = \mathscr{G}_{n,p}$ is a graph on n vertices with independent equiprobable edges of probability p. That is, each pair of vertices in the graph has equal probability p of being adjacent.

The next example illustrates the role of graph sparseness in the algorithmic comparisons. For fixed values of n and k, the time required to count k-cycles in an n-vertex graph depends on graph density for the CombiTarjan and nilpotent adjacency matrix methods, but is essentially constant for Bax's algorithm.

Example 12.2 Mean run times over 20 trials of counting $\lfloor n/2 \rfloor$-cycles in simple graphs are compared in Fig. 12.2 (top). In Fig. 12.2 (bottom), the number of vertices is fixed at 10, and edge existence probability varies from $p = 0.1$ to $p = 0.5$.

The next theorem describes a class of random graphs for which the nilpotent adjacency matrix method is more efficient than $\mathscr{O}(2^n \mathrm{poly}(n))$.

Theorem 12.4 *Let $q \in (0, 1)$ be fixed. Let n be a positive integer, and let $3 \leq k \leq n$. Let $\mathscr{G}_{n,p}$ be a homogeneous random graph on n vertices with independent equiprobable edges of probability $p \leq \frac{q}{k-1}$. Then, the average-case complexity of enumerating cycles of length less than or equal to k in $\mathscr{G}_{n,p}$ using the nilpotent adjacency matrix method is $\mathscr{O}(n^3(1+q)^n)$.*

Proof As in the proof of Theorem 12.3, the result is obtained by considering the number of nonzero coefficients in $\mathscr{A}^k$. To consider the average-case complexity, we consider expected numbers of nonzero coefficients according to the probability model indicated. In particular, the average number of blade multiplications performed in computing

$$\langle v_i | \mathscr{A}^k | v_j \rangle = \sum_{\ell=1}^{n} \langle v_i | \mathscr{A}^{k-1} | v_\ell \rangle \langle v_\ell | \mathscr{A} | v_j \rangle$$

is the product of the expected numbers of nonzero coefficients in the canonical expansions of $\langle v_i | \mathscr{A}^{k-1} | v_\ell \rangle$ and $\langle v_\ell | \mathscr{A} | v_j \rangle$.

Let $G = \mathscr{G}_{n,p}$ be a homogeneous random graph on n vertices with independent equiprobable edges of nonzero probability $p \leq \frac{q}{k-1}$ for fixed $q \in (0, 1)$.

Claim Let $n \geq 3$, and let $2 \leq k \leq n$. For any $1 \leq i, j \leq n$, the expected number of nonzero coefficients in the canonical expansion of $\langle v_i | \mathscr{A}^k | v_j \rangle$ satisfies the following inequality:

$$\mathbb{E}\big(\sharp\{\text{nonzero coefficients}\}\big) \leq q^{k-1} \binom{n-1}{k-1}. \tag{12.14}$$

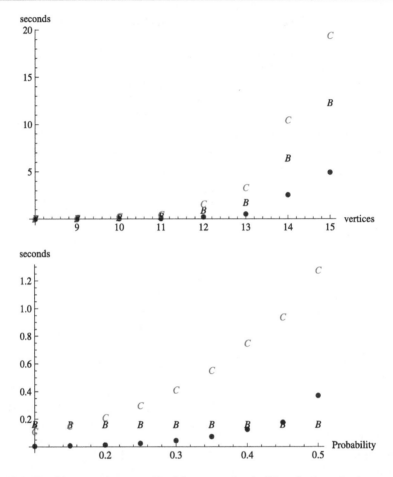

Fig. 12.2 *Top*: Mean run times over 20 trials enumerating $\lfloor n/2 \rfloor$-cycles in randomly generated n-vertex graphs having equiprobable edges ($p = 0.25$). *Bottom*: 20-Run mean run times of counting 5-cycles in 10-vertex graphs with edge probabilities running from $p = 0.1$ to $p = 0.5$. Plotmarkers: B—Bax, C—CombiTarjan, •—$\mathbb{R}\Psi_n$

Proof of Claim By Theorem 12.2, the expected number of nonzero coefficients in the canonical expansion of $\langle v_i | \mathscr{A}^k | v_j \rangle$ is equal to the expected number of k-vertex subsets $I \subseteq V$ such that there exists a k-step walk from v_i to $v_j \in I$ visiting each vertex of I exactly once when $v_i \notin I$ and revisiting v_i exactly once when $v_i \in I$.

The special case $k = 2$ is treated first. The expected number of nonzero coefficients in the canonical expansion of $\langle v_i | \mathscr{A}^2 | v_j \rangle$ is equal to the expected degree of v_i when $i = j$ and equal to the expected number of two step walks on distinct vertices $v_i \to v_\ell \to v_j$ when $i \neq j$; i.e.,

$$\mathbb{E}(\sharp\{\text{nonzero coefficients}\}) = \begin{cases} p(n-1) & i = j, \\ p^2(n-2) & \text{otherwise.} \end{cases} \quad (12.15)$$

The desired inequality for $k = 2$ is then established by observing that $p < q$, whence

$$p^2(n - 2) < p(n - 1) < q(n - 1) = q^{k-1}(n - 1). \tag{12.16}$$

Turning now to the more general case $2 < k \leq n$, the expected number of vertex sets I on which k-walks $v_i \to v_j$ exist with no repeated vertices except possibly v_i at an intermediate step is determined by partitioning the collection of these walks into two classes: (i) walks on k edges and (ii) walks on $k - 1$ edges (in which case, vertex v_i is revisited on the second step).

Unless otherwise indicated, k-walks will refer only to those walks $w : v_i \to v_j$ with no revisited vertex except possibly v_i exactly once at an intermediate step.

Note that the total number of k-walks $w : v_i \to v_j$ in K_n revisiting no vertex except possibly v_i at an intermediate step is given by

$$W = (k - 1)! \binom{n - 1}{k - 1} \tag{12.17}$$

since these walks are specified by ordered k-tuples of vertices with v_j in the kth position. Hence, $k - 1$ intermediate vertices visited are chosen from $V \setminus \{v_j\}$ with $(k - 1)!$ possible permutations.

Denote as Class I those walks on k independent equiprobable edges. Class I walks either revisit no vertices or may revisit v_i at some step other than the second step. Let W_1 denote the total number of these walks in K_n. Denote as Class II those walks on $k - 1$ independent equiprobable edges. Class II walks revisit vertex v_i at the second step. Let W_2 denote the number of Class II walks in K_n.

Note first that $W = W_1 + W_2$. Given a homogeneous random graph $G = \mathscr{G}_{n,p}$, it is now evident that

$$\begin{aligned}
\mathbb{E}\big(\sharp\{k\text{-walks } w : v_i &\to v_j \text{ in } G\}\big) \\
&= \mathbb{E}\big(\sharp\{\text{Class I } k\text{-walks } w : v_i \to v_j \text{ in } G\}\big) \\
&\quad + \mathbb{E}\big(\sharp\{\text{Class II } k\text{-walks } w : v_i \to v_j \text{ in } G\}\big).
\end{aligned} \tag{12.18}$$

When a collection of k-walks $v_i \to v_j$ exists on k independent equiprobable edges,

$$\begin{aligned}
\mathbb{E}\big(\sharp\{\text{Class I } k\text{-walks } w : v_i \to v_j\}\big) &= \sum_{\text{Class I } k\text{-walks } w:v_i \to v_j} \mathbb{P}(w \text{ exists}) \\
&= p^k W_1. \tag{12.19}
\end{aligned}$$

Similarly,

$$\mathbb{E}\big(\sharp\{\text{Class II } k\text{-walks } w : v_i \to v_j\}\big) = p^{k-1} W_2. \tag{12.20}$$

Together, (12.18), (12.19), and (12.20) give

$$\mathbb{E}\big(\sharp\{k\text{-walks } w : v_i \to v_j \text{ in } G\}\big) = p^k W_1 + p^{k-1} W_2$$
$$\leq p^{k-1}(W_1 + W_2) = p^{k-1} W. \qquad (12.21)$$

The expected number of vertex subsets supporting these walks therefore satisfies the following inequality:

$$\mathbb{E}\big(\sharp\{I : \exists\, k\text{-walk } w : v_i \to v_j\}\big) \leq p^{k-1} W. \qquad (12.22)$$

With the assumption $p \leq \frac{q}{k-1}$ for fixed $q > 0$ and substitution of W from (12.18), one thereby obtains

$$\mathbb{E}\big(\sharp\{I : \exists k\text{-walk } w : v_i \to v_j\}\big) \leq \frac{q^{k-1}}{(k-1)^{k-1}} W \leq \frac{q^{k-1}}{(k-1)!} W$$
$$= q^{k-1} \binom{n-1}{k-1}. \qquad (12.23)$$

This completes the proof of the claim. $\square$

By considering the expected number of nonzero coefficients in the canonical expansion of $\langle v_i | \mathscr{A}^{k-1} | v_j \rangle$ for $3 \leq k \leq n$, it is now evident that the expected number of blade multiplications performed in computing $\langle v_i | \mathscr{A}^{k-1} \mathscr{A} | v_j \rangle$ is bounded above by

$$\sum_{\ell=1}^{n} q^{k-2} \binom{n-1}{k-2} p = npq^{k-2} \binom{n-1}{k-2} \leq q^{k-1} \frac{n}{k-1} \binom{n-1}{k-2}$$
$$= q^{k-1} \binom{n}{k-1}. \qquad (12.24)$$

Hence, the expected number of blade multiplications performed in computing the matrix product $\mathscr{A}^k = \mathscr{A}^{k-1} \mathscr{A}$ is bounded above by $n^2 q^{k-1} \binom{n}{k-1}$. Applying this result recursively, the average number of blade multiplications required to compute $\mathscr{A}^k$ is found to be bounded above by $n^2 \sum_{\ell=1}^{k-1} q^\ell \binom{n}{\ell}$.

Observing that

$$\sum_{\ell=1}^{k-1} q^\ell \binom{n}{\ell} \leq \sum_{\ell=0}^{n} q^\ell \binom{n}{\ell} = (1+q)^n, \qquad (12.25)$$

cycle enumeration is of average-case complexity $\mathscr{O}(n^2(1+q)^n)$ in terms of blade operations required. Recalling the $\mathscr{O}(n)$ complexity of blade operations thus completes the proof. $\square$

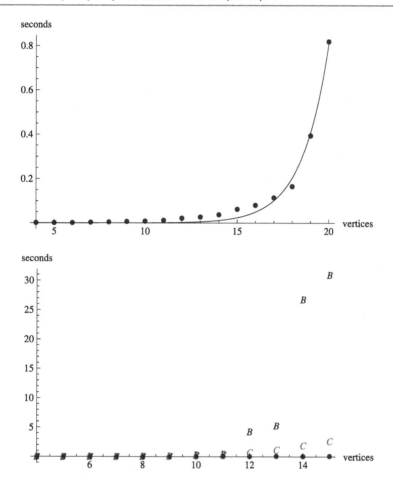

Fig. 12.3 *Top*: Counting cycles of random length $k \in \{3, \ldots, \max(\{3, n/2\})\}$ in n-vertex graphs with edge probability $p = q/(k-1)$. The continuous curve is $y = cn^3(1+q)^n$, where $q = 0.7$ and $c = 2.4862 \cdot 10^{-9}$, obtained by least squares method. *Bottom*: Counting cycles of length $k \in \{3, \ldots, \max(3, \lfloor n/2 \rfloor)\}$ in n-vertex graphs with edge probability $p = q/(k-1)$, where $q = 0.7$. Plotmarkers: B—Bax, C—CombiTarjan, •—$\mathbb{R}\Psi_n$

Example 12.3 Average computation times (over 200 trials) of enumerating cycles of length $k \in \{3, \ldots, \max(\{3, n/2\})\}$ in homogeneous random graphs satisfying the conditions of Theorem 12.4 with constant $q = 0.7$ are depicted in Fig. 12.3 (top). Also plotted is the curve $y = cn^3(1+q)^n$ with $c = 2.4862 \cdot 10^{-9}$ determined by least squares.

Example 12.4 A comparison of average computation times (over 20 trials) for the three methods of enumerating cycles of length $k \in \{3, \ldots, \max(\{3, n/2\})\}$ in homogeneous random graphs satisfying the conditions of Theorem 12.4 with constant $q = 0.7$ are depicted in Fig. 12.3 (bottom).

As the next theorem shows, the fixed cycle length case is very well-behaved in terms of complexity.

Theorem 12.5 *For fixed $k \in \mathbb{N}$, the worst-case complexity of enumerating k-cycles in an n-vertex graph by the nilpotent adjacency matrix method is $\mathcal{O}(n^{\alpha+k-1})$.*

Proof The case $k = 3$ is clear from the special case in the proof of Theorem 12.4. When $k > 3$, the maximum number of nonzero coefficients in the canonical expansion of $\langle v_i | \mathscr{A}^{k-1} | v_j \rangle$ is $\binom{n-1}{k-2}$. Asymptotically, $\binom{n-1}{k-2} \approx \frac{(n-1)^{k-2}}{(k-2)!} = \mathcal{O}(n^k)$. Hence, computing $\mathscr{A}^k$ requires computing at most

$$n^\alpha \sum_{\ell=0}^{k-2} \binom{n-1}{\ell} = \mathcal{O}(n^\alpha n^{k-2}) = \mathcal{O}(n^{k+(\alpha-2)}) \tag{12.26}$$

blade products. $\square$

12.5 Implementation Notes

In general, coding the geometric product (or any noncommutative operation) in MATHEMATICA is most reliably done using one of MATHEMATICA's undefined symbols. However, since the product in $\mathbb{R}\Psi_n$ is commutative, a much more efficient implementation is possible. The nilpotent adjacency matrix approach was implemented herein by overloading the Times operator of MATHEMATICA to facilitate multiplication of blades from $\mathbb{R}\Psi_n$. Figure 12.4 details MATHEMATICA code for implementing the multiplication in $\mathbb{R}\Psi_n$.

Once the $\mathbb{R}\Psi_n$ multiplication is implemented, the algorithm for counting cycles is very straightforward. The corresponding code is seen in Fig. 12.5.

Bax's algorithm is implemented using the formula obtained in Exercise 12.1. Counting all k-cycles in a graph G having adjacency matrix A is accomplished by computing the quantity

$$k\sharp\{k\text{-cycles in } G\} = \mathrm{Tr}\left(\sum_{\substack{S \subseteq V \\ |S| \leq k, S \neq \emptyset}} \binom{n-|S|}{k-|S|}(-1)^{k-|S|}(A_S)^k \right). \tag{12.27}$$

The corresponding code is seen in Fig. 12.6.

The CombiTarjan method was implemented by first extracting all k-vertex subgraphs and summing recovered numbers of Hamiltonian cycles on them. The code appears in Fig. 12.7.

The comparisons seen in Fig. 12.1, Fig. 12.3, and Fig. 12.2 were generated as follows: For any given trial, a random simple graph is first generated by constructing a random symmetric binary matrix. The corresponding nilpotent adjacency matrix is then constructed.

```
(* Define Rψ multiplication *)

Unprotect[ζ];
ζ = Symbol["ζ"];

Unprotect[Times];
ζ_a ζ_b := If[Length[a ∩ b] > 0, 0, ζ_a∪b];
Unprotect[Power];
(x_ /; ! FreeQ[x, ζ_a ])^n_Integer :=
  Module[{y, f}, y = Expand[x];
    Switch[EvenQ[n], True, If[n == 0, Return[1], Composition[Expand][Distribute[f[x, y]] /. f → Times]^(n/2)],
      False, If[n == 1, Return[x], Composition[Expand][Distribute[f[y, y]] /. f → Times]^((n-1)/2) x]]];

Protect[Power];
Protect[Times];
Unprotect[Expand];
Expand[x_ /; ! FreeQ[x, ζ_]] := DeleteCases[Distribute[x, Plus, Times], 0.` ζ_];
Protect[Expand];
```

Fig. 12.4 MATHEMATICA code defining $\mathbb{R}\Psi_n$ multiplication by overloading the Times operator

```
In[111]:= (* Count k-cycles in simple graph with nilpotent adjacency matrix A *)
          NAMCycles[A_, k_] := (Simplify[Tr[Expand[MatrixPower[A, k]]]] / k) /. {ζ_ → 1};
```

Fig. 12.5 MATHEMATICA code for counting k-cycles via nilpotent adjacency matrix method

```
(* Count k cycles using approach of E.T. Bax *)
BaxKCycleCount[A_, k_] := Module[{list, Kset, verts, nulltable, Ksub, ell},
  list = Table[i, {i, 1, Length[A]}];
  verts = Length[A];
  Kset = Subsets[list, {1, k}];
  nulltable = Table[0, {i, 1, verts}, {j, 1, verts}];
  Return[Sum[(-1)^(k-Length[Kset[[ell]]]) Binomial[n - Length[Kset[[ell]]], k - Length[Kset[[ell]]]]
    MatrixPower[(A ReplacePart[nulltable, CartesianProduct[Kset[[ell]], Kset[[ell]]] → 1]), k],
    {ell, 1, Length[Kset]}]];]
```

Fig. 12.6 MATHEMATICA code for counting k-cycles via algorithm of Bax

```
(* Count k-cycles in graph using Combinatorica's HamiltonianCycle algorithm *)
KCycleCount[A_, k_] := Module[{list, Kset},
  list = Table[i, {i, 1, Length[A]}];
  Kset = Subsets[list, {k}];
  Return[Sum[Length[HamiltonianCycle[FromAdjacencyMatrix[A[[Kset[[ell]], Kset[[ell]]]]], All]],
    {ell, 1, Length[Kset]}]]]
```

Fig. 12.7 MATHEMATICA implementation of "CombiTarjan" approach

The MATHEMATICA system cache was cleared before counting by each method. The system time was stored in a variable, the appropriate method was called, and the subsequent system time was stored in another variable. Relevant data were then appended to a table. Test points were incorporated after each method to ensure that all three methods were returning the same results.

12.6   Conclusion

Leslie Valiant first defined the complexity class $\sharp$P when dealing with the complexity of counting solutions to NP decision problems [8]. The problem of deciding whether or not a graph contains a Hamiltonian cycle is one such decision problem [3].

Generally, the problem of counting cycles in finite graphs is known to be $\sharp P$-complete, and no polynomial-time algorithm is known to exist for solving this problem. Algorithms with $\mathcal{O}(2^n \text{poly}(n))$ time complexity and $\mathcal{O}(\text{poly}(n))$ space complexity are known to exist (see [1, 4]).

The time complexity of Tarjan's algorithm is proportional to the number of cycles contained in the graph. Unlike Bax's algorithm, which simply counts cycles, Tarjan's algorithm actually lists the cycles. As a result, the storage and time complexity of Tarjan's algorithm are potentially factorial rather than exponential.

The nilpotent adjacency matrix method, like Bax's algorithm, counts cycles without listing them, providing an advantage over Tarjan's approach in terms of storage-complexity (assuming the counting of cycles is all that is required). Unlike Bax's algorithm, which has time complexity $\mathcal{O}(2^n \text{poly}(n))$ in all cases, the average-case complexity is significantly less when dealing with "suitably-sparse" graphs. The storage complexity of the nilpotent adjacency matrix approach lies between Bax and Tarjan, since it is proportional to the number of vertex subsets supporting cycles in the graph.

As illustrated by the experimental results, the nilpotent adjacency matrix method often has computational advantages over other classical algorithms. Much work remains to be done in characterizing theoretical complexity on various families of graphs.

12.7   Exercises

12.1 Let A be the adjacency matrix of a simple graph G on n vertices as seen in Theorem 12.1. Show that for any vertex v_i in G and positive integer $k \leq n$,

$$\sharp\{k\text{-cycles at } v_i\} = \left[\sum_{\substack{S \subseteq V \\ |S| \leq k, S \neq \emptyset}} \binom{n - |S|}{k - |S|} (-1)^{k-|S|} (A_S)^k \right]_{ii}. \tag{12.28}$$

12.2 For positive integer n, let $Z = \{\mathbf{e}_{2i-1} \wedge \mathbf{e}_{2i} : 1 \leq i \leq n\}$ as in (12.1). Prove that the pair $(Z, \wedge)$ forms an abelian semigroup.

12.3 Let $u \in \mathbb{R}\Psi_n$ be an element of the form $u = a_1\zeta_1 + a_2\zeta_2$, where a_1 and a_2 are nonzero scalars. Show that $u^2 \neq 0$ and $u^3 = 0$.

Fig. 12.8 A simple graph on 14 vertices

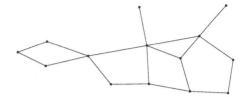

12.4 Suppose that n and k are positive integers satisfying $1 < k \leq n$ and $u \in \mathbb{R}\Psi_n$ is an element of the form $u = \sum_{i=1}^{k} a_i \zeta_i$ where $a_i \neq 0$ for $i = 1, \ldots, k$. Prove the following:

  i. $u^k = k! a_1 \cdots a_k \zeta_{\{1,\ldots,k\}}$
  ii. $u^m = 0$ for all $m > k$.

12.5 Construct a nilpotent adjacency matrix $\mathscr{A}$ for the graph appearing in Fig. 12.8 and count the 7-cycles contained therein by computing $\mathscr{A}^7$. Verify your result by hand as well as by applying the Bax and CombiTarjan algorithms.

References

1. Bax, E.: Algorithms to count paths and cycles. Inf. Process. Lett. **52**, 249–252 (1994)
2. Coppersmith, D., Winograd, S.: Matrix multiplication via arithmetic progressions. J. Symb. Comput. **9**, 251–280 (1990)
3. Karp, R.M.: Reducibility among combinatorial problems. In: Complexity of Computer Computations, pp. 85–103. Plenum, New York (1972)
4. Karp, R.: Dynamic programming meets the principle of inclusion and exclusion. Oper. Res. Lett. **1**, 49–51 (1982)
5. Schott, R., Staples, G.S.: Computational complexity reductions using Clifford algebras. In: Bayro-Corrochano, E., Scheuermann, G. (eds.) Geometric Algebra Computing for Engineering and Computer Science, pp. 431–453. Springer, Berlin (2010)
6. Schott, R., Staples, G.S.: Reductions in computational complexity using Clifford algebras. Adv. Appl. Clifford Algebras **20**, 121–140 (2010)
7. Tarjan, R.: Enumeration of the elementary circuits of a directed graph. SIAM J. Comput. **2**, 211–216 (1973)
8. Valiant, L.: The complexity of computing the permanent. Theor. Comput. Sci. **8**, 189–201 (1979)

Part V
Applications of Line Geometry

Geometric algebra provides a natural setting for encoding the geometry of 3D lines, unifying and extending earlier representations such as Plücker coordinates. This is immediately applicable to fields in which lines play the role of basic elements of expression, such as projective geometry, inverse kinematics of robots with translational joints, and visibility analysis.

Line Geometry in Terms of the Null Geometric Algebra over $\mathbb{R}^{3,3}$, and Application to the Inverse Singularity Analysis of Generalized Stewart Platforms

13

Hongbo Li and Lixian Zhang

Abstract

In this chapter, the classical line geometry is modeled in $\mathbb{R}_{3,3}$, where lines are represented by null vectors, and points and planes by null 3-blades. The group of 3D special projective transformations $SL(4)$ when acting upon points in space induces a Lie group isomorphism, with $SO(3, 3)$ acting upon lines.

As an application of the use of the $\mathbb{R}_{3,3}$ model of line geometry, this chapter analyzes the inverse singularity analysis of generalized Stewart platforms, using vectors of $\mathbb{R}^{3,3}$ to encode the force and torque wrenches to classify their singular configurations.

13.1 Introduction

H. Grassmann (1844) and J. Plücker (1865) are the co-founders of line geometry [5]. A line in space is the extension of two points. For two points with homogeneous coordinates (x_0, x_1, x_2, x_3) and (y_0, y_1, y_2, y_3), the line they extend can be represented by the outer product of their homogeneous coordinates. In coordinate form, this outer product has the following *Plücker coordinates*:

$$(l_{01}, l_{02}, l_{03}, l_{23}, l_{31}, l_{12}), \tag{13.1}$$

where $l_{ij} = x_i y_j - x_j y_i$. If we denote $\mathbf{x} = (x_1, x_2, x_3)^T$ and $\mathbf{y} = (y_1, y_2, y_3)^T$, then the line $\mathbf{xy}$ has the Plücker coordinates

H. Li (✉) · L. Zhang
Key Laboratory of Mathematics Mechanization, Academy of Mathematics and Systems Science, Chinese Academy of Sciences, Beijing 100190, P.R. China
e-mail: hli@mmrc.iss.ac.cn

L. Zhang
e-mail: shadowfly12@126.com

L. Dorst, J. Lasenby (eds.), *Guide to Geometric Algebra in Practice*,
DOI 10.1007/978-0-85729-811-9_13, © Springer-Verlag London Limited 2011

$$(\mathbf{l}, \bar{\mathbf{l}}) := (x_0 \mathbf{y} - y_0 \mathbf{x}, \mathbf{x} \times \mathbf{y}). \tag{13.2}$$

In affine geometry, $(\mathbf{l}, \bar{\mathbf{l}})$ describes an affine line if and only if $\mathbf{l} \neq 0$. When this is satisfied, then the direction of the line is its point at infinity, also called its *ideal point*: $(0, \mathbf{l})$. Let $(1, \mathbf{x})$ be an affine point on the line; then $\bar{\mathbf{l}} = \mathbf{x} \times \mathbf{l}$ is the moment vector. Obviously, we have $\mathbf{l} \cdot \bar{\mathbf{l}} = 0$ or, in the coordinate form,

$$l_{01} l_{23} + l_{02} l_{31} + l_{03} l_{12} = 0. \tag{13.3}$$

This equality is called the *Grassmann–Plücker relation* of the line. It states that all lines are points on a quadratic surface in the 5D projective space of 6-tuples of homogeneous coordinates. *Plücker's theorem* states that the converse is also true: if a point in the 5D projective space is on the *Klein quadric* defined by (13.3), then it must be the Plücker coordinates of a spatial line.

Classical line geometry studies invariant properties of line complexes under 3D projective, affine, or Euclidean transformations. Nowadays it has important applications in computer aided design, geometric modeling, scientific visualization, computer aided manufacturing, and robotics. The design of efficient algorithms involving lines can be greatly simplified if it is based on the right geometric model.

For example, a ruled surface is simply a curve of lines, whose study is much easier in line geometry. In studying developable surfaces, line geometry contributes to simplifying the computing of medial axes, rational curves with rational offsets, and cyclographic mapping. Line congruences arise in collision problems in five-axis milling, and rational congruences of line complexes are related to geometric optics. Line geometry not only provides tools for visualization, but also has interesting links to planar and spherical motions, rational curves on quadratic surfaces, and problems of surveying [4, 12, 15, 18, 22–24].

Various algebraic structures can be introduced to Plücker coordinates in geometric computing. Study (1903) considered introducing a dual element ε, which is a nilpotent element that commutes with everything, such that a line $(\mathbf{l}, \bar{\mathbf{l}})$ can be described by dual vector $\mathbf{l} + \varepsilon \bar{\mathbf{l}}$. If there is another line $\mathbf{m} + \varepsilon \bar{\mathbf{m}}$, let $\mathbf{l}, \mathbf{m}$ be unit vectors, and let

$$\theta = \angle(\mathbf{l}, \mathbf{m}),$$

$\mathbf{n} =$ unit vector along $\mathbf{l} \times \mathbf{m}$,

$\delta =$ signed distance between the two lines along direction $\mathbf{n}$,

$\bar{\mathbf{n}} =$ moment vector of the common perpendicular of the two lines.

$$\tag{13.4}$$

Then

$$\begin{cases} (\mathbf{l} + \varepsilon \bar{\mathbf{l}}) \cdot (\mathbf{m} + \varepsilon \bar{\mathbf{m}}) = \cos(\theta + \varepsilon \delta), \\ (\mathbf{l} + \varepsilon \bar{\mathbf{l}}) \times (\mathbf{m} + \varepsilon \bar{\mathbf{m}}) = (\mathbf{n} + \varepsilon \bar{\mathbf{n}}) \sin(\theta + \varepsilon \delta). \end{cases} \tag{13.5}$$

The angle, distance, and common perpendicular of the two lines can all be read from their inner product and cross product.

Fig. 13.1 Stewart platform

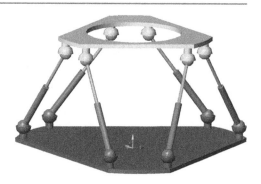

In this chapter, we introduce a nondegenerate inner product structure to the Plücker coordinates of lines and convert them to null vectors of a 6D real vector space of signature $(3, 3)$. After classifying all 2-blades and 3-blades generated by null vectors in the 6D space $\mathbb{R}^{3,3}$ and presenting their geometric interpretations in line geometry, we point out that all points and planes in space can be represented by null 3-blades, i.e., 3-blades with completely degenerate inner product structure. On one hand, any *special projective transformation* in space, i.e., special linear transformation in the 4D vector space realizing the 3D projective space, induces a special orthogonal transformation in $\mathbb{R}^{3,3}$; on the other hand, any element in $SO(3, 3)$ induces a special projective transformation in the 3D projective space of null 3-blades representing projective points in space. We further point out that all dualities in 3D projective geometry, i.e., linear mappings between the 4D vector space representing projective points and the dual 4D vector space representing projective planes, can be realized by orthogonal transformations of determinant -1 in $\mathbb{R}^{3,3}$ via their actions upon lines in space.

Hence, the Lie group isomorphism between $SL(4)$ and $SO(3, 3)$ is realized via the $\mathbb{R}^{3,3}$ model of line geometry. Since $SO(3, 3)$ can be covered by $Spin(3, 3)$, and the latter has an algebraic representation in $\mathbb{R}_{3,3}$, we can use $\mathbb{R}_{3,3}$ to construct not only all kinds of special projective transformations, but a hierarchy of advanced projective invariants.

As a specific application of the classification of blades generated by null vectors in $\mathbb{R}^{3,3}$, we consider the problem of analyzing the inverse singular configurations of generalized Stewart platforms. The topic is important in that in an inverse singular configuration, the end-effector may still possess certain degrees of freedom after all the actuators are locked, which may incur some unexpected damages such as collapse [2, 11, 14, 17, 20, 29].

Parallel robots have various advantages over serial robots, such as high accuracy, high payload-to-weight ratio, and high rigidity. They also have a few drawbacks, the most important of which is failure in or close to a singular configuration. A famous parallel manipulator is a *Stewart platform* (or *Gough–Stewart platform*) [25], as shown in Fig. 13.1. It is a 6-dof parallel manipulator composed of a static platform and a moving platform, and controlled by six distance constraints between six point pairs.

Fig. 13.2 Generalized
Stewart platforms actuated by
Left: six distance constraints
between three points and
three lines; *Right*: 6 distance
constraints between 6 lines
and 3 planes

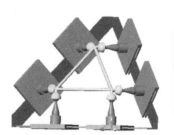 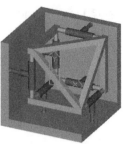

Historically, line geometry is closely related to machine mechanism and kine-matics, because lines are supports of forces and moments [1, 26, 28]. The theory of screws [27] is closely related to line geometry. Some previous works on singu-larity analysis of a Stewart platform include [3, 13, 19, 21]. In particular, Merlet [19] described the singularity configurations as degeneracies of the line complexes spanned by the six lines representing the linear actuators of the platform. This ap-proach has the advantage of avoiding complicated computing of the Jacobian ma-trix.

A natural idea to generalize the 6-dof parallel structure of a Stewart platform is to replace some of the point-pair distance actuators by distance and angle actuators between pairs of linear objects such as points, lines, and planes. Gao et al. [10] pro-posed six new types of limb actuators by distances between point/point, point/line, point/plane, line/line, line/plane, and plane/plane, and three types of limb actuators by angles between line/line, line/plane, and plane/plane. A 6-dof parallel manipu-lator controlled by such constraints is called a *generalized Stewart platform* (GSP). Figure 13.2 shows two GSPs containing limb actuators by point/line and line/plane distances.

For a GSP, the actuator of a limb corresponds to either a force or a couple, which is unanimously referred to as a *driving wrench*. We deduce by the virtual work principle the following new conclusion: *for any GSP, the inverse Jacobian is the transpose of the matrix composed of the Plücker coordinates of the driving wrenches and common constraint wrenches of the six limbs.* An inverse singularity occurs when the rank of the wrench matrix is less than six. While for a classical Stewart platform the matrix is a 6×6 square, for a GSP, the matrix often has more than six columns. Its singularities are dramatically different from those of a classical Stewart platform.

This chapter is organized as follows. Section 13.2 introduces line geometry by formulating it in the null geometric algebra over $\mathbb{R}^{3,3}$. Section 13.3 relates inverse singularity with degeneracy of wrench matrix. Section 13.4 classifies the inverse singularities of some typical GSPs.

13.2   Line Geometry with Null Geometric Algebra

In the projective space, all projective lines form a Grassmann variety. In the Grassmann space $\bigwedge(\mathscr{V}^4)$ over the 4D vector space $\mathscr{V}^4$ realizing 3D projective geometry, all bivectors form a 6D vector subspace, denoted by $\bigwedge^2(\mathscr{V}^4)$. A bivector $\mathbf{A}$ represents a projective line if and only if it satisfies the so-called *Grassmann–Plücker relation*,[1]

$$\mathbf{A} \wedge \mathbf{A} = 0. \tag{13.6}$$

If we define the following symmetric bilinear function in $\bigwedge^2(\mathscr{V}^4)$:

$$\mathbf{A} \cdot \mathbf{B} := [\mathbf{AB}], \tag{13.7}$$

where the bracket is in the Grassmann algebra $\bigwedge(\mathscr{V}^4)$, then the function defines a nondegenerate real inner product of signature $(3, 3)$.

This property can be proved as follows. Let $\mathbf{e}_0, \mathbf{e}_1, \mathbf{e}_2, \mathbf{e}_3$ be a basis of $\mathscr{V}^4$. Then the $\mathbf{e}_{ij} = \mathbf{e}_i \wedge \mathbf{e}_j$ for $0 \leq i < j \leq 3$ are the induced basis of $\bigwedge^2(\mathscr{V}^4)$. For any nonzero element $\mathbf{A} \in \bigwedge^2(\mathscr{V}^4)$, at least one of its coordinates, say its coordinate a_{01} in $\mathbf{e}_{01}$, is nonzero. Then $[\mathbf{Ae}_{23}] = a_{23} \neq 0$. This proves that if $\mathbf{A} \cdot \mathbf{B} = 0$ for all $\mathbf{B} \in \bigwedge^2(\mathscr{V}^4)$, then $\mathbf{A} = 0$. The inner product is thus nondegenerate. Since $\mathbf{e}_{01}, \mathbf{e}_{02}, \mathbf{e}_{03}$ are pairwise orthogonal to each other, and they are all basis elements, the 6D inner-product space has a 3D subspace that is *null*: the subspace spanned by $\mathbf{e}_{01}, \mathbf{e}_{02}, \mathbf{e}_{03}$ has the property that any two vectors in it are orthogonal to each other. This property leads to the conclusion that the *signature* of the inner product is $(3, 3)$. Henceforth we denote the 6D inner product space by $\mathbb{R}^{3,3}$.

By Plücker's theorem, any nonzero vector in $\mathbb{R}^{3,3}$ represents a projective line if and only if it is null. The representation of a line is unique up to scale, i.e., is *homogeneous*. The geometry of projective lines can thus be described by the geometric algebra generated by null vectors in $\mathbb{R}^{3,3}$, called the *null geometric algebra* of spatial lines.

First consider the geometric meaning of the inner product between two null vectors in $\mathbb{R}^{3,3}$. Let $\mathbf{e}_0, \mathbf{e}_1, \mathbf{e}_2, \mathbf{e}_3$ be a basis of the 3D affine space $\mathscr{V}^4$ such that $\mathbf{e}_0$ represents the origin, and $\mathbf{e}_1, \mathbf{e}_2, \mathbf{e}_3$ are points at infinity.

Notation In this section, *the outer product in $\bigwedge(\mathscr{V}^4)$ has to be denoted by juxtaposition of elements*, and the outer product in $\bigwedge(\mathbb{R}^{3,3})$ is denoted by the wedge symbol.

Consider a line $\mathbf{A}$ passing through point $\mathbf{e}_0 + \mathbf{x}$ and point at infinity $\mathbf{l}$, and a second line $\mathbf{B}$ passing through point $\mathbf{e}_0 + \mathbf{y}$ and point at infinity $\mathbf{m}$, where $\mathbf{x}, \mathbf{y}, \mathbf{l}, \mathbf{m} \in \mathbb{R}^3 \subset \mathscr{V}^4$. Then

---

[1] *Editorial note*: See also Chap. 14.

$$\mathbf{A} = (\mathbf{e}_0 + \mathbf{x})\mathbf{l}, \qquad \mathbf{B} = (\mathbf{e}_0 + \mathbf{y})\mathbf{m}, \tag{13.8}$$

so

$$\mathbf{A} \cdot \mathbf{B} = (\mathbf{e}_0 \mathbf{l}) \cdot (\mathbf{y}\mathbf{m}) + (\mathbf{x}\mathbf{l}) \cdot (\mathbf{e}_0 \mathbf{m}) = [\mathbf{l}\mathbf{y}\mathbf{m}] + [\mathbf{m}\mathbf{x}\mathbf{l}] = \big[\mathbf{l}\mathbf{m}(\mathbf{x} - \mathbf{y})\big], \tag{13.9}$$

where the bracket is with respect to $\mathbf{e}_1\mathbf{e}_2\mathbf{e}_3$ in $\bigwedge(\mathbb{R}^3)$.

When $\mathbf{B}$ is a line at infinity, let $\mathbf{B} = \mathbf{m}_1\mathbf{m}_2$ where $\mathbf{m}_1, \mathbf{m}_2 \in \mathbb{R}^3$, and let $\mathbf{A}$ be an affine line as in (13.8); then

$$\mathbf{A} \cdot \mathbf{B} = (\mathbf{e}_0\mathbf{l}) \cdot (\mathbf{m}_1\mathbf{m}_2) = [\mathbf{l}\mathbf{m}_1\mathbf{m}_2]. \tag{13.10}$$

Given two directed lines $\mathbf{A} = (\mathbf{e}_0 + \mathbf{x}, \mathbf{l})$ and $\mathbf{B} = (\mathbf{e}_0 + \mathbf{y}, \mathbf{m})$, the *signed volume* of $\mathbf{A}$ and $\mathbf{B}$ is defined as the signed volume of the parallelepiped formed by vectors $\mathbf{x} - \mathbf{y}, \mathbf{l}, \mathbf{m}$:

$$V_{\mathbf{A},\mathbf{B}} := [\mathbf{x}\mathbf{l}\mathbf{m}] + [\mathbf{y}\mathbf{m}\mathbf{l}] = \big[(\mathbf{x} - \mathbf{y})\mathbf{l}\mathbf{m}\big]. \tag{13.11}$$

It is symmetric in $\mathbf{A}$ and $\mathbf{B}$. We have thus proved the following conclusion:

> When $\mathbf{A}, \mathbf{B}$ are affine lines, then $\mathbf{A} \cdot \mathbf{B}$ is their signed volume. When one is a line at infinity, then $\mathbf{A} \cdot \mathbf{B}$ is the signed volume of the parallelepiped formed by their components at infinity, a 1D direction and a 2D direction.

In particular, two lines intersect if and only if their representative null vectors are orthogonal. If the lines are written in the Plücker coordinate form, $\mathbf{A} = (\mathbf{f}, \bar{\mathbf{f}})$ and $\mathbf{B} = (\mathbf{g}, \bar{\mathbf{g}})$, then

$$\mathbf{A} \cdot \mathbf{B} = \mathbf{f} \cdot \bar{\mathbf{g}} + \mathbf{g} \cdot \bar{\mathbf{f}}. \tag{13.12}$$

Next we classify all 2-spaces and 3-spaces of $\mathbb{R}^{3,3}$ spanned by null vectors, i.e., 2D and 3D linear extensions of null vector generators. There are only two kinds of such 2-spaces, $\mathbb{R}^{1,1,0}$ and $\mathbb{R}^{0,0,2}$. Below all points, lines, and planes are projective ones in space.

- $\mathbb{R}^{1,1,0}$ (2-space with metric $\mathrm{diag}(1, 1, 0)$): Its 1D null subspaces represent a pair of noncoplanar lines, as shown in Fig. 13.3 (left).
- $\mathbb{R}^{0,0,2}$ (2-space with metric $\mathrm{diag}(0, 0, 2)$): Its 1D null subspaces represent all the lines incident to a fixed point and a fixed plane, as shown in Fig. 13.3 (right). All such lines form a 1D algebraic variety, called a *1D concurrent pencil of lines* or a *single-wheel pencil*. The point of concurrency is called the *center*.

In $\mathbb{R}^{3,3}$, there are three kinds of 3-spaces spanned by null vectors: $\mathbb{R}^{1,2,0}$ or $\mathbb{R}^{2,1,0}$, $\mathbb{R}^{1,1,1}$, and $\mathbb{R}^{0,0,3}$.

Fig. 13.3 *Left*: null vectors in $\mathbb{R}^{1,1,0}$, a pair of noncoplanar lines. *Right*: null vectors in $\mathbb{R}^{0,0,2}$, a 1D pencil of lines concurrent at point **o** and on plane **aoc**

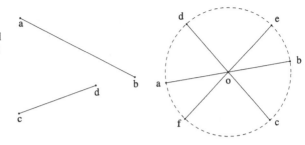

Fig. 13.4 Null vectors in $\mathbb{R}^{1,2,0}$ or $\mathbb{R}^{2,1,0}$: a 1D regulus pencil

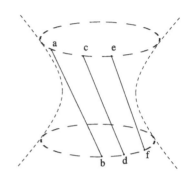

Fig. 13.5 Null vectors in $\mathbb{R}^{1,1,1}$: a 1D couple-wheel pencil; the axis is line **ab**

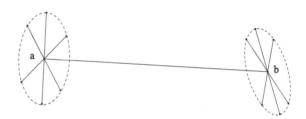

- $\mathbb{R}^{1,2,0}$ or $\mathbb{R}^{2,1,0}$ (3-space with metric diag$(1,2,0)$ or diag$(2,1,1)$): Its 1D null subspaces represent a *1D regulus pencil of lines*, i.e., a 1-parameter family of straight-line generators of a hyperboloid of one sheet, as shown in Fig. 13.4.
- $\mathbb{R}^{1,1,1}$ (3-space with metric diag$(1,1,1)$): Its 1D null subspaces represent two 1D concurrent pencils of lines sharing a unique common line, as shown in Fig. 13.5. Such a pencil is called a *1D couple-wheel pencil*; the common line is called the *axis*.
- $\mathbb{R}^{0,0,3}$ (3-space with metric diag$(0,0,3)$): Its 1D null subspaces represent either all the lines concurrent at the point, called a *2D concurrent pencil of lines*, or equivalently, the *point of concurrency*, or all the lines lying on the same plane, called a *2D coplanar pencil of lines*, or equivalently, the *supporting plane* of the lines. Figure 13.6 shows both cases.

We see that in the $\mathbb{R}^{3,3}$ model of line geometry, points and planes are both represented by null 3-blades. Algebraically they cannot be distinguished from each other. We need to introduce an *affine structure* to make the distinction.

Fig. 13.6 Null vectors in $\mathbb{R}^{0,0,3}$. *Left*: a 2D concurrent pencil, or a point. *Right*: a 2D coplanar pencil, or a plane

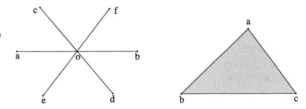

In the Grassmann algebra $\bigwedge(\mathscr{V}^n)$ over $\mathscr{V}^n$, for an r-blade $\mathbf{A}_r$ and an s-blade $\mathbf{B}_s$, their *0th-level intersection* refers to $\mathbf{A}_r \vee \mathbf{B}_s$. Their *ith-level intersection* [16] refers to

$$\mathbf{A}_r \vee^{(i)} \mathbf{B}_s := \sum_{(n-s-i,r+s-n+i) \vdash \mathbf{A}_r} [\mathbf{A}_{r(1)} \mathbf{B}_s] \mathbf{A}_{r(2)}, \tag{13.13}$$

if their jth-level intersection is zero for all $0 \le j < i$. Here $(n-s-i, r+s-n+i) \vdash \mathbf{A}_r$ denotes bipartitioning the r vectors whose outer product equals $\mathbf{A}_r$ into two subsequences of lengths $n-s-i$ and $r+s-n+i$, with $\mathbf{A}_{r(1)}$ denoting the outer product of the first subsequence, and $\mathbf{A}_{r(2)}$ denoting the outer product of the second subsequence. The bracket is set up upon the $(n-i)$D subspaces spanned by vectors in $\mathbf{A}_r$ and $\mathbf{B}_s$.

Now return to line geometry. Fix a null 3-space of $\mathbb{R}^{3,3}$, and let $\mathbf{I}_3$ be its 3-blade representation. Define the *plane at infinity* to be the set of lines in space whose representative null vectors are in $\mathbf{I}_3$. A vector $x \in \mathscr{V}^4$ denotes an affine point in 3D affine geometry if and only if for any vector $\mathbf{X} \in \mathbf{I}_3$, we have $\mathbf{xX} \ne 0$ in $\bigwedge(\mathscr{V}^4)$. Then $\mathbf{I}_3$ introduces a 3D affine structure to the underlying 4D vector space $\mathscr{V}^4$ of 3D projective geometry.

The following properties can be easily established. Let $\mathbf{A}_3$ be a null 3-blade of $\bigwedge(\mathbb{R}^{3,3})$ linearly independent of $\mathbf{I}_3$.

- If $\mathbf{A}_3 \vee \mathbf{I}_3 \ne 0$, then $\mathbf{A}_3$ represents an affine point.
- If $\mathbf{A}_3 \vee \mathbf{I}_3 = 0$ but $\mathbf{A}_3 \vee^{(1)} \mathbf{I}_3 \ne 0$, then $\mathbf{A}_3$ represents an affine plane.
- If both $\mathbf{A}_3 \vee \mathbf{I}_3 = 0$ and $\mathbf{A}_3 \vee^{(1)} \mathbf{I}_3 = 0$, then $\mathbf{A}_3$ represents a point at infinity.

Let $\mathbf{A}_3, \mathbf{B}_3$ be points (including points at infinity), and let $\mathbf{A}_3^*, \mathbf{B}_3^*$ be planes (including the plane at infinity). Then $\mathbf{A}_3 \vee^{(1)} \mathbf{B}_3$ is the line connecting the two points, and $\mathbf{A}_3^* \vee^{(1)} \mathbf{B}_3^*$ is the line of intersection of the two planes. A line represented by null vector $\mathbf{X}$ passes through point $\mathbf{A}_3$ if and only if $\mathbf{X} \wedge \mathbf{A}_3 = 0$; a line $\mathbf{X}$ is on plane $\mathbf{A}_3^*$ if and only if $\mathbf{X} \wedge \mathbf{A}_3^* = 0$.

Consider the relationship between a point $\mathbf{x} = \mathbf{e}_0 + x_1 \mathbf{e}_1 + x_2 \mathbf{e}_2 + x_3 \mathbf{e}_3$ and a plane passing through a point $\mathbf{y} = \mathbf{e}_0 + y_1 \mathbf{e}_1 + y_2 \mathbf{e}_2 + y_3 \mathbf{e}_3$. Without loss of generality, assume that the plane has normal direction $\mathbf{e}_3$. In $\bigwedge(\mathbb{R}^{3,3})$, the point has the representation

$$\mathbf{A}_3 = \mathbf{xe}_1 \wedge \mathbf{xe}_2 \wedge \mathbf{xe}_3, \tag{13.14}$$

and the plane has the representation

$$\mathbf{B}_3 = \mathbf{ye}_1 \wedge \mathbf{ye}_2 \wedge \mathbf{e}_1 \mathbf{e}_2. \tag{13.15}$$

So with respect to pseudoscalar $\mathbf{e}_{01} \wedge \mathbf{e}_{02} \wedge \mathbf{e}_{03} \wedge \mathbf{e}_{12} \wedge \mathbf{e}_{13} \wedge \mathbf{e}_{23}$,

$$\mathbf{A}_3 \vee \mathbf{B}_3 = (x_3 - y_3)^2 \tag{13.16}$$

is the squared distance between the point and the plane, or in affine terms, the squared volume of the parallelepiped formed by vectors $\mathbf{x} - \mathbf{y}, \mathbf{e}_1, \mathbf{e}_2$. Similarly, we get

$$\mathbf{A}_3 \cdot \mathbf{B}_3 = \mathbf{A}_3 \vee \mathbf{B}_3 = (x_3 - y_3)^2. \tag{13.17}$$

Now fix another null 3-space of $\mathbb{R}^{3,3}$ such that it forms a direct sum decomposition with the plane at infinity $\mathbf{I}_3$. Let $\mathbf{J}_3$ be a null 3-blade representing this 3D null subspace, such that if $\mathbf{I}_3 = \mathbf{E}_1 \wedge \mathbf{E}_2 \wedge \mathbf{E}_3$, then $\mathbf{J}_3 = -\mathbf{E}_1^* \wedge \mathbf{E}_2^* \wedge \mathbf{E}_3^*$, where $\mathbf{E}_i^* \cdot \mathbf{E}_j = \delta_{ij}$ for $1 \leq i, j \leq 3$. The $\mathbf{E}_i^*$ are called the *reciprocal basis* of the $\mathbf{E}_j$. Geometrically, $\mathbf{J}_3$ represents an affine point. It is called the *origin* of the affine space.

For example, let $\mathbf{e}_0, \mathbf{e}_1, \mathbf{e}_2, \mathbf{e}_3$ be a basis of the 3D affine space $\mathscr{V}^4$ such that $\mathbf{e}_0$ represents the origin, and $\mathbf{e}_1, \mathbf{e}_2, \mathbf{e}_3$ are an orthonormal basis of the plane at infinity. In $\bigwedge(\mathscr{V}^4)$, denote

$$\mathbf{e}_{ij} := \mathbf{e}_i \mathbf{e}_j. \tag{13.18}$$

Then in $\bigwedge(\mathbb{R}^{3,3})$, we can choose

$$\mathbf{I}_3 = \mathbf{e}_{12} \wedge \mathbf{e}_{13} \wedge \mathbf{e}_{23}, \qquad \mathbf{J}_3 = \mathbf{e}_{01} \wedge \mathbf{e}_{02} \wedge \mathbf{e}_{03}. \tag{13.19}$$

The decomposition

$$\mathbb{R}^{3,3} = \mathbf{I}_3 \oplus \mathbf{J}_3 \tag{13.20}$$

is called a *symplectification* of $\mathbb{R}^{3,3}$. Let $\mathbf{E}_1, \mathbf{E}_2, \mathbf{E}_3$ be a basis of $\mathbf{I}_3$, and let $\mathbf{E}_1^*, \mathbf{E}_2^*, \mathbf{E}_3^*$ be the corresponding reciprocal basis of $\mathbf{J}_3$. Define the *symplectic form* of the symplectification as

$$\mathbf{K}_2 = \mathbf{E}_1 \wedge \mathbf{E}_1^* + \mathbf{E}_2 \wedge \mathbf{E}_2^* + \mathbf{E}_3 \wedge \mathbf{E}_3^*. \tag{13.21}$$

It can be proved that $\mathbf{K}_2$ is invariant under different choices of the basis elements.

Notation In this section, the duality of $\mathbf{A} \in \bigwedge(\mathbf{I}_3)$ in $\mathbf{I}_3$ is denoted by $\mathbf{A}^{\mathbf{I}_3}$; the duality of $\mathbf{B} \in \bigwedge(\mathbf{J}_3)$ in $\mathbf{J}_3$ is denoted by $\mathbf{B}^{\mathbf{J}_3}$. The duality of $\mathbf{C}$ in $\bigwedge(\mathscr{V}^4)$ is denoted by $\mathbf{C}^\dagger$.

Consider a point $\mathbf{e}_0 + \mathbf{x} = \mathbf{e}_0 + x_1 \mathbf{e}_1 + x_2 \mathbf{e}_2 + x_3 \mathbf{e}_3$. Its null 3-blade representation is

$$\begin{aligned}\mathbf{A} &= (\mathbf{e}_0 + \mathbf{x})\mathbf{e}_1 \wedge (\mathbf{e}_0 + \mathbf{x})\mathbf{e}_2 \wedge (\mathbf{e}_0 + \mathbf{x})\mathbf{e}_3 \\ &= \mathbf{J}_3 - (\mathbf{e}_0\mathbf{x}) \wedge \mathbf{K}_2 + (\mathbf{e}_0\mathbf{x}) \wedge (\mathbf{e}_0\mathbf{x})^{\dagger \mathbf{I}_3},\end{aligned} \tag{13.22}$$

where $(\mathbf{e}_0\mathbf{x})^{\dagger \mathbf{I}_3}$ denotes $((\mathbf{e}_0\mathbf{x})^{\dagger})^{\mathbf{I}_3}$. This is a quadratic mapping from $\mathbb{R}^3$ to the space of null 3-blades representing affine points.

When $\mathbf{x}$ tends to infinity, (13.22) represents a point at infinity. So the point at infinity $\mathbf{x}$ has the null 3-blade representation

$$(\mathbf{e}_0\mathbf{x}) \wedge (\mathbf{e}_0\mathbf{x})^{\dagger \mathbf{I}_3}. \tag{13.23}$$

Consider an affine plane with 2D direction $\mathbf{L}_2 \in \bigwedge^2(\mathbb{R}^3)$ and passing through a point $\mathbf{x}$. Its 3-blade representation in $\bigwedge(\mathcal{V}^4)$ is $(\mathbf{e}_0 + \mathbf{x})\mathbf{L}_2$. When the plane does not pass through the origin $\mathbf{e}_0$, then $\mathbf{x}\mathbf{L}_2 \neq 0$. Rescale $\mathbf{L}_2$ so that the 3-blade representation of the plane in $\bigwedge(\mathcal{V}^4)$ becomes

$$\mathbf{e}_1\mathbf{e}_2\mathbf{e}_3 + \mathbf{e}_0\mathbf{L}_2. \tag{13.24}$$

Let $\mathbf{L}_2 = y_1\mathbf{e}_2\mathbf{e}_3 - y_2\mathbf{e}_1\mathbf{e}_3 + y_3\mathbf{e}_1\mathbf{e}_2$. The intersections of plane (13.24) with planes $\mathbf{e}_0\mathbf{e}_1\mathbf{e}_2, \mathbf{e}_0\mathbf{e}_1\mathbf{e}_3, \mathbf{e}_0\mathbf{e}_2\mathbf{e}_3$ are respectively

$$\begin{aligned}
\mathbf{P}_3 &= -y_2\mathbf{e}_{01} + y_1\mathbf{e}_{02} + \mathbf{e}_{12}, \\
\mathbf{P}_2 &= -y_3\mathbf{e}_{01} + y_1\mathbf{e}_{03} + \mathbf{e}_{13}, \\
\mathbf{P}_1 &= -y_3\mathbf{e}_{02} + y_2\mathbf{e}_{03} + \mathbf{e}_{23}.
\end{aligned} \tag{13.25}$$

So the null 3-blade representation of the plane is

$$\mathbf{B} = \mathbf{P}_3 \wedge \mathbf{P}_2 \wedge \mathbf{P}_1 = \mathbf{I}_3 - \mathbf{L}_2 \wedge \mathbf{K}_2 + \mathbf{L}_2 \wedge \mathbf{L}_2^{\dagger \mathbf{J}_3}. \tag{13.26}$$

When $\mathbf{L}_2$ tends to infinity, (13.26) represents a plane passing through the origin. So the plane passing through the origin with 2D direction $\mathbf{L}_2$ has the null 3-blade representation

$$\mathbf{L}_2 \wedge \mathbf{L}_2^{\dagger \mathbf{J}_3}. \tag{13.27}$$

A linear transformation f in $\mathcal{V}^4$ induces another linear transformation $\underline{f}$ in $\bigwedge^2(\mathcal{V}^4)$, and the mapping

$$\sqcap: f \in GL(4) \mapsto \underline{f} \in GL(6) \tag{13.28}$$

is a group monomorphism. Let $\mathbf{A}_1, \mathbf{A}_2, \ldots, \mathbf{A}_6$ be bivectors in $\bigwedge(\mathcal{V}^4)$. They are also vectors in $\mathbb{R}^{3,3}$. Then $[\underline{f}(\mathbf{A}_1)\underline{f}(\mathbf{A}_2)\ldots\underline{f}(\mathbf{A}_6)] = \det(f)^3[\mathbf{A}_1\mathbf{A}_2\ldots\mathbf{A}_6]$. So $\sqcap$ is also a group monomorphism from $SL(4)$ to $SL(6)$.

By the definition of the inner product in $\mathbb{R}^{3,3}$, any special linear transformation in $\mathcal{V}^4$ induces a special orthogonal transformation in $\mathbb{R}^{3,3}$. After some mathematical reasoning, it can be proved that $\sqcap: SL(4) \longrightarrow SO(3, 3)$ is a Lie group isomorphism.

Let $SL^-(4)$ be the set of linear transformations in $\mathcal{V}^4$ of determinant -1. Let $Sproj(3) = SL(4) \cup SL^-(4)$. Then $Sproj(3)$ is a Lie group. Similarly, let $O(3,3) = SO(3,3) \cup SO^-(3,3)$, where $SO^-(3,3)$ are the orthogonal transformations in $\mathbb{R}^{3,3}$ of determinant -1. Both $Sproj(3)$ and $O(3,3)$ have two connected components, and their connected components containing the identity are isomorphic under $\sqcap$. However, any element of $SL^-(4)$ maps points to points, maps planes to planes, but reverses the sign of the inner product in $\mathbb{R}^{3,3}$. On the contrary, any element of $SO^-(3,3)$ keeps the sign of the inner product in $\mathbb{R}^{3,3}$ but interchanges points and planes in space. Thus, (13.28) does not provide an isomorphism between $Sproj(3)$ and $O(3,3)$.

Now take a retrospect at the $\mathbb{R}^{3,3}$ model of line geometry. When we identify $O(3,3)$ as the geometric transformation group and check for its action on the null vectors in $\mathbb{R}^{3,3}$, we find that for two linearly dependent null vectors, their relative scale (or ratio) is preserved by $O(3,3)$. Indeed, this scale measures the signed length of the vector representing the direction of the line. A vector with a spatial line as its support is called a *spear*. In fact, $O(3,3)$ describes the volume-preserving spear geometry in space.

With the spin group representation of the orthogonal transformations, all special projective transformations can be classified by their spinor generators. This chapter has no room left for further discussion.

13.3 Inverse Singularity Analysis by Wrench Matrix

In screw theory [1], any nonzero vector in $\mathbb{R}^{3,3}$ is called a *screw* when describing the geometry of lines, called a *twist* when describing Euclidean motion, and called a *wrench* in statics. The general form of a nonzero vector in $\mathbb{R}^{3,3}$ is of the form

$$\mathbf{S} = (\lambda \mathbf{s}, \lambda \mathbf{r} \times \mathbf{s} + \mu \mathbf{s}), \tag{13.29}$$

where $\mathbf{s}$ is a unit vector in $\mathbb{R}^3$, $(\mathbf{s}, \mathbf{r} \times \mathbf{s})$ are the Plücker coordinates of a line with direction $\mathbf{s}$ and passing through point $\mathbf{r}$, and λ, μ are scalars that cannot be zero simultaneously. The scalar $h = \mu/\lambda$ is called the *pitch* of the screw (or twist, or wrench).

In statics, a force acting on a rigid body is a *force wrench* $\mathbf{F} = (\mathbf{f}, \mathbf{r} \times \mathbf{f})$, where $\mathbf{r}$ points to the supporting line of the force. A system of forces acting upon a rigid body can be described by a wrench consisting of a certain force and a torque whose supporting plane is perpendicular to the force. If (13.29) represents a wrench, $\mathbf{s}$ is the direction of the force in the wrench, and λ is the scale of the force. When $\lambda = 0$, then $h = \infty$, and the wrench is the sum of two forces of the same magnitude but of opposite direction. Its action leads to a rotation about an axis along the direction of $\mathbf{f}$. Such a wrench is called a *pure torque*. When $h = 0$, the wrench is a pure force.

An infinitesimal Euclidean motion in space can be described by a vector in the Lie algebra of the 3D Euclidean group, say 6D vector $\mathbf{S} = (\mathbf{s}, \bar{\mathbf{s}} + \mathbf{t})$, where $(\mathbf{s}, \bar{\mathbf{s}})$ is a spear on the axis of the infinitesimal rotation and generating the latter, and $\mathbf{t}$ is

Fig. 13.7 *Left*: a limb of
stretchable length L with two
ball joints. *Right*: the 7 twists
of the limb

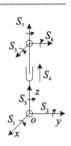

a vector generating the infinitesimal translation. By Charles' theorem, **s** and **t** are
linearly dependent for any infinitesimal Euclidean motion.

A rigid body is said to receive a *twist* about a screw if it rotates uniformly about
the screw, and at the same time translates uniformly along the screw through a distance equal to the product of the pitch and the angle of rotation. If (13.29) represents
a twist, then **s** is the direction of the rotation axis, **r** points to the axis, λ is the angle
of rotation, and μ is the distance of translation.

In (13.29), when $\lambda = 0$, then $h = \infty$, and the motion is a pure translation along
the axis. When $\mu = 0$, the twist reduces to a pure rotation around the axis. A twist
generating an infinitesimal motion is called an *infinitesimal motion twist* or *velocity
twist*.

The *virtual work* done by a wrench **W** to a twist **S** is their inner product in $\mathbb{R}^{3,3}$.
A wrench and a twist are said to be *reciprocal* to each other if they have zero virtual
work. If the work done by a wrench of a mechanism to a velocity twist is zero, the
wrench is called a *constraint* of the mechanism. If a wrench is reciprocal to all twists
of a mechanism, it is called a *common constraint* of the mechanism.

The $\mathbb{R}^{3,3}$ model of line geometry provides vector representation to the driving
wrenches of 1-dof kinematic pairs such as revolute pairs, prismatic pairs, and screw
pairs. It provides bivector representation to the driving wrenches of 2-dof kinematic
pairs such as cylindrical pairs.

In $\mathbb{R}^{3,3}$, the common constraints are the orthogonal complement of the kinematic
screw (twist) system of a mechanism [8, 9, 19]. Figure 13.7 is a typical example of a
limb used in parallel manipulators. It has a stretchable length L, and its two ends are
ball joints. Let $\mathbf{e}_1, \mathbf{e}_2, \mathbf{e}_3$ be an orthonormal basis of the space such that $\mathbf{e}_3$ is along
the limb. The kinematic screw system of the limb is composed of the following,
where 0 denotes the zero vector in $\mathbb{R}^3$:

1. The three infinitesimal rotation generators of the base ball joint:

$$\mathbf{S}_1 = (\mathbf{e}_1, 0)^T, \qquad \mathbf{S}_2 = (\mathbf{e}_2, 0)^T, \qquad \mathbf{S}_3 = (\mathbf{e}_3, 0)^T. \tag{13.30}$$

2. The infinitesimal translation generator along the shaft of the limb:

$$\mathbf{S}_4 = (0, \mathbf{e}_3)^T. \tag{13.31}$$

Fig. 13.8 *Left*: a limb of stretchable length L and variable angle θ between two revolute joint axes. *Right*: the 6 twists of the limb

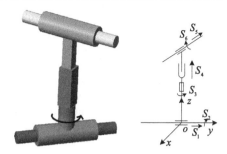

3. The three infinitesimal rotation generators of the moving ball joint:

$$\mathbf{S}_5 = (\mathbf{e}_1, L\mathbf{e}_2)^T, \qquad \mathbf{S}_6 = (\mathbf{e}_2, -L\mathbf{e}_1)^T, \qquad \mathbf{S}_7 = (\mathbf{e}_3, 0)^T. \qquad (13.32)$$

Since the above seven vectors span $\mathbb{R}^{3,3}$, their orthogonal complement is zero, and the limb does not have any common constraint.

Traditionally, inverse singularities of a Stewart platform are defined to be the singularities of the Jacobian mapping the velocity of the end-effector to the joint velocities. In [6], it is shown that the columns of the Jacobian matrix are zero-pitch or infinite-pitch wrenches (i.e., pure forces or pure torques) acting upon the moving platform. The inverse singularities can thus be interpreted as configurations where the lines of actions are linearly dependent [7].

By (13.30) to (13.32), for a Stewart platform, in the course of motion there does not occur any common constraint. For a GSP, however, things are quite different. Figure 13.8 is a GSP limb connecting two lines where either the distance L or the angle θ between the two lines is used to drive the mechanism. Let $\mathbf{e}_1, \mathbf{e}_2, \mathbf{e}_3$ be an orthonormal basis where $\mathbf{e}_2$ is along the lower revolute joint axis, and $\mathbf{e}_3$ is along the shaft of the limb. Let the direction of the upper revolute joint axis be $\mathbf{e}_4 = (\cos\theta, \sin\theta, 0)$. The kinematic screw system of the limb is composed of the following:

1. The infinitesimal rotation generator and translation generator of the base revolute joint:

$$\mathbf{S}_1 = (0, \mathbf{e}_2)^T, \qquad \mathbf{S}_2 = (\mathbf{e}_2, 0)^T. \qquad (13.33)$$

2. The infinitesimal rotation generator and translation generator of the limb shaft:

$$\mathbf{S}_4 = (0, \mathbf{e}_3)^T, \qquad \mathbf{S}_3 = (\mathbf{e}_3, 0)^T. \qquad (13.34)$$

3. The infinitesimal rotation generator and translation generator of the upper revolute joint:

$$\mathbf{S}_6 = (\mathbf{e}_4, L\mathbf{e}_5)^T, \qquad \mathbf{S}_5 = (0, \mathbf{e}_4)^T, \qquad (13.35)$$

where $\mathbf{e}_5 = \mathbf{e}_3 \times \mathbf{e}_4 = (-\sin\theta, \cos\theta, 0)$.

When $\cos\theta = 0$, i.e., the two revolute joint axes are parallel to each other, then $\mathbf{e}_4 = \pm\mathbf{e}_2$ and $\mathbf{e}_5 = \mp\mathbf{e}_1$. So $(\mathbf{e}_1, 0)^T$ is not spanned by the above six vectors in $\mathbb{R}^{3,3}$; since $(0, \mathbf{e}_1)^T$ spans the orthogonal complement of them, it is the common constraint of the limb. It is a pure torque that should result in a virtual rotation of some part of the limb around an axis in the direction of $\mathbf{e}_1$ but fails to make it due to the perpendicularity of the whole structure to $\mathbf{e}_1$.

The occurrence of common constraints can compensate for the linear degeneracy of driving wrenches to make a configuration nonsingular. If we stick to the original Jacobian map from the velocity of the end-effector to the joint velocities, we get wrong conclusions on the singularity of the configuration. For a GSP, we need to consider the following more general Jacobian map $\mathscr{J}$.

In a parallel manipulator, let there be all together n driving wrenches and linearly independent common constraint wrenches. For a GSP, $n \geq 6$. Each driving wrench $\mathbf{P}_i$ when executed leads to a virtual displacement q_i along the wrench. Let the end-effector velocity be denoted by vector $\mathbf{X}^\# \in \mathbb{R}^{3,3}$, where "#" is the linear transformation in $\mathbb{R}^{3,3}$ changing $\mathbf{X} = (\mathbf{x}, \mathbf{y})$ to $(\mathbf{y}, \mathbf{x})$ for any $\mathbf{x}, \mathbf{y} \in \mathbb{R}^3$. Let

$$\mathscr{Q} = (q_1, q_2, \ldots, q_6, \underbrace{0, 0, \ldots, 0}_{n-6})^T, \tag{13.36}$$

where the $n - 6$ zeroes denote the virtual displacements of the $n - 6$ linearly independent common constraints. Now the parallel manipulator is in singular configuration if and only if the following matrix $\mathscr{J}$ defined for arbitrary $\mathbf{X}$ is of rank less than 6:

$$\mathscr{J}\mathbf{X} = \mathscr{Q}. \tag{13.37}$$

Below we prove that $\mathscr{J}$ can be represented by the matrix of driving wrenches and common constraint wrenches. This conclusion holds not only for GSPs, but for all parallel robots of at most 6 degrees of freedom.

Each driving wrench or common constraint wrench is a vector $\mathbf{P}_i$ in $\mathbb{R}^{3,3}$. The *wrench matrix* of the system refers to the following $6 \times n$ matrix:

$$\mathscr{W} = [\mathbf{P}_1 \ \mathbf{P}_2 \ \ldots \ \mathbf{P}_n]. \tag{13.38}$$

In a force statical equilibrium of the end-effector, the external force and torque wrench $\mathbf{E}$ equals the sum of the driving wrenches and common constraint wrenches:

$$\mathbf{E} = \mathscr{W}\mathscr{F}, \tag{13.39}$$

where $\mathscr{F}$ is an nD vector whose components are the force scales of the driving wrenches and common constraint wrenches.

The virtual work done by $\mathbf{E}$ equals the virtual work done by $\mathscr{F}$, so

$$\mathscr{F}^T \delta\mathscr{Q} = \mathbf{E} \cdot \delta\mathbf{X}^\# = \mathbf{E}^T \delta\mathbf{X}, \tag{13.40}$$

Fig. 13.9 A limb driven by line/plane angle

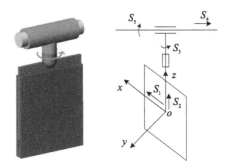

where the dot symbol denotes the inner product in $\mathbb{R}^{3,3}$. By (13.37), (13.39), and (13.40), $\mathscr{F}^T \mathscr{J} \delta\mathbf{X} = \mathscr{F}^T \mathscr{W}^T \delta\mathbf{X}$ for arbitrary $\mathscr{F}$ and $\delta\mathbf{X}$; hence

$$\mathscr{J} = \mathscr{W}^T. \tag{13.41}$$

A GSP is in singular configuration if and only if its driving wrenches and constraint wrenches span a linear subspace of dimension less than 6.

13.4 Singular Configurations of GSPs

According to the driving parameters of a GSP, all GSPs can be divided into four classes: (1) 3D3A: 3 distance control parameters and 3 angle control parameters, (2) 4D2A, (3) 5D1A, (4) 6D. There cannot be more than three angle control parameters due to the fact that a rigid body needs at most three angle constraints to determine its orientation.

Figure 13.9 illustrates a limb of fixed length L and driven by line/plane angle θ. Let $\mathbf{e}_3$ be along the shaft, and $\mathbf{e}_2$ be normal to the plane. The upper revolute line is in direction $\mathbf{e}_4 = (\cos\theta, \sin\theta, 0)$. Let $\mathbf{e}_5 = (-\sin\theta, \cos\theta, 0)$. The kinematic screw system of the limb is the following:
1. The infinitesimal translation generators of the base plate:

$$\mathbf{S}_1 = (0, \mathbf{e}_1)^T, \qquad \mathbf{S}_2 = (0, \mathbf{e}_3)^T. \tag{13.42}$$

2. The infinitesimal rotation generator of the limb shaft:

$$\mathbf{S}_3 = (\mathbf{e}_3, 0)^T. \tag{13.43}$$

3. The infinitesimal translation generator and rotation generator of the upper revolute axis:

$$\mathbf{S}_4 = (0, \mathbf{e}_4)^T, \qquad \mathbf{S}_5 = (\mathbf{e}_4, L\mathbf{e}_5)^T. \tag{13.44}$$

Fig. 13.10 A limb driven by
plane/plane angle

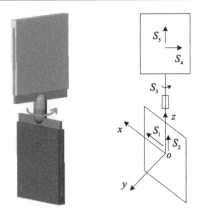

The above five twists span a 5D subspace of $\mathbb{R}^{3,3}$ when $\sin\theta \neq 0$, i.e., when the upper revolute axis is not parallel to the base plate. Its 1D orthogonal complement is spanned by a vector $(0, \mathbf{e}_5)^T$, which represents the common constraint of the limb. When $\theta = 0 \mod \pi$, the orthogonal complement is 2D, and is spanned by vectors $(0, \mathbf{e}_2)^T$ and $(\mathbf{e}_2, -L\mathbf{e}_1)^T$.

Figure 13.10 shows a limb of fixed length L and driven by plane/plane angle θ. Let $\mathbf{e}_3$ be along the limb shaft, and $\mathbf{e}_2$ be normal to the plane. Let $\mathbf{e}_4 = (\cos\theta, \sin\theta, 0)^T$. The kinematic screw system of the limb is the following:

1. The infinitesimal translation generators of the base plate:

$$\mathbf{S}_1 = (0, \mathbf{e}_1)^T, \qquad \mathbf{S}_2 = (0, \mathbf{e}_3)^T. \tag{13.45}$$

2. The infinitesimal rotation generator of the limb shaft:

$$\mathbf{S}_3 = (\mathbf{e}_3, 0)^T. \tag{13.46}$$

3. The infinitesimal translation generators of the upper plate:

$$\mathbf{S}_4 = (0, \mathbf{e}_4)^T, \qquad \mathbf{S}_5 = \mathbf{S}_2. \tag{13.47}$$

When $\sin\theta \neq 0$, i.e., when the two plates are not parallel to each other, then the above five twists span a 4D subspace of $\mathbb{R}^{3,3}$, whose 2D orthogonal complement is spanned by vectors $(0, \mathbf{e}_1)^T$ and $(0, \mathbf{e}_2)^T$, which represent two linearly independent common constraints of the limb. When the two plates are parallel, the twists span a 3D subspace, whose 3D orthogonal complement has a third vector generator $(\mathbf{e}_2, 0)^T$.

We have computed the common constraints of all the limbs used in GSPs, four of which have been shown in this chapter so far. The following list summarizes the number of linearly independent common constraint wrenches of a limb in GSP:

$\# = 0$ (1) point to point (or point to line, or point to plane) distance drive; (2) non-parallel line/line distance or angle drive.

Fig. 13.11 A 5D1A-type
GSP

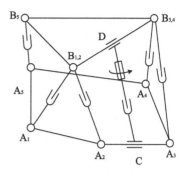

$\# = 1$ (1) line to parallel line distance or angle drive; (2) line to plane distance drive; (3) line to nonparallel plane angle drive.
In the above three cases, the common constraint wrench is a null vector in $\mathbb{R}^{3,3}$ representing a pure torque.
$\# = 2$ (1) line to parallel plane angle drive; (2) plane/plane distance drive; (3) non-parallel plane/plane angle drive.
In the first case, the common constraint wrenches form a 2D subspace $\mathbb{R}^{1,1}$ in $\mathbb{R}^{3,3}$, whose two null 1-subspaces represent a pure force and a pure torque, respectively. In the latter two cases, the common constraint wrenches form a 2D null subspace whose null 1-subspaces represent pure torques.
$\# = 3$ parallel plane/plane angle drive.
In this case, the common constraint wrenches form a 3D subspace $\mathbb{R}^{1,1,1}$ in $\mathbb{R}^{3,3}$, where the unique null 1-subspace orthogonal to the whole $\mathbb{R}^{1,1,1}$ represents a pure force, and the two null 2-subspaces both represent pure torques.

Below consider the 5D1A-type GSP shown in Fig. 13.11. It is actuated by five distance constraints between point pairs and one angle constraint between two lines. There are only three ball joints linking the moving platform: $B_{1,2}$, $B_{3,4}$, and B_5. Lines $B_{1,2}B_{3,4}$ and A_2A_3 are connected by a revolute limb S of stretchable length. This GSP has the following kinds of singularities:

Singularity type 1   The five lines supporting the driving wrenches of the limbs between point pairs are linearly independent, while the driving wrench and possible common constraint wrench of the limb S between two lines are both within the 5D subspace spanned by the five lines.

Figure 13.12 shows such a singular configuration. Line A_5B_5 meets line $B_{1,2}B_{3,4}$, and lines $A_2A_3 \parallel B_{1,2}B_{3,4}$. The five limbs between point pairs all have their supporting lines intersecting line $B_{1,2}B_{3,4}$ but are otherwise arbitrary, so they form a 5D subspace of $\mathbb{R}^{3,3}$ with degenerate inner product.

Let $\mathbf{e}_1$ be along $B_{1,2}B_{3,4}$, and $\mathbf{e}_2$ be along S. Let $B_{1,2}$ be the origin of the coordinate system. Now limb S has a driving torque $(0, \mathbf{e}_2)^T$ and a common constraint wrench $(0, \mathbf{e}_3)^T$, and both are orthogonal to $(\mathbf{e}_1, 0)^T$, i.e., both when taken as lines intersect line $B_{1,2}B_{3,4}$. So all seven wrenches have their supporting lines intersecting a common line, i.e., their representative null vectors orthogonal to the same null vector in $\mathbb{R}^{3,3}$. Hence the configuration is singular.

Fig. 13.12 A configuration of singularity type 1

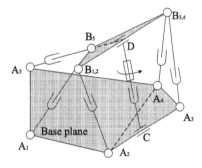

Fig. 13.13 A configuration of singularity type 3

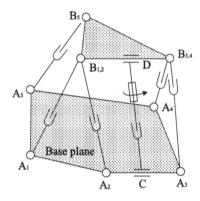

Singularity type 2 The five lines between point pairs are linearly dependent, and the two lines connected by the angle-driven limb S are not parallel.

Singularity type 3 The five lines between point pairs span a 4D linear subspace, the two lines of limb S are parallel, and the 4D subspace has nonempty intersection with the 2D subspace spanned by the driving wrench and common constraint wrench of limb S.

Figure 13.13 shows such a singular configuration. All points except for B_5 lie in the same plane, and $A_2A_3 \parallel B_{1,2}B_{3,4}$. Since lines $A_1B_{1,2}$, $A_2B_{1,2}$, $A_3B_{3,4}$, $A_4B_{3,4}$ are in the same plane, and the line at infinity of the plane supports the common constraint wrench of S, the rank of the five lines is 3, and the rank of the 6×7 wrench matrix is 5.

Singularity 4 The five lines between point pairs span a linear subspace of dimension at most 3.

Figure 13.14 shows another type of 5D1A GSP in a configuration where the matrix of driving wrenches is singular but the whole wrench matrix is not, and hence the configuration is nonsingular.

In the configuration, lines A_5B_5, $A_2B_{1,2}$, A_3B_3, A_4B_4 intersect at the same point O, and $A_3A_4 \parallel B_3B_4$. Let $\mathbf{e}_1$ be along B_3B_4, and $\mathbf{e}_2$ be along the shaft of the limb S connecting two lines. Let O be the origin of the coordinate system. Then

Fig. 13.14 A nonsingular configuration whose matrix of driving wrenches is singular

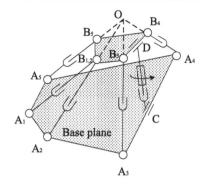

limb S has a driving torque $(0, \mathbf{e}_2)^T$ and a common constraint wrench $(0, \mathbf{e}_3)^T$, and limbs $A_5 B_5$, $A_2 B_{1,2}$, $A_3 B_3$, $A_4 B_4$ span a null 3-space of $\mathbb{R}^{3,3}$ with basis $(\mathbf{e}_1, 0)^T$, $(\mathbf{e}_2, 0)^T$, and $(\mathbf{e}_3, 0)^T$. So to limb $A_1 B_{1,2}$, let $\mathbf{s}$ be its direction, and let $\mathbf{r}$ be the vector from O to $B_{1,2}$. Then as long as the three vectors $\mathbf{r}, \mathbf{s}, \mathbf{e}_1$ are linearly independent, the wrench matrix is nonsingular. On the contrary, the matrix of driving wrenches is always singular.

13.5 Conclusion

The $\mathbb{R}^{3,3}$ model of line geometry is ideal for analyzing line complexes and designing line-based surfaces. Spinor representation of 3D special projective transformations casts new light on geometric construction and decomposition of projective transformations. This model thus carries the hope of many GAers in improving the performance of GA in projective geometry.

13.6 Exercises

13.1 Investigate which lines are represented by the nonline element $E = e_{03} \wedge e_{12}$ in $\mathbb{R}^{3,3}$ if the original 3D space is Euclidean. Probe with $L = (e_0 + \mathbf{d})\mathbf{u}$ (where $\mathbf{d} \cdot \mathbf{u} = 0$) in two ways, related dually: solve $L \cdot E = 0$, and solve $L \wedge E = 0$.

13.2 (*Continued from previous*): You should have found $e_3 \cdot (\mathbf{u}(1 \mp \mathbf{d}^*))$, where $*$ denotes the dual in the 3D space. Interpret these sets of lines geometrically.

13.3 Following the hints at the end of Sect. 13.2, show that $\exp(\frac{1}{2}\tau e_{12} \wedge e_{31})$ represents the translation versor for a translation over τe_1.

13.4 Following the hints at the end of Sect. 13.2, show that $\exp(\frac{1}{2}\gamma e_{01} \wedge e_{23})$ represents the scaling versor for a scaling relative to the origin by e^γ along the e_1-direction.

13.5 (*Continued from previous*): Provide versors for the remaining projective transformations.

References

1. Ball, R.: The Theory of Screws: A Study in the Dynamics of a Rigid Body. Hodges, Foster (1876)
2. Basu, D., Ghosal, A.: Singularity analysis of platform-type multi-loop spatial mechanism. Mech. Mach. Theory **32**, 375–389 (1997)
3. Ben-Horin, P., Shoham, M.: Singularity analysis of a class of parallel robots based on Grassmann–Cayley algebra. Mech. Mach. Theory **41**, 958–970 (2006)
4. Busemann, H.: Projective Geometry and Projective Metrics. Academic Press, New York (1953)
5. Cayley, A.: On the six coordinates of a line. Trans. Camb. Philos. Soc. **5**, 290–323 (1869)
6. Collins, C., Long, G.: Singularity analysis of an in-parallel hand controller for force-reflected teleoperation. IEEE J. Robot. Autom. **11**, 661–669 (1995)
7. Dandurand, A.: The rigidity of compound spatial grids. Struct. Topol. **10**, 41–56 (1984)
8. Fang, Y., Tsai, L.: Structure synthesis of a class of 4-DoF and 5-DoF parallel manipulators with identical limb structures. Int. J. Robot. Res. **21**, 799–810 (2002)
9. Featherstone, R.: Robot Dynamics Algorithms. Springer, Berlin (1987)
10. Gao, X., Lei, D., Liao, Q., Zhang, G.: Generalized Stewart–Gough platforms and their direct kinematics. IEEE Trans. Robot. Autom. **21**, 141–151 (2005)
11. Gosselin, C., Angeles, J.: Singularity analysis of closed-loop kinematic chains. IEEE Trans. Robot. Autom. **6**, 281–290 (1990)
12. Hodge, W., Pedoe, D.: Methods of Algebraic Geometry, vol. 1. Cambridge University Press, Cambridge (1952)
13. Huang, Z., Chen, L., Li, W.: The singularity principle and property of Stewart parallel manipulator. J. Robot. Syst. **20**, 163–176 (2003)
14. Hunt, K.: Kinematic Geometry of Mechanisms. Oxford University Press, Oxford (1978)
15. Jenner, W.: Rudiments of Algebraic Geometry. Oxford University Press, Oxford (1963)
16. Li, H.: Invariant Algebras and Geometric Reasoning. World Scientific, Singapore (2008)
17. Long, G.: Use of the cylindroid for the singularity analysis of rank 3 robot manipulator. Mech. Mach. Theory **32**, 391–404 (1997)
18. Maxwell, E.: General Homogeneous Coordinates in Spaces of Three Dimensions. Cambridge University Press, Cambridge (1951)
19. Merlet, J.: Singular configurations of parallel manipulators and Grassmann geometry. Int. J. Robot. Res. **8**, 45–56 (1989)
20. Merlet, J.: Parallel Robots. 2nd edn. Springer, Heidelberg (2006)
21. Park, F., Kim, J.: Singularity analysis of closed kinematics chains. J. Mech. Des. **121**, 32–38 (1999)
22. Pottmann, H.: Computational Line Geometry. Springer, Heidelberg (2001)
23. Semple, J., Roth, L.: Introduction to Algebraic Geometry. Oxford University Press, Oxford (1949)
24. Sommerville, D.: Analytic Geometry of Three Dimensions. Cambridge University Press, Cambridge (1934)
25. Stewart, D.: A platform with six degrees of freedom. Proc. Inst. Mech. Eng. **180**, 371–378 (1965)
26. Study, E.: Geometrie der Dynamen. Leipzig (1903)
27. Woo, L., Freudenstein, F.: Application of line geometry to theoretical kinematics and the kinematic analysis of mechanical systems. J. Mech. **5**, 417–460 (1970)
28. Yang, A.: Calculus of screws. In: Spillers, W. (ed.) Basic Questions of Design Theory. Elsevier, Amsterdam (1974)
29. Zlatanov, D., Fenton, R., Benhabib, B.: Identification and classification of the singular configurations of mechanisms. Mech. Mach. Theory **33**, 743–760 (1998)

A Framework for n-Dimensional Visibility Computations

14

Lilian Aveneau, Sylvain Charneau, Laurent Fuchs, and Frederic Mora

Abstract

This chapter introduces global visibility computation using Grassmann Algebra. Visibility computation is a fundamental task in computer graphics, as in many other scientific domains. While it is well understood in two dimensions, this does not remain true in higher-dimensional spaces.

Grassmann Algebra allows to think about visibility at a high level of abstraction and to design a framework for solving visibility problems in any n-dimensional space for $n \geq 2$. Contrary to Stolfi's framework which allows only the representation of geometric lines, its algebraic nature deals means general applicability, with no exceptional cases.

This chapter shows how the space of lines can be defined as a projective space over the bivector vector space. Then line classification, a key point for the visibility computation, is achieved using the exterior product. Actually, line classification turns out to be equivalent to point vs. hyperplane classification relative to a nondegenerate bilinear form. This ensures it is well defined and computationally robust.

Using this, the lines stabbing an n-dimensional convex face are characterized. This set of lines appears to be the intersection of the decomposable bivectors set (i.e., bivectors that represent a line) and a convex polytope. Moreover, this

L. Aveneau (✉) · S. Charneau · L. Fuchs
XLIM/SIC, CNRS, University of Poitiers, Poitiers, France
e-mail: lilian.aveneau@xlim.fr

S. Charneau
e-mail: sylvain.charneau@xlim.fr

L. Fuchs
e-mail: laurent.fuchs@xlim.fr

F. Mora
XLIM/SIC, CNRS, University of Limoges, Limoges, France
e-mail: frederic.mora@xlim.fr

L. Dorst, J. Lasenby (eds.), *Guide to Geometric Algebra in Practice*,
DOI 10.1007/978-0-85729-811-9_14, © Springer-Verlag London Limited 2011

convex polytope is proved to be minimal. This property allows useful algorithmic improvements.

To illustrate the use of our framework in practice, we present the computation of soft shadows for three-dimensional illuminated scenes.

14.1 Problem Statement

14.1.1 About Visibility

Visibility is a fundamental problem in computer graphics. All rendering algorithms aim at simulating the light transfer in a virtual environment, which strongly depends on the mutual visibility of each element in the scene. This is clearly illustrated by the following well-known rendering equation:

$$L(x, \omega) = E(x, \omega) + \int_y \rho(x, \omega, x \to y) L(x, x \to y) \frac{\cos \theta_x \cos \theta_y}{|x - y|^2} V(x, y) \, dy.$$

The radiance L leaving a point x in the direction ω is the sum of the emitted radiance E at x, plus the reflected light as the sum of the incoming radiance from all the points y in the scene, according to the surface property ρ and the incident angles. In this equation, $V(x, y)$ is the visibility function, whose value is 1 if x and y are mutually visible and 0 otherwise.

As a consequence, the accuracy of the visibility solution has a direct impact on the quality of the result. This explains why visibility is a central question. And it goes beyond the scope of computer graphics: Other domains, such as electromagnetism or acoustics, for instance, derive algorithms to simulate wave propagation.

There are many visibility problems. The simplest one is between two points. A classical solution uses a visibility ray, which works in any dimension where such a ray approach is available [9]. But visibility queries can be more complicated. For example: "What parts of the scene can be seen from this point?" or "What parts of the scene can be seen from this region?" In the latter case, the visibility problem becomes very complex. Contrary to the point-to-point visibility query, it is not sufficient to answer "It is visible" or "It is invisible." The challenge is to compute the whole visibility set, i.e., a global visibility information between two continuous sets of points. This implies to study all the discontinuities in the visibility that may occur because of occluders lying between the sets of points. Visibility discontinuities, sometimes called visibility events, happen at the occluder boundaries. They are the frontiers where the visibility changes.

For simplifying global visibility problems, a common approach consists of first to sample the continuous sets of points and then to perform successive point-to-point visibility queries. However, this sampling step introduces noise, altering the quality of the result. Increasing the sampling density helps to minimize the problem but may badly affect the computation time. In addition, notice that a sampling strategy may be unusable. Considering the following problem: "Prove that two continuous sets of points are not mutually visible," an infinite number of samples would be required!

This illustrates the need for algorithms able to solve *exactly* any global visibility problem: On the one hand, it ensures high quality results in applications; on the other hand, it is the only way to solve some visibility problems.

14.1.2 The Dimension Problem

Global visibility problems take place in line-space. For example, the visibility of two continuous sets of points corresponds to the lines intersecting the two sets without intersecting their occluders. Thus, visibility discontinuities correspond to lines incident to occluder boundaries. As a consequence, the complexity of a visibility problem is strongly related to the dimension of its underlying line-space.

In a two-dimensional space, the line-space is also two-dimensional. Global visibility in 2D has been studied for convex objects through the visibility complex [16] and used in different applications such as radiosity computation [14]. Using another line parameterization, Bittner et al. [4] focus on the visibility from a region in the plane.

In a three-dimensional space, the line-space is not three-dimensional, but of dimension 5. As a consequence, visibility problems are much more difficult to apprehend, and the generalization of two-dimensional visibility algorithms is not possible. So, dedicated algorithms were proposed. F. Durand has developed the 3D visibility complex [7], a data structure that encodes the global visibility by tracking all the discontinuities generated by the vertices, edges, and faces of a polygonal environment. This data structure illustrates the complexity of the three-dimensional visibility but is not practicable due to robustness issues. The visibility skeleton [6] is a derivative of the 3D visibility complex. It is a multipurpose visibility tool, but it does not encode all the visibility data.

A line in 3D has four degrees of freedom, but a 4D parameterization is not possible without singularities (for instance, we can consider the bounding sphere of the scene; then any line intersects the sphere in two different points, and since the sphere is a surface of degree 2, then a line can be defined using four parameters; however, singularities remain at the poles: the azimuthal coordinate can take any value, leading to different coordinates describing a same line). This can make algorithms sensitive to numerical stability. To avoid this problem, other approaches use the Plücker line parameterization. The Plücker space is a five-dimensional projective space embedding all the 3D lines in a four-dimensional manifold. It is useful to group lines according to the objects they intersect. Pellegrini [15] uses this formalism to find upper bounds on geometric problems involving three-dimensional lines. In the Plücker space, lines stabbing a sequence of convex polygons can be represented as a convex polytope set. This property is used by Teller [18] for computing visibility through a sequence of *portals* or convex transparent polygons. Nirenstein [13] and Bittner [3] take into account occlusion to compute from-region visibility, further improved by Haumont [8] and Mora [10, 11].

The first practicable global visibility algorithms in three-dimensional space are quite recent. This area of research is still being investigated. If it is quite difficult to apprehend visibility in a three-dimensional space, it is worse in a four-dimensional space, e.g., in dynamic environments. At present, we are not aware of any practicable algorithms dedicated to 4D space.

14.1.3 Toward a Global Visibility Framework

This brief overview highlights several difficulties. At first, visibility algorithms are dependent on the geometrical space dimension. The gap of complexity, for example, from two-dimensional to three-dimensional space, prevents a general approach. In addition, since global visibility is expressed in a line space, the parameterization choice greatly affects the algorithm design, properties, and robustness.

Geometry algebra gives the opportunity to think about visibility at a higher level of abstraction. It allows one to analyze problems and to design their solutions regardless of the dimension of space, using a single approach.

In this chapter, we propose a global visibility framework based on a n-dimensional line space, defined using Grassmann Algebra [5]. While this is a classical definition of lines in mathematics, it remains uncommon in computer graphics. Thanks to this formalism, we prove a major theorem on the representation of a set of lines by a convex polytope. Next, we propose a generalization of Mora's work [10] into an n-dimensional visibility framework. Finally, as an application, we explain how it can be used to compute very high quality soft shadows in the rendering of artificial scenes.

14.2   Line Spaces

For computing visibility between objects, let us denote $\mathfrak{G}_n$ the n-dimension geometrical space of the geometric objects. It is embedded into the projective space $\mathbb{P}^n$. As $\mathbb{P}^n$ is built from $\mathbb{R}^{n+1}$, linear subspaces of $\mathbb{P}^n$ can be represented by elements of $\bigwedge(\mathbb{R}^{n+1})$.

14.2.1 n-Dimensional Lines

Whatever the dimension of the space it belongs to, a line is a one-dimensional subspace. It expresses a dependency between two distinct points. So, we can formulate the following definition:

Definition 14.1 An n-dimensional line, passing through two projective points A and B of $\mathfrak{G}_n$ with respective 1-vector coordinates a and b, is represented by the exterior product $a \wedge b$.

Example 14.1 As an example, let us consider in two dimensions the line going through the points of homogeneous coordinates $(1, 0, 1)$ and $(2, 1, 1)$. Using the exterior product, it follows that the expression of this line in $\bigwedge(\mathbb{R}^3)$ is $\mathbf{e_0} \wedge \mathbf{e_1} - \mathbf{e_1} \wedge \mathbf{e_2} - \mathbf{e_2} \wedge \mathbf{e_1}$, where the 1-vectors $(\mathbf{e_0}, \mathbf{e_1}, \mathbf{e_2})$ form the basis of $\bigwedge^1(\mathbb{R}^3)$.

Example 14.2 In computer graphics, the three-dimensional lines are most known using Plücker coordinates. In fact, they can be retrieved using Definition 14.1. Using again a homogeneous notation, a point P is denoted by four coordinates using the vector: $p = (x, y, z, w)^T$. The line going through A and B with respective coordinates $(x_a, y_a, z_a, 1)$ and $(x_b, y_b, z_b, 1)$ is known as [15]

$$\begin{pmatrix} x_b - x_a \\ y_b - y_a \\ z_b - z_a \\ y_a z_b - y_b z_a \\ z_a x_b - z_b x_a \\ x_a y_b - x_b y_a \end{pmatrix}.$$

With our definition, the same line through A and B is defined, using the Grassmann exterior product, as $\Pi_{AB} = a \wedge b$—see Exercise 14.1. Then, it is quite easy to show that Plücker coordinates are coordinates in $\bigwedge^2(\mathbb{R}^4)$.

14.2.2 From Line to Line Space

The elements of $\bigwedge^k(\mathbb{R}^{n+1})$ for $k \leq n$ are homogeneous: If $\mathcal{K} \in \bigwedge^k(\mathbb{R}^{n+1})$ represents a subspace of $\mathbb{R}^{n+1}$, then $\mathcal{K}' = \lambda K$ for $\lambda \in \mathbb{R}^*$ represents the same subspace. Hence, $\mathbb{P}^{\binom{n+1}{k}-1} = \mathbb{P}(\bigwedge^k(\mathbb{R}^{n+1}))$ is the space of the 1-subspaces of $\bigwedge^k(\mathbb{R}^{n+1})$. Each point of $\mathbb{P}^{\binom{n+1}{k}-1}$ represents a unique linear manifold of $\mathfrak{G}_n$. This leads to the following definition of the line space $\mathfrak{L}_n$ of $\mathfrak{G}_n$:

Definition 14.2 The space of lines of $\mathfrak{G}_n$, denoted by $\mathfrak{L}_n$, is the projective space $\mathbb{P}(\bigwedge^2(\mathbb{R}^{n+1}))$.

From this definition, the line space is a projective space of dimension $\binom{n+1}{2} - 1$. With $\mathfrak{G}_2$, the line space is of dimension $\binom{3}{2} - 1 = 2$, while with $\mathfrak{G}_3$, it corresponds to the classical Plücker space of dimension 5. This is directly related to the dimension problem, as presented in Sect. 14.1.2.

Considering again Example 14.1, the line passing through the points with homogeneous coordinates $(1, 0, 1)$ and $(2, 1, 1)$ has coordinates $(1, -1, -1)$ using the basis $(\mathbf{e_0} \wedge \mathbf{e_1}, \mathbf{e_1} \wedge \mathbf{e_2}, \mathbf{e_2} \wedge \mathbf{e_0})$ of $\bigwedge^2(\mathbb{R}^3)$.

14.2.3 About the Grassmannian

From previous works on visibility computations [13, 19], it is well known that three-dimensional Plücker lines do not fill all the linear space $\mathcal{L}_3$. The mapping of three-dimensional lines to the Plücker space is not surjective. Indeed, a line must pass through at least two distinct points and then is represented as a 2-blade. They are all located on the Grassmannian (manifold) $G^{\mathbb{R}}(2, 4)$ in the linear space $\bigwedge^2(\mathbb{R}^4)$, defined as the set of all 2-subspaces of $\mathbb{R}^4$. It is also the set of all decomposable bivectors, or 2-blades. In dimension n, the set of *geometric* lines are located on the Grassmannian $G^{\mathbb{R}}(2, n + 1)$ within the linear space $\bigwedge^2(\mathbb{R}^{n+1})$.

For deciding if a given point of $\mathcal{L}_n$ is on the Grassmannian, and so is a geometric line, it is sufficient to verify that it is a decomposable vector. The following theorem gives an easy way to solve this.

Theorem 14.1 *Let $\bigwedge(\mathbb{R}^{n+1})$ be the Grassmann algebra. A nonzero bivector M from $\bigwedge(\mathbb{R}^{n+1})$ is a 2-blade—a decomposable vector—if and only if $M \wedge M = 0$.*

Proof If M is a 2-blade, then there exist two linearly independent vectors m_1 and m_2 from $\mathbb{R}^{n+1}$ such that $M = m_1 \wedge m_2 \neq 0$. Then, $M \wedge M = m_1 \wedge m_2 \wedge m_1 \wedge m_2$, and using the antisymmetry property of the exterior product, $M \wedge M = 0$.

Now, assuming that M is not decomposable, by definition it can be written as a finite sum of linearly independent 2-blades: $\sum_{i=1}^p M_i$, $\binom{n+1}{2} \geq p \geq 2$. It follows that

$$M \wedge M = \left(\sum_{i=1}^p M_i \right) \wedge \left(\sum_{j=1}^p M_j \right) \tag{14.1}$$

$$= 2 \sum_{i \leq j}^p M_i \wedge M_j. \tag{14.2}$$

Since all these terms are linearly independent 4-vectors, $M \wedge M \neq 0$.   □

This theorem helps for computing the intersection between a set of lines in $\mathcal{L}_n$ and the Grassmannian, for instance, to decide if it contains at least one geometric line or can be dropped in future computations.

14.2.4 Line Orientation

Previous works on visibility computation use line orientation as a key element. In Grassmann algebra, it is expressed using the exterior product, which expresses the dependency between vector subspaces.

Property 14.1 Let M and M' be two projective linear subspaces of $\mathfrak{G}_n$. Their intersection is nonempty if and only if $M \wedge M' = 0$.

This property can be applied to a bivector L and an $(n-1)$-vector F of $\bigwedge(\mathbb{R}^{n+1})$, allowing us to check if a line is incident to the boundary of an occluder, i.e., an $(n-2)$-variety or flat. For instance, as proposed in Exercise 14.3, in three dimensions it is possible to check that the Plücker relation corresponds to the test $L \wedge H$, where H is also a line since $n - 1 = 2$. A particular case of this first property is the following:

Property 14.2   Let M and M' be respectively a k-vector and an $(n-k+1)$-vector, representing two projective linear subspaces of $\mathfrak{G}_n$. Their relative orientation is denoted by the sign of $\lambda \in \mathbb{R}$ using the exterior product $M \wedge M' = \lambda I$, where I is the pseudo-scalar in $\bigwedge(\mathbb{R}^{n+1})$.

For instance, in two dimensions, it allows us to check if a directed line turn clockwise or counterclockwise with respect to a point. This property is essential for computing the line stabbing a convex $(n-1)$ face, as presented in Sect. 14.3.1.

In order to work with lines in $\mathfrak{L}_n$, we need a similar product using only bivectors. Moreover, we want a robust solution, working with geometric lines, but also with any bivector. In fact, it is possible to express Property 14.2 into the line space, using only bivectors and the inner product. This is based on the duality between $\bigwedge^2(\mathbb{R}^{n+1})$ and $\bigwedge^{n-1}(\mathbb{R}^{n+1})$.

Theorem 14.2   *Let δ be an isomorphism from $\bigwedge^2(\mathbb{R}^{n+1})$ to $\bigwedge^{n-1}(\mathbb{R}^{n+1})$ such that $\delta(l) = l \rfloor I_{n+1}$, where $\rfloor$ denotes the left contraction, and I_{n+1} is the pseudoscalar over $\bigwedge(\mathbb{R}^{n+1})$. Then, the inner product between any bivectors L_1 and L_2 is equivalent to check the orientation of the line L_1 with a $(n-2)$-flat:*

$$L_1 \cdot L_2 \equiv L_1 \wedge \delta(L_2)$$

up to an identification of pseudoscalars and scalars in $\bigwedge(\mathbb{R}^{n+1})$.[1]

From this result, two immediate and important properties follow:
1. Since the inner product is nondegenerate, there is no singularity. It asserts that the line orientation test is always defined, whatever the line and the flat are.
2. The duality does not depend on the Grassmannian: The previous property is also valid for nondecomposable bivectors and $(n-1)$-vectors, or equivalently, when they do not respectively represent a line and an $(n-2)$-flat.

The first property plays a fundamental role in our visibility framework. Firstly, it explains that the visibility computations work for all dimensions and all configurations, without any singularity, ensuring the generality of this approach. Secondly,

---

[1]*Editorial note*: The δ is a dualization in $\bigwedge(\mathbb{R}^{n+1})$. If one would have a geometric algebra for $\mathbb{R}^{n+1}$, this would be proportional to multiplication by the pseudoscalar, see Chap. 21 and their 2D example below. In a nonmetric Grassmann algebra, the dualization is necessarily introduced somewhat more abstractly.

it is computationally simple, as is it reduced to evaluations of a particular scalar product of vectors of dimension $\binom{n+1}{2}$, and this ensures its robustness.

The n-dimensional proof of this theorem is left to the reader, but we illustrate it in two dimensions. Let $(\mathbf{e}_0, \mathbf{e}_1, \mathbf{e}_2)$ be an orthogonal basis of $\mathbb{R}^3$, and $\mathbf{e}_0 \wedge \mathbf{e}_1 \wedge \mathbf{e}_2$ be the pseudoscalar for $\bigwedge(\mathbb{R}^3)$. Let $(\mathbf{e}_1 \wedge \mathbf{e}_2, \mathbf{e}_2 \wedge \mathbf{e}_0, \mathbf{e}_0 \wedge \mathbf{e}_1)$ be a basis of $\bigwedge^2(\mathbb{R}^3)$. We know that any two-dimensional line can be represented using three coordinates in such a basis. The left contraction is used to define the isomorphism $\delta : \bigwedge^2(\mathbb{R}^3) \mapsto \bigwedge^1(\mathbb{R}^3)$ as:

$$\delta(\mathbf{e}_1 \wedge \mathbf{e}_2) = (\mathbf{e}_1 \wedge \mathbf{e}_2)\rfloor(\mathbf{e}_0 \wedge \mathbf{e}_1 \wedge \mathbf{e}_2) = \mathbf{e}_0,$$

$$\delta(\mathbf{e}_2 \wedge \mathbf{e}_0) = (\mathbf{e}_2 \wedge \mathbf{e}_0)\rfloor(\mathbf{e}_0 \wedge \mathbf{e}_1 \wedge \mathbf{e}_2) = \mathbf{e}_1,$$

$$\delta(\mathbf{e}_0 \wedge \mathbf{e}_1) = (\mathbf{e}_0 \wedge \mathbf{e}_1)\rfloor(\mathbf{e}_0 \wedge \mathbf{e}_1 \wedge \mathbf{e}_2) = \mathbf{e}_2.$$

So, any two-dimensional line can be mapped to a 1-vector using δ, and conversely with δ^{-1}. Without loss of generality, let $A : (\alpha_0, \alpha_1, \alpha_2)$ and $B : (\beta_0, \beta_1, \beta_2)$ be two lines, i.e., bivectors. Using the anticommutativity of the exterior product, it follows that

$$A \wedge \delta(B) = (\alpha_0 \mathbf{e}_1 \wedge \mathbf{e}_2 + \alpha_1 \mathbf{e}_2 \wedge \mathbf{e}_0 + \alpha_2 \mathbf{e}_0 \wedge \mathbf{e}_1) \wedge (\beta_0 \mathbf{e}_0 + \beta_1 \mathbf{e}_1 + \beta_2 \mathbf{e}_2)$$

$$= (\alpha_0 \beta_0 + \alpha_1 \beta_1 + \alpha_2 \beta_2)\mathbf{e}_0 \wedge \mathbf{e}_1 \wedge \mathbf{e}_2$$

$$\equiv A \cdot B.$$

Using any isomorphism between $\mathbb{R}^3$ and $\bigwedge^3(\mathbb{R}^3)$, the obtained expression can be recognized as a classical inner product in $\bigwedge^2(\mathbb{R}^3)$.

14.2.5 Dual Line Representation

Theorem 14.2 has a nice and useful interpretation in the line-space $\mathfrak{L}_n$.

As a classical result of vector algebra, in a vector space E of dimension m, it is well known that the set of vectors orthogonal to any other vector x (the set of v such that $v \cdot x = 0$, where $\cdot$ is the inner product) describes a vector subspace of dimension $m - 1$ in E, i.e., a hyperplane. Transposed to our problem, this classical result means that each $(n - 1)$-vector (i.e., $\delta(L_2)$ in the theorem statement) can be dually associated to a hyperplane in $\bigwedge^2(\mathbb{R}^{n+1})$ (the set of bivectors L orthogonal to L_2 in $\bigwedge^2(\mathbb{R}^{n+1})$, i.e., such that $L \cdot L_2 = 0$), which corresponds to a projective hyperplane in the line-space $\mathfrak{L}_n$. In dimension 2, it can be remarked that some similarities exist between this model of line-space $\mathfrak{L}_2$ with the well-known dual plane where lines map to points and conversely points map to lines. This reveals that classifying a line L against an $(n - 2)$-flat F, by computing the sign of the product $L \wedge F$ for the bivector L and the $(n - 1)$-vector F, can always be seen as determining in which half-space is the point L of $\mathfrak{L}_n$, according to the oriented hyperplane $\delta^{-1}F$ associated to F in $\mathfrak{L}_n$. The product is zero if and only if L is a point on the hyperplane F^*.

This interpretation will be particularly helpful in Sect. 14.3.2, to give a geometrical significance to the global visibility computation and representation, which only makes sense in the line-space.

14.3   Visibility in $\mathfrak{L}_n$

14.3.1   Lines Stabbing a Convex $(n-1)$-Face

The following theorem unambiguously characterizes the set of lines stabbing a convex face in any dimension n.

Theorem 14.3 *Let $\mathbf{F}$ be a convex $(n-1)$-face in $\mathfrak{G}_n$, supported by the hyperplane $\mathcal{H}_F$ (i.e., an $(n-1)$-flat in $\mathfrak{G}_n$) and bounded by the $(n-2)$-flats f_i for $i \in [1, \ldots, r]$. The flats f_i have two orientations such that for any line L, $L \cap \mathcal{H}_F \neq L$, L stabs $\mathbf{F}$ if and only if one of the following two properties is verified:*

$$\forall i \in [1, \ldots, r], \quad L \wedge f_i \geq 0, \tag{14.3}$$
$$\forall i \in [1, \ldots, r], \quad L \wedge f_i \leq 0. \tag{14.4}$$

Let $\mathscr{S}_{\mathbf{F}}$ be the set of lines stabbing the face $\mathbf{F}$.

The proof of the theorem is based on the following remarks. Firstly, $\mathbf{F}$ is a convex polytope, delimited by the flats f_i and restricted to the hyperplane $\mathcal{H}_F$ in $\mathfrak{G}_n$. Secondly, when a line L does not lie on $\mathcal{H}_F$, it has one and only one intersection point with $\mathcal{H}_F$, at infinity if L is parallel to $\mathcal{H}_F$. Thirdly, L stabs the face $\mathbf{F}$ if and only if its intersection point with $\mathcal{H}_F$ is in $\mathbf{F}$.

Proof The line L corresponds to a 2-vector, and the hyperplane $\mathcal{H}_F$ to an n-vector in $\bigwedge(\mathbb{R}^{n+1})$. Since $n + 2 > n + 1$, then $L \wedge \mathcal{H}_F = 0$. So, there is always an intersection between L and $\mathcal{H}_F$, either of dimension 1 (a point) or 2 (L itself). This allows us to propose the following lemma:

Lemma 14.1 *Every line L in $\mathfrak{G}_n$ intersects $\mathcal{H}_F$ in a unique point, except if L lies in $\mathcal{H}_F$.*

Let $P_\cap$ be the intersection of L with $\mathcal{H}_F$. Assuming that L does not lie on $\mathcal{H}_F$, $P_\cap$ is a nonzero 1-vector. Obviously, L stabs $\mathbf{F}$ if and only if $P_\cap$ is inside $\mathbf{F}$. Let $P_{\not\cap}$ be any other point on L out of $\mathcal{H}_F$ such that $P_{\not\cap} \wedge \mathcal{H}_F > 0$, i.e., $P_{\not\cap}$ is in the positive half-space of $\mathcal{H}_F$. We can write $L = P_\cap \wedge P_{\not\cap}$. It follows that

$$\forall i \in [i, \ldots, r], \quad P_\cap \wedge P_{\not\cap} \wedge f_i = P_\cap \wedge (P_{\not\cap} \wedge f_i).$$

By hypothesis, $P_{\not\cap}$ is not incident to f_i, and the n-vector $P_{\not\cap} \wedge f_i \neq 0$ represents a hyperplane in $\mathfrak{G}_n$. The sign of the pseudoscalar $P_\cap \wedge (P_{\not\cap} \wedge f_i)$ indicates in which half-space of the hyperplane $P_{\not\cap} \wedge f_i$ the point $P_\cap$ is.

Let $\mathscr{P}$ be the polytope generated by $\mathbf{F}$ and the vertex $P_{\widehat{\eta}}$. Since $\mathbf{F}$ is convex, so is $\mathscr{P}$: It is the intersection between $\mathscr{H}_F^+$ and the positive half-spaces associated to the hyperplanes $P_{\widehat{\eta}} \wedge f_i$, $i \in [1, \ldots, r]$, for a *particular* but consistent orientation of them. By hypothesis, $P_{\widehat{\eta}} \wedge \mathscr{H}_F$ is a nonzero pseudoscalar. The orientation of each hyperplane $P_{\widehat{\eta}} \wedge f_i$ can be determined from any point $P_{\mathbf{F}}$ into $\mathbf{F}$, such that $P_{\mathbf{F}} \wedge (P_{\widehat{\eta}} \wedge f_i)$ are only positively oriented pseudoscalars. Since $P_{\mathbf{F}} \wedge (P_{\widehat{\eta}} \wedge f_i) = -P_{\widehat{\eta}} \wedge P_{\mathbf{F}} \wedge f_i$ and $P_{\mathbf{F}} \wedge f_i$ generates the flat $\mathscr{H}_F$, $P_{\mathbf{F}} \wedge f_i$ has an opposite orientation than the one of $\mathscr{H}_F$ since $\forall i \in [1, \ldots, r]$, $P_{\mathbf{F}} \wedge f_i = \lambda F$, $\lambda < 0$. This shows that the orientations of the flats f_i can be determined uniquely from those of $\mathscr{H}_F$ and the position of $P_{\widehat{\eta}}$ relatively to $\mathscr{H}_F$.

The orientation of the flats f_i allows us to state that a point P of $\mathfrak{G}_n$ is in the polytope $\mathscr{P}$ if and only if $P \wedge \mathscr{H}_F \geq 0$ and $P \wedge (P_{\widehat{\eta}} \wedge f_i) > 0$ $\forall i \in [1, \ldots, r]$. In particular, this concerns every point in $\mathbf{F}$, including $P_{\cap}$. Then, $P_{\cap}$ is a point of $\mathscr{P}$ if and only if:

$$P_{\cap} \wedge (P_{\widehat{\eta}} \wedge f_i) \geq 0 \quad \forall i \in [1, \ldots, r].$$

In other words, L stabs the face $\mathbf{F}$ if and only if

$$L \wedge f_i \geq 0 \quad \forall i \in [1, \ldots, r].$$

Considering the opposite line $L' = P_{\widehat{\eta}} \wedge P_{\cap}$, obviously it stabs $\mathbf{F}$ if and only if

$$L' \wedge f_i \leq 0 \quad \forall i \in [1, \ldots, r].$$

This result does not depend on $P_{\widehat{\eta}}$ with respect to the sign of the half-space associated to the hyperplane $\mathscr{H}_F$; it remains valid for all lines L not in $\mathscr{H}_F$.

Changing the orientation of some but not all of the flats f_i makes the previous result false. On the contrary, changing the orientation of all of them—or equivalently the orientation of $\mathscr{H}_F$ or the position of $P_{\widehat{\eta}}$ relatively to $\mathscr{H}_F$—only interchanges all signs in the in-equations, and so the result remains true. This means there are only two valid orientations for the flats f_i. $\qquad\square$

Theorem 14.3 has a fundamental consequence: It gives an algebraic method for determining whether or not a line stabs a given face in any n-dimensional space. This method has no singularity, since our algebraic framework also handles flats at infinity. However, the lines lying on the hyperplane $\mathscr{H}_F$ are excluded, since they cannot properly "stab" $\mathbf{F}$. But this distinction does not impact the visibility computation, both from a theoretical and an algorithmic point of view.

Moreover, this theorem has a useful interpretation in $\mathfrak{L}_n$. It first reveals what the visibility computation through faces relies on, geometrically, and then indicates what kind of data-structures and algorithms can be used to compute the visibility in practice. The following section aims to explain this interpretation.

14.3.2 Convex Cells and Visibility Events in the Line-Space

14.3.2.1 Interpretation in $\mathfrak{L}_n$ and Consequences

Theorem 14.3 has a suitable geometrical meaning in $\mathfrak{L}_n$. By duality, every flat f_i bounding the face **F** can be associated to an unique hyperplane f_i^* in $\mathfrak{L}_n$ $\forall i \in [1, \ldots, r]$. Then, by choosing a positive orientation for the flats, Theorem 14.3 implies that $\mathscr{S}_\mathbf{F}$, the set of lines stabbing **F** in $\mathfrak{L}_n$, is the intersection between $G^{\mathbb{R}}(2, n+1)$ and the convex polytope defined as the intersection of the positive half-spaces delimited by the hyperplanes f_i^*. This is a useful result, as convex polytopes have the following well-known properties in computational geometry:

- They have multiple representations: A hyperplane set, a vertex set, and a face lattice.
- It can easily be determined if a point is either inside or outside a polytope.
- It can be easily determined if two polytopes intersect each other.
- Boolean operations are expressed as geometrical computations, such as split, intersections, etc.

Nevertheless, a single face **F** is not sufficient to define a polytope: The hyperplanes f_i^*, $i \in [1, \ldots, r]$, delimit a region in $\mathfrak{L}_n$ partially bounded by infinity. This is stated by the dimension of $\mathfrak{L}_n$ (see Sect. 14.2.3): While at least $2n - 1$ hyperplanes are required to define a simplex in $\mathfrak{L}_n$, an $(n-1)$-face can only have n facets in general (i.e., independent) positions, for instance, those of an $(n-1)$-simplex in $\mathfrak{G}_n$. As a consequence of this closure by the infinity in $\mathfrak{L}_n$, it becomes impossible to determine a convex-hull representation of the polytope $\mathscr{S}_\mathbf{F}$.

While this interpretation leads to some interesting properties, it also illustrates the fundamental role of the Grassmannian $G^{\mathbb{R}}(2, n+1)$ and its embedding line-space $\mathfrak{L}_n$. Indeed, the polytope representing the lines stabbing some faces also contains points outside the Grassmannian. Then, representing the lines stabbing faces in $\mathfrak{G}_n$ by a convex polytope in $\mathfrak{L}_n$ is only possible by considering the whole line-space $\mathfrak{L}_n$, but not the Grassmannian $G^{\mathbb{R}}(2, n+1)$ alone. This mainly explains why the Stolfi framework [17], which only represents points located on the Grassmannian, is not suitable for computing the global visibility. On the contrary, by enabling computations on nondecomposable multivectors, geometric algebras make the global visibility computation sum up to boolean operations on convex polytopes in $\mathfrak{L}_n$.

14.3.2.2 Global Visibility in $\mathfrak{G}_n$ as Convex Cells in $\mathfrak{L}_n$

Extending this representation to two or more faces is straightforward. In the example depicted in Fig. 14.1, **A**, **B**, and **O** are three edges in $\mathfrak{G}_2$, with bounding vertices $i \in [1, \ldots, 6]$. In $\mathfrak{G}_2$, these vertices are associated to hyperplanes that subdivide the line-space into cells, grouping together the lines stabbing the same edges. Figure 14.1 shows two such cells: $\mathscr{P}_\mathbf{AB}$ representing lines stabbing **A** and **B** but missing **O**; and $\mathscr{P}_\mathbf{AOB}$ representing lines stabbing **A**, **O**, and **B**. It must be noticed that $\mathscr{P}_\mathbf{AB}$ completely describes the global visibility between **A** and **B** by taking into account the occlusion by **O**. This example shows that visibility in $\mathfrak{G}_2$ can be described in $\mathfrak{L}_2$ by a set of convex polytopes, obtained using Theorem 14.3.

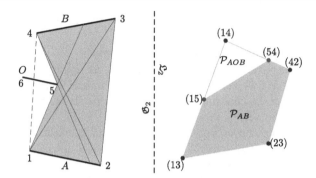

Fig. 14.1 (Color online) Visibility computation and representation. *Left*: In $\mathfrak{G}_2$, the edges **A** and **B** are partly hidden by **O**. *Right*: In $\mathfrak{L}_2$, the lines stabbing the three edges are the convex cell $\mathcal{P}_{\mathbf{AOB}}$, while the lines stabbing **A** and **F**B but not **O** are represented by the convex cell $\mathcal{P}_{AB}$. These two cells or polytopes are obtained from the intersection of the positive half-spaces associated to the six vertices bounding the edges

Since Theorem 14.3 does not depend on the geometric space dimension, it can be applied to compute the visibility in $\mathfrak{G}_n$: Visibility through some faces in $\mathfrak{G}_n$ can always be represented by a subdivision of $\mathfrak{L}_n$ in cells which group together the lines stabbing the same faces. The boundary of the cells are then the hyperplanes associated to the $(n-2)$-flats which bound the objects in $\mathfrak{G}_n$.

However, we show in Sect. 14.4.2 that some special configurations prevent grouping two or more faces in only one convex cell or polytope in $\mathfrak{L}_n$. The distinction and description of these degenerate cases come as a part of the proof of the minimal polytope solution. They give a precise understanding of how lines in the line-space subdivision are grouped together.

14.3.2.3 Visibility Events in $\mathfrak{L}_n$

According to Durand [6], a visual event is defined as the locus where visibility changes in $\mathfrak{G}_n$. This notion is central in many approaches concerning visibility computation, since both the visibility modification and knowledge of topology are sufficient to fully describe the visibility. In practice, a visual event appears as a line tangent to a finite number of geometrical objects. The degree of freedom gives supplementary information, leading to the k-visual event notion.

As depicted in Fig. 14.1, vertex 5 is a locus with important visibility variations. The red lines 15 and 54 are two examples of visual events that separate the visibility for all the lines passing through 5. In $\mathfrak{L}_2$, they become two 1-vectors that form a part of the cells $\mathcal{P}_{AOB}$ and $\mathcal{P}_{AB}$, as they lie on a common 2-vector, the dual of 5. Obviously, this is a general rule: The visual events are located on the cell boundaries. It comes from the visual event definition and Theorem 14.3.

In n dimensions, only the *real* visual events are of interest, so the whole cell boundary is not interesting. In $\mathfrak{G}_n$, the visual events are located on the Grassmannian too. Then, a k-visual event is a k-submanifold located at the intersection between $G^{\mathbb{R}}(2, n+1)$ and a cell in $\mathfrak{L}_n$. This shows that the visibility is fully described using a partition in $\mathfrak{L}_n$.

14.4 The Minimal Polytope

14.4.1 Minimal Polytope Interest

All the previous approaches fail to give the minimal set of lines stabbing two convex faces in $\mathfrak{G}_n$ for $n > 2$. The Grassmann algebra allows us to define and to compute the minimal polytope enclosing this set of lines. This is a key for our visibility framework, as it ensures computation efficiency.

Let us enumerate some properties and goals of a minimal polytope representation from both the theoretical and practical points of view:

1. It procures a vertex representation of the polytope containing the lines stabbing two faces. This is useful for applications needing to split polytopes, to detect collisions between them or to classify them according to some hyperplanes in the line-space.
2. By splitting the minimal polytope with hyperplanes, such a vertex representation can be extended for representing lines stabbing more than two faces.
3. The minimal polytope is a general solution, in any dimension, to the open problem stated in three dimensions [13]. It is also the most appropriate to avoid the splittings leading to polytopes that do not represent any line in $\mathfrak{G}_n$, i.e., that do not intersect the Grassmannian in $\mathfrak{L}_n$.
4. From the polytope vertices, all the faces in the polytope boundary (edges, hyperplanes, ...), and their incidences can be computed.
5. It unveils the case where a single polytope cannot be used to represent the visibility through two polygons. These *degenerate* cases appear in previous three-dimensional works [13] and are generalized in $\mathfrak{G}_n$ in this chapter.

14.4.2 The Minimal Polytope for Two Convex Faces

Let $\mathbf{A}$ and $\mathbf{B}$ be two convex $(n-1)$-faces in $\mathfrak{G}_n$, and $a_1, \ldots, a_q$ and $b_1, \ldots, b_r$ their respective vertices.

Definition 14.3 The minimal polytope, denoted $\mathscr{M}_\mathbf{A}^\mathbf{B}$, represents the set of lines $\mathscr{S}_\mathbf{A}^\mathbf{B}$ stabbing $\mathbf{A}$ and $\mathbf{B}$ in $\mathfrak{L}_n$. It is the convex polytope with the following properties:

1. $\mathscr{S}_\mathbf{A}^\mathbf{B} \subseteq \mathscr{M}_\mathbf{A}^\mathbf{B}$.
2. $\mathscr{M}_\mathbf{A}^\mathbf{B} \cap G^{\mathbb{R}}(2, n+1) \subseteq \mathscr{S}_\mathbf{A}^\mathbf{B}$.
3. If $\mathscr{P}_\mathbf{A}^\mathbf{B}$ is a convex set in $\mathfrak{L}_n$ such that $\mathscr{S}_\mathbf{A}^\mathbf{B} \subseteq \mathscr{P}_\mathbf{A}^\mathbf{B}$, then $\mathscr{M}_\mathbf{A}^\mathbf{B} \subseteq \mathscr{P}_\mathbf{A}^\mathbf{B}$.

Properties 1 and 2 mean that the polytope $\mathscr{M}_\mathbf{A}^\mathbf{B}$ is a representation of $\mathscr{S}_\mathbf{A}^\mathbf{B}$ in $\mathfrak{L}_n$, i.e., a line L stabs $\mathbf{A}$ and $\mathbf{B}$ if and only if its representation in $\mathfrak{L}_n$ is contained in $\mathscr{M}_\mathbf{A}^\mathbf{B}$. The third property indicates that $\mathscr{M}_\mathbf{A}^\mathbf{B}$ is the minimal polytope: There does not exist another convex polytope representing $\mathscr{S}_\mathbf{A}^\mathbf{B}$ and contained in $\mathscr{M}_\mathbf{A}^\mathbf{B}$.

The following theorem gives a computational characterization of the minimal polytope for two faces in some canonical configurations and indicates the nonexistence of any polytope for the other configurations.

Theorem 14.4 *Let $\mathcal{H}_A$ and $\mathcal{H}_B$ be respectively the supporting planes of the faces* **A** *and* **B**. *If $\mathcal{H}_A$ and $\mathcal{H}_B$ do not respectively intersect the faces* **B** *or* **A**, *or only on their boundary, then the minimal polytope $\mathcal{M}_A^B$ is the convex hull of the lines $L_{ij} = a_i \wedge b_j$, $(i, j) \in [1, \ldots, q] \times [1, \ldots, r]$ from the vertices of* **A** *to the ones of* **B**. *Otherwise, the set of lines stabbing* **A** *and* **B** *cannot be represented by any convex polytope in $\mathfrak{L}_n$.*

14.4.3 Proof of the Minimal Polytope Solution

To prove Theorem 14.4, we consider the two $(n-1)$-faces **A** and **B**, with respective vertices $a_1, \ldots, a_q$ and $b_1, \ldots, b_r$. We suppose that these faces are supported by the hyperplanes $\mathcal{H}_A$ and $\mathcal{H}_B$, and bounded by the $(n-2)$-flats $f_1^a, \ldots, f_s^a$ and $f_1^b, \ldots, f_t^b$, respectively.

The proof is decomposed into three steps:

1. If the polytope $\mathcal{M}_A^B$ exists, then it is minimal.
2. If the hyperplanes $\mathcal{H}_A$ and $\mathcal{H}_B$ do not intersect the faces **B** and **A**, respectively, or only their boundary, then:
   a. $\mathcal{S}_A^B \subset \mathcal{M}_A^B$;
   b. $\mathcal{M}_A^B \cap G^{\mathbb{R}}(2, n+1) \subset \mathcal{S}_A^B$.
3. If the polytope $\mathcal{M}_A^B$ is not defined, then the lines $\mathcal{S}_A^B$ cannot be represented by only one convex polytope.

14.4.3.1 If the Polytope $\mathcal{M}_A^B$ Exists, then It Is Minimal

Let $\mathcal{M}_A^B$ be the polytope defined as the convex hull of the vertices $a_i \wedge b_j$, according to Theorem 14.4. Assuming that this polytope represents lines $\mathcal{S}_A^B$ stabbing **A** and **B**, its vertices, i.e., the points in $\mathfrak{L}_n$ associated to the lines $a_i \wedge b_j$ $\forall i \in [1, \ldots, q], j \in [1, \ldots, r]$, are in $\mathcal{S}_A^B$.

Let $\mathcal{P}$ be a convex polytope strictly contained in $\mathcal{M}_A^B$. Obviously, any convex polytope containing all the vertices of $\mathcal{M}_A^B$ also contains their convex hull $\mathcal{M}_A^B$. Then, it follows that $\mathcal{P}$ does not contain at least one of the vertices of $\mathcal{M}_A^B$. Since those vertices are in $\mathcal{S}_A^B$, we deduce that $\mathcal{P}$ does not represent all the lines stabbing **A** and **B**, proving that $\mathcal{M}_A^B$ is minimal.

14.4.3.2 Proof of $\mathcal{S}_A^B \subseteq \mathcal{M}_A^B$

Let us assume that $\mathcal{H}_A$ and $\mathcal{H}_B$ do not intersect the inside of **A** or **B**, respectively. The set $\mathcal{S}_A^B$ in $\mathfrak{G}_n$ contains lines defined by any couple of points on **A** and **B** such that the point of **A** is not on $\mathcal{H}_B$, and conversely the point of **B** is not on $\mathcal{H}_A$.

Let $a \in \mathbf{A}$ and $b \in \mathbf{B}$ be two such points. Since **A** and **B** are convex, then the homogeneous representation of a and b in $\mathcal{G}_{n+1}$ can be represented by combinations of the vertices of **A** and **B**, respectively. For instance,[2]

---

[2]Using homogeneous coordinates, the sum of the coefficients does not need to be normalized to unity, as it is usually done in computational geometry.

$$a = \sum_{i=1}^{q} \alpha_i a_i, \quad \alpha_i \geq 0 \ \forall i \in [1, \ldots, q], \quad \text{and}$$

$$b = \sum_{j=1}^{r} \beta_j b_i, \quad \beta_j \geq 0 \ \forall j \in [1, \ldots, r].$$

So, the line $D = a \wedge b$ is:

$$D = a \wedge b$$
$$= \left(\sum_{i=1}^{q} \alpha_i a_i \right) \wedge \left(\sum_{j=1}^{r} \beta_j b_i \right)$$
$$= \sum_{i=1}^{q} \sum_{j=1}^{r} \alpha_i \beta_j a_i \wedge b_j$$
$$= \sum_{i \in [1,\ldots,q], j \in [1,\ldots,r]} \gamma_{ij} \, a_i \wedge b_j.$$

By hypothesis, since $\alpha_i \geq 0$ and $\beta_j \geq 0$, we have $\gamma_{ij} = \alpha_i \beta_j \geq 0$. This shows that any line D in $\mathscr{S}_A^B$ is a convex combination of the points $a_i \wedge b_j$ in $\mathfrak{L}_n \ \forall i \in [1, \ldots, q]$ and $j \in [1, \ldots, r]$. These points are precisely the vertices of $\mathscr{M}_A^B$. This proves that $\mathscr{S}_A^B$ is contained in $\mathscr{M}_A^B$.

14.4.3.3 Proof of $\mathscr{M}_A^B \cap G^{\mathbb{R}}(2, n+1) \subseteq \mathscr{S}_A^B$

By hypothesis, since $\mathscr{H}_A$ (resp. $\mathscr{H}_B$) does not split the inside of **B** (resp. **A**), all the vertices b_j, $j \in [1, \ldots, r]$ (resp. a_i, $i \in [1, \ldots, q]$) are in a same half-space delimited by $\mathscr{H}_A$ (resp. $\mathscr{H}_B$).

From this remark and Theorem 14.3, it can deduced that there is a unique orientation of the flats f_i^a, $i \in [1, \ldots, s]$, and f_j^b, $j \in [1, \ldots, t]$, verifying the following inequalities:

$$a_i \wedge b_j \wedge f_k^a > 0 \quad \forall (i, j, k) \in [1, \ldots, q] \times [1, \ldots, r] \times [1, \ldots, s],$$
$$a_i \wedge b_j \wedge f_l^b > 0 \quad \forall (i, j, l) \in [1, \ldots, q] \times [1, \ldots, r] \times [1, \ldots, t].$$

Let $D = \sum_{i \in [1,\ldots,q], j \in [1,\ldots,r]} \gamma_{ij} a_i \wedge b_j$, $\gamma_{ij} \geq 0$ for all (i, j) in $[1, \ldots, q] \times [1, \ldots, r]$, be any point inside $\mathscr{M}_A^B$. It follows that:

$$\forall k \in [1, \ldots, s], \quad D \wedge f_k^a = \sum_{i \in [1,\ldots,q], j \in [1,\ldots,r]} \gamma_{ij} a_i \wedge b_j \wedge f_k^a,$$
$$\forall l \in [1, \ldots, t], \quad D \wedge f_l^b = \sum_{i \in [1,\ldots,q], j \in [1,\ldots,r]} \gamma_{ij} a_i \wedge b_j \wedge f_l^b.$$

Since all γ_{ij}, $a_i \wedge b_j \wedge f_k^a$ and $a_i \wedge b_j \wedge f_l^b$ are positive scalars or pseudoscalars, we have:

Fig. 14.2 Degenerate case in $\mathfrak{G}_2$, where it is not possible to determine an orientation of the boundary of both the faces **A** and **B**, in order to characterize consistently all the lines stabbing the two polygons: the lines l^+ and l^- need opposite orientations

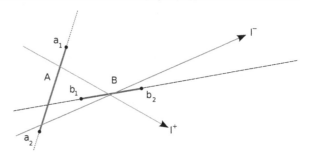

$$D \wedge f_k^a \geq 0 \quad \forall k \in [1, \ldots, s],$$

$$D \wedge f_l^b \geq 0 \quad \forall l \in [1, \ldots, t].$$

Let us assume that D is in the Grassmannian $G^{\mathbb{R}}(2, n+1)$ and $\mathcal{H}_A$ and $\mathcal{H}_B$. Then, by Theorem 14.3, D is in $\mathscr{S}_A^B$. By the hypothesis, D lies on one of the hyperplanes $\mathcal{H}_A$ and $\mathcal{H}_B$ if and only if it is incident to the $(n-2)$-flat f_i, defined as the intersection of the two hyperplanes $\mathcal{H}_A$ and $\mathcal{H}_B$ in $\mathfrak{G}_n$. This $(n-2)$-flat corresponds in $\mathfrak{L}_n$ to a hyperplane f_i^* that bounds the polytope $\mathcal{M}_A^B$. Thus, the lines incident to f_i can be easily excluded from the polytope $\mathcal{M}_A^B$ by considering it *open* on the boundary corresponding to the hyperplane f_i^*.

Since the previous results are proved for any point in $\mathcal{M}_A^B$ and on the Grassmannian $G^{\mathbb{R}}(2, n+1)$, we deduce that $\mathcal{M}_A^B \cap G^{\mathbb{R}}(2, n+1) \subseteq \mathscr{S}_A^B$.

14.4.3.4 When the Hyperplane $\mathcal{H}_B$ or $\mathcal{H}_A$ Intersects the Inside of A or B

Assuming that the hyperplane $\mathcal{H}_B$ intersects the inside of **A**, there are at least two vertices a_{i_1} and a_{i_2} of **A** which are in the two opposite half-spaces delimited by $\mathcal{H}_B$ (see Fig. 14.2).

Let b be a point of **B**, and let $D_1 = a_{i_1} \wedge b$ and $D_2 = a_{i_2} \wedge b$ be two lines in $\mathscr{S}_A^B$. Assuming that the flats f_i^a, $i \in [1, \ldots, s]$, are correctly oriented if $D_1 \wedge f_i^a$ are only positive pseudoscalars for all $i \in [1, \ldots, s]$, the $D_2 \wedge f_i^a$ are also positive, and conversely. This comes from the pseudoscalar sign which only depends on which half-space delimited by $\mathcal{H}_A$ the point b lies (see the proof of Theorem 14.3).

On the contrary, supposing the flats f_j^b for $j \in [1, \ldots, t]$ correctly oriented, since the vertices a_{i_1} and a_{i_2} do not lie on the same half-space, according to $\mathcal{H}_B$, if $D_1 \wedge f_j^b$ is a positive pseudoscalar, then $D_2 \wedge f_j^b$ will be a negative pseudoscalar, and conversely.

Reversing the orientation of one of the two lines, for instance, D_2, is not a solution: The pseudoscalars $D_1 \wedge f_i^a$ and $-D_2 \wedge f_i^a$ still have an opposite sign.

This proves that it is not possible to point the flats f_i^a and f_j^b, $i \in [1, \ldots, s]$ and $j \in [1, \ldots, t]$, such that the classification against those flats of all the lines stabbing both **A** and **B** only results in positive pseudoscalars. In other words, it is not possible to group together the lines $\mathscr{S}_A^B$ in only one convex polytope. $\qquad\square$

14.4.3.5 Dealing with Degenerate Cases

In this chapter, we talk about a degenerate case for two faces when at least one of the two faces has an intersection with the hyperplane that extends the second face. From this definition, two different kinds of degenerate cases can be specified: Firstly, when the intersection is limited to a boundary part of a face; secondly, when the intersection also concerns the inside of a face.

From Theorem 14.4 we know that there exists a minimal polytope representing the lines stabbing the two faces for the first kind of degeneracy, whereas there is not for the second one. However, this latter case can always be transformed in the former one, splitting the two faces along their intersection with the hyperplane supporting the opposite one. This split allows us to divide the initial degenerate configuration in two or four configurations of the first type, depending on whether one or both faces are split.

14.5 An Application Example: Soft Shadows Computation

14.5.1 The n-Dimensional Visibility Framework Implementation

As presented in previous sections, the set of lines intersecting two convex $n - 1$ faces A and B in $\mathfrak{G}_n$ can be represented as an $\binom{n+1}{2}$-dimensional convex polytope P_{AB} in $\mathbb{P}^{\binom{n+1}{2}}$. Denoting by O_i, $1 \le i \le m$, the m occluding $(n - 1)$-faces, the visibility between A and B is

$$P_{AB} - \bigcup_{i=1}^{m} P_{O_i} = P_{AB} - \bigcup_{i=1}^{m} P_{AO_i} = P_{AB} - \bigcup_{i=1}^{m} P_{O_i B}.$$

This can be computed using Computational Solid Geometry operations: Each polytope P_{AO_i} (or $P_{O_i B}$) has to be subtracted from P_{AB}. All n-dimensional CSG operations can be implemented using Binary Space Partitioning trees [12]. The core of this method requires to split an n-dimensional convex polytope against an $(n - 1)$-dimensional hyperplane. Two different approaches can be used:

1. An enumeration algorithm such as [1] can solve the linear system induced both by the splitting hyperplane and the bounding polytope hyperplanes (the so-called H-representation). However, such an approach is prone to numerical errors, especially in higher dimensions as noticed by Bittner [3], whose method relies on a similar algorithm.
2. Bajaj et al. [2] propose a more robust method relying on the relative position of a point and a hyperplane. Nirenstein [13] or Mora [11] use this algorithm. We also choose this technique because robustness is crucial in image synthesis. In particular, even a small error always leads to a blatant visual artifact.

As a result, this allows us to implement the n-dimensional visibility framework whatever $n \ge 2$ is, contrary to previous works which are only correct in 2D or 3D space. In addition, our framework takes advantage of the minimal polytope theorem to optimize CSG computations, whereas previous works construct nonminimal polytopes, increasing the vertex number and thus the complexity of the CSG operations.

The visibility framework can be considered as a black box and easily plugged into any applications that need to perform visibility queries.

14.5.2 Soft Shadow Computations

In computer graphics, soft shadows are very important to render realistic pictures, because they unveil the relative positions of the objects in the scene. But it is a difficult problem, since it requires to compute the visibility of an area light source from any point in the scene, which is very time consuming. In this section, we explain how the visibility framework can be used to solve exactly the visibility of an area light source and to speed up the computation.

14.5.2.1 Application Overview

We consider a 3D environment made of convex polygons and precompute their visibility with an area light source L. Denoting T a polygon in the scene, this leads to compute, for each pair (L, T), a 6D BSP tree whose inner nodes are 5D projective hyperplanes corresponding to the duals of occluders' edges and whose leaves are polytopes representing a visible or invisible set of lines. Such a tree is an exact and coherent representation of the visibility of L from any point on T. As a consequence, it is used during the rendering step to query the visibility of L for each point on T visible from the camera. A simple algorithm to perform such a query is presented in [10]. It provides an exact polygonal subdivision of the visible parts of L from a given point. This result is then used to compute the direct illumination received by the point.

We compare our approach to the solution commonly used in production rendering software: A stratified sampling of the area light source. In this case, the visibility of L from a given point is evaluated by shooting shadow rays toward each sample on L. The quality of the result increases with the number of samples and the computation time.

14.5.2.2 Results

All tests are run on an Intel Core 2 Duo at 2.4 GHz with 3 Gb of memory. For comparison purposes, all pictures are rendered at 800×600 on one thread without anti-aliasing. The comparison method uses 256 samples per area light source, since this number is usually considered sufficient for producing quality results. The ray tracer is an implementation of [20], taking advantage of SSE instructions to trace four rays at a time.

Figure 14.3 presents the pictures. The first scene, Eagle, is a model with a moderate shadow complexity, while the second scene, Panther, is a more complex case. Despite the high number of samples used by the comparison method, noise remains in soft shadows as illustrated by the close-ups. Using our visibility framework, the soft shadows quality is optimal, whatever the zooming is, since the visibility queries

Fig. 14.3 The *left column* presents the pictures with soft shadows computed using our exact visibility framework. The *right column* presents the differences' images of the same pictures computed using a classical sampling strategy. The *close-ups* underline the significant differences in soft shadows

Table 14.1 Result details for the two test scenes, *Eagle* and *Panther*, with one area light source. The *first column* gives the number of polygons in a scene. The *second column* presents the total number of inner nodes for all precomputed BSP trees. The *third column* indicates the time spent for precomputing all BSP trees. The *fourth column* gives the time spent in soft shadows computation using our framework, whereas the *last column* gives this time using the comparison method

	Polygons	BSP-tree	Ptime	Rtime	Ctime
Eagle	5520	19 154	9 min 34 s	2.9 s	1 min 24 s
Panther	12 993	47 684	54 min 12 s	4.1 s	1 min 45 s

are exact. It is worth underlying that we were not able to precompute correctly the visibility on the Panther scene using a nonminimal polytope like in [13] or [10]: Because of numerical instabilities, errors occur in the visibility data, leading to visual artifacts in soft shadows. Using the minimal polytope, we avoid to perform useless CSG operations, improving robustness.

Table 14.1 presents the computation details. The size of the BSP-trees illustrates their compactness and ability to efficiently encode the visibility data. The precomputation times are significant since CSG operations in high dimensions are time consuming. However the method remains practicable, and it does not depend on the point of view. As a consequence, it can be computed once then stored into files to be reused later. Finally, the time spent in soft shadows computation during the rendering step clearly shows the efficiency of the visibility framework. Indeed, the visibility queries used on the BSP trees depend on their average depth and compactness. Thus, the benefit from the precomputation step is really important.

In this application, our visibility framework manages to reconcile accuracy and efficiency, often considered as two opposite qualities in computer graphics.

14.6 Exercises

14.1 Prove that Plücker's coordinates correspond to the coordinates of a bivector in $\bigwedge^2(\mathbb{R}^4)$.

14.2 From the parametric equation of a line (i.e., $tP + (1 - t)Q$, where P and Q are n-dimensional points), find a 2-blade that represents it.

14.3 The Plücker relation between two lines expressed with their six Plücker coordinates $(\Pi_0, \ldots, \Pi_5)$ and $(\Delta_0, \ldots, \Delta_5)$ is $\Pi_0 \Delta_3 + \Pi_1 \Delta_4 + \Pi_2 \Delta_5 + \Pi_3 \Delta_0 + \Pi_4 \Delta_1 + \Pi_5 \Delta_2$. Show that it is equivalent to the inner product in $\mathfrak{L}_3$.

14.4 Let A, B, and C be three Euclidean points in $\mathfrak{G}_3$, with respective coordinates $(1, 0, 0)$, $(2, 1, 1)$, and $(1, 0, 2)$. Let F be the triangle (A, B, C). Let P, Q, and R be three Euclidean points in $\mathfrak{G}_3$ with respective coordinates $(0, 1, 1)$, $(2, 0, 2)$, and $(4, 0, 4)$. Do the lines (PQ) and (RP) stab the face F? Same question for the lines (PR) and (QP), but without any new computations.

14.5 Find a bivector that is not decomposable, i.e., that is not a 2-blade. Show that this bivector cannot represent a line into the space of the geometric objects. (Hint: consider dimension 4.)

14.6 Prove Theorem 14.2. Notice that the duality cannot be expressed easily directly in $\bigwedge(\mathbb{R}^{n+1})$. The left contraction allows us to express it, and so the difficulty only resides in dimension n.

14.7 Consider two faces A and B. Show that any line that is outside the minimal polytope cannot cross A and B.

References

1. Avis, D., Fukuda, K.: Reverse search for enumeration. Discrete Appl. Math. **65**, 21–46 (1996)
2. Bajaj, C.L., Pascucci, V.: Splitting a complex of convex polytopes in any dimension. In: Proceedings of the Twelfth Annual Symposium on Computational Geometry (ISG '96), pp. 88–97, ACM, New York (1996)
3. Bittner, J.: Hierarchical techniques for visibility computation. PhD thesis, Department of Computer Science and Engineering, Czech Technical University in Prague (2002)
4. Bittner, J., Prikryl, J., Slavík, P.: Exact regional visibility using line space partitioning. Comput. Graph. **27**(4), 569–580 (2003)
5. Dorst, L., Fontijne, D., Mann, S.: Geometric Algebra for Computer Science: An Object-Oriented Approach to Geometry. The Morgan Kaufmann Series in Computer Graphics. Morgan Kaufmann, San Francisco (2007)
6. Durand, F., Drettakis, G., Puech, C.: The visibility skeleton: a powerful and efficient multipurpose global visibility tool. In: Proceedings of the 24th Annual Conference on Computer Graphics and Interactive Techniques, SIGGRAPH '97, pp. 89–100 (1997)
7. Durand, F., Drettakis, G., Puech, C.: The 3d visibility complex. ACM Trans. Graph. **21**, 176–206 (2002)
8. Haumont, D., Makinen, O., Nirenstein, S.: A low dimensional framework for exact polygon-to-polygon occlusion queries. In: Rendering Techniques, pp. 211–222 (2005)
9. Möller, T., Trumbore, B.: Fast minimum storage ray–triangle intersection. J. Graph. GPU Game Tools **2**(1), 21–28 (1997)
10. Mora, F., Aveneau, L.: Fast exact direct illumination. In: Proceedings of the Computer Graphics International 2005, pp. 191–197 (2005)
11. Mora, F., Aveneau, L., Mériaux, M.: Coherent exact polygon-to-polygon visibility. In: WSCG'05, pp. 87–94 (2005)
12. Naylor, B.F., Amanatides, J., Thibault, W.C.: Merging BSP trees yields polyhedral set operations. In: SIGGRAPH, pp. 115–124 (1990)
13. Nirenstein, S., Blake, E.H., Gain, J.E.: Exact from-region visibility culling. In: Rendering Techniques, pp. 191–202 (2002)
14. Orti, R., Durand, F., Rivière, S., Puech, C.: Using the visibility complex for radiosity computation. In: WACG, pp. 177–190 (1996)
15. Pellegrini, M.: Ray shooting and lines in space. In: Goodman, J.E., O'Rourke, J. (eds.) Handbook of Discrete and Computational Geometry, 2nd edn., pp. 839–856. Chapman & Hall/CRC Press, Boca Raton (2004)
16. Pocchiola, M., Vegter, G.: The visibility complex. Int. J. Comput. Geom. Appl. **6**(3), 279–308 (1996)
17. Stolfi, J.: Oriented Projective Geometry: A Framework for Geometric Computations. Academic Press, San Diego (1991)

18. Teller, S.J.: Computing the antipenumbra of an area light source. In: Proceedings of the 19th Annual Conference on Computer Graphics and Interactive Techniques, SIGGRAPH '92, pp. 139–148. ACM, New York (1992)
19. Teller, S.J., Séquin, C.H.: Visibility preprocessing for interactive walkthroughs. In: Proceedings of the 18th Annual Conference on Computer Graphics and Interactive Techniques, SIGGRAPH '91, pp. 61–70. ACM, New York (1991)
20. Wald, I., Slusallek, P., Benthin, C., Wagner, M.: Interactive rendering with coherent ray tracing. Comput. Graph. Forum **20**(3), 153–164 (2001)

Part VI
Alternatives to Conformal Geometric Algebra

The 3D conformal geometric algebra $\mathbb{R}_{4,1}$ is five-dimensional and often feels like a slight overkill for the description of rigid body motion and other limited geometries. This part presents several four-dimensional alternatives for the applications we saw in Part I.

On the Homogeneous Model of Euclidean Geometry   15

Charles Gunn

Abstract

We attach the degenerate signature $(n, 0, 1)$ to the dual Grassmann algebra of projective space to obtain a real Clifford algebra which provides a powerful, efficient model for Euclidean geometry. We avoid problems with the degenerate metric by constructing an algebra isomorphism between the Grassmann algebra and its dual that yields non-metric meet and join operators. We focus on the cases of $n = 2$ and $n = 3$ in detail, enumerating the geometric products between k-blades and m-blades. We identify sandwich operators in the algebra that provide all Euclidean isometries, both direct and indirect. We locate the spin group, a double cover of the direct Euclidean group, inside the even subalgebra of the Clifford algebra, and provide a simple algorithm for calculating the logarithm of group elements. We conclude with an elementary account of Euclidean kinematics and rigid body motion within this framework.

15.1   Introduction

The work presented here was motivated by the desire to integrate the work of Study [28] on dual quaternions into a Clifford algebra setting. The following exposition introduces the modern mathematical structures—projective space, exterior algebra, and Cayley–Klein metrics—required to imbed the dual quaternions as the even subalgebra of a particular Clifford algebra, and shows how the result can be applied to Euclidean geometry, kinematics, and dynamics. Those interested in more details, exercises, and background material are referred to an extended on-line version [13].

C. Gunn (✉)
DFG-Forschungszentrum Matheon, MA 8-3, Technisches Universität Berlin,
Str. des 17. Juni 136, 10623 Berlin, Germany
e-mail: gunn@math.tu-berlin.de

L. Dorst, J. Lasenby (eds.), *Guide to Geometric Algebra in Practice*,   297
DOI 10.1007/978-0-85729-811-9_15, © Springer-Verlag London Limited 2011

15.2   The Grassmann Algebra(s) of Projective Space

Real projective n-space $\mathbb{R}P^n$ is obtained from the $(n+1)$-dimensional Euclidean vector space $\mathbb{R}^{n+1}$ by introducing an equivalence relation on vectors $\mathbf{x}, \mathbf{y} \in \mathbb{R}^{n+1} \setminus \{\mathbf{0}\}$ defined by: $\mathbf{x} \sim \mathbf{y} \iff \mathbf{x} = \lambda \mathbf{y}$ for some $\lambda \neq 0$. That is, points in $\mathbb{R}P^n$ correspond to lines through the origin in $\mathbb{R}^{n+1}$.

Grassmann Algebra   The Grassmann, or exterior, algebra $\bigwedge(\mathbb{R}^n)$, is generated by the outer (or exterior) product $\wedge$ applied to the vectors of $\mathbb{R}^n$. The outer product is an alternating bilinear operation. The product of a k- and m-vector is a $(k+m)$-vector when the operands are linearly independent subspaces. An element that can be represented as a wedge product of k 1-vectors is called a simple k-vector, or k-*blade*. The k-blades generate the vector subspace $\bigwedge^k(\mathbb{R}^n)$, whose elements are said to have grade k. This subspace has dimension $\binom{n}{k}$, and hence the total dimension of the exterior algebra is 2^n. $\bigwedge^n(\mathbb{R}^n)$ is one-dimensional, generated by a single element $\mathbf{I}$, sometimes called the pseudo-scalar.

Simple and Nonsimple Vectors   A k-blade represents the subspace of $\mathbb{R}^n$ spanned by the k vectors which define it. Hence, the exterior algebra contains within it a representation of the subspace lattice of $\mathbb{R}^n$. For $n > 3$, there are also k-vectors which are not blades and do not represent a subspace of $\mathbb{R}^n$. Such vectors occur as bivectors when $n = 4$ and play an important role in the discussion of kinematics and dynamics, see Sect. 15.6.

Projectivized Exterior Algebra   The exterior algebra can be projectivized using the same process defined above for the construction of $\mathbb{R}P^n$ from $\mathbb{R}^{n+1}$ but applied to the vector spaces $\bigwedge^k(\mathbb{R}^{n+1})$. This yields the projectivized exterior algebra $W := \mathbf{P}(\bigwedge(\mathbb{R}^{n+1}))$. The operations of $\bigwedge(\mathbb{R}^{n+1})$ carry over to W, since, roughly speaking, "Projectivization commutes with outer product." The difference lies in how the elements and operations are projectively interpreted. The k-blades of W correspond to $(k-1)$-dimensional subspaces of $\mathbb{R}P^n$. All multiples of the same k-blade represent the same projective subspace and differ only by intensity [31, §16–17]. 1-blades correspond to points; 2-blades to lines; 3-blades to planes, etc.

Dual Exterior Algebra   The dual algebra $W^* := \mathbf{P}(\bigwedge \mathbb{R}^{(n+1)*})$ is formed by projectivizing the exterior algebra of the dual vector space $(\mathbb{R}^{n+1})^*$. Details can be found in the excellent Wikipedia article [32], based on [7]. W^* is the alternating algebra of k-multilinear forms and is naturally isomorphic to W; again, the difference lies in how the elements and operations are interpreted. Like W, W^* represents the subspace structure of $\mathbb{R}P^n$, but turned on its head: 1-blades represent projective hyperplanes, while n-blades represent projective points. The outer product $\mathbf{a} \wedge \mathbf{b}$ corresponds to the *meet* rather than *join* operator. In order to distinguish the two outer products of W and W^*, we write the outer product in W as $\vee$, and leave the outer product in W^* as $\wedge$. These symbols match closely the affiliated operations of join (union $\cup$) and meet (intersection $\cap$), respectively.

15.2.1 Remarks on Homogeneous Coordinates

We use the terms *homogeneous* model and *projective* model interchangeably, to denote the projectivized version of Grassmann (and, later, Clifford) algebra.

The projective model allows a certain freedom in specifying results within the algebra. In particular, when the calculated quantity is a subspace, then the answer is only defined up to a non-zero scalar multiple. In some literature, this fact is represented by always surrounding an expression x in square brackets $[x]$ when one means "the projective element corresponding to the vector space element x." We do not adhere to this level of rigor here, since in most cases the intention is clear.

Some of the formulas introduced below take on a simpler form which take advantage of this freedom, but they may appear unfamiliar to those used to working in the more strict vector-space environment. On the other hand, when the discussion later turns to kinematics and dynamics, then this projective equivalence is no longer strictly valid. Different representatives of the same subspace represent weaker or stronger instances of a velocity or momentum (to mention two possibilities). In such situations terms such as *weighted* point or "point with intensity" will be used. See [31, Book III, Chap. 4].

15.2.2 Equal Rights for W and W^*

From the point of view of representing $\mathbb{R}P^n$, W and W^* are equivalent. There is no a priori reason to prefer one to the other. Every geometric element in one algebra occurs in the other, and any configuration in one algebra has a dual configuration in the other obtained by applying the Principle of Duality [9] to the configuration. We refer to W as a *point-based* algebra and W^* as a *plane-based* algebra.

Depending on the context, one or the other of the two algebras may be more useful. Here are some examples:

- *Joins and meets.* W is the natural choice to calculate subspace joins and W^* to calculate subspace meets. See (15.2) and (15.3).
- *Spears and axes.* Lines appear in two aspects: as spears (bivectors in W) and axes (bivectors in W^*). See Fig. 15.2 and its discussion.
- *Euclidean geometry.* W^* is the correct choice to use for modeling Euclidean geometry. See Sect. 15.3.2.
- *Reflections in planes.* W^* has advantages for kinematics, since it naturally allows building up rotations as products of reflections in planes. See Sect. 15.4.2.

Bases and Isomorphisms for W and W^* Our treatment differs from other approaches (for example, Grassmann–Cayley algebras) in explicitly maintaining both algebras on an equal footing rather than expressing the wedge product in one in terms of the wedge product of the other (as in the Grassmann–Cayley *shuffle* product) [23, 26]. To switch back and forth between the two algebras, we construct an algebra isomorphism that, given an element of one algebra, produces the element of the second algebra which corresponds to the same geometric entity of $\mathbb{R}P^n$. We show how this works for the case of interest $n = 3$.

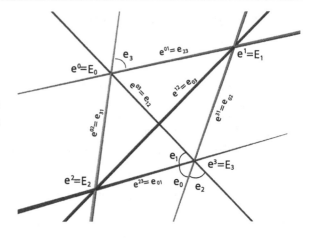

Fig. 15.1 Fundamental tetrahedron with dual labeling. Entities in W have superscripts; entities in W^* have subscripts. Planes are identified by labeled *angles* of two spanning lines. A representative sampling of equivalent elements is shown

The Isomorphism J Each weighted subspace S of $\mathbb{R}P^3$ corresponds to a unique element S_W of W and to a unique element S_{W^*} of W^*. We seek a bijection $\mathbf{J}: W \to W^*$ such that $\mathbf{J}(S_W) = S_{W^*}$. If we have found $\mathbf{J}$ for the basis k-blades, then it extends by linearity to multivectors. To that end, we introduce a basis for $\mathbb{R}^4$ and extend it to a basis for W and W^* so that $\mathbf{J}$ takes a particularly simple form. Refer to Fig. 15.1.

The Canonical Basis A basis $\{e^0, e^1, e^2, e^3\}$ of $\mathbb{R}^4$ corresponds to a coordinate tetrahedron for $\mathbb{R}P^3$, with corners occupied by the basis elements.[1] Use the same names to identify the elements of $P(\bigwedge^1(\mathbb{R}^4))$ which correspond to these projective points. Further, let $\mathbf{I}^0 := e^0 \vee e^1 \vee e^2 \vee e^3$ be the basis element of $P(\bigwedge^4(\mathbb{R}^4))$, and 1^0 be the basis element for $P(\bigwedge^0(\mathbb{R}^4))$. Let the basis for $P(\bigwedge^2(\mathbb{R}^4))$ be given by the six edges of the tetrahedron,

$$\{e^{01}, e^{02}, e^{03}, e^{12}, e^{31}, e^{23}\}$$

where $e^{ij} := e^i \vee e^j$ represents the oriented line joining e^i and e^j.[2] Finally, choose a basis $\{E^0, E^1, E^2, E^3\}$ for $P(\bigwedge^3(\mathbb{R}^4))$ satisfying the condition that $e^i \vee E^i = \mathbf{I}^0$. This corresponds to choosing the ith basis 3-vector to be the plane opposite the ith basis 1-vector in the fundamental tetrahedron, oriented in a consistent way.

We repeat the process for the algebra W^*, writing indices as subscripts. Choose the basis 1-vector e_i of W^* to represent the same plane as E^i. That is, $\mathbf{J}(E^i) = e_i$. Let $\mathbf{I}_0 := e_0 \wedge e_1 \wedge e_2 \wedge e_3$ be the pseudoscalar of the algebra. Construct bases for grade-0, grade-2, and grade-3 using the same rules as above for W (i.e., replacing subscripts by superscripts). The results are represented in Table 15.1.

---

[1] We use superscripts for W and subscripts for W^* since W^* will be the more important algebra for our purposes.

[2] Note that the orientation of e^{31} is reversed; this is traditional since Plücker introduced these line coordinates.

Table 15.1 Comparison of W and W^*

Feature	W	W^*
0-Vector	Scalar $\mathbf{1}^0$	Scalar $\mathbf{1}_0$
Vector	Point $\{e^i\}$	Plane $\{e_i\}$
Bivector	"Spear" $\{e^{ij}\}$	"Axis" $\{e_{ij}\}$
Trivector	Plane $\{E^i\}$	Point $\{E_i\}$
4-Vector	$\mathbf{I}^0$	$\mathbf{I}_0$
Outer product	Join $\vee$	Meet $\wedge$

Given this choice of bases for W and W^*, examination of Fig. 15.1 makes clear that, on the basis elements, $\mathbf{J}$ takes the following simple form:

$$\mathbf{J}(e^i) := E_i, \quad \mathbf{J}(E^i) := e_i, \quad \mathbf{J}(e^{ij}) := e_{kl} \tag{15.1}$$

where in the last equation, $(ijkl)$ is an even permutation of (0123).[3]

Furthermore, $\mathbf{J}(\mathbf{1}^0) = \mathbf{I}_0$ and $\mathbf{J}(\mathbf{I}^0) = \mathbf{1}_0$ since these grades are one-dimensional. To sum up: the map $\mathbf{J}$ is grade-reversing and, considered as a map of coordinate-tuples, it is the identity map on all grades except for bivectors. What happens for bivectors? In W, consider $\mathbf{e}^{01}$, the joining line of points $\mathbf{e}^0$ and $\mathbf{e}^1$ (refer to Fig. 15.1). In W^*, the same line is $\mathbf{e}_{23}$, the intersection of the only two planes which contain both of these points, $\mathbf{e}_2$ and $\mathbf{e}_3$. See Fig. 15.2. Since $\mathbf{J}^{-1}$ is obtained from the definition of $\mathbf{J}$ by swapping superscripts and subscripts, we can consider $\mathbf{J} : W \leftrightarrow W^*$ as a defined on both algebras, with $\mathbf{J}^2$ the identity. The full significance of $\mathbf{J}$ will only become evident after metrics are introduced (Sect. 15.3.3). We now show how to use $\mathbf{J}$ to define meet and join operators valid for both W and W^*.

Projective Join and Meet   Knowledge of $\mathbf{J}$ allows equal access to join and meet operations. We define a meet operation $\wedge$ for two blades $A, B \in W$ by

$$A \wedge B = \mathbf{J}(\mathbf{J}(A) \wedge \mathbf{J}(B)) \tag{15.2}$$

and extend by linearity to the whole algebra. There is a similar expression for the join $\vee$ operation for two blades $A, B \in W^*$:

$$A \vee B := \mathbf{J}(\mathbf{J}(A) \vee \mathbf{J}(B)) \tag{15.3}$$

We turn now to another feature highlighting the importance of maintaining W and W^* as equal citizens.

---

[3]*Editorial note*: The reader may find the alternative coordinate-free construction in Sect. 18.3 enlightening.

Fig. 15.2 A line in its dual nature as spear, or point range; and as axis, or plane pencil

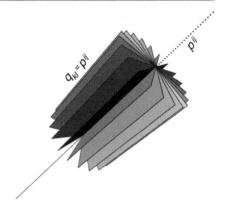

There Are No Lines, Only Spears and Axes!   Given two points $\mathbf{x}$ and $\mathbf{y} \in W$, the condition that a third point $\mathbf{z}$ lies in the subspace spanned by the 2-blade $\mathbf{l} := \mathbf{x} \vee \mathbf{y}$ is that $\mathbf{x} \vee \mathbf{y} \vee \mathbf{z} = 0$, which implies that $\mathbf{z} = \alpha\mathbf{x} + \beta\mathbf{y}$ for some α, β not both zero. In projective geometry, such a set is called a *point range*. We prefer the more colorful term *spear*. Dually, given two planes $\mathbf{x}$ and $\mathbf{y} \in W^*$, the condition that a third plane $\mathbf{z}$ passes through the subspace spanned by the 2-blade $\mathbf{l} := \mathbf{x} \wedge \mathbf{y}$ is that $\mathbf{z} = \alpha\mathbf{x} + \beta\mathbf{y}$. In projective geometry, such a set is called a *plane pencil*. We prefer the more colorful term *axis*.

Within the context of W and W^*, lines exist only in one of these two aspects: of spear—as bivector in W—and axis—as bivector in W^*. This naturally generalizes to nonsimple bivectors: there are pointwise bivectors (in W), and planewise bivectors (in W^*). Many of the important operators of geometry and dynamics we will meet below, such as the polarity on the metric quadric (Sect. 15.3.1) and the inertia tensor of a rigid body (Sect. 15.6.2.2), map $\langle W \rangle_2$ to $\langle W^* \rangle_2$ and hence map spears to axes and vice-versa. Having both algebras on hand preserves the qualitative difference between these dual aspects of the generic term "line."

We now proceed to describe how to introduce metric relations.

15.3   Clifford Algebra for Euclidean Geometry

The outer product is antisymmetric, so $\mathbf{x} \wedge \mathbf{x} = 0$. However, in geometry there are important bilinear products which are symmetric. We introduce a real-valued *inner product* on pairs of vectors $\mathbf{x} \cdot \mathbf{y}$ which is a real-valued symmetric bilinear map. Then, the geometric product on 1-vectors is defined as the sum of the inner and outer products:

$$\mathbf{xy} := \mathbf{x} \cdot \mathbf{y} + \mathbf{x} \wedge \mathbf{y}$$

How this definition can be extended to the full exterior algebra is described elsewhere [10, 15]. The resulting algebraic structure is called a *real Clifford algebra*. It is fully determined by its signature, which describes the inner product structure.

The signature is a triple of integers (p, n, z) where $p + n + z$ is the dimension of the underlying vector space, and p, n, and z are the numbers of positive, negative, and zero entries along the diagonal of the quadratic form representing the inner product. We denote the corresponding Clifford algebra constructed on the point-based Grassmann algebra as $\mathbf{P}(\mathbb{R}_{p,n,z})$ and that based on the plane-based Grassmann algebra as $\mathbf{P}(\mathbb{R}^*_{p,n,z})$.

The discovery and application of signatures to create different sorts of metric spaces within projective space goes back to a technique invented by Arthur Cayley and developed by Felix Klein [20]. The so-called *Cayley–Klein construction* provides models of the three standard metric geometries (hyperbolic, elliptic, and Euclidean)—along with many others!—within projective space. This work provides the mathematical foundation for the inner product as it appears within the homogeneous model of Clifford algebra. Since the Cayley–Klein construction for Euclidean space has some subtle points, is relatively sparsely represented in the current literature, and is crucial to what follows, we describe it below.

15.3.1 The Cayley–Klein Construction

For simplicity, we focus on the case $n = 3$. To obtain metric spaces inside $\mathbb{R}P^3$, begin with a symmetric bilinear form Q on $\mathbb{R}^4$. The quadric surface associated to Q is then defined to be the points $\{\mathbf{x} \mid Q(\mathbf{x}, \mathbf{x}) = 0\}$. For nondegenerate Q, a distance between points A and B can be defined by considering the cross ratio of the four points A, B, and the two intersections of the line AB with Q. Such a Q is characterized by its signature, there are two cases of interest for $n = 3$: $(4, 0, 0)$ yielding elliptic geometry and $(3, 1, 0)$ yielding hyperbolic geometry. These are point-based metrics; they induce an inner product on planes, which, as one can show, is identical to the original signatures. By interpolating between these two cases, one is led to the degenerate case in which the quadric surface collapses to a plane, or to a point. In the first case, one obtains Euclidean geometry; the plane is called the *ideal* plane. The signature breaks into two parts: for points, it is $(1, 0, 3)$, and for planes, it is $(3, 0, 1)$. The distance function for Euclidean geometry is based on a related limiting process. For details, see [20] or [13]. *Warning*: in the projective model, the signature $(n, 0, 0)$ is called the *elliptic* metric, and *Euclidean* metric refers to these degenerate signatures.

Polarity on the Metric Quadric   For a Q and a point $\mathbf{P}$, define the set $\mathbf{P}^\perp := \{\mathbf{X} \mid Q(\mathbf{X}, \mathbf{P}) = 0\}$. When $\mathbf{P}^\perp$ is a plane, it is called the *polar plane* of the point. For a plane $\mathbf{a}$, there is also an associated *polar point* defined analogously using the "plane-based" metric. Points and planes with such polar partners are called *regular*. In the Euclidean case, the polar plane of every finite point is the ideal plane; the polar point of a finite plane is the ideal point in the normal direction to the plane. Ideal points and the ideal plane are not regular and have no polar partner. The polar plane of a point is important since it can be identified with the tangent space of the point when the metric space is considered as a differential manifold. Many of the peculiarities

of Euclidean geometry may be elegantly explained due to the degenerate form of
the polarity operator. In the Clifford algebra setting, this polarity is implemented by
multiplication by the pseudoscalar.

Free Vectors and the Euclidean Metric   As mentioned above, the tangent space
at a point is the polar plane of the point. Every Euclidean point shares the same polar
plane, the ideal plane. In fact, the ideal points (points of the ideal plane) can be identified with Euclidean free vectors. A model for Euclidean geometry should handle
both Euclidean points and Euclidean free vectors. This is complicated by the fact
that free vectors have a natural signature $(3, 0, 0)$. However, since the limiting process (in Cayley–Klein) that led to the degenerate point metric $(1, 0, 3)$ only effects
the nonideal points, it turns out that the original nondegenerate metric, restricted to
the ideal plane, yields the desired signature $(3, 0, 0)$. As we will see in Sect. 15.4.1
and Sect. 15.5.2, the model presented here is capable of mirroring this subtle fact.

15.3.2 A Model for Euclidean Geometry

As noted above, the Euclidean inner product has signature $(1, 0, 3)$ on points and
$(3, 0, 1)$ on planes. If we attach the first signature to W, we have the following
relations for the basis 1-vectors:

$$\left(e^0\right)^2 = 1; \qquad \left(e^1\right)^2 = \left(e^2\right)^2 = \left(e^3\right)^2 = 0$$

It is easy to see that these relations imply that, for all basis trivectors E_i, $E_i^2 = 0$. But
the trivectors represent planes, and the signature for the planewise Euclidean metric
is $(3, 0, 1)$, not $(0, 0, 4)$. Hence, we cannot use W to arrive at Euclidean space. If
instead, we begin with W^* and attach the plane-wise signature $(3, 0, 1)$, we obtain:

$$\left(e_0\right)^2 = 0; \qquad \left(e_1\right)^2 = \left(e_2\right)^2 = \left(e_3\right)^2 = 1$$

It is easy to check that this inner product, when extended to the higher grades, produces the proper behavior on the trivectors, since only $E_0 = e_1 e_2 e_3$ has nonzero
square, producing the pointwise signature $(0, 1, 3)$ (equivalent to the signature
$(1, 0, 3)$). Hence, W^* is the correct choice for constructing a model of Euclidean
geometry.

Counterspace   What space *does* one obtain by attaching the signature $(3, 0, 1)$
to W? One obtains a different metric space, sometimes called polar-Euclidean space
or counterspace. Its metric quadric is a *point* along with all the *planes passing
through it* (dual to the Euclidean ideal *plane* and all the *points lying in it*).[4] See
[8, pp. 71ff.], for a related discussion.

---

[4]Blurring the distinction between these two spaces may have led some authors to incorrect conclusions about the homogeneous model of Euclidean geometry, see [21, p. 11].

We retain W, the point-based algebra, solely as a Grassmann algebra, primarily for calculating the join operator. All Euclidean metric operations are carried out in W^*. Or equivalently, we attach the metric $(0, 0, 4)$ to W, forcing all inner products to zero. Due to the more prominent role of W^*, the basis element for scalar and pseudoscalar in W^* will be written without index as $\mathbf{1}$ and $\mathbf{I}$; we may even omit $\mathbf{1}$ when writing scalars, as is common in the literature.

15.3.3 J, Metric Polarity, and the Regressive Product

We can now appreciate better the significance of $\mathbf{J} : W \to W^*$. Consider the map $\mathit{\Pi} : W \to W$ defined analogously to $\mathbf{J}$ in (15.1):

$$\mathit{\Pi}(e^i) := E^i, \quad \mathit{\Pi}(E^i) := e^i, \quad \mathit{\Pi}(e^{ij}) := e^{kl} \tag{15.4}$$

$\mathit{\Pi}$ is the same as $\mathbf{J}$, but interpreted as a map to W instead of W^*. It is easy to see that $\mathit{\Pi}$ is the polarity on the elliptic metric quadric with signature $(4, 0, 0)$. Many authors (see [15]) define the meet operation between two blades $A, B \in W$ (also known since Grassmann as the *regressive* product) via $\mathit{\Pi}(\mathit{\Pi}(A) \wedge \mathit{\Pi}(B))$, where $\wedge$ is the exterior product in W. One can define a similar join operator in W^*. We prefer to use $\mathbf{J}$ for this purpose (see (15.2)) since it provides a *projective* solution for a *projective* (incidence) problem, and it is useful on its own (see, for example, Sect. 15.6.2.2), while $\mathit{\Pi}$, being a foreign entity, must always appear in the second power so that it has no side-effects. To distinguish the two approaches, we suggest calling $\mathit{\Pi}$ the *metric polarity* and $\mathbf{J}$, the *duality* operator, consistent with mathematical literature. For an n-dimensional discussion and proof, see Appendix 1 of [13].

15.4   The Euclidean Plane via $\mathbf{P}(\mathbb{R}^*_{2,0,1})$

Due to the combination of unfamiliar concepts involved in the algebras $\mathbf{P}(\mathbb{R}^*_{n,0,1})$—notably the dual construction and the degenerate metric—we begin our study with the Clifford algebra for the Euclidean plane: $\mathbf{P}(\mathbb{R}^*_{2,0,1})$. Then, when we turn to the 3D case, we can focus on the special challenges which it presents, notably the existence of nonsimple bivectors. A basis for the full algebra of $\mathbf{P}(\mathbb{R}^*_{2,0,1})$ is given by

$$\{\mathbf{1} := \mathbf{1}_0, \mathbf{e}_0, \mathbf{e}_1, \mathbf{e}_2, \ \mathbf{E}_0 := \mathbf{e}_1\mathbf{e}_2, \ \mathbf{E}_1 := \mathbf{e}_2\mathbf{e}_0, \ \mathbf{E}_2 := \mathbf{e}_0\mathbf{e}_1, \ \mathbf{I} := \mathbf{e}_0\mathbf{e}_1\mathbf{e}_2\}$$

with the relations $\{\mathbf{e}_0^2 = 0; \ \mathbf{e}_1^2 = \mathbf{e}_2^2 = 1\}$.

Consequences of Degeneracy   The pseudoscalar $\mathbf{I}$ satisfies $\mathbf{I}^2 = 0$. Hence, $\mathbf{I}^{-1}$ is not defined. Many standard formulas of geometric algebra are, however, typically stated using $\mathbf{I}^{-1}$ [10, 15], since that can simplify things for nondegenerate metrics. As explained in Sect. 15.2.1, many formulas remain projectively valid when $\mathbf{I}^{-1}$ is replaced by $\mathbf{I}$; in such cases this is the solution we adopt.

Table 15.2 Geometric product in $\mathbf{P}(\mathbb{R}^*_{2,0,1})$

	1	e_0	e_1	e_2	E_0	E_1	E_2	I
1	1	e_0	e_1	e_2	E_0	E_1	E_2	I
e_0	e_0	0	E_2	$-E_1$	I	0	0	0
e_1	e_1	$-E_2$	1	E_0	e_2	I	$-e_0$	E_1
e_2	e_2	E_1	$-E_0$	1	-1	e_0	I	E_2
E_0	E_0	I	$-e_2$	e_1	-1	$-E_2$	E_1	$-e_0$
E_1	E_1	0	I	$-e_0$	E_2	0	0	0
E_2	E_2	0	e_0	I	$-E_1$	0	0	0
I	I	0	E_1	E_2	$-e_0$	0	0	0

Notation We denote 1-vectors with bold small letters and 2-vectors with bold capital letters. We will use the term *ideal* to refer to geometric elements contained in projective space but not in Euclidean space. Then e_0 is the *ideal* line of the plane, e_1 is the line $x = 0$, and e_2 the line $y = 0$. E_0 is the origin $(1, 0, 0)$, while E_1 and E_2 are the *ideal* points in the x- and y-directions, respectively. Points and lines which are not ideal are called *finite*, or *Euclidean*.

We write the natural embedding of a Euclidean *position* $\mathbf{x} = (x, y)$ as $\mathbf{i}(\mathbf{x}) = E_0 + xE_1 + yE_2$. A Euclidean *vector* $\mathbf{v} = (x, y)$ corresponds to an ideal point (see Sect. 15.3.1); we denote its embedding with the same symbol $\mathbf{i}(\mathbf{v}) = xE_1 + yE_2$. We sometimes refer to such an element as a *free* vector. Conversely, a bivector $wE_0 + xE_1 + yE_2$ with $w \neq 0$ corresponds to the Euclidean point $(\frac{x}{w}, \frac{y}{w})$. We refer to w as the *intensity* or *weight* of the bivector, and we write $\underline{\mathbf{A}}$ to refer to $\mathbf{i}^{-1}(\mathbf{A})$. The line $ax + by + c = 0$ maps to the 1-vector $ce_0 + ae_1 + be_2$. A line is Euclidean if and only if $a^2 + b^2 \neq 0$.

The multiplication table is shown in Table 15.2. Inspection of the table reveals that the geometric product of a k- and l-vector yields a product that involves at most two grades. When these two grades are $|k - l|$ and $k + l$, we can write the geometric product for two arbitrary blades $\mathbf{A}$ and $\mathbf{B}$ as

$$\mathbf{A}\mathbf{B} = \mathbf{A} \cdot \mathbf{B} + \mathbf{A} \wedge \mathbf{B}$$

where $\cdot$ is the generalized inner product, defined to be $\langle \mathbf{A}\mathbf{B} \rangle_{|k-l|}$ [15]. The only exception is $(l, k) = (2, 2)$ where the grades $|k - l| = 0$ and $|k - l| + 2 = 2$ occur. Following [15], we write the grade-2 part as

$$\mathbf{A} \times \mathbf{B} := \langle \mathbf{A}\mathbf{B} \rangle_2 = \frac{1}{2}(\mathbf{A}\mathbf{B} - \mathbf{B}\mathbf{A})$$

where $\mathbf{A}$ and $\mathbf{B}$ are bivectors. This is called the *commutator* product. Since all vectors in the algebra are blades, the above decompositions are valid for the product of any two vectors in our algebra.

Fig. 15.3 A selection of the geometric products between various k-blades. Points and lines are assumed to be normalized. Ideal points are drawn as *vectors*, distances indicated by *norms*

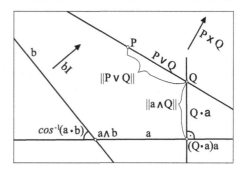

15.4.1 Enumeration of Various Products

We want to spend a bit of time now investigating the various forms which the geometric product takes in this algebra. For this purpose, define two arbitrary 1-vectors **a** and **b** and two arbitrary bivectors **P** and **Q** with

$$\mathbf{a} = a_0 \mathbf{e}_0 + a_1 \mathbf{e}_1 + a_2 \mathbf{e}_2, \quad \text{etc.}$$

These coordinates are of course not intrinsic, but they can be useful in understanding how the Euclidean metric is working in the various products. See the companion diagram in Fig. 15.3.

1. *Norms.* It is often useful to normalize vectors to have a particular intensity. There are different definitions for each grade:
   - 1-*vectors.* $\mathbf{a}^2 = \mathbf{a} \cdot \mathbf{a} = a_1^2 + a_2^2$. Define the *norm* of **a** to be $\|\mathbf{a}\| := \sqrt{\mathbf{a} \cdot \mathbf{a}}$. Then $\frac{\mathbf{a}}{\|\mathbf{a}\|}$ is a vector with norm 1, defined for all vectors except $\mathbf{e}_0$ and its multiples. In particular, all Euclidean lines can be normalized to have norm 1. Note that when **a** is normalized, then so is $-\mathbf{a}$. These two lines represents opposite *orientations* of the line.[5]
   - 2-*vectors.* $\mathbf{P}^2 = \mathbf{P} \cdot \mathbf{P} = p_0^2 \mathbf{E}_0^2 = -p_0^2$. Define the *norm* of **P** to be p_0 and write it $\|\mathbf{P}\|$. Note that this can take positive or negative values, in contrast to $\sqrt{\mathbf{P} \cdot \mathbf{P}}$. Then $\frac{\mathbf{P}}{\|\mathbf{P}\|}$ is a bivector with norm 1, defined for all bivectors except where $p_0 = 0$, that is, ideal points. In particular, all Euclidean points can be normalized to have norm 1. This is also known as *dehomogenizing*.
   - 3-*vectors.* Define $\mathbf{S} : \mathbf{P}(\bigwedge^3 \mathbb{R}^{3*}) \to \mathbf{P}(\bigwedge^0 \mathbb{R}^{3*})$ by $\mathbf{S}\alpha\mathbf{I} = \alpha\mathbf{1}$. This gives the scalar magnitude of a pseudoscalar in relation to the basis pseudoscalar **I**. We sometimes write $\frac{1}{\mathbf{I}}(\alpha\mathbf{I})$ for the same. In a nondegenerate metric, the same can be achieved by multiplication by $\mathbf{I}^{-1}$.
2. *Inverses.* $\mathbf{a}^{-1} = \frac{\mathbf{a}}{\mathbf{a} \cdot \mathbf{a}}$ and $\mathbf{P}^{-1} = \frac{-\mathbf{P}}{\mathbf{P} \cdot \mathbf{P}}$ for Euclidean **a** and **P**.
3. *Euclidean distance.* For normalized **P** and **Q**, $\|\mathbf{P} \vee \mathbf{Q}\|$ is the Euclidean distance between **P** and **Q**.

---

[5]Orientation is an interesting topic which lies outside the scope of this article.

4. *Free vectors.* For an ideal point $\mathbf{V}$ (that is, a free vector) and *any* normalized Euclidean point $\mathbf{P}$, $\|\mathbf{V}\|_\infty := \|\mathbf{V} \vee \mathbf{P}\| = \sqrt{v_1^2 + v_2^2}$ is the length of $\mathbf{V}$. Then $\frac{\mathbf{V}}{\|\mathbf{V}\|_\infty}$ is normalized to have length 1.

5. $\mathbf{a} \wedge \mathbf{P} = (a_0 p_0 + a_1 p_2 + a_2 p_2)\mathbf{I}$ vanishes only if $\mathbf{a}$ and $\mathbf{P}$ are incident. Otherwise, when $\mathbf{a}$ and $\mathbf{P}$ are normalized, it is equal to the signed distance of the point to the line times the pseudoscalar $\mathbf{I}$.

6. $\mathbf{P} \cdot \mathbf{a} = (p_2 a_1 - p_1 a_2)\mathbf{e}_0 + p_0 a_2 \mathbf{e}_1 - p_0 a_1 \mathbf{e}_2$ is a line which passes through $\mathbf{P}$ and is perpendicular to $\mathbf{a}$. Reversing the order changes the orientation of the line.

7. $\mathbf{a} \wedge \mathbf{b} =: \mathbf{T}$ is the intersection point of the lines $\mathbf{a}$ and $\mathbf{b}$. For normalized $\mathbf{a}$ and $\mathbf{b}$, $\|\mathbf{T}\| = \sin\alpha$ where α is the angle between the lines. Reversing the order reverses the orientation of the resulting point.

8. $\mathbf{a} \cdot \mathbf{b} = \cos\alpha$ for normalized vectors $\mathbf{a}$ and $\mathbf{b}$. Which of the two possible angles is being measured here depends on the orientation of the lines.

9. $\mathbf{P} \vee \mathbf{Q}$ is the joining line of $\mathbf{P}$ and $\mathbf{Q}$.

10. $\mathbf{P} \times \mathbf{Q} =: \mathbf{T}$ is the ideal point in the direction perpendicular to the direction of the line $\mathbf{P} \vee \mathbf{Q}$.

11. $\mathbf{aI} = a_1 \mathbf{E}_1 + a_2 \mathbf{E}_2$ is the polar point of the line $\mathbf{a}$: the ideal point in the perpendicular direction to the line $\mathbf{a}$. All lines parallel to $\mathbf{a}$ have the same polar point.

12. $\mathbf{PI} = p_0 \mathbf{e}_0$ is the polar line of the point $\mathbf{P}$: for finite points, the ideal line, weighted by the intensity of $\mathbf{P}$. Ideal points have no polar line.

13. $\mathbf{I}^2 = 0$. This is equivalent to the degeneracy of the metric. Notice that this fact has no effect on the validity of the above calculations.[6]

For a variety of exercises, see [13].

15.4.2 Euclidean Isometries via Sandwich Operations

One of the most powerful aspects of Clifford algebras for metric geometry is the ability to realize isometries as sandwich operations of the form[7]

$$\mathbf{X} \to \mathbf{g}\mathbf{X}\mathbf{g}^{-1}$$

where $\mathbf{X}$ is any geometric element of the algebra, and $\mathbf{g}$ is a specific geometric element, unique to the isometry. $\mathbf{g}$ is in general a *versor*, that is, it can be written as the product of 1-vectors [15]. Let us explore whether this works in $\mathbf{P}(\mathbb{R}_{2,0,1}^*)$.

---

[6]In fact, the validity of most of the above calculations *requires* that $\mathbf{I}^2 = 0$.

[7]The presence of a minus sign (or "the factor $(-1)^{nk}$") in some literature arises from the fact that the desired reflection is in the hyperplane orthogonal to the 1-vector appearing in the sandwich. Since 1-vectors represent hyperplanes here, in the dual algebra, no such correction factor is required.

Reflections   Let $\mathbf{a} := \mathbf{e}_0 - \mathbf{e}_1$ (the line $x = 1$), and $\mathbf{P}$ a normalized point $\mathbf{E}_0 + x\mathbf{E}_1 + y\mathbf{E}_2$. Simple geometric reasoning shows that reflection in the line $\mathbf{a}$ sends the point (x, y) to the point $(2 - x, y)$. Alternatively, the reader is encouraged to verify the missing steps of the following computation:

$$\mathbf{P}' := \mathbf{aPa}^{-1} = \mathbf{aPa} = \cdots$$
$$= (x - 2)\mathbf{E}_1 - \mathbf{E}_0 - y\mathbf{E}_2$$
$$= \mathbf{E}_0 + (2 - x)\mathbf{E}_1 + y\mathbf{E}_2$$

This algebra element corresponds to the Euclidean point $(2 - x, y)$, so the sandwich operation *is* the desired reflection in the line $\mathbf{a}$. We leave it as an exercise for the interested reader to carry out the same calculation for a general line.

Direct Isometries   By well-known results in plane geometry, the composition of two reflections yields a rotation around the common point of the two lines. Translating this into the language of the Clifford algebra, the composition of reflections in lines $\mathbf{a}$ and $\mathbf{b}$ will look like:

$$\mathbf{P}' = \mathbf{b}(\mathbf{aPa})\mathbf{b}$$
$$= \mathbf{T}\mathbf{P}\tilde{\mathbf{T}}$$

where we write $\mathbf{T} := \mathbf{ba}$, and $\tilde{\mathbf{T}}$ is the reversal of $\mathbf{T}$.

Note that the intersection of the two lines will be fixed by the resulting isometry. There are two cases: the point is ideal, or it is Euclidean. In the case of an ideal point, the two lines are parallel, and the composition is a translation. Let us look at an example.

Retaining $\mathbf{a}$ as above, define the normalized line $\mathbf{b} := 2\mathbf{e}_0 - \mathbf{e}_1$, the line $x = 2$. By simple geometric reasoning, the composition "reflect first in $\mathbf{a}$, then in $\mathbf{b}$" *should* be the translation $(x, y) \rightarrow (x + 2, y)$. Defining $\mathbf{T} := \mathbf{ba}$, the sandwich operator looks like $\mathbf{T}\mathbf{P}\tilde{\mathbf{T}}$. Calculate the product $\mathbf{T} = 1 - \mathbf{E}_2$ and $\mathbf{T}\mathbf{P}\tilde{\mathbf{T}} = \mathbf{E}_0 + (2 + x)\mathbf{E}_1 + \mathbf{E}_2$. This shows that $\mathbf{T}\mathbf{P}\tilde{\mathbf{T}}$ is the desired translation operator. One can generalize the above to show that a translation by the vector (x_0, y_0) is given by the sandwich operation $\mathbf{T}\mathbf{P}\tilde{\mathbf{T}}$ where

$$\mathbf{T} := 1 + \frac{1}{2}(y_0\mathbf{E}_1 - x_0\mathbf{E}_2) \tag{15.5}$$

It is interesting to note that $\mathbf{T}\mathbf{P}$ and $\mathbf{P}\tilde{\mathbf{T}}$ are both translations of $(x, y) \rightarrow (x + 1, y)$, so one does not need a sandwich to implement translations, but for simplicity of representation, we continue to do so.

Rotations   Similar remarks apply to rotations. A rotation around a normalized point $\mathbf{R}$ by an angle θ is given by $\mathbf{T} = \cos(\frac{\theta}{2}) + \sin(\frac{\theta}{2})\mathbf{R}$. This can be checked by substituting into (15.5) and multiplying out. We will explore a method for constructing such rotators using the exponential function in the next section. See [13] for a more detailed discussion, including constructions of glide reflections and point reflections.

15.4.3 Spin Group, Exponentials, and Logarithms

We have seen above in (15.5) that Euclidean rotations and translations can be represented by sandwich operations in $\mathbf{P}(\mathbb{R}_{2,0,1}^*)$, in fact, in the even subalgebra $\mathbf{P}(\mathbb{R}_{2,0,1}^{*+})$.

Definition 15.1 The *spin group* **Spin(2, 0, 1)** consists of elements $\mathbf{g}$ of the even subalgebra $\mathbf{P}(\mathbb{R}_{2,0,1}^{*+})$ such that $\mathbf{g}\tilde{\mathbf{g}} = 1$. An element of the spin group is called a *rotor*.

Write $\mathbf{g} = s\mathbf{1} + \mathbf{M}$ where $\mathbf{M} = \langle\mathbf{g}\rangle_2 = m_0\mathbf{E}_0 + m_1\mathbf{E}_1 + m_2\mathbf{E}_2$. Then $\mathbf{g}\tilde{\mathbf{g}} = s^2 + m_0^2 = 1$. There are two cases.

- $m_0 \neq 0$, so $\mathbf{M}$ is a Euclidean point. Then there exists $\theta \neq 0$ such that $s = \cos(\theta)$ and $m_0 = \sin(\theta)$, yielding $\mathbf{g} = \cos(\theta) + \sin(\theta)\mathbf{N}$, where $\mathbf{N} = \frac{\mathbf{M}}{\sin\theta}$ is a normalized point, hence $\mathbf{N}^2 = -1$. Thus, the formal exponential $e^{t\mathbf{N}}$ can be evaluated to yield:

$$e^{t\mathbf{N}} = \sum_{i=0}^{\infty} \frac{(t\mathbf{N})^i}{i!}$$
$$= \cos(t) + \sin(t)\mathbf{N}$$

Hence, the rotor $\mathbf{g}$ can be written as an exponential: $\mathbf{g} = e^{\theta\mathbf{N}}$.

- $m_0 = 0$, so $\mathbf{M}$ is an ideal point. Then we can assume that $s = 1$ (if $s = -1$, take the element $-\mathbf{g}$ with the same sandwich behavior as $\mathbf{g}$). Also, $\mathbf{M} = m_1\mathbf{E}_1 + m_2\mathbf{E}_2$. Again, the formal exponential $e^{t\mathbf{M}}$ can be evaluated to yield:

$$e^{t\mathbf{M}} = \sum_{i=0}^{\infty} \frac{(t\mathbf{M})^i}{i!}$$
$$= 1 + t\mathbf{M}$$

So in this case, too, $\mathbf{g} = e^{\mathbf{M}}$ has an exponential form.
The above motivates the following definitions:

Definition 15.2 A *rotator* is a rotor whose bivector part is a Euclidean point. A *translator* is a rotor whose bivector part is an ideal point.

Definition 15.3 The *logarithm* of a translator $\mathbf{g} = s\mathbf{1} + \mathbf{M} \in \mathbf{P}(\mathbb{R}_{2,0,1}^*)$ is $\mathbf{M}$, since $e^{\mathbf{M}} = \mathbf{g}$.

Definition 15.4 Given a rotator $\mathbf{g} = s\mathbf{1} + m_0\mathbf{E}_0 + m_1\mathbf{E}_1 + m_2\mathbf{E}_2 \in \mathbf{Spin(2,0,1)}$. Define $\theta := \tan^{-1}(m_0, s)$ and $\mathbf{N} := \frac{\mathbf{M}}{\|\mathbf{M}\|}$. Then the *logarithm* of $\mathbf{g}$ is $\theta\mathbf{N}$, since $e^{\theta\mathbf{N}} = \mathbf{g}$.

Lie Groups and Lie Algebras The above remarks provide a realization of the two-dimensional Euclidean direct isometry group $se(2)$ and its Lie algebra $se(2)$ within $\mathbf{P}(\mathbb{R}^{*+}_{2,0,1})$. The Spin group $\mathbf{Spin(2,0,1)}$ forms a double cover of $SE(2)$ since the rotors g and $-g$ represent the same isometry. Within $\mathbf{P}(\mathbb{R}^{*+}_{2,0,1})$, the spin group consists of elements of unit norm; the Lie algebra consists of the pure bivectors plus the zero element. The exponential map $\mathbf{X} \to e^{\mathbf{X}}$ maps the latter bijectively onto the former. This structure is completely analogous to the way the unit quaternions sit inside $\mathbf{P}(\mathbb{R}^{*+}_{3,0,0})$ and form a double cover of $SO(3)$. The full group including indirect isometries is also naturally represented in $\mathbf{P}(\mathbb{R}^{*}_{2,0,1})$ as the group generated by reflections in lines, sometimes called the *Pin* group.

15.4.4 Guide to the Literature

There is a substantial literature on the four-dimensional even subalgebra $\mathbf{P}(\mathbb{R}^{*+}_{2,0,1})$ with basis $\{\mathbf{1}, \mathbf{E}_0, \mathbf{E}_1, \mathbf{E}_2\}$. In an ungraded setting, this structure is known as the *planar quaternions*. The original work appears to have been done by Study [27, 28]; this was subsequently expanded and refined by Blaschke [4]. Study's parameterization of the full planar Euclidean group as "quasi-elliptic" space is worthy of more attention. Modern accounts include [22].

15.5 $\mathbf{P}(\mathbb{R}^{*}_{3,0,1})$ and Euclidean Space

The extension of the results in the previous section to the three-dimensional case $\mathbf{P}(\mathbb{R}^{*}_{3,0,1})$ is mostly straightforward. Many of the results can be carried over virtually unchanged. The main challenge is due to the existence of nonsimple bivectors; in fact, most bivectors are *not* simple! (See Sect. 15.5.1 below.) This means that the geometric interpretation of a bivector is usually *not* a simple geometric entity, such as a spear or an axis, but a more general object known in the classical literature as a *linear line complex*, or *null system*. Such entities are crucial in kinematics and dynamics; we will discuss them below in more detail.

Notation As a basis for the full algebra, we adopt the terminology for the exterior algebra W^* in Section 15.2.2, interpreted as a *plane-based* algebra. We add an additional basis 1-vector satisfying $\mathbf{e}_3^2 = 1$. $\mathbf{e}_0$ now represents the ideal *plane* of space, and the other basis vectors represent the coordinate planes. $\mathbf{E}_0$ is the origin of space, while $\mathbf{E}_1 = \mathbf{e}_0\mathbf{e}_3\mathbf{e}_2$ is the ideal point in the x-direction, similarly for $\mathbf{E}_2$ and $\mathbf{E}_3$. The bivector $\mathbf{e}_{01}$ is the ideal line in the $x = 0$ plane, and similarly for $\mathbf{e}_{02}$ and $\mathbf{e}_{03}$. $\mathbf{e}_{23}, \mathbf{e}_{31}$, and $\mathbf{e}_{12}$ are the x-, y-, and z-axes, respectively. We use $\mathbf{i}$ again to denote the embedding of Euclidean points, lines, and planes, from $\mathbb{R}P^3$ into the Clifford algebra.

We continue to denote 1-vectors with bold small Roman letters $\mathbf{a}$; trivectors will be denoted with bold capital Roman letters $\mathbf{P}$; and bivectors will be represented with bold capital Greek letters $\boldsymbol{\Xi}$.[8]

We leave the construction of a multiplication table as an exercise. Once again, most of the geometric products of two vectors obey the pattern $AB = A \cdot B + A \wedge B$. Two new exceptions involve the product of a bivector with another bivector, and with a trivector:

$$\boldsymbol{\Xi}\boldsymbol{\Phi} = \boldsymbol{\Xi} \cdot \boldsymbol{\Phi} + \boldsymbol{\Xi} \times \boldsymbol{\Phi} + \boldsymbol{\Xi} \wedge \boldsymbol{\Phi} \qquad (15.6)$$

$$\boldsymbol{\Xi}\mathbf{P} = \boldsymbol{\Xi} \cdot \mathbf{P} + \boldsymbol{\Xi} \times \mathbf{P} \qquad (15.7)$$

Here, as before, the commutator product $A \times B := \frac{1}{2}(AB - BA)$.

We now describe in more detail the nature of bivectors. We work in W^*, since that is the foundation of the metric. As a result, even readers familiar with bivectors from a *point-based* perspective will probably benefit from going through the following *plane-based* development.

15.5.1 Properties of Bivectors

We begin with a simple bivector $\boldsymbol{\Xi} := \mathbf{a} \wedge \mathbf{b}$ where $\mathbf{a}$ and $\mathbf{b}$ are two planes with coefficients $\{a_i\}$ and $\{b_i\}$. The resulting bivector has the coefficients

$$p_{ij} := a_i b_j - a_j b_i \quad \left(ij \in \{01, 02, 03, 12, 31, 23\}\right)$$

These are the plane-based Plücker coordinates for the intersection line (*axis*) of $\mathbf{a}$ and $\mathbf{b}$. Clearly $\boldsymbol{\Xi} \wedge \boldsymbol{\Xi} = 0$. Conversely, if

$$\boldsymbol{\Xi} \wedge \boldsymbol{\Xi} = 2(p_{01}p_{23} + p_{02}p_{31} + p_{03}p_{12})\mathbf{I} = 0$$

for a bivector $\boldsymbol{\Xi}$, the bivector is simple [16].

Given a second axis $\boldsymbol{\Phi} = \mathbf{c} \wedge \mathbf{d}$, the condition $\boldsymbol{\Xi} \wedge \boldsymbol{\Phi} = 0$ implies they have a plane in common, or, equivalently, they have a point in common. For general bivectors,

$$\boldsymbol{\Xi} \wedge \boldsymbol{\Phi} = (p_{01}q_{23} + p_{02}q_{31} + p_{03}q_{12} + p_{12}q_{03}$$
$$+ p_{31}q_{02} + p_{23}q_{01})\mathbf{I} \qquad (15.8)$$

The parenthesized expression is called the Plücker inner product of the two lines and is written $\langle \boldsymbol{\Xi}, \boldsymbol{\Phi} \rangle_P$. With this inner product, the space of bivectors $P(\bigwedge^2(\mathbb{R}^4)^*)$ is the Cayley–Klein space $\mathfrak{B} := \mathbf{P}(\mathbb{R}^{3,3})$, and the space of lines is the quadric surface

---

[8]A convention apparently introduced by Klein, see [18].

$\mathbf{L}^{3,3} = \{\varXi \mid \langle \varXi, \varXi \rangle_P = 0\} \subset \mathfrak{B}$. When (15.8) vanishes, the two bivectors are said to be *in involution*.

Null System   A line which is in involution with a given bivector $\varXi$ is called a *null line of $\varXi$*. Through every point and in every plane of space, lies a line pencil of null lines. In the case of a nonsimple $\varXi$, this sets up an polarity[9] between the points and planes of space called the *null polarity* determined by $\varXi$: the null plane of a point is the plane in which the null lines of the point lie, and vice-versa. Section 15.5.2 shows how the null plane and null point can be expressed in the Clifford algebra. We will meet the null system again in Sect. 15.6 since it is fundamental to understanding rigid body mechanics.

Metric Properties of Bivectors   Write the bivector $\varXi$ as the sum of two simple bivectors $\varXi = \varXi_\infty + \varXi_o$:

$$\varXi_\infty := p_{01}\mathbf{e}_{01} + p_{02}\mathbf{e}_{02} + p_{03}\mathbf{e}_{03}$$
$$\varXi_o := p_{12}\mathbf{e}_{12} + p_{31}\mathbf{e}_{31} + p_{23}\mathbf{e}_{23}$$

This is the unique decomposition of $\varXi$ as the sum of a line lying in the ideal plane ($\varXi_\infty$) and a Euclidean part ($\varXi_o$). We sometimes write $\varXi = (\varXi_\infty; \varXi_o)$. $\varXi_o \wedge \varXi_\infty = 0 \iff \varXi$ is simple. $\varXi_o$ is invariant under Euclidean translations, while $\varXi_\infty$ is not (exercise). We say that a bivector is *ideal* if $\varXi_o = 0$; otherwise it is *Euclidean*. $\varXi_o$ is a line through the origin, whose direction is given by the ideal point $\mathbf{Q} := e_0\varXi_o = p_{23}\mathbf{E}_1 + p_{31}\mathbf{E}_2 + p_{12}\mathbf{E}_3$. We call $\mathbf{Q}$ the *direction vector* of the bivector.

Guide to the Literature   [17] and [19] are older, classical treatments. [30] is an excellent introduction to Study's approach to the subject. [24] is a modern treatment providing many useful details. See also Chap. 13 in this volume.

15.5.2 Enumeration of Various Products

All the products described in Sect. 15.4.1 have counter-parts here, obtained by leaving points alone and replacing lines by planes. We leave it as an exercise to the reader to enumerate them. Here we focus on the task of enumerating the products that involve bivectors. For that purpose, we extend the definition of $\mathbf{a}$, $\mathbf{b}$, $\mathbf{P}$, and $\mathbf{Q}$ to have an extra coordinate and introduce two arbitrary bivectors, which may or may not be simple:

1. *Inner product.* $\varXi \cdot \varPhi = -(p_{12}g_{12} + p_{31}g_{13} + p_{23}g_{23}) = -\cos(\alpha)$ where α is the angle between the direction vectors of the two bivectors (see Sect. 15.5.1 above). $\varXi \cdot \varPhi$ is a symmetric bilinear form on bivectors, called the *Killing* form. We sometimes write $\varXi \cdot \varPhi = \langle \varXi, \varPhi \rangle_k$. Note that just as in the 2D case, the ideal elements play no role in this inner product. This angle formula is only valid for Euclidean bivectors.

---

[9]A *polarity* is an involutive projectivity that swaps points and planes.

2. *Norm.* There are two cases:
   a. *Euclidean bivectors.* For Euclidean Ξ, define the norm $\|\Xi\| = \sqrt{-\Xi \cdot \Xi}$. Then $\frac{\Xi}{\|\Xi\|}$ has norm 1; we call it a normalized Euclidean bivector.
   b. *Ideal bivectors.* As in Sect. 15.4.1, we get the desired norm on an ideal line Ξ by joining the line with *any* Euclidean point $\mathbf{P}$ and taking the norm of the plane: $\|\Xi\|_\infty = \|\Xi \vee \mathbf{P}\|$. We normalize ideal bivectors with respect to this norm.
3. *Distance.* Verify that the Euclidean distance of two normalized points $\mathbf{P}$ and $\mathbf{Q}$ is still given by $\|\mathbf{P} \vee \mathbf{Q}\|$, and the norm of an ideal point $\mathbf{V}$ (i.e., vector length) is given by $\|\mathbf{V} \vee \mathbf{P}\|$ where $\mathbf{P}$ is any normalized Euclidean point.
4. *Inverses.* For Euclidean Ξ, define $\Xi^{-1} = \frac{\Xi}{\Xi \cdot \Xi}$. Inverses are unique.
5. $\Xi \wedge \Phi = \langle \Xi, \Phi \rangle_P \mathbf{I}$ is the Plücker inner product times $\mathbf{I}$. When both bivectors are simple, this is proportional to the Euclidean distance between the two lines they represent (exercise).
6. *Commutator.* $\Xi \times \Phi$ is a bivector which is in involution to both Ξ and Φ (exercise). We will meet this later in the discussion of mechanics (Sect. 15.6) as the Lie bracket.
7. *Null point.* $\mathbf{a} \wedge \Xi$ for simple Ξ is the intersection point of Ξ with the plane $\mathbf{a}$; in general, it is the *null point* of the plane with respect Ξ.
8. *Null plane.* $\mathbf{P} \vee \Xi$ for simple Ξ is the joining plane of $\mathbf{P}$ and Ξ; in general, it is the *null plane* of the point with respect to Ξ.
9. $\mathbf{a} \cdot \Xi$ for simple Ξ is a plane containing Ξ whose intersection with $\mathbf{a}$ is perpendicular to Ξ.
10. $\Xi\mathbf{I} = (p_{23}\mathbf{e}_{01} + p_{31}\mathbf{e}_{02} + p_{12}\mathbf{e}_{03})$ is the polar bivector of the bivector Ξ. It is an ideal line which is orthogonal (in the elliptic metric of the ideal plane, see Sect. 15.3.1) to the direction vector of Ξ. $(\Xi_\infty; \Xi_o)\mathbf{I} = (\Xi_o; 0)$.

15.5.3 Dual Numbers

We call a number of the form $a + b\mathbf{I}$ for $a, b \in \mathbf{R}$ a *dual number*, after Study [28]. Dual numbers are similar to complex numbers, except that $\mathbf{I}^2 = 0$ rather than $i^2 = -1$. We will need some results on dual numbers to calculate rotor logarithms below. Dual numbers commute with other elements of the Clifford algebra. Given a dual number $\mathbf{z} = a + b\mathbf{I}$, we say that $\mathbf{z}$ is *Euclidean* if $a \neq 0$; otherwise $\mathbf{z}$ is *ideal*. Define the *conjugate* $\bar{\mathbf{z}} = a - b\mathbf{I}$. Then $\mathbf{z}\bar{\mathbf{z}} = a^2$. Define the *norm* $\|a + b\mathbf{I}\| := \sqrt{\mathbf{z}\bar{\mathbf{z}}} = a$. For Euclidean $\mathbf{z}$, define the inverse $(a + b\mathbf{I})^{-1} = \frac{1}{a^2}(a - b\mathbf{I})$. The inverse is the unique dual number $\mathbf{w}$ such that $\mathbf{z}\mathbf{w} = 1$. Given a Euclidean dual number $a + b\mathbf{I}$, define $c = \sqrt{a}$ and $d = \frac{b}{2\sqrt{a}}$. Then $\mathbf{w} := c + d\mathbf{I}$ satisfies $\mathbf{w}^2 = \mathbf{z}$, and we write $\mathbf{w} = \sqrt{\mathbf{z}}$.

Dual Analysis   Just as one can extend real power series to complex power series with reliable convergence properties, power series with a dual variable have well-behaved convergence properties. See [28] for a proof. In particular, the power series for $\cos(x + y\mathbf{I})$ and $\sin(x + y\mathbf{I})$ have the same radii of convergence as their real counterparts. One can use the addition formulae for cos and sin to show that:

$$\cos(x + y\mathbf{I}) = \cos x - \sin x (y\mathbf{I})$$
$$\sin(x + y\mathbf{I}) = \sin x + \cos x (y\mathbf{I})$$

The Axis of a Bivector   Working with Euclidean bivectors is simplified by identifying a special line, the *axis*, the unique Euclidean line in the linear span of $\boldsymbol{\Xi}_\infty$ and $\boldsymbol{\Xi}_o$. The axis $\boldsymbol{\Xi}_x$ is defined by $\boldsymbol{\Xi}_x = (a + b\mathbf{I})\boldsymbol{\Xi}$ for a dual number $a + b\mathbf{I}$. In fact, one can easily check that the choice $a : b = -2\langle \boldsymbol{\Xi}, \boldsymbol{\Xi}\mathbf{I}\rangle_P : \langle \boldsymbol{\Xi}, \boldsymbol{\Xi}\rangle_P$ yields the desired simple bivector. We usually normalize so that $\boldsymbol{\Xi}_x^2 = -1$. The axis appears later in the discussion of Euclidean isometries in Sect. 15.5.5, since most isometries are characterized by a unique invariant axis.

15.5.4 Reflections, Translations, Rotations, and ...

The results of Sect. 15.4.2 can be carried over without significant change to 3D:
1. For a 1-vector **a**, the sandwich operation $\mathbf{P} \to \mathbf{aPa}$ is a Euclidean reflection in the plane represented by **a**.
2. For a pair of 1-vectors **a** and **b** such that $\mathbf{g} := \mathbf{ab}$, $\mathbf{P} \to \mathbf{gP\tilde{g}}$ is a Euclidean isometry. There are two cases:
   a. When $\langle \mathbf{g}\rangle_2$ is Euclidean, it is a rotation around the line represented by $\langle \mathbf{g}\rangle_2$ by twice the angle between the two planes.
   b. When $\langle \mathbf{g}\rangle_2$ is an ideal line $p_{01}\mathbf{e}_{01} + p_{02}\mathbf{e}_{02} + p_{03}\mathbf{e}_{03}$, it is a translation by the vector $(x, y, z) = 2(p_{01}, p_{02}, p_{03})$.

A rotor responsible for a translation (rotation) is called, as before, a *translator (rotator)*. There are however other direct isometries in Euclidean space besides these two types.

Definition 15.5   A *screw motion* is an isometry that can be factored as a rotation around a line $\boldsymbol{\Xi}$ followed by a translation in the direction of $\boldsymbol{\Xi}$. $\boldsymbol{\Xi}$ is called the *axis* of the screw motion.

Like the linear line complex, a screw motion has no counterpart in 2D. In fact, $\langle \mathbf{g}\rangle_2$ is a nonsimple bivector $\Longleftrightarrow$ **g** is the rotor of a screw motion. To show this, we need to extend 2D results on rotors.

15.5.5 Rotors, Exponentials and Logarithms

As in Sect. 15.4.3, the spin group **Spin**$(3, 0, 1)$ is defined to consist of all elements **g** of the even subalgebra $\mathbf{P}(\mathbb{R}_{3,0,1}^{*+})$ such that $\mathbf{g\tilde{g}} = 1$. A group element is called a *rotor*. In this section we seek the logarithm of a rotor **g**. Things are complicated by the fact that the even subalgebra includes the pseudo-scalar **I**. Dual numbers help overcome this difficulty.

Write $\mathbf{g} = s_r + s_d\mathbf{I} + \boldsymbol{\Xi}$. Then

$$\mathbf{g}\tilde{\mathbf{g}} = s_r^2 + 2s_rs_d\mathbf{I} - \boldsymbol{\Xi}^2 = 1$$

Suppose $\boldsymbol{\Xi}^2$ is real. Then $\boldsymbol{\Xi}$ is simple, $s_d = 0$, and $\boldsymbol{\Xi}^2 = s_r^2 - 1$. If $\boldsymbol{\Xi}^2 < 0$, then find real λ such that $\boldsymbol{\Xi}_N := \lambda\boldsymbol{\Xi}$ satisfies $\boldsymbol{\Xi}_N^2 = -1$ and evaluate the formal exponential $e^{t\boldsymbol{\Xi}_N}$ as before (Sect. 15.4.3) to yield

$$e^{t\boldsymbol{\Xi}_N} = \cos(t) + \sin(t)\boldsymbol{\Xi}_N \tag{15.9}$$

We can use this formula to derive exponential and logarithmic forms for rotations as in the 2D case (exercise). If $\boldsymbol{\Xi}^2 = 0$, $\boldsymbol{\Xi}$ is ideal, and the rotor is a translator, similar to the 2D case (exercise). This leaves the case $s_d \neq 0$. Let $\boldsymbol{\Phi} = (a+b\mathbf{I})\boldsymbol{\Xi}$ be the axis of $\boldsymbol{\Xi}$ (see Sect. 15.5.3 above). Since the axis is Euclidean, $a \neq 0$, and the inverse $c + d\mathbf{I} := (a+b\mathbf{I})^{-1}$ exists:

$$(c+d\mathbf{I})\boldsymbol{\Phi} = \boldsymbol{\Xi} \tag{15.10}$$

Replace the real parameter t in the exponential with a dual parameter $t + u\mathbf{I}$, replace $\boldsymbol{\Xi}$ with $\boldsymbol{\Phi}$, substitute $\boldsymbol{\Phi}^2 = -1$, and apply the results above on dual analysis:

$$e^{(t+u\mathbf{I})\boldsymbol{\Phi}} = \cos(t+u\mathbf{I}) + \sin(t+u\mathbf{I})\boldsymbol{\Phi} \tag{15.11}$$
$$= \cos(t) - u\sin(t)\mathbf{I} + \big(\sin(t) + u\cos(t)\mathbf{I}\big)\boldsymbol{\Phi} \tag{15.12}$$

We seek values of t and u such that $\mathbf{g}$ equals the RHS of (15.12):

$$g = s_r + s_d\mathbf{I} + (c+d\mathbf{I})\boldsymbol{\Phi} = \cos(t) - u\sin(t)\mathbf{I} + \big(\sin(t) + u\cos(t)\mathbf{I}\big)\boldsymbol{\Phi} \tag{15.13}$$

We solve for t and u, doing our best to avoid numerical problems that might arise from $\cos(t)$ or $\sin(t)$ alone:

$$t = \tan^{-1}(c, s_r)$$
$$u = \begin{cases} \frac{d}{\cos(t)} & \text{if } |\cos(t)| > |\sin(t)| \\ \frac{-s_d}{\sin(t)} & \text{otherwise} \end{cases}$$

Definition 15.6 Given a rotor $\mathbf{g} \in Sp(3,0,1)$ with nonsimple bivector part, the bivector $(t+u\mathbf{I})\boldsymbol{\Phi}$ defined above is the *logarithm* of **g**.

Theorem 15.1 *Let $(t+u\mathbf{I})\boldsymbol{\Phi}$ be the logarithm of the rotor* $\mathbf{g}$ *($= s_r + s_d\mathbf{I} + \boldsymbol{\Xi}$) with $s_d \neq 0$. Then $u \neq 0$, $t \neq 0$, and **g** represents a screw motion along the axis $\boldsymbol{\Phi}$ consisting of rotation by angle $2t$ and translation by distance $2u$.*

Proof $\boldsymbol{\Phi}$ commutes with $\boldsymbol{\varXi}$ and with $\mathbf{g}$ (exercise). Hence $\mathbf{g}\boldsymbol{\Phi}\tilde{\mathbf{g}} = \boldsymbol{\Phi}$ is fixed by the sandwich. Write the sandwich operation on an arbitrary blade $\mathbf{x}$ as the composition of a translation followed by a rotation:

$$\mathbf{g}\mathbf{x}\tilde{\mathbf{g}} = e^{(t+u\mathbf{I})\boldsymbol{\Phi}}\mathbf{x}e^{-(t+u\mathbf{I})\boldsymbol{\Phi}}$$
$$= e^{t\boldsymbol{\Phi}}\left(e^{u\mathbf{I}\boldsymbol{\Phi}}\mathbf{x}e^{-u\mathbf{I}\boldsymbol{\Phi}}\right)e^{-t\boldsymbol{\Phi}}$$

This makes clear the decomposition into a translation through distance $2u$ (exercise), followed by a rotation around $\boldsymbol{\Phi}$ through an angle $2t$. One sees that the translation and rotation commute by reversing the order. $\qquad\square$

We have succeeded in showing that the bivector of a rotor is nonsimple if and only if the associated isometry is a nondegenerate screw motion. This result closes our discussion of $\mathbf{P}(\mathbb{R}^*_{3,0,1})$. For a fuller discussion, see [13].

15.6   Case Study: Rigid Body Motion

The remainder of the article shows how to model Euclidean rigid body motion using the Clifford algebra structures described above. It begins by showing how to use the Clifford algebra $\mathbf{P}(\mathbb{R}^*_{3,0,1})$ to represent Euclidean motions and their derivatives. Dynamics is introduced with Newtonian particles, which are collected to construct rigid bodies. The inertia tensor of a rigid body is derived as a positive definite quadratic form on the space of bivectors. Equations of motion in the force-free case are derived. In the following, we represent velocity states by $\boldsymbol{\varOmega}$, momentum states by $\boldsymbol{\varPi}$, and forces by $\boldsymbol{\varDelta}$. Due to space limitations, results are compressed. For a fuller discussion, see [13].

15.6.1 Kinematics

Definition 15.7 A *Euclidean motion* is a C^1 path $g : [0, 1] \to Spin(3, 0, 1)$ with $g(0) = \mathbf{1}$.

Theorem 15.2 $\tilde{\mathbf{g}}\dot{\mathbf{g}}$ *is a bivector.*

Proof $\tilde{\mathbf{g}}\dot{\mathbf{g}}$ is in the even subalgebra. For a bivector X, $\tilde{X} = -X$; for scalars and pseudoscalars, $\tilde{X} = X$. Hence it suffices to show $\widetilde{\tilde{\mathbf{g}}\dot{\mathbf{g}}} = -\tilde{\mathbf{g}}\dot{\mathbf{g}}$:

$$\tilde{\mathbf{g}}\mathbf{g} = 1$$
$$(\tilde{\mathbf{g}}\mathbf{g})^{\cdot} = 0$$
$$\dot{\tilde{\mathbf{g}}}\mathbf{g} + \tilde{\mathbf{g}}\dot{\mathbf{g}} = 0$$
$$\widetilde{\tilde{\mathbf{g}}\dot{\mathbf{g}}} + \tilde{\mathbf{g}}\dot{\mathbf{g}} = 0$$
$$\widetilde{\tilde{\mathbf{g}}\dot{\mathbf{g}}} = -\tilde{\mathbf{g}}\dot{\mathbf{g}} \qquad\square$$

Fig. 15.4 Two examples of a
point **P**, its null plane
(**P** ∨ $\varXi$), and the null plane's
polar point

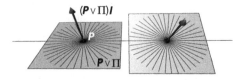

Define $\varOmega := \dot{\mathbf{g}}(0)$; by the theorem, $\varOmega$ is a bivector. We call $\varOmega$ a *Euclidean velocity state*. For a point **P**, the motion **g** induces a path $\mathbf{P}(t)$, the *orbit* of the point **P**, given by $\mathbf{P}(t) = \mathbf{g}(t)\mathbf{P}\tilde{\mathbf{g}}(t)$. Taking derivatives of both sides and evaluating at $t = 0$ yields:

$$\dot{\mathbf{P}}(t) = \dot{\mathbf{g}}(t)\mathbf{P}\tilde{\mathbf{g}}(t) + \mathbf{g}(t)\mathbf{P}\dot{\tilde{\mathbf{g}}}(t)$$
$$\dot{\mathbf{P}}(t) = \dot{\mathbf{g}}(t)\mathbf{P}\tilde{\mathbf{g}}(t) + \mathbf{g}(t)\mathbf{P}\dot{\tilde{\mathbf{g}}}(t)$$
$$\dot{\mathbf{P}}(0) = \varOmega\mathbf{P} - \mathbf{P}\varOmega$$
$$= 2(\varOmega \times \mathbf{P})$$

The last step follows from the definition of the commutator product of bivectors. In this formula we can think of **P** as a normalized Euclidean point which is being acted upon by the Euclidean motion **g**. From Sect. 15.5.2 we know that $\varOmega \times \mathbf{P}$ is a ideal point, that is, a free vector. We sometimes use the alternative form $\varOmega \times \mathbf{P} = (\varOmega \vee \mathbf{P})\mathbf{I}$ (exercise). The vector field vanishes wherever $\varOmega \vee \mathbf{P} = 0$. This only occurs if $\varOmega$ is a line and **P** lies on it. The picture is consistent with the knowledge, gained above, that in this case $e^{t\varOmega}$ generates a rotation (or translation) with axis $\varOmega$. Otherwise, the motion is an instantaneous screw motion around the axis of $\varOmega$, and no points remain fixed.

Null Plane Interpretation   In the formulation $\dot{\mathbf{P}} = 2(\varOmega \vee \mathbf{P})\mathbf{I}$, we recognize the result as the polar point (with respect to the Euclidean metric) of the null plane of **P** (with respect to $\varOmega$). See Fig. 15.4. Thus, the vector field can be considered as the *composition* of two simple polarities: first the null polarity on $\varOmega$, then the metric polarity on the Euclidean quadric. This leads to the somewhat surprising result that regardless of the metric used, the underlying null polarity remains the same.

15.6.2 Dynamics

With the introduction of forces, our discussion moves from the kinematic level to the dynamic one. We begin with a treatment of statics. We then introduce Newtonian particles, build rigid bodies out of collection of such particles, and state and solve the equations of motions for these rigid bodies.[10]

---

[10]*Editorial note*: This approach to dynamics may be compared to Chap. 1 and Chap. 18 elsewhere in this volume.

3D Statics Traditional statics represents 3D a single force F as a pair of 3-vectors (V, M), where $V = (v_x, v_y, v_z)$ is the direction vector, and $M = (m_x, m_y, m_z)$ is the moment with respect to the origin (see [12, Chap. 2]). The resultant of a system of forces F_i is defined to be the sum of the corresponding direction vectors V_i and moment vectors M_i. The forces are in equilibrium if both terms of the resultant are zero.

If $\mathbf{P}$ is a normalized point on the line carrying the force, define $H(F) := \mathbf{P} \vee \mathbf{i}(V)$. We call $H(F)$ the *homogeneous form* of the force and verify that

$$H(F) = m_x \mathbf{E}_{01} + m_y \mathbf{E}_{02} + m_z \mathbf{E}_{03} + v_z \mathbf{E}_{12} + v_y \mathbf{E}_{31} + v_x \mathbf{E}_{23}$$

If F is the resultant of a force system $\{F_i\}$, then $H(F) = \sum_i H(F_i)$. Hence, a system of forces $\{F_i\}$ is the null force $\iff \sum_i H(F_i) = 0$. Furthermore, $H(F)$ is an ideal line $\iff$ the system of forces reduces to a force-couple, and $H(F)$ is a simple Euclidean bivector $\iff$ F represents a single force. Notice that the intensity of a bivector is significant, since it is proportional to the strength of the corresponding force. For this reason, we sometimes say forces are represented by *weighted* bivectors.

15.6.2.1 Newtonian Particles

The basic object of Newtonian mechanics is a particle P with mass m located at the point represented by a trivector $\mathbf{R}$. Stated in the language of ordinary Euclidean vectors, Newton's law asserts that the force F acting on P is $F = m\ddot{\underline{\mathbf{R}}}$.

Definition 15.8 The *spear* of the particle is $\Lambda := \mathbf{R} \vee \dot{\mathbf{R}}$.

Definition 15.9 The *momentum state* of the particle is $\Pi := m\Lambda$.

Definition 15.10 The *velocity state* of the particle is $\Gamma := \Lambda \mathbf{I}$.

Definition 15.11 The *kinetic energy* E of the particle is

$$E := \frac{m}{2} \|\dot{\mathbf{R}}\|_\infty^2 = -\frac{m}{2} \Lambda \cdot \Lambda = -\frac{1}{2} \mathsf{S}\Gamma \wedge \Pi \tag{15.14}$$

Remarks Since we can assume that $\mathbf{R}$ is normalized, $\dot{\mathbf{R}}$ is an ideal point. Π is a weighted bivector whose weight is proportional to the mass and the velocity of the particle. Γ is ideal, corresponding to the fact that the particle's motion is *translatory*. Up to the factor m, Γ is the polar line of Π with respect to the Euclidean metric. It is straightforward to verify that the linear and angular momentum of the particle appear as Π_o and Π_∞, respectively, and that the definition of kinetic energy agrees with the traditional one (exercise). The second and third equalities in (15.14) are also left as an exercise.

We consider only force-free systems. The extension to include external forces is straightforward but lies outside the scope of this introduction.

Theorem 15.3 *If $F = 0$, then Λ, Π, Γ, and E are conserved quantities.*

Proof $F = 0$ implies $\ddot{\mathbf{R}} = 0$. Then:
- $\dot{\Lambda} = (\dot{\mathbf{R}} \vee \dot{\mathbf{R}} + \mathbf{R} \vee \ddot{\mathbf{R}}) = 0$
- $\dot{\Pi} = m\dot{\Lambda} = 0$
- $\dot{\Gamma} = (\dot{\Lambda})\mathbf{I} = 0$
- $\dot{E} = \frac{1}{2}(\mathbf{S}\dot{\Gamma} \wedge \Pi + \mathbf{S}\Gamma \wedge \dot{\Pi}) = 0.$ □

Inertia Tensor of a Particle   Assume the particle is "governed by" a Euclidean motion $\mathbf{g}$ with associated Euclidean velocity state $\Omega := \dot{\mathbf{g}}(0)$. Then Π, Γ, and E depend on Ω as follows:

$$\dot{\mathbf{R}} = 2(\Omega \times \mathbf{R}) \tag{15.15}$$

$$\Pi = m\big(\mathbf{R} \vee 2(\Omega \times \mathbf{R})\big) \tag{15.16}$$

$$\Gamma = \big(\mathbf{R} \vee 2(\Omega \times \mathbf{R})\big)\mathbf{I} \tag{15.17}$$

$$E = -\frac{m}{2}\mathbf{S}\big((\mathbf{R} \vee 2(\Omega \times \mathbf{R}))\mathbf{I}\big) \wedge \big(\mathbf{R} \vee 2(\Omega \times \mathbf{R})\big) \tag{15.18}$$

$$= -\mathbf{S}\Omega \wedge \Pi \tag{15.19}$$

The step from (15.18) to (15.19) is described in more detail in [13].

Define a real-valued bilinear operator $\mathbf{A}$ on pairs of bivectors:

$$\mathbf{A}(\Omega, \Xi) := -\frac{m}{2}\mathbf{S}\big((\mathbf{R} \vee 2(\Omega \times \mathbf{R}))\mathbf{I}\big) \wedge \big(\mathbf{R} \vee 2(\Xi \times \mathbf{R})\big) \tag{15.20}$$

$$= \frac{m}{2}\big(\mathbf{R} \vee 2(\Omega \times \mathbf{R})\big) \cdot \big(\mathbf{R} \vee 2(\Xi \times \mathbf{R})\big) \tag{15.21}$$

where the step from (15.20) to (15.21) can be deduced from Sect. 15.5.2. (15.21) shows that $\mathbf{A}$ is symmetric since $\cdot$ on bivectors is symmetric: $\Lambda \cdot \Delta = \Delta \cdot \Lambda$. We call $\mathbf{A}$ the *inertia tensor* of the particle, since $E = \mathbf{A}(\Omega, \Omega) = -\mathbf{S}\Omega \wedge \Pi$. We overload the operator and write $\Pi = \mathbf{A}(\Omega)$ to indicate the polar relationship between Π and Ω. We will construct the inertia tensor of a rigid body out of the inertia tensors of its particles below.

15.6.2.2 Rigid Body Motion
Begin with a finite set of mass points P_i; for each, derive the velocity state Γ_i, the momentum state Π_i, and the inertia tensor $\mathbf{A}_i$.[11] Such a collection of mass points is called a *rigid body* when the Euclidean distance between each pair of points is constant.

---

[11] We restrict ourselves to the case of a finite set of mass points, since extending this treatment to a continuous mass distribution presents no significant technical problems; summations have to be replaced by integrals.

Extend the momenta and energy to the collection of particles by summation:

$$\Pi := \sum \Pi_i = \sum \mathbf{A}_i(\Omega) \tag{15.22a}$$

$$E := \sum E_i = \sum \mathbf{A}_i(\Omega, \Omega) \tag{15.22b}$$

Since for each single particle, these quantities are conserved when $F = 0$, this is also the case for the aggregate Π and E defined here.

We introduce the inertia tensor A for the body:

Definition 15.12 $\mathbf{A} := \sum \mathbf{A}_i$.

Then $\Pi = \mathbf{A}(\Omega)$ and $E = \mathbf{A}(\Omega, \Omega)$; neither formula requires a summation over the particles: the shape of the rigid body has been encoded into $\mathbf{A}$. One can proceed traditionally and diagonalize the inertia tensor by finding the center of mass and moments of inertia (see [2]). Due to space constraints, we omit the details. Instead, we sketch how to integrate the inertia tensor more tightly into the Clifford algebra framework.

Clifford Algebra for Inertia Tensor   We define a Clifford algebra $\mathbf{C_A}$ based on $P(\bigwedge^2 \mathbb{R}^{4*})$ by attaching the quadratic form $\mathbf{A}$ as the inner product.[12] We denote the pseudoscalar of this alternative Clifford algebra by $\mathbf{I_A}$, and the inner product of bivectors by $\langle \cdot, \cdot \rangle_{\mathbf{A}}$. We use the same symbols to denote bivectors in W^* as 1-vectors in $\mathbf{C_A}$. Bivectors in W are represented by 5-vectors in $\mathbf{C_A}$. Multiplication by $\mathbf{I_A}$ swaps 1-vectors and 5-vectors in $\mathbf{C_A}$; we use $\mathbf{J}$ (lifted to $\mathbf{C_A}$) to convert 5-vectors back to 1-vectors as needed. The following theorem, which we present without proof, shows how to obtain Π directly from $\mathbf{I_A}$ in this context:

Theorem 15.4 *Given a rigid body with inertia tensor* $\mathbf{A}$ *and velocity state* Ω, *the momentum state* $\Pi = \mathbf{A}(\Omega) = \mathbf{J}(\Omega \mathbf{I_A})$.

Conversely, given a momentum state Π, we can manipulate the formula in the theorem to deduce:

$$\Omega = \mathbf{A}^{-1}(\Pi) = \left(\mathbf{J}(\Pi)\mathbf{I_A}^{-1} \right)$$

In the sequel we denote the polarity on the inertia tensor by $\mathbf{A}(\Omega)$ and $\mathbf{A}^{-1}(\Pi)$, leaving open whether the Clifford algebra approach indicated here is followed.

---

[12]It remains to be seen if this approach represents an improvement over the linear algebra approach, which could also be maintained in this setting.

15.6.2.3 The Euler Equations for Rigid Body Motion

In the absence of external forces, the motion of a rigid body is completely determined by its momentary velocity state or momentum state at a given moment. How can one compute this motion? First we need a few facts about coordinate systems.

The following discussion assumes that we observe a system as it evolves in time. All quantities are then potentially time dependent; instead of writing $\mathbf{g}(t)$, we continue to write $\mathbf{g}$ and trust the reader to bear in mind the time-dependence.

We use the subscripts X_s and X_c[13] to distinguish whether the quantity X belongs to the space or the body coordinate system. The conservation laws of the previous section are generally valid only in the space coordinate system, for example, $\dot{\boldsymbol{\Pi}}_s = 0$. On the other hand, the inertia tensor will be constant only with respect to the body coordinate system, so, $\boldsymbol{\Pi}_c = \mathbf{A}(\boldsymbol{\Omega}_c)$. When we consider a Euclidean motion $\mathbf{g}$ as being applied to the body, then the relation between body and space coordinate systems for *any* element $\mathbf{X} \in \mathbf{P}(\mathbb{R}^*_{3,0,1})$, with respect to a motion $\mathbf{g}$, is given by the sandwich operator:

$$\mathbf{X}_s = \mathbf{g}\mathbf{X}_c\tilde{\mathbf{g}}$$

Definition 15.13 The *velocity in the body* $\boldsymbol{\Omega}_c := \tilde{\mathbf{g}}\dot{\mathbf{g}}$, and the *velocity in space* $\boldsymbol{\Omega}_s := \mathbf{g}\boldsymbol{\Omega}_c\tilde{\mathbf{g}}$.

Definition 15.14 The *momentum in the body* $\boldsymbol{\Pi}_c := \mathbf{A}(\boldsymbol{\Omega}_c)$, and the *momentum in space* $\boldsymbol{\Pi}_s := \mathbf{g}\boldsymbol{\Pi}_c\tilde{\mathbf{g}}$.

We derive a general result for a time-dependent element (of arbitrary grade) in these two coordinate systems:

Theorem 15.5 *For time-varying* $\mathbf{X} \in \mathbf{P}(\mathbb{R}^*_{3,0,1})$ *subject to the motion* $\mathbf{g}$ *with velocity in the body* $\boldsymbol{\Omega}_c$,

$$\dot{\mathbf{X}}_s = \mathbf{g}\big(\dot{\mathbf{X}}_c + 2(\boldsymbol{\Omega}_c \times \mathbf{X}_c)\big)\tilde{\mathbf{g}}$$

Proof

$$\dot{\mathbf{X}}_s = \dot{\mathbf{g}}\mathbf{X}_c\tilde{\mathbf{g}} + \mathbf{g}\dot{\mathbf{X}}_c\tilde{\mathbf{g}} + \mathbf{g}\mathbf{X}_c\dot{\tilde{\mathbf{g}}}$$
$$= \mathbf{g}(\tilde{\mathbf{g}}\dot{\mathbf{g}}\mathbf{X}_c + \dot{\mathbf{X}}_c + \mathbf{X}_c\dot{\tilde{\mathbf{g}}}\mathbf{g})\tilde{\mathbf{g}}$$
$$= \mathbf{g}(\boldsymbol{\Omega}_c\mathbf{X}_c + \dot{\mathbf{X}}_c + \mathbf{X}_c\tilde{\boldsymbol{\Omega}}_c)\tilde{\mathbf{g}}$$
$$= \mathbf{g}(\dot{\mathbf{X}}_c + \boldsymbol{\Omega}_c\mathbf{X}_c - \mathbf{X}_c\boldsymbol{\Omega}_c)\tilde{\mathbf{g}}$$
$$= \mathbf{g}\big(\dot{\mathbf{X}}_c + 2(\boldsymbol{\Omega}_c \times \mathbf{X}_c)\big)\tilde{\mathbf{g}}$$

---

[13]From *corpus*, Latin for body.

The next-to-last equality follows from the fact that for bivectors, $\tilde{\boldsymbol{\Omega}} = -\boldsymbol{\Omega}$; the last equality is the definition of the commutator product. ☐

We will be interested in the case $\mathbf{X}_c$ is a bivector. In this case, $\mathbf{X}_c$ and $\boldsymbol{\Omega}_c$ can be considered as Lie algebra elements, and $2(\boldsymbol{\Omega}_c \times \mathbf{X}_c)$ is called the *Lie bracket*. It expresses the change in one ($\mathbf{X}$) due to an instantaneous Euclidean motion represented by the other ($\boldsymbol{\Omega}$).

15.6.2.4 Solving for the Motion

Since $\boldsymbol{\Omega}_c = \tilde{\mathbf{g}}\dot{\mathbf{g}}$, $\dot{\mathbf{g}} = \mathbf{g}\boldsymbol{\Omega}_c$, a first-order ODE. If we had a way of solving for $\boldsymbol{\Omega}_c$, we could solve for $\mathbf{g}$. If we had a way of solving for $\boldsymbol{\Pi}_c$, we could apply Theorem 15.4 to solve for $\boldsymbol{\Omega}_c$. So, how to solve for $\boldsymbol{\Pi}_c$?

We apply the corollary to the case of force-free motion. Then $\dot{\boldsymbol{\Pi}}_s = 0$: the momentum of the rigid body in space is constant. By Theorem 15.5,

$$0 = \dot{\boldsymbol{\Pi}}_s = \mathbf{g}\big(\dot{\boldsymbol{\Pi}}_c + 2(\boldsymbol{\Omega}_c \times \boldsymbol{\Pi}_c)\big)\tilde{\mathbf{g}} \tag{15.23}$$

The only way the RHS can be identically zero is that the expression within the parentheses is also identically zero, implying:

$$\dot{\boldsymbol{\Pi}}_c = 2\boldsymbol{\Pi}_c \times \boldsymbol{\Omega}_c$$

Use the inertia tensor to convert velocity to momentum yields a differential equation purely in terms of the momentum:

$$\dot{\boldsymbol{\Pi}}_c = 2\boldsymbol{\Pi}_c \times \mathbf{A}^{-1}(\boldsymbol{\Pi}_c)$$

When the inner product is written out in components, one arrives at the well-known Euler equations for the motion [2, p. 143].

The complete set of equations for the motion g are given by the pair of first-order ODEs:

$$\dot{\mathbf{g}} = \mathbf{g}\boldsymbol{\Omega}_c \tag{15.24}$$

$$\dot{\boldsymbol{\Pi}}_c = 2\boldsymbol{\Pi}_c \times \boldsymbol{\Omega}_c \tag{15.25}$$

where $\boldsymbol{\Omega}_c = \mathbf{A}^{-1}(\boldsymbol{\Pi}_c)$. When written out in full, this gives a set of 14 first-order linear ODEs. The solution space is 12 dimensions; the extra dimensions corresponds to the normalization $\mathbf{g}\tilde{\mathbf{g}} = 1$. At this point the solution continues as in the traditional approach, using standard ODE solvers. Our experience is that the cost of evaluating (15.24) is no more expensive than traditional methods. For an extension to the presence of external forces, see [13].

Comparison The projective Clifford algebra approach outlined here exhibits several advantages over other approaches to rigid body motion. The representation of kinematic and dynamic states as bivectors rather than as pairs of ordinary 3-vectors (linear and angular velocity, momentum, etc.) provides a framework free of the special cases which characterize the split approach (for example, translations are rotations around an ideal line, a force couple is a single force carried by an ideal line, etc.). The Clifford algebra product avoids cumbersome matrix formulations and, as seen in Theorem 15.5, yields formulations which are valid for points, lines, and planes uniformly. The inertia tensor of a rigid body can be represented as a separate (positive definite) Clifford algebra defined on the space of bivectors. Finally, the treatment of Newtonian particles reveals an underlying velocity–momentum polarity in bivector space analogous to that of rigid bodies.

15.7 Guide to the Literature

The work presented here, in its conceptual basis, is derived from [28]. This seminal book worked out in impressive detail the structure which in modern form appears as the even subalgebra $\mathbf{P}(\mathbb{R}^{*+}_{3,0,1})$. Study avoided using the term quaternion; the structure he described has nonetheless become known as the *dual quaternions*. His parameter ϵ maps to the pseudoscalar $\mathbf{I}$. A full description of the correspondence between the two systems lies outside the scope of this work, nor is the full scope of [28] reflected in material presented here. [33] gives an excellent survey of the historical development that led up to Study's work. Möbius, Plücker, Hamilton, and Klein were Study's most important predecessors; he and Ball [3] had a relationship based on mutual appreciation. Weiss, a student of Study's, wrote a concise introduction [30] to Study's investigations which can be recommended. For beginners, [6] provides a simpler introduction to Study's approach.

Study's contribution in the direction of mechanics was developed further in [29] and [5]. The former was hampered by the awkward matrix notation required for stating transformation laws. These, on the other hand, are "built in" to the Clifford algebra approach and simplify the approach considerably. [5] concentrated on the non-Euclidean case.

The modern legacy of this work is varied. Some modern literature, such as [12], use *spatial vectors* to model rigid body motion; these are 6D vectors equivalent to our bivectors, but developed within a linear algebra framework reminiscent of [29]. While the dual quaternions have developed a following in robotics and other applied areas [22], their embedding as the even subalgebra of $\mathbf{P}(\mathbb{R}^*_{3,0,1})$ has not received widespread acceptance. Modern sources using $\mathbf{P}(\mathbb{R}^*_{3,0,1})$ include [25] and [26]. An advanced treatment of the structure of spin groups in degenerate Clifford algebras can be found in [1]. Much contemporary work in mechanics which uses Clifford algebra methods applied to physics and engineering uses the conformal model [11, 14, 23] rather than the homogeneous model presented here. Our handling of rigid body motion owes much to the spirit of [2].

Study himself appears to be aware of the possibility of extending his work within a more comprehensive algebraic structure. He remarked at the end of [28] (p. 595, translated by the author):

> The elementary geometric theory, that hovers thus before us, will surpass the construction possibilities of the quaternions to the same degree that the geometry of dynamen [linear line complexes] surpasses the addition of vectors. The accessory analytic machinery will consist of a system of compound quantities, with eight, or better yet, with *sixteen* units. {Study's italics!}

It seems likely that $\mathbf{P}(\mathbb{R}^*_{3,0,1})$ is in fact the 16-dimensional algebraic realizations of Study's prophetic inkling.

15.8 The Homogeneous Model: A Serious Alternative

How does the theory developed so far—geometric, kinematic, dynamic—compare with alternative models for the same purposes? In many respects—for example, geometric expressiveness, efficiency of computation, fidelity of isometry group representation, calculation of rotor logarithms, and succinctness of kinematic and dynamic models—the homogeneous theory is a serious competitor to the other models presented in this book. For application areas with purely Euclidean content it may have advantages over higher dimensional models with regard to computational efficiency, ease of representation, and numerical integration. The excellent results obtained with this model are not surprising, as the previous section shows that the mathematicians involved with rigid body motion in the projective model (Klein, Clifford, Study, and others) were the same ones who laid the foundations for modern Clifford algebras.

15.9 Non-Euclidean Extension

The work presented here is part of a forthcoming dissertation which includes also the non-Euclidean geometries of elliptic and hyperbolic space. A full theory of these non-Euclidean spaces has been developed in the same dual projectivized Clifford algebra setting used for the present Euclidean treatment, based on the Cayley–Klein construction (Sect. 15.3.1), and including kinematics and rigid body motion. Space considerations made a detailed inclusion here impractical. Readers interested in learning more can consult [13] (Sect. 7). Chapter 18 in this volume also addresses this topic, taking a different perspective.

15.10 Conclusion

We have established that $\mathbf{P}(\mathbb{R}^*_{n,0,1})$ is a model for Euclidean geometry. By using a mixture of projective, ideal, and properly Euclidean elements, we have avoided the problems traditionally associated with degenerate metrics. The result provides a complete and compact representation of Euclidean geometry, kinematics and rigid body dynamics. We hope that this work will stimulate others working in these fields to consider the homogeneous model as a practical solution to their problem domain and to deepen and extend these initial results.

15.11 Exercises

15.1 In analogy to Table 15.2, construct the multiplication table for $\mathbf{P}(\mathbb{R}^*_{3,0,1})$. Define a norm for points, vectors, and planes in analogy to those given in Sect. 15.4.1. Translate the products $\mathbf{a} \cdot \mathbf{b}, \mathbf{P} \cdot \mathbf{Q}, \mathbf{a} \wedge \mathbf{P}, \mathbf{a} \cdot \mathbf{P}, \mathbf{P} \vee \mathbf{Q}, \mathbf{P} \times \mathbf{Q}, \mathbf{aI}$, and $\mathbf{PI}$ from Sect. 15.4.1 into this setting.

15.2 *Projecting a point onto a line, and* vice-versa. For Euclidean $\mathbf{P}$ and Euclidean simple $\boldsymbol{\Xi}$, show:
1. $\mathbf{Q} := (\mathbf{P} \cdot \boldsymbol{\Xi})\boldsymbol{\Xi}^{-1}$ is the point of $\boldsymbol{\Xi}$ closest to $\mathbf{P}$. $\mathbf{Q} \vee \mathbf{P}$ is the line through $\mathbf{P}$ cutting $\boldsymbol{\Xi}$ at right angles.
2. $(\mathbf{P} \cdot \boldsymbol{\Xi})\mathbf{P}^{-1}$ is a line parallel to $\boldsymbol{\Xi}$ passing through $\mathbf{P}$.

15.3 *The common normal of two lines.* Let $\boldsymbol{\Xi}$ and $\boldsymbol{\Phi}$ be two simple Euclidean bivectors.
1. Show that $\boldsymbol{\Theta} := \boldsymbol{\Xi} \times \boldsymbol{\Phi}$ is a bivector which is in involution to both $\boldsymbol{\Xi}$ and $\boldsymbol{\Phi}$.
2. Show that the axis $\boldsymbol{\Theta}_x$ of $\boldsymbol{\Theta}$ is a line perpendicular to both $\boldsymbol{\Xi}$ and $\boldsymbol{\Phi}$.
3. Calculate the points where $\boldsymbol{\Theta}_x$ intersects $\boldsymbol{\Xi}$ and $\boldsymbol{\Phi}$. (Hint: consider where the plane spanned by $\mathbf{N}_{\boldsymbol{\Theta}} := \mathbf{e}_0 \boldsymbol{\Theta}_x$ (the direction vector of $\boldsymbol{\Theta}_x$) and $\boldsymbol{\Phi}$ cuts $\boldsymbol{\Xi}$.)

15.4 *3D rotations and translations.* Handle the case $s_d = 0$ ($\mathbf{g} = s_r + \boldsymbol{\Xi}$) from Sect. 15.5.5 to obtain exponential forms for 3D rotations and translations. Define the corresponding logarithms. What does the case $s_r = s_d = 0$ represent? (Hint: $\cos\left(\frac{\pi}{2}\right) = 0$.)

15.5 Find the rotator corresponding to a rotation of $\frac{\pi}{3}$ radians around the line through the origin and the point $(1, 1, 1)$.

15.6 From the proof of Theorem 15.1:
1. Show that $\boldsymbol{\Phi}$ commutes with $\boldsymbol{\Xi}$ and with $\mathbf{g}$.
2. Define $\mathbf{T} := -2u\mathbf{N}_{\boldsymbol{\Phi}}$, where $\mathbf{N}_{\boldsymbol{\Phi}}$ is the direction vector of the axis $\boldsymbol{\Phi}$. Show that the translation part of $\mathbf{g}$ moves a point $\mathbf{X}$ to $\mathbf{X} + \mathbf{T}$, hence a distance of $2u$. [Hint: Expand $(e^{u\mathbf{I}\boldsymbol{\Phi}}\mathbf{X}e^{-u\mathbf{I}\boldsymbol{\Phi}})$ using $(\mathbf{I}\boldsymbol{\Phi})^2 = 0$, and $\boldsymbol{\Phi}^2 = -1$ implies $\|\mathbf{I}\boldsymbol{\Phi}\|_\infty = 1$.]

References

1. Ablamowicz, R.: Structure of spin groups associated with degenerate Clifford algebras. J. Math. Phys. **27**, 1–6 (1986)

2. Arnold, V.I.: Mathematical Methods of Classical Physics. Springer, New York (1978), Appendix 2
3. Ball, R.: A Treatise on the Theory of Screws. Cambridge University Press, Cambridge (1900)
4. Blaschke, W.: Ebene Kinematik. Teubner, Leipzig (1938)
5. Blaschke, W.: Nicht-euklidische Geometrie und Mechanik. Teubner, Leipzig (1942)
6. Blaschke, W.: Analytische Geometrie. Birkhäuser, Basel (1954)
7. Bourbaki, N.: Elements of Mathematics, Algebra I. Springer, Berlin (1989)
8. Conradt, O.: Mathematical Physics in Space and Counterspace. Verlag am Goetheanum, Goetheanum (2008)
9. Coxeter, H.M.S.: Projective Geometry. Springer, New York (1987)
10. Dorst, L., Fontijne, D., Mann, S.: Geometric Algebra for Computer Science. Morgan Kaufmann, San Francisco (2009)
11. Doran, C., Lasenby, A.: Geometric Algebra for Physicists. Cambridge University Press, Cambridge (2003)
12. Featherstone, R.: Rigid Body Dynamics Algorithms. Springer, Berlin (2007)
13. Gunn, C.: On the homogeneous model for Euclidean geometry: extended version. http://arxiv.org/abs/1101.4542 (2011)
14. Hestenes, D.: New tools for computational geometry and rejuvenation of screw theory. In: Bayro-Corrochano, E.J., Scheuermann, G. (eds.) Geometric Algebra Computing: In Engineering and Computer Science, pp. 3–35. Springer, Berlin (2010)
15. Hestenes, D., Sobczyk, G.: Clifford Algebra to Geometric Calculus. Fundamental Theories of Physics. Reidel, Dordrecht (1987)
16. Hitchin, N.: Projective geometry. http://people.maths.ox.ac.uk/hitchin/hitchinnotes/Projective_geometry/Chapter_3_Exterior.pdf (2003)
17. Jessop, C.M.: A Treatise on the Line Complex. Chelsea, New York (1969). Original 1903, Cambridge
18. Klein, F.: Über Liniengeometrie und metrische Geometrie. Math. Ann. 2, 106–126 (1872)
19. Klein, F.: Vorlesungen Über Höhere Geometrie. Chelsea, New York (1927)
20. Klein, F.: Vorlesungen Über Nicht-euklidische Geometrie. Chelsea, New York (1949). Original 1926, Berlin
21. Li, H.: Invariant Algebras and Geometric Algebra. World Scientific, Singapore (2008)
22. McCarthy, J.M.: An Introduction to Theoretical Kinematics. MIT Press, Cambridge (1990)
23. Perwass, C.: Geometric Algebra with Applications to Engineering. Springer, Berlin (2009)
24. Pottmann, H., Wallner, J.: Computational Line Geometry. Springer, Berlin (2001)
25. Selig, J.: Clifford algebra of points, lines, and planes. Robotica 18, 545–556 (2000)
26. Selig, J.: Geometric Fundamentals of Robotics. Springer, Berlin (2005)
27. Study, E.: Von den bewegungen und umlegungen. Math. Ann. 39, 441–566 (1891)
28. Study, E.: Geometrie der Dynamen. Teubner, Leipzig (1903)
29. von Mises, R.: Die Motorrechnung Eine Neue Hilfsmittel in der Mechanik. Z. Rein Angew. Math. Mech. 4(2), 155–181 (1924)
30. Weiss, E.A.: Einführung in die Liniengeometrie und Kinematik. Teubner, Leipzig (1935)
31. Whitehead, A.N.: A Treatise on Universal Algebra. Cambridge University Press, Cambridge (1898)
32. Wikipedia. http://en.wikipedia.org/wiki/Exterior_algebra
33. Ziegler, R.: Die Geschichte Der Geometrischen Mechanik im 19. Jahrhundert. Franz Steiner Verlag, Stuttgart (1985)

A Homogeneous Model for Three-Dimensional Computer Graphics Based on the Clifford Algebra for $\mathbb{R}^3$

16

Ron Goldman

Abstract

We construct a homogeneous model for Computer Graphics using the Clifford Algebra for $\mathbb{R}^3$. To incorporate points as well as vectors within this model, we employ the odd-dimensional elements of this graded eight-dimensional algebra to represent mass-points by exploiting the pseudoscalars to represent mass. The even-dimensional elements of this Clifford Algebra are isomorphic to the quaternions, which operate on the odd-dimensional elements by sandwiching. Along with the standard sandwiching formulas for rotations and reflections, this paradigm allows us to use sandwiching to compute perspective projections.

16.1 Introduction

Although the everyday visual world is three-dimensional, present-day Computer Graphics typically uses four coordinates to represent points and vectors, and 4×4 matrices to represent the standard transformations in the graphics pipeline [4]. Four coordinates and 4×4 matrices are necessary in order to represent perspective projection using matrix multiplication because perspective projection is not a linear transformation in three dimensions; rather in three dimensions, perspective projection is a rational linear (i.e., a projective) transformation. Therefore in contemporary Computer Graphics a fourth coordinate is introduced to represent the denominators introduced by perspective projections.

The four coordinates (x, y, z, w) correspond to the affine point in three dimensions located at the point with rectangular coordinates $(x/w, y/w, z/w)$ provided that $w \neq 0$. The fourth coordinate w can be interpreted as a weight or mass (possibly negative) associated with this affine point. Thus the natural domain for everyday Computer Graphics is not $\mathbb{R}^3$, but rather $\mathbb{R}^4$: three of the dimensions are spatial, and the fourth dimension is due to the mass.

R. Goldman (✉)
Department of Computer Science, Rice University, Houston, TX 77005, USA
e-mail: rng@rice.edu

L. Dorst, J. Lasenby (eds.), *Guide to Geometric Algebra in Practice*,
DOI 10.1007/978-0-85729-811-9_16, © Springer-Verlag London Limited 2011

Fig. 16.1 Mass-points
represented as vectors in four
dimensions. Here O
represents the origin for the
affine points in R^3, and Ω
represents the zero vector in
R^3 which is also the origin
in R^4

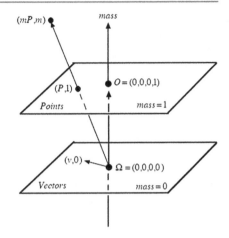

These four coordinates are sometimes called *homogeneous coordinates*, but in order not to confuse these coordinates with the homogeneous coordinates for the points in three-dimensional projective space, we prefer to call the objects in this four-dimensional vector space *mass-points*. The three-dimensional vectors reside in this space as objects with zero mass, while the affine points are embedded in this space as mass-points with unit mass (see Fig. 16.1). For a more detailed explanation of the four-dimensional vector space of mass-points, see [5].

A homogeneous model for Clifford Algebra that deals directly with mass-points has indeed been constructed. This model is called the *Conformal Model* and is a *Clifford Algebra* for $\mathbb{R}^5$ [1, 2, 9]. Thus this algebra is a vector space with $2^5 = 32$ dimensions. This 32-dimensional algebra has many convenient properties. For example, not only lines and planes but also circles and spheres are included as primitive objects in this model; moreover all conformal transformations on $\mathbb{R}^3$ including inversions in the sphere are represented in this algebra. Nevertheless, the dimension of this algebra seems excessively high and the algebra itself unnecessarily complicated for contemporary Computer Graphics. The purpose of this chapter is to develop a simpler, lower-dimensional Clifford Algebra for investigating the four-dimensional space of mass-points.

Our goal is to make four principal contributions:

1. To show how to apply the standard Clifford Algebra for $\mathbb{R}^3$ to represent mass-points by taking advantage of the pseudoscalars to represent mass;
2. To present novel ways to understand the effect of sandwiching on points and vectors in three dimensions based on insights from the algebra and geometry of complex numbers and quaternions. Thus this chapter is a companion to [6];
3. To develop more intuitive proofs of the sandwiching formulas for rotations and reflections in three dimensions by studying simple rotations in four dimensions;
4. To demonstrate how to use sandwiching to compute perspective projections.

This chapter is organized in the following fashion. We begin in Sect. 16.2 with a brief review of the standard model for the Clifford Algebra of $\mathbb{R}^3$. Here we establish our sign conventions and review some basic formulas. In Sect. 16.3 we introduce the operators and operands—mass-points and quaternions—by exploiting the pseudoscalars to represent mass. Section 16.4 is devoted to studying the action of the unit quaternions on the space of mass-points, and Sect. 16.5 shows how to apply the sandwiching operators to compute rotation, reflection, and perspective projection on points as well as on vectors. Most of the results in Sects. 16.5.1 and 16.5.2 for rotation and reflection will already be known to readers familiar with Clifford Algebra, but our point of view is quite different from the standard approach of versors and rotors, and the material on perspective projection in Sect. 16.5.3 appears to be completely new. We close in Sect. 16.6 with a short summary our principal results along with a brief discussion of the main limitations of our model.

16.2   The Standard Model of the Clifford Algebra for Three Dimensions

The Clifford Algebra associated with the three-dimensional vector space $\mathbb{R}^3$ is an eight-dimensional vector space. We shall endow R^3 with the usual dot product and let e_1, e_2, e_3 be an orthonormal basis for $\mathbb{R}^3$. Then the eight canonical generators for the Clifford Algebra of $\mathbb{R}^3$ are denoted by the products:

1	scalars
e_1, e_2, e_3	vectors
e_1e_2, e_2e_3, e_3e_1	bivectors $\cdots$
$e_1e_2e_3$	pseudoscalars.

The formal algebra of this eight-dimensional vector space is defined by the following rules:
   i. multiplication is associative;
  ii. multiplication distributes through addition;
 iii. 1 is the identity for multiplication;
  iv.

$$e_1^2 = e_2^2 = e_3^2 = -1; \tag{16.1}$$

  v.

$$e_2e_1 = -e_1e_2, \qquad e_3e_2 = -e_2e_3, \qquad e_1e_3 = -e_3e_1. \tag{16.2}$$

The minus sign on the right-hand side of Eq. (16.1) is somewhat arbitrary. This sign is chosen here for convenience so that $(e_1e_2e_3)^2 = 1$. This minus sign also has the following ramifications.

Consider the Clifford product of two arbitrary vectors $u = u_1e_1 + u_2e_2 + u_3e_3$ and $v = v_1e_1 + v_2e_2 + v_3e_3$. Since multiplication distributes through addition, it follows from Eqs. (16.1) and (16.2) that

$$uv = (u_1e_1 + u_2e_2 + u_3e_3)(v_1e_1 + v_2e_2 + v_3e_3)$$
$$= -(u_1v_1 + u_2v_2 + u_3v_3)$$
$$+ (u_1v_2 - u_2v_1)e_1e_2 + (u_2v_3 - u_3v_2)e_2e_3 + (u_3v_1 - u_1v_2)e_3e_1.$$

Let

$$u \wedge v = (u_1v_2 - u_2v_1)e_1e_2 + (u_2v_3 - u_3v_2)e_2e_3 + (u_3v_1 - u_1v_3)e_3e_1.$$

Then

$$uv = -u \cdot v + u \wedge v. \tag{16.3}$$

Notice the minus sign, instead of a plus sign, adjacent to the dot product.

Duality is quite convenient with our choice of signs. Consider the pseudoscalar

$$O = -e_1e_2e_3. \tag{16.4}$$

By Eqs. (16.1) and (16.2),

$$Oe_1 = e_2e_3 = e_1O, \qquad Oe_2 = e_3e_1 = e_2O,$$
$$Oe_3 = e_1e_2 = e_3O, \tag{16.5}$$
$$Oe_1e_2 = e_3 = e_1e_2O, \qquad Oe_2e_3 = e_1 = e_2e_3O,$$
$$Oe_3e_1 = e_2 = e_3e_1O. \tag{16.6}$$

In particular,

$$O(u \wedge v) = u \times v = (u \wedge v)O, \tag{16.7}$$
$$O(u \times v) = u \wedge v = (u \times v)O. \tag{16.8}$$

Thus in this algebra, duality is mediated by multiplication with the pseudoscalar O.

The pseudoscalar O shares many properties with the scalar 1. For example, it follows easily from Eqs. (16.5) and (16.6) that O commutes with every element p of the Clifford Algebra, that is,

$$Op = pO. \tag{16.9}$$

Moreover, by Eqs. (16.1) an (16.2),

$$O^2 = 1. \tag{16.10}$$

16.3   Operands and Operators: Mass-Points and Quaternions

The Clifford Algebra of $\mathbb{R}^3$ is a real eight-dimensional vector space. This vector space splits conveniently into two four-dimensional subspaces: one consisting of the even-dimensional elements, which we shall see shortly are isomorphic to the quaternions, and one consisting of the odd-dimensional elements, which we shall identify with the mass-points.

16.3.1 Odd Order: Mass-Points

The ambient geometric space underlying contemporary Computer Graphics is the four-dimensional vector space of mass-points [5]: three of the dimensions are spatial; the fourth dimension is due to the mass. Let MP denote this four-dimensional space of mass-points. We are going to represent the space of mass-points MP as a four-dimensional subspace of the eight-dimensional Clifford Algebra for $\mathbb{R}^3$. Since the fourth dimension is mass-like rather than spatial, we should expect that our representation for the fourth dimension would be qualitatively somewhat different from our representation for the three spatial dimensions.

In the Clifford Algebra associated to $\mathbb{R}^3$, every vector v can be expressed uniquely in terms of our fixed orthonormal basis:

$$v = v_1 e_1 + v_2 e_2 + v_3 e_3.$$

But in the standard geometric interpretation of the Clifford Algebra for $\mathbb{R}^3$, there is no way to represent points, let alone mass-points. Therefore we shall now adopt a nonstandard interpretation.

We need to represent one more dimension, the dimension corresponding to mass. For this purpose, we will adopt the pseudoscalars. We shall represent the mass m by scalar multiples of the pseudoscalar

$$O = -e_1 e_2 e_3,$$

and we shall identify the pseudoscalar O with the point (or more precisely with the vector in four dimensions representing the point) at the origin of a three-dimensional coordinate system. Note that with this interpretation O represents a point, not a vector in three dimensions; the origin of the coordinate system is not the same as the zero vector (see Fig. 16.1).[1]

Classically the pseudoscalar $e_1 e_2 e_3$ is used to represent an element of unit volume. Here we are going to use $-e_1 e_2 e_3$ to represent a point. Denote by $Cl(R^3)$ the Clifford Algebra for R^3. Letting the pseudoscalar $-e_1 e_2 e_3$ represent a point is essentially equivalent to invoking a vector space isomorphism

---

[1]*Editorial note*: The reader may find the geometrical view of Gunn (in Chap. 15, this volume) enlightening: the basis vectors represent normal vectors of coordinate planes, and the point at the origin is then the trivector representing the intersection of those three coordinate planes.

$$T : Cl(R^3)^{\text{odd}} \to MP,$$

where

$$T(e_1) = e_1, \qquad T(e_2) = e_2, \qquad T(e_3) = e_3, \qquad T(-e_1 e_2 e_3) = O.$$

The minus sign is inserted in $-e_1 e_2 e_3$ to make the signs turn out right when we multiply the pseudoscalars by other elements of the Clifford Algebra. For example, by Eqs. (16.7) and (16.8),

$$O(u \wedge v) = u \times v = (u \wedge v)O,$$
$$O(u \times v) = u \wedge v = (u \times v)O.$$

We shall often abuse notation and use the symbol O to represent both the pseudoscalar $-e_1 e_2 e_3$ in the Clifford Algebra $Cl(R^3)$ and the origin of the coordinate system in MP. The correct interpretation will be clear from the context.

Now every mass-point p can be represented uniquely in the Clifford Algebra for R^3 as the sum of a vector v and the fixed point O times a mass m, that is,

$$p = mO + v$$

(see Fig. 16.1). This formula means that p has mass m and is located at the point $p/m = O + v/m$, provided that $m \neq 0$. We shall write

$$mO + v \equiv O + v/m$$

to indicate that the mass-point $mO + v$ is located at the affine point $O + v/m$.

To summarize: we shall use the odd-dimensional elements of the Clifford Algebra to represent mass-points. Elements of dimension one are vectors; elements of dimension three have mass. The sum of an element of dimension one and an element of dimension three is a mass-point.

An algebra is only an algebra. The formal rules of the Clifford Algebra are fixed. But the geometric interpretation we assign to elements of this algebra is completely up to us, constrained only by consistency and applicability.

16.3.2 Even Order: Quaternions

The algebra generated by the even-dimensional basis elements $\{1, e_1 e_2, e_2 e_3, e_3 e_1\}$ of the Clifford Algebra for R^3 is isomorphic to the algebra of quaternions. Indeed, let

$$i = e_2 e_3, \qquad j = e_3 e_1, \qquad k = e_1 e_2. \tag{16.11}$$

Then by Eqs. (16.1) and (16.2) it is easy to verify that

$$i^2 = j^2 = k^2 = -1, \tag{16.12}$$

$$ij = k, \qquad jk = i, \qquad ki = j, \tag{16.13}$$

which is the standard algebra of quaternion multiplication. We shall use the classical notation H to denote the quaternion algebra, represented here by the even-dimensional elements of the Clifford Algebra.

Let π represent multiplication in the Clifford Algebra $Cl(R^3)$. Then by Eqs. (16.1) and (16.2),

$$\pi : Cl(R)^{\text{even}} \oplus Cl(R^3)^{\text{odd}} \to Cl(R^3)^{\text{odd}}$$

or equivalently

$$\pi : H \oplus MP \to MP.$$

Thus we can think of the even-dimensional elements of the Clifford Algebra, the quaternions, as acting by multiplication on the odd-dimensional elements of the Clifford Algebra, the mass-points. Notice that H acts on MP on both the left and the right, but these actions are not identical since Clifford multiplication is not commutative. We are going to exploit this noncommutativity in Sect. 16.5 to develop sandwiching formulas for rotation, reflection, and perspective projection.

16.4   Decomposing Mass-Points into Two Complementary Planes

Relative to any fixed bivector b spanned by two direction vectors,[2] the four-dimensional space of mass-points can be decomposed into two complementary planes: the plane of vectors determined by b, and the two-dimensional subspace of the four-dimensional space of mass-points determined by the pseudoscalar O and the vector bO. To avoid confusion between the bivector b and the plane determined by the bivector b, we will denote by $b^{\|}$ the plane of vectors determined by the bivector b. Similarly, we will denote by $b^{\perp}$ the mass-points in the plane determined by O and bO. Note that by Eq. (16.7), the vector bO is perpendicular to the plane $b^{\|}$, because in three dimensions every bivector b can be written as $u \wedge v$ for some vectors u, v. Since O and bO are linearly independent (they have different grades), and both lie outside the plane $b^{\|}$, clearly $MP \cong b^{\|} \oplus b^{\perp}$. Here and elsewhere we use

---

[2]*Editorial note*: Since this chapter uses the algebra $\mathbb{R}_3$, bivectors from its 3-D space $\mathbb{R}^3$ are always also 2-blades, but the author prefers to use the term 'bivector'. In contrast, he describes the planes in the 4-D representational space consistently as 'planes', giving their spanning 4-D vectors but not representing them algebraically.

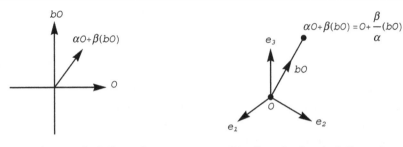

(a) a plane of vectors in 4-dimensions   (b) a line of points in 3-dimensions

Fig. 16.2 (a) The plane of vectors (mass-points) $b^{\perp}$ spanned by $\{O, bO\}$ in four dimensions is equivalent to (b) the line of affine points in three dimensions passing through the point O in the direction of the vector bO. This line is actually two-dimensional, but the second dimension is mass-like not spatial, so this dimension is not visible in (b)

the term *plane* to denote any two-dimensional subspace of the space of mass-points, rather than a physical plane in R^3.

Indeed, although $b^{\parallel}$ and $b^{\perp}$ both represent planes in four dimensions, the geometric interpretations of these two planes in terms of mass-points in three dimensions are markedly different: the plane $b^{\parallel}$ represents a plane of vectors in three dimensions, but in three dimensions, the plane $b^{\perp}$ represents a line of points.

Consider first the plane $b^{\parallel}$ spanned by the vectors $\{u, v\}$, where $b = u \wedge v$. Since u and v are linearly independent vectors in R^3, the vectors

$$w = \alpha u + \beta v$$

spanned by $\{u, v\}$ represent a two-dimensional plane of vectors in R^3.

In contrast, consider the plane $b^{\perp}$ spanned by $\{O, bO\}$. In three dimensions, O represents a point, not a vector. Thus the mass-points

$$P = cO + sbO$$

spanned by $\{O, bO\}$ actually represent a line in three dimensions: the line through the point O in the direction of the vector bO (see Fig. 16.2). In fact,

$$P \equiv O + \frac{s}{c}bO,$$

that is, P is a mass-point with mass c on the line $P(t) = O + tbO$. (If $c = 0$, then P is a vector sbO parallel to the line $P(t)$.) The plane $b^{\perp}$ spanned by $\{O, bO\}$ does have two dimensions, but only one dimension is spatial; the other dimension, the coefficient of O, represents mass, not length.

We are now going to study the geometric effect of multiplying an arbitrary mass-point p in MP by a unit quaternion

$$q(b, \theta) = \cos(\theta) + \sin(\theta)b$$

in H, where b is a unit bivector, that is, where b represents a planar segment with unit area. We shall proceed by investigating the geometric effects of this multiplication in the two complementary planes $b^{\parallel}$ and $b^{\perp}$.

16.4.1 Action of $q(b, \theta)$ on $b^{\parallel}$

Let v be a unit vector in the plane represented by the bivector b. Then the vectors v, bv are an orthonormal basis for $b^{\parallel}$. This observation can be proved in the following fashion.

Lemma 16.1 *Let b be a bivector, and let v be a vector in the plane $b^{\parallel}$. Then*
  i. $bv = bO \times v$,
 ii. $vb = v \times bO$.

Proof By Eqs. (16.10), (16.9), and (16.3),

$$bv = (bv)O^2 = \big((bO)v\big)O = \big(-(bO \cdot v) + (bO \wedge v)\big)O.$$

But

$$bO \cdot v = 0$$

because bO is orthogonal to the plane $b^{\parallel}$. By Eq. (16.7),

$$(bO \wedge v)O = bO \times v.$$

Therefore,

$$bv = bO \times v.$$

Similarly,

$$vb = v \times bO. \qquad \qquad \square$$

Corollary 16.1 *Let b be a bivector, and let v be a vector in the plane $b^{\parallel}$. Then*
  i. *bv is a vector in the plane $b^{\parallel}$.*
 ii. *$bv \perp v$.*
iii. *$|bv| = |b||v|$.*
 iv. *$vb = -bv$.*
Thus if $|b| = |v| = 1$, then v, bv is an orthonormal basis for $b^{\parallel}$.

Fig. 16.3 The plane
$b^\perp = \text{span}\{O, bO\}$ in the
space of mass-points MP
(*left*) and the plane
$b^\| = \text{span}\{v, bv\}$ in the space
of mass-points MP (*right*)

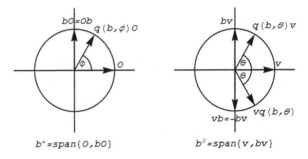

Proof These results follow immediately from Lemma 16.1.   □

Let v be a vector in $b^\|$. Using the orthonormal basis v, bv for $b^\|$, we can now analyze the effect of multiplication by $q(b, \theta)$ on vectors v in the plane $b^\|$.

Corollary 16.2 *The effect of multiplication by $q(b, \theta)$ on the left on vectors v in the plane $b^\|$ is just counterclockwise rotation through the angle θ in the plane $b^\|$. Moreover, since*

$$vb = -bv,$$

multiplication by $q(b, \theta)$ on the right on vectors in $b^\|$ rotates these vectors clockwise through the angle θ in the plane $b^\|$.

Proof It is enough to prove this result for unit vectors v in $b^\|$. Now

$$q(b, \theta)v = \big(\cos(\theta) + \sin(\theta)b\big)v = \cos(\theta)v + \sin(\theta)(bv).$$

Therefore, since by Corollary 16.1 v, bv is an orthonormal basis for $b^\|$, multiplication by $q(b, \theta)$ on the left rotates the vector v counterclockwise by the angle θ in the plane $b^\|$ (see Fig. 16.3, right). Similarly, since $vb = -bv$, multiplication by $q(b, \theta)$ on the right rotates the vector v clockwise by the angle θ in the plane $b^\|$.   □

16.4.2 Action of $q(b, \theta)$ on $b^\perp$

Multiplication by O is a vector space isomorphism from the subspace of the quaternions H spanned by $\{1, b\}$ to the subspace of the mass-points MP spanned by $\{O, bO\}$. Hence to understand how the unit quaternions $q(b, \theta)$ act on $b^\perp$, we first need to understand how multiplication works on the subspace of quaternions spanned by $\{1, b\}$. We begin with the following lemma.

Lemma 16.2 $b^2 = -|b|^2$.

Proof In three dimensions, every bivector b is a blade, that is, $b = u \wedge v$ for some choice of vectors u, v. Moreover, by definition,

$$|b| = |u \wedge v| = \text{area}(u, v) = |u \times v|.$$

Now by Eqs. (16.7), (16.9), and (16.10),

$$b^2 = b^2 O^2 = (bO)(bO) = -(bO) \cdot (bO) = -|u \times v|^2 = -|b|^2. \qquad \square$$

Here is the key observation. By Lemma 16.2,

$$|b| = 1 \Rightarrow b^2 = -1.$$

Therefore the quaternion plane spanned by $\{1, b\}$ is isomorphic to the complex plane.

Now we know how multiplication works in the complex plane:

$$e^{i\theta} \cdot e^{i\phi} = e^{i(\theta+\phi)} = e^{i\phi} \cdot e^{i\theta}.$$

Thus multiplication is commutative, and angles add. Therefore we have analogous results for multiplication in the plane $b^{\perp}$.

Lemma 16.3 *Let b be a unit bivector. Then*
i. $q(b, \theta)q(b, \phi) = q(b, \theta + \phi)$.
ii. $q(b, \theta)q(b, \phi) = q(b, \phi)q(b, \theta)$.

Proof To prove i, we simply apply the fact that $b^2 = -1$ and use the trigonometric identities for the sine and the cosine of the sum of two angles:

$$\begin{aligned}
q(b, \theta)q(b, \phi) &= \big(\cos(\theta) + \sin(\theta)b\big)\big(\cos(\phi) + \sin(\phi)b\big) \\
&= \big(\cos(\theta)\cos(\phi) - \sin(\theta)\sin(\phi)\big) \\
&\quad + \big(\sin(\theta)\cos(\phi) + \cos(\theta)\sin(\phi)\big)b \\
&= \cos(\theta + \phi) + \sin(\theta + \phi)b \\
&= q(b, \theta + \phi).
\end{aligned}$$

ii. Follows immediately from i. $\qquad \square$

A complex number $e^{i\theta}$ acts on vectors in the plane R^2 by rotating the vectors by the angle θ. Similarly, we would like to *interpret* the action of the quaternions $q(b, \theta)$ on vectors in the plane $b^{\perp}$ as rotation by the angle θ. But although we started with a metric on R^3, the plane $b^{\perp}$ is not yet endowed with a metric, since $O \in b^{\perp}$, but $O \notin R^3$. Thus to interpret multiplication by $q(b, \theta)$ as rotation in $b^{\perp}$, we need to extend our metric from R^3 to $MP \cong R^3 \oplus \text{span}\{O\}$. One natural way to extend the metric from R^3 to MP is to set $O \perp R^3$ and let $\|O\| = 1$ (see Fig. 16.1). Now in analogy with Corollary 16.2, we have the following result.

Corollary 16.3 *The effect of multiplication by $q(b, \theta)$ on the mass-points in the plane $b^\perp$ is just counterclockwise rotation through the angle θ in the plane $b^\perp$. Moreover, since*

$$Ob = bO,$$

multiplication by $q(b, \theta)$ on the left and the right has the same effect on the elements in $b^\perp$.

Proof It is enough to consider mass-points p represented by four-dimensional vectors of unit length in the plane $b^\perp = \text{span}\{O, bO\}$. Since O, bO is an orthonormal basis for $b^\perp$, there is an angle ϕ such that

$$p = \cos(\phi)O + \sin(\phi)(bO) = q(b, \phi)O$$

(see Fig. 16.3, left). Therefore, by Lemma 16.3,

$$\begin{aligned} q(b, \theta)p &= q(b, \theta)q(b, \phi)O = q(b, \theta + \phi)O \\ &= \cos(\theta + \phi)O + \sin(\theta + \phi)(bO). \end{aligned}$$

Thus the effect of multiplication by $q(b, \theta)$ on the mass-points in the plane $b^\perp = \text{span}\{O, bO\}$ is just counterclockwise rotation through the angle θ. □

16.4.3 Sandwiching

To facilitate our future discussions, we introduce the following notation. Let q be a quaternion, and let p be a mass-point. Then
- $L_q(p) = qp$ left multiplication by q,
- $R_q(p) = pq$ right multiplication by q,
- $T_q(p) = qpq = L_q R_q((p))$ sandwiching p between two copies of q,
- $S_q(p) = qpq^* = L_q(R_{q^*}(p))$ sandwiching p between q and q^*.

Here for each quaternion $q = c_1 + c_2e_1e_2 + c_3e_2e_3 + c_4e_3e_1$, the conjugate q^* is defined by $q^* = c_1 - c_2e_1e_2 - c_3e_2e_3 - c_4e_3e_1$. In particular, if $q(b, \theta) = \cos(\theta) + \sin(\theta)b$, then

$$q^*(b, \theta) = \cos(\theta) - \sin(\theta)b = q(b, -\theta). \tag{16.14}$$

The functions $L_q(p)$ and $R_q(p)$ are linear transformations on the vector space of mass-points because multiplication distributes through addition. The sandwiching operators $T_q(p)$ and $S_q(p)$ are also linear transformations because they are composites of linear transformations.

We now summarize the geometric effects of left and right multiplication by the unit quaternions

$$q(b, \theta) = \cos(\theta) + \sin(\theta)b$$

on vectors in the plane $b^{\parallel}$ and on mass-points in the complementary plane $b^{\perp}$.

Proposition 16.1 (Left Multiplication)

i. $L_{q(b,\theta)}(v) = q(b, \theta)v$ *rotates vectors* $v \in b^{\parallel}$ *by the angle* θ *in the plane* $b^{\parallel}$.

ii. $L_{q(b,\theta)}(p) = q(b, \theta)p$ *rotates mass-points* $p \in b^{\perp}$ *by the angle* θ *in the plane* $b^{\perp}$.

Proof These results follow immediately from Corollaries 16.2 and 16.3.   □

Proposition 16.2 (Right Multiplication)

i. $R_{q(b,\theta)}(v) = vq(b, \theta)$ *rotates vectors* $v \in b^{\parallel}$ *by the angle* $-\theta$ *in the plane* $b^{\parallel}$.

ii. $R_{q(b,\theta)}(p) = pq(b, \theta)$ *rotates mass-points* $p \in b^{\perp}$ *by the angle* θ *in the plane* $b^{\perp}$.

Proof These results also follow immediately from Corollaries 16.2 and 16.3.   □

By Propositions 16.1 and 16.2 both left and right multiplication by the unit quaternion $q(b, \theta)$ represent double isoclinic rotations in the four-dimensional space of mass-points, that is, there are two mutually orthogonal planes in four dimensions where vectors are rotated by the same angle θ [8]. For left multiplication, the rotations in both planes are counterclockwise, but for right multiplication, the rotation in one plane is counterclockwise, while the rotation in the other plane is clockwise; thus left and right multiplication by the unit quaternions $q(b, \theta)$ generate left and right screws in the four-dimensional space of mass-points [3]. To generate simple rotations—rotations in a single plane—in the four-dimensional space of mass-points, we need to use sandwiching.

Proposition 16.3 (Sandwiching by $S_{q(b,\theta)}$)

i. $S_{q(b,\theta)}(v) = q(b, \theta)vq^*(b, \theta)$ *rotates vectors* $v \in b^{\parallel}$ *by the angle* 2θ *in the plane* $b^{\parallel}$.

ii. $S_{q(b,\theta)}(p) = q(b, \theta)pq^*(b, \theta)$ *is the identity on mass-points* $p \in b^{\perp}$.

Proof These results follow immediately from Propositions 16.1 and 16.2 and Eq. (16.14).   □

Proposition 16.4 (Sandwiching by $T_{q(b,\theta)}$)

i. $T_{q(b,\theta)}(v) = q(b, \theta)vq(b, \theta)$ *is the identity on vectors* $v \in b^{\parallel}$.

ii. $T_q(b, \theta)(p) = q(b, \theta)pq(b, \theta)$ *rotates mass-points* $p \in b^{\perp}$ *by the angle* 2θ *in the plane* $b^{\perp}$.

Proof These results follow immediately from Propositions 16.1 and 16.2.   □

16.5 Rotation, Reflection, and Perspective Projection

Applying sandwiching to compute rotation and reflection on vectors is a well-known technique in Clifford Algebra. Since our model of Clifford Algebra includes points as well as vectors, here we shall also extend these standard results on rotation and reflection from vectors to points. Our proofs, however, are more intuitive than the standard proofs, since we shall take advantage of what we already know about the simple effects of the sandwiching maps $T_{q(b,\theta)}$ and $S_{q(b,\theta)}$ on the planes $b^{\parallel}$ and $b^{\perp}$. Using sandwiching to compute perspective projection seems to be new. We are able to employ sandwiching to perform perspective projection only because we have adopted a rather unconventional interpretation of the pseudoscalars and because in addition to the classical sandwiching maps $S_{q(b,\theta)}$, we also have available the sandwiching maps $T_{q(b,\theta)}$.

16.5.1 Rotation

Rotations in three dimensions are typically specified by an axis of rotation and an angle of rotation. But in R^3 planes are dual to vectors, so instead of specifying an axis of rotation, we can specify a plane of rotation, a plane perpendicular to the axis of rotation. Rotation mostly occurs in this rotation plane, since a vector is rotated around an axis of rotation by rotating its orthogonal projection in the plane of rotation. In Clifford Algebra, planes are represented by bivectors. Therefore here we shall specify a rotation by a bivector b and an angle θ. The plane of rotation is the plane $b^{\parallel}$, and the axis of rotation is the vector bO.

To rotate a vector v around the axis vector bO, we shall decompose v into two components $v_{\parallel}$ and $v_{\perp}$ such that

$$v = v_{\parallel} + v_{\perp},$$

where

$$v_{\parallel} = \text{component of } v \text{ in } b^{\parallel},$$
$$v_{\perp} = \text{component of } v \text{ perpendicular to } b^{\parallel}$$

(see Fig. 16.4(a)). Since $v_{\perp}$ is perpendicular to the plane of rotation $b^{\parallel}$, the vector $v_{\perp}$ is not altered by rotation. Hence, after rotation,

$$v_{\perp}^{\text{new}} = v_{\perp}.$$

Thus, to compute the effect of rotation on v, we need only compute the effect of rotation on $v_{\parallel}$. Now $v_{\parallel}$ lies in the plane $b^{\parallel}$; moreover, we showed in Sect. 16.4.1 that $bv_{\parallel}$ is perpendicular to $v_{\parallel}$, has the same length as $v_{\parallel}$, and

Fig. 16.4 (a) Decomposing a vector v into components parallel ($v_\parallel$) and perpendicular ($v_\perp$) to the plane of rotation (*left*); and (b) rotation in the plane of rotation (*right*)

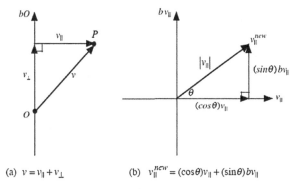

(a) $v = v_\parallel + v_\perp$ (b) $v_\parallel^{new} = (\cos\theta)v_\parallel + (\sin\theta)bv_\parallel$

also lies in the plane $b^\parallel$ (see Fig. 16.3, right). Hence, after rotation by the angle θ,

$$v_\parallel^{new} = (\cos\theta)v_\parallel + (\sin\theta)bv_\parallel$$

(see Fig. 16.4(b)). Therefore, by linearity, after rotation,

$$v^{new} = v_\perp^{new} + v_\parallel^{new} = v_\perp + (\cos\theta)v_\parallel + (\sin\theta)bv_\parallel. \qquad (16.15)$$

To rotate points, let O be a point on the axis of rotation. The choice of the origin is arbitrary, so we shall identify O with the origin of our coordinate system. Since $P = O + (P - O)$, by linearity, after rotation,

$$P^{new} = O^{new} + (P - O)^{new} = O + (P - O)^{new}. \qquad (16.16)$$

Now let us interpret these results in the four-dimensional space of mass-points. In the space of mass-points, the axis line is represented by the plane of $b^\perp$ spanned by $\{O, bO\}$, since every mass-point

$$p = cO + sbO \equiv O + \frac{s}{c}bO$$

in this plane in four dimensions lies on the line through the point O in the direction of the vector bO in three dimensions (see Sect. 16.4). The plane of vectors in three dimensions perpendicular to the axis vector bO is represented by the plane $b^\parallel$ in four dimensions.

To rotate points or vectors in three dimensions around the line through the point O parallel to the axis vector bO, we need to keep the plane $b^\perp$ fixed (since the point O and the axis vector bO are fixed by rotation) and to rotate vectors in the plane $b^\parallel$ by the angle θ (since these vectors are perpendicular to the axis vector in three dimensions). But by Proposition 16.3 these results are precisely the effects of the sandwiching map $S_{q(b,\theta/2)}$. Thus we are led to the following theorem.

Theorem 16.1 (Sandwiching with Conjugates Rotates Points and Vectors in 3-D)
Let
- $b = a$ *unit bivector,*
- $v = a$ *vector in three dimensions,*
- $P = a$ *point in three dimensions.*

Then
i. $S_{q(b,\theta/2)}(v) = q(b,\theta/2)vq^*(b,\theta/2)$ *rotates v by the angle θ around the axis bO.*
ii. $S_{q(b,\theta/2)}(P) = q(b,\theta/2)Pq^*(b,\theta/2)$ *rotates P by the angle θ around the line passing through the point O parallel to the vector bO.*

Proof To prove i, let $v_{\parallel}$ be the component of v in $b^{\parallel}$, and let $v_{\perp}$ be the component of v perpendicular to $b^{\parallel}$. Then, by Proposition 16.3,
- $S_{q(b,\theta/2)}(v_{\perp}) = q(b,\theta/2)v_{\perp}q^*(b,\theta/2) = v_{\perp}$,
- $S_{q(b\theta/2)}(v_{\parallel}) = q(b,\theta/2)v_{\parallel}q^*(b,\theta/2) = \cos(\theta)v_{\parallel} + \sin(\theta)bv_{\parallel}$ rotates $v_{\parallel}$ by the angle θ in the plane $b^{\parallel}$.

Therefore, since $S_{q(b,\theta/2)}$ is a linear transformation and $v = v_{\perp} + v_{\parallel}$,

$$S_{q(b,\theta/2)}(v) = S_{q(b,\theta/2)}(v_{\perp}) + S_{q(b,\theta/2)}(v_{\parallel}) = v_{\perp} + \cos(\theta)v_{\parallel} + \sin(\theta)bv_{\parallel}.$$

Hence by Eq. (16.15) sandwiching has the same effect on v as rotating v in three dimensions by the angle θ around the axis vector bO.

To prove ii, observe that since P is a point in affine space,

$$P = O + (P - O).$$

But
- $S_{q(b,\theta/2)}(O) = O$ (Proposition 16.3),
- $S_{q(b\theta/2)}(P - O)$ rotates the vector $P - O$ by the angle θ around the axis bO (part i).

Therefore, since $S_{q(b,\theta/2)}$ is a linear transformation and $P = O + (P - O)$,

$$S_{q(b,\theta/2)}(P) = O + S_{q(b,\theta/2)}(P - O).$$

Hence, by part i and Eq. (16.16), sandwiching has the same effect on P as rotating P in three dimensions by the angle θ around the line passing through the point O parallel to the vector bO. $\qquad\square$

16.5.2 Mirror Image

Suppose that we want to mirror a vector v in a plane specified by a bivector b. To compute the effect of this reflection on the vector v, we once again decompose v into two components $v_{\parallel}$ and $v_{\perp}$ where
- $v_{\parallel} = $ component of v parallel to $b^{\parallel}$,
- $v_{\perp} = $ component of v perpendicular to $b^{\parallel}$.

Since $v_{\parallel}$ lies in the mirror plane, the vector $v_{\parallel}$ is not altered by reflection. Hence, after reflection,

$$v_{\parallel}^{\text{new}} = v_{\parallel}.$$

Thus to compute the effect of reflection on v, we need only compute the effect of reflection on $v_{\perp}$. But reflection simply reverses the direction of $v_{\perp}$, so

$$v_{\perp}^{\text{new}} = -v_{\perp}.$$

Therefore, by linearity, after reflection,

$$v^{\text{new}} = v_{\parallel} - v_{\perp}. \tag{16.17}$$

To find the mirror image of points in a plane, let O be a point in the mirror plane. Again since the choice of the origin is arbitrary, we shall identify O with the origin of our coordinate system. Since $P = O + (P - O)$, by linearity, after reflection,

$$P^{\text{new}} = O^{\text{new}} + (P - O)^{\text{new}} = O + (P - O)^{\text{new}}. \tag{16.18}$$

To reflect vectors in three dimensions in the plane $b^{\parallel}$, we need to keep vectors in the plane $b^{\parallel}$ fixed and negates vectors perpendicular to the plane $b^{\parallel}$. But notice that by Proposition 16.4 the maps $T_{q(b,\theta)}$ are the identity on vectors in the plane $b^{\parallel}$. Therefore, to compute the mirror image of vectors in the plane $b^{\parallel}$, we need only find an angle θ for which $T_{q(b,\theta)}$ maps the normal vector bO to $-bO$ (or equivalently rotates bO by the angle π). Since by Proposition 16.4 the map $T_{q(b,\theta)}$ rotates mass-points in the plane $b^{\perp} = \text{span}\{O, bO\}$ by the angle 2θ, the angle we seek is $\theta = \pi/2$. But

$$q(b, \pi/2) = \cos(\pi/2) + \sin(\pi/2)b = b.$$

Therefore we are led directly to the following result.

Theorem 16.2 (Sandwiching with Unit Bivectors Reflects Vectors in 3-D) *Let*
- *$b = $ a unit bivector,*
- *$v = $ a vector in three dimensions.*

Then
- *$T_b(v) = bvb = -S_b(v)$ is the mirror image of v in the plane $b^{\parallel}$.*

Proof Let $v_{\parallel}$ be the component of v in $b^{\parallel}$, and let $v_{\perp}$ be the component of v perpendicular to $b^{\parallel}$. Since $b = \cos(\pi/2) + \sin(\pi/2)b = q(b, \pi/2)$, it follows by Proposition 16.4 that
- $T_b(v_{\perp}) = bv_{\perp}b = q(b, \pi/2)v_{\perp}q(b, \pi/2) = T_{q(b,\pi/2)}(v_{\perp}) = -v_{\perp}$,
- $T_b(v_{\parallel}) = bv_{\parallel}b = q(b, \pi/2)v_{\parallel}q(b, \pi/2) = T_{q(b,\pi/2)}(v_{\parallel}) = v_{\parallel}$.

Therefore, since T_b is a linear transformation and $v = v_{\parallel} + v_{\perp}$,

$$T_b(v) = T_b(v_{\parallel}) + T_b(v_{\perp}) = v_{\parallel} - v_{\perp}.$$

Hence by Eq. (16.17) sandwiching with b has the same effect on v as reflecting v in three dimensions in the plane $b^{\|}$. $\square$

Since by Theorem 16.1 the sandwiching maps $S_{q(b,\theta/2)}$ can be used to rotate points as well as vectors in three dimensions around lines through the origin O, Theorem 16.2 seems to invite us to use the sandwiching maps T_b to compute the mirror image of points in planes $b^{\|}$ passing through the origin O. But recall that unlike the transformations $S_{q(b,\theta/2)}$, the maps T_b are not the identity on the plane $b^{\perp}$ in four dimensions spanned by $\{O, bO\}$. Indeed, even though the map T_b is the identity on the vectors v in the plane $b^{\|}$, we find that

$$T_b(O) = bOb = -O \neq O.$$

Therefore $T_b(P)$ is not the identity on the affine plane in three dimensions through the point O is perpendicular to the vector bO. Thus, perhaps contrary to intuition, sandwiching a point P with the unit bivector b is not the mirror image in three dimensions of the point P in the plane $b^{\|}$ passing through the origin O. Rather by linearity,

$$T_b(P) = T_b\big(O + (P - O)\big) = T_b(O) + T_b(P - O) = -O + b(P - O)b,$$

so

$$T_b(P) \equiv O - b(P - O)b = O + b(P - O)b^* = S_{q(b,\pi/2)}(P).$$

Hence, by Theorem 16.2, sandwiching a point P between two copies of a bivector b rotates the point P through the angle π around the line passing through the point O parallel to the vector bO.

Thus sandwiching with b does not reflect points P in the plane $b^{\|}$ passing through the origin O. Nevertheless we can compute the mirror image of points in this plane by sandwiching using the following approach.

Theorem 16.3 (Sandwiching $P - 2O$ with the Bivector b Reflects P in the Plane $b^{\|}$ Passing Through the Point O) *Let*
- $b = $ *a unit bivector,*
- $P = $ *a point in three dimensions.*
Then
- $T_b(P - 2O) = b(P - 2O)b$ *is the mirror image of the point P in the plane $b^{\|}$ passing through the point O.*

Proof Clearly,

$$P - 2O = (P - O) - O.$$

Now

- $T_b(-O) = -bOb = -b^2O = O,$
- $T_b(P - O) = b(P - O)b.$

Therefore, since T_b is a linear transformation and $P - 2O = (P - O) - O,$

$$T_b(P - 2O) = O + b(P - O)b,$$

which by Theorem 16.3 and Eq. (16.18) is the mirror image of the point P in the plane $b^{\|}$ passing through the point O.   □

If we introduce rectangular coordinates, then $P = (p_1, p_2, p_3, 1)$ and $O = (0, 0, 0, 1)$, so

$$P - 2O = (p_1, p_2, p_3, -1).$$

Thus, by Theorem 16.3, to find the mirror image of a point P in the plane $b^{\|}$ passing through the point O, we simply negate the fourth coordinate, the mass, of P, and sandwich the resulting mass-point between two copies of b.

Notice that in Theorems 16.2 and 16.3, to compute reflections, we sandwich points and vectors between two copies of the bivector b rather than, as traditional in most presentations of Clifford Algebra, between two copies of the normal vector bO. That is, we invoke simple rotations in four dimensions represented by bivectors instead of reflections in three dimensions represented by versors. Of course, these results are equivalent because by Eqs. (16.9) and (16.10)

$$bvb = (bO)v(bO).$$

Nevertheless, our focus is on simple rotations in four dimensions as the primary operators, whereas the traditional approach emphasizes reflections (versors) in three dimensions as the primary operators. In our approach, reflections in three dimensions are special rotations in four dimensions; in the traditional approach, rotations in three dimensions (rotors) are products of reflections (versors) in three dimensions.

Next we shall display the full power of our approach by using our understanding of simple rotations in four dimensions to represent perspective projections in three dimensions.

16.5.3 Perspective Projection

So far we have investigated the maps $T_{q(b,\theta)}(v)$ that sandwich the vector v between two copies of the unit quaternion $q(b, \theta)$ only when $q(b, \theta) = b$, that is, only for $\theta = \pi/2$. By Theorem 16.2, $T_b(v)$ is the mirror image of the vector v in the plane $b^{\|}$. We are now going to study the maps $T_{q(b,\theta)}$ for $\theta \neq \pi/2$.

By Proposition 16.4, $T_{q(b,\theta)}$ is the identity on the plane $b^{\|}$. Therefore we should expect that the map $T_{q(b,\theta)}$ represents some kind of projection into the plane $b^{\|}$. We shall now show that the maps $T_{q(b,\theta)}$ can be used to compute perspective projections in three dimensions onto a plane parallel to $b^\parallel$. The angle θ parameterizes the distance $d = |\csc(\theta)| \geq 1$ from the eye point to the plane of projection. When $\theta = \pi$, the eye recedes to infinity, and the map $T_{q(b,-\theta/2)}(v) = T_{-b}(v)$ computes the mirror image of the vector v in the plane $b^\parallel$ (see Sect. 16.5.2).

The following proposition is stated and proved only for a special position of the eye point and the plane of projection. In Theorem 16.4 we shall generalize this result to arbitrary positions of the eye point and the plane of projection.

Proposition 16.5 (Sandwiching Vectors to the Eye with $q(b, -\theta)$ Gives Perspective) *Suppose that $0 < \theta < \pi$, and let*
- $b = a$ *unit bivector,*
- $E(b, \theta) = O + (\cot(\theta) - \csc(\theta))bO = $ *eye point,*
- $P = a$ *point in three dimensions.*
Then
- $T_{q(b,-\theta/2)}(P - E) = q(b, -\theta/2)(P - E)q(b, -\theta/2)$ *is a mass-point, where:*
  - *the point is located at the perspective projection of the point P from the eye point $E(b, \theta)$ onto the plane $b^\parallel(\theta)$ passing through the point $O + \cot(\theta)(bO) \equiv T_{q(b,-\theta)}(bO)$;*
  - *the mass is equal to $d \sin(\theta)$, where d is the distance of the point P from the plane through the eye point $E(b, \theta)$ perpendicular to the vector bO.*

Proof Let $P - E = d(bO) + v$, where d is a scalar and $v \perp bO$ (see Fig. 16.5). Since by Proposition 16.4 the map $T_{q(b,-\theta/2)}$ rotates mass-points in the plane $b^\perp = \mathrm{span}\{O, bO\}$ by the angle $-\theta$:

$$T_{q(b,-\theta/2)}\big(d(bO)\big) = d\cos(\pi/2 - \theta)O + d\sin(\pi/2 - \theta)(bO)$$
$$= d\sin(\theta)O + d\cos(\theta)(bO).$$

Moreover, again by Proposition 16.4, the map $T_{q(b,-\theta/2)}$ is the identity on vectors in the plane $b^\parallel$, so

$$T_{q(b,-\theta/2)}(v) = v.$$

Therefore, by linearity,

$$T_{q(b,-\theta/2)}(P - E) = T_{q(b,-\theta/2)}\big(d(bO) + v\big)$$
$$= d\sin(\theta)O + d\cos(\theta)(bO) + v$$
$$\equiv O + \cot(\theta)(bO) + \csc(\theta)\frac{v}{d}.$$

Since by construction

$$O + \cot(\theta)(bO) = E(b, \theta) + \csc(\theta)(bO),$$

Fig. 16.5 By similar triangles, the point $P^{\text{new}} = O + \cot(\theta)(bO) + \csc(\theta)\frac{v}{d} = E + \csc(\theta)(bO) + \csc(\theta)\frac{v}{d}$ is the perspective projection of the point P from the eye point E onto the plane $b^{\parallel}$ passing through the point $Q = O + \cot(\theta)(bO)$

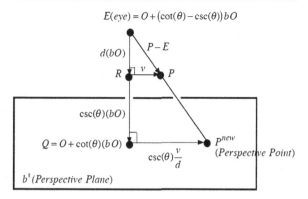

it follows by similar triangles (see Fig. 16.5) that the point corresponding to the mass-point $T_{q(b,-\theta/2)}(P - E)$ is the perspective projection of the point P from the eye point $E(b, \theta)$ onto the plane $b^{\parallel}(\theta)$ passing through the point $O + \cot(\theta)bO$. Moreover, the mass is $d\sin(\theta)$, where d is the distance of the point P from the plane through the eye point $E(b, \theta)$ perpendicular to the vector bO.   □

Notice that since the mass of the mass-point that we compute with sandwiching is a constant $(\sin(\theta))$ times the distance d from the point P to the plane through the eye E parallel to the plane of projection, we can use this sandwiching formula for perspective projection to detect hidden surfaces: if two points project to the same point, then the smaller the mass, the closer the surface point is to the eye. Thus even though we are projecting a point onto a plane in three dimensions, no information is actually lost: distance is simply converted into mass. Since sandwiching represents a simple rotation in four dimensions, this general sandwiching formula for perspective projection allows us to use 4×4 orthogonal matrices to compute perspective projection; for details, see [7].

Proposition 16.5 is stated and proved only for a special configuration of the eye point $E(b, \theta)$ and the plane of projection $b^{\parallel}(\theta)$. We shall call this position of the eye point E and the plane of projection $b^{\parallel}$ the *canonical position* for the angle θ and the unit bivector b. Now the rather astonishing observation is that a simple variant of Proposition 16.5 remains valid even if the eye point E and the plane of projection $b^{\parallel}$ are in arbitrary rather than in canonical position.

Projecting a scene from an eye point E into a plane S and then translating the projection by a vector v into the plane $S + v$ is equivalent to first translating the eye and the entire scene by the vector v and then projecting from the translated eye point $E + v$ into the translated plane $S + v$. Thus translation and projection commute. Therefore we have the following generalization of Proposition 16.5.

Theorem 16.4 (Sandwiching Vectors to the Eye with $q(b, -\theta/2)$ Gives Perspective) *Suppose that $0 < \theta < \pi$, and let*
- *E = eye point,*
- *S = plane of projection parallel to a bivector b,*

- $d = \csc(\theta) = $ *distance from the eye point to the plane of projection along the normal* bO,
- $P = $ *a point in three dimensions.*

Then

- $T_{q(b,-\theta/2)}(P - E) = q(b, -\theta/2)(P - E)q(b, -\theta/2)$ *is a mass-point, where*:
  - *the point is located at the perspective projection of the point P from the eye point E onto the plane S translated by the vector* $E(b, \theta) - E$ *to the canonical plane* $b^\parallel(\theta)$;
  - *the mass is equal to* $d \sin(\theta)$, *where d is the distance of the point P from the plane through the eye point E perpendicular to the unit normal* bO.

Proof This result follows immediately from Proposition 16.5 and the observation that translation and projection commute. □

By Theorem 16.4, if we display the points generated by the sandwiching transformation

$$T_{q(b,-\theta/2)}(P - E) = q(b, -\theta/2)(P - E)q(b, -\theta/2),$$

the scene will appear in perspective. But the scene materializes in the canonical plane rather than in the specified plane of projection. Thus for fixed values of the unit bivector b and the scalar distance $d = \csc(\theta)$, the scene always appears in the identical plane independent of the absolute position of the eye point E and the plane of projection S. Only the relative positions of E and S matter, not their absolute locations relative to a fixed coordinate system. Typically we are interested only in viewing the scene in perspective; translating the plane of perspective does not matter. Thus we can hard code the location of the viewing plane $b^\parallel(\theta)$, depending only on the values of b and θ and not worry about the absolute locations of the eye point E and the plane of projection S.[3]

16.6 Summary

In the standard approach to the Clifford Algebra for three dimensions the basic operations are reflections: every rotation (rotor) in three dimensions is the product of two reflections (versors) in three dimensions. In our homogeneous model of Clifford Algebra, rotations in three dimensions still factor into pairs of reflections in three dimensions. Nevertheless, in our homogeneous model, the basic operations are not reflections in three dimensions, but rather simple rotations in four dimensions.

---

[3]*Editorial note*: Note that this chapter gives a geometric algebra description of a perspective projection onto a plane. For a geometric algebra representation of a general projective transformation in 3-D, the reader is referred to Chap. 13 in this volume, which uses $\mathbb{R}^{3,3}$.

For any bivector b, rotations in four dimensions that leave the plane $b^\perp$ fixed correspond to rotations in three dimensions in the plane $b^\|$; rotations in four dimensions that leave the plane $b^\|$ fixed correspond to either reflections or perspective projections in three dimensions. Thus in our homogeneous model of Clifford Algebra rather than think of reflections as the basic operations, it is more natural to think of reflections in three dimensions as special rotations in four dimensions. The main strength of our technique is that this approach allows us to model perspective projections as basic operations in the Clifford Algebra.

The main weakness of our method is that our approach does not model translations. In the conformal model, every translation is the product of two reflections in parallel planes. But in our model there is no way to represent translation. Indeed, our model is not translation invariant; the pseudoscalar $O = -e_1e_2e_3$ is rotation invariant, but O is not translation invariant. In effect, we have traded the capacity to perform translations for the ability to perform perspective projections.

16.7   Exercises

16.1 Let L be an axis of rotation through the origin O parallel to the unit vector u. Using Theorem 16.1 from Sect. 16.5.1, derive the following *Formula of Rodrigues* for rotating a vector v or a point P around the axis line L through the angle θ:

$$v^{new} = (\cos\theta)v + (1 - \cos\theta)(v \cdot u)u + (\sin\theta)u \times v,$$
$$P^{new} = O + (\cos\theta)(P - O) + (1 - \cos\theta)\big((P - O) \cdot u\big)u$$
$$+ (\sin\theta)u \times (P - O).$$

[Hint: Let b be the bivector representing the plane of rotation through the origin O, perpendicular to the axis line L. Then $u = bO$.]

16.2 Let M be a mirror plane through the origin O perpendicular to the unit vector u. Using Theorem 16.2 from Sect. 16.5.2, show that the mirror image of a vector v or a point P in the plane M is given by:

$$v^{new} = v - 2(v \cdot u)u,$$
$$P^{new} = O + 2\big((P - O) \cdot u\big)u.$$

[Hint: Let b be the bivector representing the mirror plane through the origin O, perpendicular to the unit vector u. Then $u = bO$.]

16.3 Suppose that the eye is located along the z-axis at the point $E = (0, 0, 1)$, and the perspective plane S is the xy-plane. Using Proposition 16.5 from Sect. 16.5.3, show that in this case perspective projection maps the point $P = (x, y, z)$ to the point $P^{new} = (\frac{x}{1-z}, \frac{y}{1-z}, 0)$.

Acknowledgements   I would like to thank Leo Dorst and Steve Mann for reading a preliminary draft of this manuscript and providing valuable comments, criticisms, and suggestions. I would also like to thank the anonymous referees for their constructive criticisms. This work is much improved as a result of the observations of these people. Any mistakes that still remain are, of course, entirely my own.

References

1. Doran, C., Lasenby, A.: Geometric Algebra for Physicists. Cambridge University Press, Cambridge, UK (2003)
2. Dorst, L., Fontijne, D., Mann, S.: Geometric Algebra for Computer Science: An Object-Oriented Approach to Geometry. Morgan Kaufmann, Amsterdam (2007)
3. Du Val, P.: Homographies, Quaternions and Rotations. Oxford Mathematical Monographs. Clarendon, Oxford (1964)
4. Foley, J., van Dam, A., Feiner, S., Hughes, J.: Computer Graphic: Principles and Practice, 2nd edn. Addison Wesley, Reading (1990)
5. Goldman, R.: On the algebraic and geometric foundations of computer graphics. Trans. Graph. **21**, 1–35 (2002)
6. Goldman, R.N.: Understanding quaternions. Graph. Models **73**, 21–49 (2011)
7. Goldman, R.N.: Modeling perspective projections in 3-dimensions by rotations in 4-dimensions. Trans. Vis. Comput. Graph. (2010, to appear)
8. Mebius, J.E.: A matrix based proof of the quaternion representation theorem for four-dimensional rotations. http://arXiv:math/0501249v1 [math.GM] (2005)
9. Perwass, C.: Geometric Algebra with Applications in Engineering. Springer, Berlin (2009)

Rigid-Body Transforms Using Symbolic Infinitesimals

17

Glen Mullineux and Leon Simpson

Abstract

There is a requirement to be able to represent three-dimensional objects and their transforms in many applications, including computer graphics and mechanism and machine design. A geometric algebra is constructed which can model three-dimensional geometry and rigid-body transforms. The representation is exact since the square of one of the basis vectors is treated symbolically as being infinite. The non-zero, even-grade elements of the algebra represent precisely all rigid-body transforms. By allowing the transform to vary, smooth motions are obtained. This can be achieved using Bézier and B-spline combinations of even-grade elements.

17.1 Introduction

There are a number of applications, including computer graphics, machine design and robotics, where there is a need to represent physical objects and to be able to manipulate these. The objects lie in three-dimensional Euclidean space, and manipulation is often done in terms of transforms of the space. As the objects do not distort, these transformations are combinations of rotations and translations. By varying the displacements correctly, smooth spatial motions of objects can be obtained.

A number of techniques are available [3]. One well-established approach is to represent geometry in terms of vectors, using homogeneous coordinates. This allows transformations of 3D Euclidean space to be created using 4×4 matrices, which are used to multiply the vectors.

G. Mullineux (✉) · L. Simpson
Innovative Design and Manufacturing Research Centre, Department of Mechanical Engineering, University of Bath, Bath BA2 7AY, UK
e-mail: g.mullineux@bath.ac.uk

L. Simpson
e-mail: l.c.simpson@bath.ac.uk

L. Dorst, J. Lasenby (eds.), *Guide to Geometric Algebra in Practice*,
DOI 10.1007/978-0-85729-811-9_17, © Springer-Verlag London Limited 2011

If objects are required to move, this can be achieved by applying a varying transform [1, 10]. In particular, B-spline techniques can be applied to create matrix functions to generate motions [10].

The requirement of speed and stability for some computer-generated motions, particularly in the games industry, has seen a revival of interest in alternative representations. Quaternions have mathematical properties which allow them to handle smooth rotations [11, 17, 23]. In particular, they are commonly used to model mechanical systems involving revolute joints [4, 9]. They are closely related to screw displacements and Plücker coordinates [5] which have additionally been used to generate motions by extending Bézier and related techniques for smooth curves [6, 8, 16].

However, quaternions do not easily represent translations and ways to extend them have been investigated. Dual quaternions [20] can be created by introducing an additional element whose square is zero.

Quaternions lie naturally within (suitably formulated) geometric algebras. In such an algebra [2], double quaternions can be defined and translations handled as rotations dependent upon a large radius R, chosen to be sufficiently large that the error in the approximation is less than a prescribed tolerance value. An alternative approach [18] is to define the square of one of the basis vectors of the algebra to be zero (which provides the geometric algebra $\mathbb{R}_{3,0,1}$). This gives a representation of three-dimensional space together with transforms but has the disadvantage that vectors in the algebra correspond to planes (rather than points).[1]

In this chapter, interest is in obtaining a more natural representation of three-dimensional Euclidean space and its rigid-body transforms by use of a geometric algebra. To this end, an algebra, $\mathscr{G}_4$, is created with four basis vectors so that a representation of the projective space $\mathbb{RP}^3$ is obtained. It is effectively this space which is used when describing transforms in terms of 4×4 matrices.

Section 17.2 discusses the construction of $\mathscr{G}_4$. One of its basis vectors is chosen to have a square which is infinite (so that $\mathscr{G}_4$ behaves as a dual of $\mathbb{R}_{3,0,1}$), and this is achieved by defining it to be the reciprocal of a small quantity ε. This is treated symbolically through required calculations, so that the scalars involved are essentially power series in ε.

Section 17.3 discusses the representation of geometry and transforms, and shows that the non-zero even-grade elements generate isometries of three-dimensional space. These are seen to be precisely combinations of rotations and translations in Sect. 17.4, which also discusses the relation to Chasles's theorem.

Since a Bézier or B-spline combination of even-grade elements is again of even grade, this opens the possibility of creating continuously varying transforms and hence smooth motions of objects [13]. This is presented in Sect. 17.5, which also gives properties of first degree Bézier motions which form the basic step in the de Casteljau algorithm.

---

[1] *Editorial note*: However, this representation is explored in detail in Chap. 15 (this volume).

17.2 Geometric Algebra $\mathcal{G}_4$

This section considers the construction of the geometric algebra, here called $\mathcal{G}_4$, which is used to model the projective space $\mathbb{RP}^3$. The approach used extends to other dimensions, but, for simplicity, it is here restricted to representing the space which relates to three-dimensional Euclidean geometry.[2]

The starting point for the construction is a real vector space of dimension four. Suppose that a basis consists of the vectors e_0, e_1, e_2, e_3. The space is extended to allow a multiplication of the basis vectors to be defined. The new basis is defined comprising the 16 vectors e_σ, where σ is a subset of the set $\{0, 1, 2, 3\}$. Strictly speaking, σ is an ordered subset, but, if σ and τ are two such subsets with the same members, then e_σ and e_τ are taken as being identical if τ is an even permutation of σ, and $e_\tau = -e_\sigma$ if it is an odd permutation. The multiplication is then defined by saying that e_σ is the product of the e_i with $i \in \sigma$, with the ordering implied by σ preserved. Thus, for example,

$$e_{123} = e_1 e_2 e_3.$$

Note that the multiplication is not generally commutative. In particular, for original basis vectors e_i and e_j with $i \neq j$,

$$e_j e_i = e_{ji} = -e_{ij} = -e_i e_j.$$

One of the new basis elements is $e_\emptyset$ where $\emptyset$ is the empty set. The multiplication defines $e_\emptyset e_\sigma$ to be e_σ. Thus $e_\emptyset$ acts as the identity, and it is regarded as being the same as the real number 1.

The typical member a of the geometric algebra $\mathcal{G}_4$ is a linear combination,

$$a = \sum_\sigma a_\sigma e_\sigma, \qquad (17.1)$$

of the basis elements where the a_σ are scalars. Addition of such elements is carried out in the obvious way, and their multiplication is the natural extension from the definition for the basis elements.

One thing remains to be considered and that is the squares of the original basis vectors. For three of these, their squares are taken to be unity:

$$e_1^2 = e_2^2 = e_3^2 = 1.$$

The case of e_0 is special. Taking $e_0^2 = 1$ means that the algebra represents four-dimensional Euclidean space. An alternative, which yields a projective space, is to take e_0^2 to be zero [18]. However, in this representation, the original basis vectors

---

[2]*Editorial note*: In its basic definitions of geometric algebra, this chapter repeats some of the elementary constructions given in given in the tutorial (Chap. 21) in this volume. Since the anomalous element e_0 changes some of the details crucially, we kept this re-explanation.

e_1, e_2, e_3 correspond to planes rather than points which seems unnatural. (Additionally, in the presentation in [18], a form of the "Hodge star" operator needs to be introduced when some geometric objects are combined; the corresponding manipulation in $\mathcal{G}_4$ can be done entirely using the defined multiplication without the need for new operators.) So the approach used here is to try to take e_0^2 to be infinite. This is achieved by setting

$$e_0^2 = \varepsilon^{-1},$$

where ε is a (vanishingly) small real quantity.[3]

This means that ε regularly appears in manipulations of expressions and it needs to be carried through these. Once a calculation is complete, higher powers of ε can be disregarded, and this is often equivalent to setting ε equal to zero. An alternative view is to regard the scalars a_σ in the typical element, Eq. (17.1), as being a power series in ε of the form

$$a_\sigma = \sum_{i=m}^{\infty} \alpha_i \varepsilon^i, \tag{17.2}$$

where the α_i are real, and m is a finite (possibly negative) integer.

Elements whose squares are very small or large have been used elsewhere to describe motions. For example, dual quaternions [17, 20] have an element whose square is zero, and double quaternions [2, 17] approximate translations by rotations acting over a large radius. This raises the question of how large is "large". The implementation used here overcomes this by treating ε as a symbol. The coefficients a_σ are regarded as formal power series and are implemented computationally as arrays of the real numbers α_i in Eq. (17.2). Addition, subtraction and multiplication of such arrays is straightforward on a term-by-term basis. Division is achieved by use of the power series expansion of $(1 + x)^{-1}$. Entries corresponding to high powers of ε are allowed to "drop off" the end of the array as the corresponding terms in the series are small.

The definition of multiplication in $\mathcal{G}_4$ is now complete and the following are some examples:

$$(1 - e_{12})(e_0 + e_1)(1 + e_{12})$$
$$= (1 - e_{12})(e_0 + e_0 e_{12} + e_1 + e_1 e_{12})$$
$$= e_0 + e_0 e_{12} + e_1 + e_1 e_{12} - e_{12} e_0 - e_{12} e_0 e_{12} - e_{12} e_1 - e_{12} e_1 e_{12}$$
$$= e_0 + e_{012} + e_1 + e_2 - e_{012} + e_0 + e_2 - e_1$$
$$= 2(e_0 + e_2),$$

---

[3]*Editorial note*: This somewhat unusual construction may find its motivation in a limiting procedure from curved spaces to flat Euclidean space, see Chap. 18.

$$(e_0 - e_1 + e_{23})(1 + \varepsilon e_{01} + e_{123})$$
$$= e_0 + \varepsilon e_0^2 e_1 + e_0 e_{123} - e_1 - \varepsilon e_1 e_{01} - e_1 e_{123} + e_{23} + \varepsilon e_{23} e_{01} + e_{23} e_{123}$$
$$= e_0 + e_1 + e_{0123} - e_1 + \varepsilon e_0 - e_{23} + e_{23} + \varepsilon e_{0123} - e_1$$
$$= (1 + \varepsilon) e_0 - e_1 + (1 + \varepsilon) e_{0123}$$
$$\simeq e_0 - e_1 + e_{0123}.$$

Here, the symbol $\simeq$ is used to indicate that there is equality of the expressions in the limit as ε tends to zero; the expressions on either side of the symbol differ by terms involving (positive) powers of ε, and these have been ignored since they are small.

The *grade* of the basis element e_σ is the size of the subset σ. If a general element of $\mathcal{G}_4$ is a combination only of basis elements of a single grade, then that is also the grade of the element. Elements of grade 1 are called *vectors*, elements of grade 2 are *bivectors*, and those of grade 3 are *trivectors*. Note that the parity of the grade behaves naturally under multiplication. So, for example, the product of two elements of odd grade has even grade.

The only elements in $\mathcal{G}_4$ of grade 4 are scalar multiples of the basis element e_{0123}. This basis vector is called *the pseudoscalar*, and, for convenience, it is denoted by I. The set of elements of the form $\alpha + \beta I$, where α and β are scalars, is closed under addition and multiplication. An element of this form is called *a pseudoscalar*.[4]

The *reverse* of the basis element e_σ is the element obtained by reversing the order of the entries in the subset σ. The effect is to leave the element unchanged if its grade is 0, 1 or 4, and to change its sign if the grade is 2 or 3.

More generally, the reverse of the typical element a of $\mathcal{G}_4$, Eq. (17.1), is obtained by reversing each of the basis elements. The reverse of a is denoted by $\tilde{a}$. The following gives an example:

$$(e_1 + \varepsilon e_{01} + e_{23} + e_{123})^{\sim} = e_1 + \varepsilon e_{10} + e_{32} + e_{321}$$
$$= e_1 - \varepsilon e_{01} - e_{23} - e_{123}.$$

If a and b are two elements, then the reverse of their product is the product of their reverses in the other order:

$$\widetilde{(ab)} = \tilde{b}\tilde{a}.$$

An inner and outer product are now introduced. It should be noted that the definition of the outer product taken here is different to the one used by other authors (in that it does not extract the component of highest grade). The versions given here take the expressions for the products of a pair of vectors [7, 21] and use them for any pair of elements. They are used because of their simplicity (they are purely algebraic

---

[4] *Editorial note*: This deviates from the usage of the term "pseudoscalar" in an n-D algebra elsewhere in this book, where it is restricted to pure n-blades.

combinations) and the fact that they have proved useful in dealing with $\mathscr{G}_4$ (due partly to its low dimension).

The inner and outer products of two elements a and b are defined respectively by the following expressions:

$$a \cdot b = \frac{1}{2}(ab + ba),$$

$$a \wedge b = \frac{1}{2}(ab - ba).$$

These have the expected distributive properties when additive combinations of elements are used, as in the following example:

$$(e_0 - 3e_1) \wedge (e_0 + e_2) = (e_0 \wedge e_0) - 3(e_1 \wedge e_0) + (e_0 \wedge e_2) - 3(e_1 \wedge e_2)$$
$$= 3e_{01} + e_{02} - 3e_{12}.$$

The definitions allow the ordinary products of a and b to be expressed in terms of the new products:

$$ab = (a \cdot b) + (a \wedge b), \tag{17.3}$$

$$ba = (a \cdot b) - (a \wedge b). \tag{17.4}$$

Finally in this section, it is noted that $\mathscr{G}_4$ contains other well-known sets of numbers. In fact these all lie within the "finite" part of $\mathscr{G}_4$, that is the set $\mathscr{E}_3$ consisting of combinations of the eight basis elements which do not involve e_0. Since $e_\emptyset = 1$, the real numbers form a subset of $\mathscr{G}_4$. Elements of the form $\alpha + \beta e_{12}$, where α and β are scalars, form a subset isomorphic to the complex numbers with e_{12} taking the role of the square root of -1. In this case, the reverse operation is the same as complex conjugation. Lastly, the even grade elements of $\mathscr{G}_4$ which do not involve e_0 are isomorphic to the ring of quaternions with the following correspondences for the unit quaternions: $i = e_{12}$, $j = e_{31} = -e_{13}$, $k = e_{23}$, so that, as required, the product ijk becomes

$$ijk = -e_{12}e_{13}e_{23} = -e_1e_2e_1e_3e_2e_3 = -e_1e_1e_2e_2e_3e_3 = -1.$$

17.3   Geometry and Transforms

The vector $p = p_0e_0 + p_1e_1 + p_2e_2 + p_3e_3$ in $\mathscr{G}_4$ is used to represent the point in three-dimensional space whose coordinates are $(p_1/p_0, p_2/p_0, p_3/p_0)$. This assumes of course that p_0 is non-zero; if not, then what is represented is an ideal point (that is a point at infinity). Clearly any (non-zero) scalar multiple of p represents the same point. It is often convenient to assume that $p_0 = 1$. Note that p_0, and the other coefficients, may not be just simply numbers, but instead combinations of powers of ε as in Eq. (17.2). There may even be occasions when all the coefficients are multiples of a power of ε which then cancels out when division by p_0 takes place.

If p and q are two vectors in $\mathscr{G}_4$, then the bivector $p \wedge q$ represents the line joining the corresponding points [7, 12, 18]. This idea is not pursued further here except to note that it enables the distance between the two points to be found. Suppose that $p = e_0 + p_1 e_1 + p_2 e_2 + p_3 e_3$ and $q = e_0 + q_1 e_1 + q_2 e_2 + q_3 e_3$. Then

$$p \wedge q = (q_1 - p_1)e_{01} + (q_2 - p_2)e_{02} + (q_3 - p_3)e_{03}$$
$$+ (p_1 q_2 - p_2 q_1)e_{12} + (p_1 q_3 - p_3 q_1)e_{13} + (p_2 q_3 - p_3 q_2)e_{23}.$$

The inner product of $p \wedge q$ with itself can now be formed. If $d(p, q)$ is the Euclidean distance between the points corresponding to p and q, then it is found that

$$d(p, q)^2 = (q_1 - p_1)^2 + (q_2 - p_2)^2 + (q_3 - p_3)^2$$
$$= -\varepsilon(p \wedge q) \cdot (p \wedge q). \tag{17.5}$$

More generally, when the coefficients of e_0 are not necessarily unity, the relation becomes

$$d(p, q)^2 = (p \wedge q) \cdot (p \wedge q)/p \cdot q.$$

Now consider transforms. Let S be any element of even grade. A map, F_S, can be defined whose action on any $p \in \mathscr{G}_4$ is given by

$$F_S : p \mapsto \widetilde{S} p S.$$

In the case where p is a vector, its image is an element of odd grade which is equal to its own reverse. Hence the image is again a vector. So F_S maps vectors to vectors and hence induces a transform on three-dimensional space. Two examples are now given.

Example Take $S = c + s e_{12}$, where $c = \cos\theta$ and $s = \sin\theta$ for some angle θ, and let p be $e_0 + x e_1 + y e_2 + z e_3$. Manipulation shows that

$$\widetilde{S} p S = e_0 + \left[(c^2 - s^2)x - 2csy\right]e_1 + \left[2csx + (c^2 - s^2)y\right]e_2 + z e_3$$
$$= e_0 + \left[(\cos 2\theta)x - (\sin 2\theta)y\right]e_1$$
$$+ \left[(\sin 2\theta)x + (\cos 2\theta)y\right]e_2 + z e_3, \tag{17.6}$$

and the transform is a rotation through 2θ about the z-axis.

Example Now take $S = 1 + \varepsilon e_0 u$, where $u \in \mathscr{E}_3$ is a vector (not involving e_0). Again let $p = e_0 + q$ where $q = x e_1 + y e_2 + z e_3 \in \mathscr{E}_3$; then

$$\tilde{S}pS = \left(1 - \varepsilon u^2 - \varepsilon u \cdot q\right)e_0 + q + 2u - \varepsilon u q u$$
$$\simeq p + 2u. \tag{17.7}$$

The terms which are multiples of ε are ignored in the final result since they are small compared to other terms and they multiply elements within $\mathscr{E}_3$. The corresponding transform is seen to be the translation given by twice the components of u.

When $S = \alpha + \varepsilon \beta I$ is a pseudoscalar, with α non-zero, the transform it generates is the identity. This is because for any vector $p \in \mathscr{G}_4$,

$$\tilde{S}pS = (\alpha + \varepsilon \beta I)p(\alpha + \varepsilon \beta I)$$
$$= (\alpha + \varepsilon \beta I)(\alpha - \varepsilon \beta I)p$$
$$= \left(\alpha^2 - \varepsilon \beta^2\right)p, \tag{17.8}$$

which is a simply a scalar multiple of p and so represents the same point as p in three-dimensional space.

The typical even-grade element of $\mathscr{G}_4$ has the form

$$S = c + sb + e_0 v + \gamma I,$$

where c, s, γ are scalars, and b, $v \in \mathscr{E}_3$ are a bivector and vector respectively, with b normalised so that $\tilde{b}b = 1$. Then

$$\tilde{S}S = \left(c^2 + s^2\right) - 2se_0(b \cdot v) + 2c\gamma I - \varepsilon^{-1}v^2 + \varepsilon^{-1}I\gamma^2,$$

which is clearly non-finite (as ε tends to zero) unless both v and γ are multiples of ε.

So instead take the following as the typical even-grade element:

$$S = c + sb + \varepsilon e_0 v + \varepsilon \gamma I \tag{17.9}$$

and then

$$\tilde{S}S = \left(c^2 + s^2\right) - 2\varepsilon se_0(b \cdot v) + 2\varepsilon c\gamma I - \varepsilon v^2 + \varepsilon I\gamma^2$$
$$\simeq \left(c^2 + s^2\right) - 2\varepsilon se_0(b \cdot v) + 2\varepsilon c\gamma I. \tag{17.10}$$

Thus $\tilde{S}S$ is equivalent to the identity transform as in Eq. (17.8), and hence $\tilde{S}$ generates the inverse transform to S. Note that there is no need to assume that $\tilde{S}S$ is unity or indeed that it is purely scalar. It is however possible to normalise S so that this becomes true. Suppose $\tilde{S}S = \lambda + \varepsilon \mu I$ with $\lambda = c^2 + s^2 > 0$. Set

$$S_1 = \frac{1}{\sqrt{\lambda}}\left[1 - \frac{\varepsilon \mu I}{2\lambda}\right]S,$$

which is of even grade and has the property that $\widetilde{S_1} S_1 = 1$. Since S_1 is the product of S and a pseudoscalar, S_1 and S generate the same transform of three-dimensional space.

In the exceptional case where $\lambda = c^2 + s^2 = 0$, both c and s are zero, and hence $IS = \gamma - I_3 v$ where $I_3 = e_{123}$. So IS is a combination of a scalar and a bivector from $\mathscr{E}_3$ and can be normalised as before (unless $\gamma = v = 0$, in which case S is zero).

The form of the map F_S defined by S means that it preserves products. Assume for simplicity that $\widetilde{S} S = 1$, which also means that $S \widetilde{S} = 1$. Then the images of the product of two elements a and b becomes

$$F_S(ab) = \widetilde{S} ab S = \widetilde{S} a S \widetilde{S} b S = F_S(a) F_S(b),$$

which is the product of the images of the two elements.

Now suppose that p and q are two vectors in which the coefficients of e_0 are unity. Equation (17.5) says that $(p \wedge q) \cdot (p \wedge q)$ is scalar. This means that it is unchanged by F_S since $\widetilde{S} S = 1$. Then, since products are preserved by F_S,

$$(p \wedge q) \cdot (p \wedge q) = F_S\big((p \wedge q) \cdot (p \wedge q)\big)$$
$$= \big(F_S(p) \wedge F_S(q)\big) \cdot \big(F_S(p) \wedge F_S(q)\big).$$

Equation (17.5) is again used and shows that the distance between p and q is the same as that between their images. Thus F_S preserves distances and so is an isometry.

This idea is further investigated in the next section where it is seen that the isometry is a combination of a rotation and a translation and hence is a rigid-body transform.

17.4 Rotations and Translations

In this section, the typical even-grade element S given by Eq. (17.9) is investigated. Let I_3 be the basis element e_{123} and set $a = -bI_3$ so that a is a vector and $b = aI_3$. Recall that b is a unit bivector so that $\widetilde{b} b = 1$. This means that $a^2 = 1$. Since I_3 commutes with all elements of $\mathscr{E}_3$, Eq. (17.10) becomes

$$\widetilde{S} S = (c^2 + s^2) - 2\varepsilon s e_0(a \cdot v)I_3 + 2\varepsilon c\gamma I = (c^2 + s^2) - 2\varepsilon[c\gamma - s(a \cdot v)]I.$$

It is assumed that S has been normalised so that $\widetilde{S} S = 1$, and hence

$$c^2 + s^2 = 1, \tag{17.11}$$
$$s(a \cdot v) = c\gamma. \tag{17.12}$$

In particular, this means that $c = \cos\theta$ and $s = \sin\theta$ for some angle θ.

Consider the case in which v and γ are both zero, so that

$$S = c + sb = c + sa I_3. \tag{17.13}$$

Its action on the vector $p = e_0 + \lambda a$ where λ is a scalar is

$$\begin{aligned}
\widetilde{S} p S &= (c - sa I_3)(e_0 + \lambda a)(c + sa I_3) \\
&= c^2 e_0 - csae_0 I_3 + \lambda c^2 a + \lambda cs I_3 + csae_0 I_3 + s^2 e_0 - \lambda cs I_3 + \lambda s^2 a \\
&= e_0 + \lambda a = p,
\end{aligned}$$

so that p is unchanged by S. Now p represents the typical point on a line through the origin. The transform leaves fixed every point on this line, and hence it is rotation with the line as its axis. Further investigation shows that the angle of rotation is 2θ (cf. Eq. (17.6)). This is the extension of the idea that, in complex numbers, $\exp(I\phi)$ creates a rotation of the plane (through an angle ϕ). Indeed the element S given by Eq. (17.13) can be written as $S = \exp(b\theta)$.

A rotation about an axis through a general point $e_0 + p$ is now constructed where $p \in \mathscr{E}_3$. This is achieved by translating $e_0 + p$ to the origin, performing the rotation and then translating back. The element P for the translation of the origin to p is $P = 1 + \frac{1}{2} e_0 p$ (as in Eq. (17.7)), and so the rotation required needs the following even-grade element:

$$\begin{aligned}
R &= \widetilde{P}(c + sb) P \\
&= c + \frac{1}{2}\varepsilon c e_0 p + sb + \frac{1}{2}\varepsilon s e_0 bp - \frac{1}{2}\varepsilon c e_0 p \\
&\quad + \frac{1}{4}\varepsilon c p^2 - \frac{1}{2}\varepsilon s e_0 pb + \frac{1}{4}\varepsilon s pbp \\
&\simeq c + sb + \varepsilon s e_0 (b \wedge p). \tag{17.14}
\end{aligned}$$

This represents a general rotation. The typical translation is given by

$$T = 1 + \varepsilon e_0 t, \tag{17.15}$$

where t lies in $\mathscr{E}_3$, and the translation is over $2t$ (cf. Eq. (17.7)).

The two products of R and T evaluate as follows:

$$\begin{aligned}
RT &= c + sb + \varepsilon e_0 \big[s(b \wedge p) + ct + sbt \big], \\
TR &= c + sb + \varepsilon e_0 \big[s(b \wedge p) + ct + stb \big].
\end{aligned}$$

So R and T commute if t and b commute, that is if $b \wedge t = 0$, or equivalently if $a \wedge t = 0$. The last equation here is the condition that the directions represented by a and t are parallel, and these are the directions of the axis of rotation and of the translation. Assume that there is commutativity and that $t = \lambda a$ for a scalar λ. The above expression for RT becomes the following:

$$RT = (c + saI_3) + \varepsilon\left[s(a \wedge p)I + \lambda ce_0a\right] + \varepsilon\lambda sI.$$

It is now shown that any even-grade element S with $\tilde{S}S = 1$ can be written as a product $RT = TR$.

Starting with Eq. (17.9), assume first that $s \neq 0$ and set $\lambda = \gamma/s$. Vector p needs to be found such that

$$s(a \wedge p)I + \lambda ce_0a = e_0v.$$

Assume that p can be chosen with $a \cdot p = 0$. Then, by Eq. (17.3), $a \wedge p = ap$ and

$$e_0v = se_0apI_3 + \lambda ce_0a.$$

Further manipulation, using Eq. (17.12), yields

$$p = \left[\frac{v \wedge a}{s}\right]I_3.$$

The exceptional case is where $s = 0$. Equation (17.11) says that $c^2 = 1$, and without loss of generality take $c = 1$. Then Eq. (17.12) shows that $\gamma = 0$, and hence S has the form of Eq. (17.15) and represents a pure translation.

There is an alternative way of dealing with the axis. The outer product of two points (in any geometric algebra) gives the bivector which represents the line joining them. In the above, $e_0 + p$ is one point on the axis; another is $e_0 + p + a$ (which is unit distance away). So the axis is represented by the (unitised) line $\ell = (e_0 + p + a) \wedge (e_0 + p) = (a \wedge e_0) + (a \wedge p)$. Since $a = -bI_3$, manipulation shows that $a \wedge e_0 = bI$ and $a \wedge p = -(b \wedge p)\hat{I}_3$. It follows that $\varepsilon\ell I = b + \varepsilon e_0(b \wedge p)$ and so

$$R = c + s(\varepsilon\ell I),$$

which is a simpler form for R than Eq. (17.14) relating it directly to the axial line.

To summarise the results of this section, Eqs. (17.14) and (17.15) give the forms of even-grade elements representing a general rotation and translation. Further, any even-grade element S can be expressed as a product of a rotation R and a translation T which commute. As shown at the end of Sect. 17.3, this element, being of even-grade, represents an isometry. It now follows that it cannot involve a reflection and so the isometry is a rigid-body transform. The commuting of R and T implies that the direction of translation is along the axis of rotation. This is Chasles's theorem.

Two examples are now given illustrating the decomposition of an even-grade element S. In both cases S is normalised so that $\tilde{S}S = 1$.

Example $S = \frac{1}{\sqrt{2}}[1 + e_{12} + \varepsilon e_{01} - \varepsilon e_{02} + \varepsilon e_{03} + e_{0123}]$. In the previous notation, the
following values are identified: $c = s = \gamma = 1/\sqrt{2}$, $b = e_{12}$, $v = (e_1 - e_2 + e_3)/\sqrt{2}$,
$a = e_3$, $\lambda = 1$, $t = e_3$, and $p = e_1 + e_2$. Hence $S = RT = TR$, where $R = \frac{1}{\sqrt{2}}[1 +$
$e_{12} + \varepsilon e_{01} - \varepsilon e_{02}]$ and $T = 1 + \varepsilon e_{03}$.

Example $S = \frac{1}{2}[1 + e_{12} + e_{13} + e_{23} + \varepsilon e_{01} + \varepsilon e_{02} + \varepsilon e_{03} + \varepsilon e_{0123}]$. This time:
$c = \gamma = 1/2$, $s = \sqrt{3}/2$, $b = (e_{12} + e_{13} + e_{23})/\sqrt{3}$, $v = (e_1 + e_2 + e_3)/2$, $a = (e_1 -$
$e_2 + e_3)/\sqrt{3}$, $\lambda = 1/\sqrt{3}$, $t = (e_1 - e_2 + e_3)/3$, and $p = 2(-e_1 + e_3)/3$. The rotation
and translation are given by $R = \frac{1}{6}[3 + 3e_{12} + 3e_{13} + 3e_{23} + 2\varepsilon e_{01} + 4\varepsilon e_{02} + 2\varepsilon e_{03}]$
and $T = \frac{1}{3}[3 + \varepsilon e_{01} - \varepsilon e_{02} + \varepsilon e_{03}]$.

17.5   Motions

The previous section looks at even-grade elements and the rigid-body transforms
they represent. Given an object (defined in its own local coordinate system), a *pose*
is a transform which maps the body from its own local space to a position and
orientation in world space. Suppose that the pose corresponds to the even-grade
element S. If $S = S(t)$ is regarded as a function of a parameter t, then as t varies, so
does the pose, and this gives a *motion* to the body [22].

How can smooth motions be generated? One approach is via the *slerp* operation.
In its basic form, this performs an interpolation between a pair of transforms [11].
If these transforms are S_0 and S_1, they are combined to give $S(t) = S_0[\tilde{S_0}S_1]^t$. This
has the property that $S(0) = S_0$ and $S(1) = S_1$, and, for $0 \leq t \leq 1$, the motion passes
between the given poses. However the evaluation of non-integer powers requires
the use of logarithmic and exponential functions [14, 15, 19] and is not dealt with
further here. Instead attention is given to linear combinations of transforms.

The advent of computer graphics in the 1960s has led to a number of ways of
representing smooth curves and surfaces for numerical processing. Chief among
these are the Bézier and B-spline approaches [3].

Consider the case of the Bézier curve segment (similar properties apply to B-
spline segments). Suppose that a number, $n + 1$, of position vectors $\mathbf{r}_0, \mathbf{r}_1, \ldots, \mathbf{r}_n$ are
given. These are called *control points*, and they are combined to form the following
function:

$$\mathbf{r}(t) = \sum_{i=0}^{n} \binom{n}{i} t^i (1 - t)^{n-i} \mathbf{r}_i,$$

where $\binom{n}{i}$ is one of the binomial coefficients. This is a polynomial of degree n in t.
Further $\mathbf{r}(0) = \mathbf{r}_0$ and $\mathbf{r}(1) = \mathbf{r}_n$. So as t varies between 0 and 1, $\mathbf{r}(t)$ creates a
smooth curve between these end points. The arrangement of the Bézier form allows
properties of some of the other control points to be identified. In particular, the curve
segment (for $0 \leq t \leq 1$) lies within the convex hull of the control points, the initial
tangent is along the line joining $\mathbf{r}_0$ and $\mathbf{r}_1$, and the final tangent along the line joining

Fig. 17.1 Bézier quadratics:
left—curve; *right*—motion

r_{n-1} and r_n. Figure 17.1 shows, on the left, a Bézier quadratic segment with its three control points.

These ideas can be extended to motions. Suppose that $n + 1$ *control poses* are given $S_0, S_1, \ldots, S_n$. The equivalent combination is the following:

$$S(t) = \sum_{i=0}^{n} \binom{n}{i} t^i (1 - t)^{n-i} S_i.$$

This is a combination of even-grade elements and so is itself of even grade. Hence it represents a rigid-body transform (unless it happens to be zero). As expected, $S(t)$ varies smoothly between S_0 and S_1 for $0 \le t \le 1$. On the right of Fig. 17.1 is shown a quadratic Bézier motion with its three control poses. (This is called quadratic since $S(t)$ is of degree two; note however that the path followed by any point in the body is a parametric curve of degree four since $S(t)$ appears twice in the map.)

A Bézier curve has the property that it can be transformed as a whole by applying the transform to each of the control points. A similar invariance property holds for a Bézier motion. The image of a point p under the pose transform S_i is $\widetilde{S}_i p S_i$. If each of these images is transformed by element U, the new images are $\widetilde{U} \widetilde{S}_i p S_i U$. Effectively, the pose transform S_i has been replaced by $S_i U$. The new motion transform is then $SU = S(t)U$, and the typical point p transforms to $\widetilde{SU} p SU = \widetilde{U}(\widetilde{S}pS)U$. Thus the result is to transform the entire motion by U.

The case of a Bézier segment of degree one is now explored. In view of the above invariance, the control poses can be post-multiplied by $\widetilde{S}_0$ so that it can be assumed, without loss, that $S_0 = 1$, and $S(t)$ takes the form $S(t) = (1 - t) + tU$, where $U = S_1 \widetilde{S}_0$. The image of the typical point p is

$$\widetilde{S} p S = (1 - t)^2 p + 2t(1 - t)q + t^2 \widetilde{U} p U \quad \text{where } q = \frac{1}{2}[\widetilde{U}p + pU].$$

$$(17.16)$$

Here q is an element of odd degree and is equal to its own reverse, so q represents a point. This means that Eq. (17.16) is a Bézier quadratic combination of three points, namely p, q and $\widetilde{U}pU$. So any point in the body follows a quadratic curve, and this curve is planar, lying in the plane defined by the three points.

Suppose that $U = RT = TR$ is the decomposition of U as the product of a rotation and a translation, with $\widetilde{R}R = 1$. The effect of S on a point p on the axis of the rotation is now considered. In Eq. (17.16), the image of p is a combination of three terms. The first and last certainly lie on the axis. Since R fixes p, $\widetilde{R}pR = p$, and hence R commutes with p. So R commutes also with q and hence fixes it as

Fig. 17.2 General Bézier
linear motion around
cylindrical surface

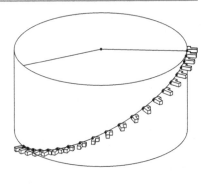

well. This means that q also lies on the axis. Thus the image of p is a combination of three points on the axis and so also lies on the axis. Hence S maps the axis onto itself and so is also the product of a rotation about that axis and a translation along it.

It follows that the path traced out by any point in the body lies on a cylinder whose axis is along the axis of rotation as suggested in Fig. 17.2. The path is elliptical, lying on a planar slice through the cylinder (and is therefore not exactly part of a helix).

One way to form Bézier curves is via the de Casteljau algorithm [3]. The significance of motions of first degree is that they form basic step in the algorithm when applied to poses. Consider the case of a Bézier cubic motion with control poses S_0, S_1, S_2, S_3. Additional poses are created according to the pattern

$$
\begin{array}{llll}
S_0 & & & \\
& S_{01} & & \\
S_1 & & S_{012} & \\
& S_{12} & & S_{0123} \\
S_2 & & S_{123} & \\
& S_{23} & & \\
S_3 & & &
\end{array}
$$

where each new pose is a linear combination of the two poses to its left, so that, for example, $S_{012} = (1 - t)S_{01} + tS_{12}$. The same value of the parameter t is used throughout. Each new pose lies on the first-degree motion interpolating the two poses which define it. An example is shown on the left in Fig. 17.3. The first-degree motions are the curves shown. The pose S_{0123} becomes the value of $S(t)$ for that value of t and hence represents a pose along the cubic motion. This motion is shown as the dashed curve on the left in the figure, and the full motion is shown on the right.

In particular, when t is close to 0 or 1, the motion is approximately the first-degree motion defined by the first or last pair of control poses. The criterion for two Bézier curve segments to join together smoothly is that their end-tangents align. The condition for two Bézier motions to join smoothly is that the next segment must start with the same cylindrical motion (defined by its first pair of control poses) with which the previous segment finishes (defined by its last pair of control poses).

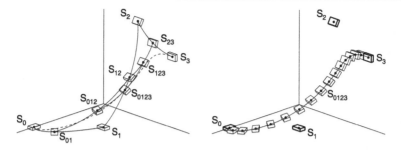

Fig. 17.3 Cubic Bézier motion: *left*—de Casteljau construction; *right*—resultant motion with control poses

17.6   Conclusions

It is possible to construct a 4D geometric algebra which represents exactly three-dimensional Euclidean space and its rigid-body transforms. To do this, the square of one of the basis vectors is treated as being infinite by defining it to be the reciprocal of a (vanishingly) small number ε. This is carried (symbolically) through any calculations and then allowed to become zero. An alternative view is to regard the scalars of the algebra as being power series in ε.

The vectors of the algebra then correspond (projectively) to the points of Euclidean space, and the even-grade elements of the algebra represent rigid-body transforms, and vice versa. This means that a (non-zero) linear combination of even-grade elements is also a transform and hence Bézier (and B-spline) combination of control poses can be used to represent smooth motions.

17.7   Exercises

17.1 The following elements of $\mathcal{G}_4$ are given: $R = e_{12}$; $T = 1 + \varepsilon e_{03}$; $S = RT$; $p = e_0 + e_1$; and $q = e_0 + e_2$. By multiplying out, show that the following are true:

$$S = e_{12} + \varepsilon e_{0123},$$
$$S = TR,$$
$$\widetilde{S}pS = (1 - \varepsilon)e_0 - (1 + \varepsilon)e_1 + 2e_3,$$
$$\widetilde{S}qS = (1 - \varepsilon)e_0 - (1 + \varepsilon)e_2 + 2e_3.$$

17.2 A even-grade element S defines the map $F_S : p \mapsto \widetilde{S}pS$ for $p \in \mathcal{G}_4$, generating a transform which is a rotation about the z-axis through a right angle followed by a translation through two units in the x direction. Show that S is $(1 + \varepsilon e_{01} - \varepsilon e_{02} + e_{12})/\sqrt{2}$ (or possibly a multiple of it by a pseudoscalar). Check that the images of the origin and points unit distance along each coordinate axis are as expected. Show that $S^4 \simeq -1$; what transform does this element create?

17.3 For an even-grade element S, let F_S denote the map $F_S : p \mapsto \widetilde{S}pS$ for $p \in \mathcal{G}_4$. Set $\alpha = \widetilde{S}S$. Check that α is a pseudoscalar and so commutes with every even-grade element of $\mathcal{G}_4$. Check also that $\alpha = S\widetilde{S}$. Show that F_S preserves the inner and outer products in the sense that $\alpha F_S(a \cdot b) = F_S(a) \cdot F_S(b)$ and $\alpha F_S(a \wedge b) = F_S(a) \wedge F_S(b)$.

17.4 Starting with $S = (1 + \varepsilon e_{01} + \varepsilon e_{02} + e_{12})/\sqrt{2}$, use the method of Sect. 17.4 to write S as a product of commuting elements R and T which represent respectively a rotation and a translation. In particular, deduce that S corresponds to a pure rotation about an axis in the z-direction through the point $e_0 + e_1 + e_2$.

17.5 What transforms are generated by the even-grade elements $S_0 = 1 + \varepsilon e_{01}$ and $S_1 = \varepsilon e_{02} + e_{12}$? Let $S(t) = (1 - t)S_0 + tS_1$ be the Bézier combination of unit degree. Show that $\widetilde{S(t)}e_0S \simeq we_0 + xe_1 + ye_2$ where $w = 1 - 2t + 2t^2$, $x = 2 - 4t$, and $y = 4t - 4t^2$. Check that $(x/2)^2 + (y/2)^2 - w^2 = 0$. Deduce that, under the motion generated by $S(t)$, the image of the origin, e_0, moves along a semicircle of radius 2 as t varies between 0 and 1.

Acknowledgements The work reported in the paper was carried within the Innovative Design and Manufacturing Research Centre at the University of Bath, and the second author is funded by a studentship provided via the Centre. The Centre is funded by the Engineering and Physical Sciences Research Council (EPSRC), and this support is gratefully acknowledged.

References

1. Belta, C., Kumar, V.: An SVD-based projection method for interpolation on $SE(3)$. IEEE Trans. Robot. Autom. **18**, 334–345 (2002)
2. Etzel, K.R., McCarthy, J.M.: Interpolation of spatial displacements using the Clifford algebra of E^4. J. Mech. Des. **121**, 39–44 (1999)
3. Farin, G.: Curves and Surfaces for CAGD: A Practical Guide, 5th edn. Morgan Kaufmann, San Francisco (2001)
4. Gan, D., Liao, Q., Wei, S., Dai, J.S., Qiao, S.: Dual quaternion-based inverse kinematics of the general spatial 7R mechanism. Proc. Inst. Mech. Eng., Part C, J. Mech. Eng. Sci. **222**, 1593–1598 (2008)
5. Ge, Q.J.: On the matrix realization of the theory of biquaternions. J. Mech. Des. **120**, 404–407 (1998)
6. Ge, Q.J., Ravani, R.: Geometric construction of Bézier motions. J. Mech. Des. **116**, 749–755 (1994)
7. González Calvet, R.: Treatise of Plane Geometry Through Geometric Algebra. Cerdanyola del Vallès (2007)
8. Hofer, M., Pottmann, H., Ravani, B.: From curve design algorithms to the design of rigid body motions. Vis. Comput. **20**, 279–297 (2004)
9. Jin, Z., Ge, Q.J.: Computer aided synthesis of piecewise rational motions for planar 2R and 3R robot arms. J. Mech. Des. **129**, 1031–1036 (2007)
10. Jüttler, B., Wagner, M.G.: Computer-aided design with spatial rational B-spline motions. J. Mech. Des. **118**, 193–201 (1996)
11. Leeney, M.: Fast quaternion slerp. Int. J. Comput. Math. **86**, 79–84 (2009)
12. Mullineux, G.: Clifford algebra of three dimensional geometry. Robotica **20**, 687–697 (2002)

13. Mullineux, G.: Modelling spatial displacements using Clifford algebra. J. Mech. Des. **126**, 420–424 (2004)
14. Özgören, M.K.: Kinematics analysis of spatial mechanical systems using exponential rotation matrices. J. Mech. Des. **129**, 1144–1152 (2007)
15. Perez-Garcia, A., McCarthy, J.M.: Kinematic synthesis of spatial serial chains using Clifford algebra exponentials. Proc. Inst. Mech. Eng., Part C, J. Mech. Eng. Sci. **220**, 953–968 (2006)
16. Purwar, A., Jin, Z., Ge, Q.J.: Rational motion interpolation under kinematic constraints of spherical 6R closed chains. J. Mech. Des. **130** 062301 (2008)
17. Röschel, O.: Rational motion design—a survey. Comput. Aided Des. **30**, 169–178 (1998)
18. Selig, J.M.: Clifford algebra of points, lines and planes. Robotica **20**, 545–556 (2000)
19. Simpson, L., Mullineux, G.: Exponentials and motions in geometric algebra. In: Vaclav, S., Hildenbrand, D. (eds.) International Workshop on Computer Graphics, Computer Vision and Mathematics (GraVisMa), pp. 9–16. Union Agency, Plzen (2009)
20. Srinivasen, L.N., Ge, Q.J.: Fine tuning of rational B-spline motions. J. Mech. Des. **120**, 46–51 (1998)
21. Vince, J.: Geometric Algebra for Computer Graphics. Springer, London (2008)
22. Wareham, R., Lasenby, J.: Mesh vertex pose and position interpolation using geometric algebra. In: Perales, F.J., Fisher, R.B. (eds.) Articulated Motion and Deformable Objects, 5th International Conference, AMDO 2008, pp. 122–131. Springer, Berlin (2008)
23. Wu, W., You, Z.: Modelling rigid origami with quaternions and dual quaternions. Proc. R. Soc. A **466**, 2155–2174 (2010)

Rigid Body Dynamics in a Constant Curvature Space and the '1D-up' Approach to Conformal Geometric Algebra

18

Anthony Lasenby

Abstract

We discuss a '1D up' approach to Conformal Geometric Algebra, which treats the dynamics of rigid bodies in 3D spaces with constant curvature via a 4D conformal representation. All equations are derived covariantly from a 4D Lagrangian, and definitions of energy and momentum in the curved space are given. Some novel features of the dynamics of rigid bodies in these spaces are pointed out, including a simple non-relativistic version of the Papapetrou force in General Relativity. The final view of ordinary translational motion that emerges is perhaps surprising, in that it is shown to correspond to precession in the 1D up conformal space. We discuss the alternative approaches to Euclidean motions and rigid body dynamics outlined by *Gunn* in Chap. 15 and *Mullineux and Simpson* in Chap. 17 of this volume, which also use only one extra dimension, and compare these with the Euclidean space limit of the current approach.

18.1 Introduction

In Chap. 1, we examined rigid body dynamics using a Conformal Geometric Algebra (CGA) approach. This adjoins two extra vectors to the 3D base space and thus allows translations, rotations and dilations in Euclidean space to be expressed via a unified rotor structure. It is also possible to add just one extra dimension and still be able to express both rotations and translations within a unified rotor structure. The 'penalty' for this is that one then needs to work in a *curved* space. This study of the curved space dynamics is interesting in itself, but also the decrease in dimension and therefore number of elements that need to be operated upon may make this a useful approach in practice, even for Euclidean space dynamics. We will see that

A. Lasenby (✉)
Cavendish Laboratory and Kavli Institute for Cosmology, University of Cambridge, Cambridge, UK
e-mail: a.n.lasenby@mrao.cam.ac.uk

L. Dorst, J. Lasenby (eds.), *Guide to Geometric Algebra in Practice*,
DOI 10.1007/978-0-85729-811-9_18, © Springer-Verlag London Limited 2011

some interesting issues of principle come forward in this approach and show how a force on a spinning body in curved space, normally only considered in a full general relativistic context and called there the Papapetrou force [6], can be derived in an elementary fashion in the non-relativistic context described here.

A '1D up' approach to Euclidean geometry is considered elsewhere in this volume, in the chapters by Gunn (Chap. 15), Goldman (Chap. 16) and Mullineux and Simpson (Chap. 17). Here we discuss the relationship with the work of Gunn, since this also has a focus on the dynamics of rigid bodies, and also comment briefly on the relationship of the curved space approach to that of Mullineux and Simpson.

18.2 The '1D up' approach

As discussed in Chap. 1, we know that in the setup we use for CGA, we adjoin to ordinary 3D space two further vectors:

$$e: \quad e^2 = +1$$
$$\bar{e}: \quad \bar{e}^2 = -1$$

and that this allows us to create[1]

$$\begin{cases} n = e + \bar{e} & \text{the point at infinity } n_\infty \\ \bar{n} = e - \bar{e} & \text{proportional to the point at the origin } n_o \end{cases}$$

Euclidean geometry is basically the geometry obtained by keeping n_∞ constant. Now as discussed e.g. in [4], it turns out we can do *spherical geometry* by instead keeping $\bar{e}$ constant and *hyperbolic geometry* by keeping e constant. Since these vectors (respectively, in the different geometries) are constant, there is the potential of a '1D up' approach where we drop them from the CGA space and work in a space of one less dimension. The same comment could be made about Euclidean geometry as well of course—why do not we just drop n_∞, since it is constant under rigid body transformations? The difference lies with the form of the translation rotors in the two cases. In the Euclidean case these look like

$$T_a = 1 + \frac{n_\infty a}{2} \tag{18.1}$$

where a is the 3D vector corresponding to the displacement, whereas in e.g. the spherical case, where we keep $\bar{e}$ constant, the translation rotor is

---

[1] *Editorial note*: The notation n and $\bar{n}$ is used in [4]. When we compare to the notation used elsewhere in this volume we find that $n = n_\infty$ and $\bar{n} = -2n_o$. Correspondingly, $e = \frac{1}{2}(n + \bar{n}) = -n_o + \frac{1}{2}n_\infty$ and $\bar{e} = \frac{1}{2}(n - \bar{n}) = n_o + \frac{1}{2}n_\infty$. Note that $\bar{n} \cdot n = 2$ corresponds to $n_o \cdot n_\infty = -1$. As a compromise between the notation in this book and [4] that avoids awkward factors, in this chapter we will use $\bar{n}$, but replace n by n_∞.

$$T_a = \frac{\lambda + ae}{\sqrt{\lambda^2 + a^2}} \tag{18.2}$$

where λ is a length scale related to the 'radius of curvature' of the space, and which is introduced to keep everything dimensionally homogeneous.[2]

Thus with e as origin and by applying the translation rotor with $a = x$, to get the representative of a general 3D position x, we will never obtain a vector including $\bar{e}$. Since it appears neither in the vectors nor the rotors, $\bar{e}$ can be dropped from the algebra. In contrast, in the Euclidean case, the invariant point at infinity appears in both the rotors and the points, and cannot be dropped, at least not in this way.

We note that the explicit form of representation in the case where we take e as the origin is

$$\hat{Y} = \left(\frac{2\lambda}{\lambda^2 + x^2}\right)x + \left(\frac{\lambda^2 - x^2}{\lambda^2 + x^2}\right)e \tag{18.3}$$

and that this satisfies the (covariant) normalisation $\hat{Y}^2 = 1$.

All the advantages of translations and rotations being unified in a rotor structure are still present in this treatment, and since we are working in a constant curvature space, the notion of a 'rigid body' displacement still makes sense. Specifically a rigid-body translation or rotation maintains the geodesic distance between two points as being constant, and this is guaranteed by the rotor structure, since for points represented by 4D vectors $\hat{Y}_1$ and $\hat{Y}_2$, this distance is a function of $\hat{Y}_1 \cdot \hat{Y}_2$ and the rotors preserve this dot product (see [4]). For an example of a rigid translation in 2D hyperbolic space, see Fig. 18.1.

An interesting point of using such an approach for rigid bodies is that the counting of degrees of freedom is now exactly right. Translations and rotations in 3D have altogether six d.o.f., and this matches the number of *bivectors*, and therefore generators of motion, in 4D (i.e. in 1D up).

This works generally—to describe rotation in m dimensions, we need $\frac{1}{2}m(m-1)$ components to describe rotation, whilst position needs another m. Therefore the total is $\frac{1}{2}m(m+1)$, which is the same as the number of bivector (and therefore independent rotor) components in $m+1$ dimensions.

A further motivation for working in such spaces effectively comes from *physics*: does the notion of the dynamics of a rigid body even make sense in a curved space? This may seem esoteric, but it should not be forgotten that the current space we are living in is probably curved. It is *certainly* curved near gravitating objects, but also, on cosmological scales it is clear that we are heading asymptotically towards a final *de Sitter* phase of the universe and this has constant curvature everywhere at all positions and times.

---

[2]Those interested might like to know that if we extend this work to the space–time, rather than the purely spatial case, then the positive curvature version of the space we obtain is called *de Sitter* space, and λ is related to the usual *cosmological constant*, Λ (as measured in inverse metres squared), by $\Lambda = 12/\lambda^2$. (See e.g. [3].)

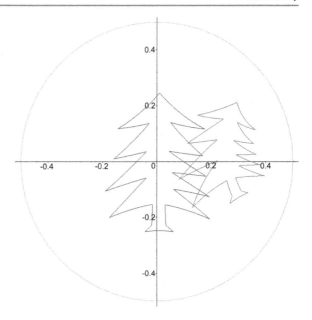

Fig. 18.1 A rigid body translation in 2D hyperbolic space. The representation used is the Poincaré disc, and the *circle* is its boundary. The geodesic distance between each pair of points is maintained in the motion

We now show that the CGA, in the 1D-up approach, provides a self-consistent and sensible framework in which to study this question. This question as to whether dynamics of a rigid body can be formulated sensibly in a constant curvature space has a long history (see e.g. [2] for a first attempt), but it is not clear to the present author that the particular equations and solutions given here have ever been given before.

In a final section of this chapter, we ask the question of whether, by taking the limit as the curvature scale $\lambda \to \infty$ at the end of the calculation, we can recover a description of *Euclidean* space motion which preserves the advantages of a conformal GA description but is operating with only one extra dimension compared to 3D. As discussed above, we then compare and contrast this approach with those taken by *Gunn* (Chap. 15) and *Mullineux and Simpson* (Chap. 17) in this volume, seeking an overall consensus on the usefulness of this approach.

18.2.1 Equations and Solutions for Rigid Body Motion in Spherical Space

For definiteness, we will work in spherical space, so that the metric is $(+, +, +, +)$. Note that in this space the pseudoscalar $I = e_1e_2e_3e$ squares to $+1$ and anticommutes with all vectors.

In setting up a Lagrangian for this case, there are a number of subtleties we must deal with, and to begin with we start with the much simpler case of the motion of a single point particle.

18.2.1.1 Point Particle Motion in Curved Space

One of the forms of Lagrangian that can be used for the free non-relativistic motion of a point particle in ordinary 3D space is

$$\mathscr{L} = \frac{1}{2} m \dot{x}^2 \tag{18.4}$$

where x is the 3D position. This is obviously just the standard kinetic energy. This form generalises to a *curved* space, with interval and metric given by

$$ds^2 = g_{\mu\nu} dx^\mu dx^\nu \tag{18.5}$$

as

$$\mathscr{L} = \frac{1}{2} m g_{\mu\nu} \dot{x}^\mu \dot{x}^\nu \tag{18.6}$$

where $\dot{x}^\mu \equiv \frac{dx^\mu}{d\tau}$, and τ is an affine parameter along the path. (Note that this Lagrangian differs from the more usual one that would be used in General Relativity, namely $\mathscr{L} = m\sqrt{g_{\mu\nu}(dx^\mu/ds)(dx^\nu/ds)}$, but nevertheless successfully reproduces the desired geodesics. We use it here since it has an obvious connection with non-relativistic energy.)

We can find the metric in our curved 3D spherical base space via the route sketched out in Sect. 3 of [3], which can be easily adapted from the de Sitter case considered there to the positive curvature 3-space considered here. This yields:

$$ds^2 = \frac{\lambda^4}{(\lambda^2 + r^2)^2} \left(dx^2 + dy^2 + dz^2 \right) \tag{18.7}$$

where $r^2 = x^2 + y^2 + z^2$.

We could go on to use (18.6) to find the geodesics in this metric, which would be the paths followed by free point-particles. However, we would like to understand how these paths can be derived from an equivalent formulation in the '1D up' conformal space, since this will hopefully show us the route to the correct rigid-body Lagrangian in this space.

This topic is discussed in [5], and the net result is that if we equip the '1D up' space with a conformal metric in which the conformal factor is the simple function $1/Y^2$, where Y is the position vector in the conformal space, then the geodesics in this space will (after projecting down) be geodesics in the 3D curved base space as well.

This means that the point particle Lagrangian equivalent to (18.6) we should take in this space is

$$\mathscr{L} = \mathscr{L}(Y, \dot{Y}) = \frac{1}{2} m \frac{\lambda^2}{4} \frac{\dot{Y}^2}{Y^2} \tag{18.8}$$

where we have inserted $\lambda^2/4$ to keep the dimensions as those of energy (the $1/4$ will be explained later), and the ˙ derivative from here on corresponds to choosing the ordinary non-relativistic time as the affine parameter.

Now the Euler–Lagrange equations we get from (18.8) are fairly easy to work out. We get the neat form

$$\ddot{Y} = \frac{2Y\cdot\dot{Y}\dot{Y} - Y\dot{Y}^2}{Y^2} = \frac{\dot{Y}Y\dot{Y}}{Y^2} \tag{18.9}$$

An important step now is to note that if Y and $\dot{Y}$ are orthogonal at some given time, then they remain orthogonal thereafter. This follows from

$$\frac{d}{dt}(Y\cdot\dot{Y}) = \dot{Y}\cdot\dot{Y} + Y\cdot\ddot{Y} = 2\frac{(Y\cdot\dot{Y})^2}{Y^2} \tag{18.10}$$

where we have used (18.9) for $\ddot{Y}$. We can thus specialise our Y representation to the case we are expecting in the '1D up approach', namely that our representation points have unit norm, and this will be preserved along the geodesics. The geodesic equation is now, for normalised $\hat{Y}$,

$$\ddot{\hat{Y}} = -\dot{\hat{Y}}^2\hat{Y} \tag{18.11}$$

and we see that the acceleration is now radial. In fact we can interpret this equation as simply the centripetal acceleration $-v^2/r$ necessary to move on the 3D spherical 'surface'! From now on, to avoid a multiplicity of 'hats', we will henceforth assume that all Y points are normalised.

An interesting feature of the geodesic motion is that the specific angular momentum

$$L = Y \wedge \dot{Y} = Y\dot{Y} \tag{18.12}$$

is automatically conserved (which is not surprising), but also that L simultaneously acts both as a generator of the motion, via

$$\dot{Y} = L\cdot Y \tag{18.13}$$

and as the '1D up' conformal representation of the 'd-line' along which the motion happens. Specifically, for any two points on such a d-line, with representative 4D points Y_1 and Y_2, then L is a multiple of $Y_1 \wedge Y_2$. This dual role is discussed further in [4]. Furthermore the action of L on the velocity is to give the acceleration:

$$\ddot{Y} = L\cdot\dot{Y} \tag{18.14}$$

We thus start to build up a picture of the motion as being represented by a particle with three axes sticking out of it, Y, $\dot{Y}$ and $\ddot{Y}$. The Y axis tells us where the (actual) particle is, the $\dot{Y}$ axis what its velocity is and the $\ddot{Y}$ axis its acceleration. In this trivial case of free motion, the $\ddot{Y}$ axis is not distinct from the Y axis, but it would be in the presence of forces, and in this case, the generator L would no longer be constant.

18.2.1.2 Rigid Body Motion in Curved Space

We can now generalise these considerations to the case of a rigid body. To avoid complications with volume elements in the curved space, we are going to restrict attention to a rigid body consisting of an assemblage of point particles. We thus write as Lagrangian

$$\mathscr{L} = \sum_i \frac{1}{2} m_i \frac{\lambda^2}{4} \frac{\dot{Y}_i}{Y_i^2} \tag{18.15}$$

where of course here, since we want the Lagrangian to work properly, we mean the full Ys, not just unit vector equivalents.

What now takes the place of the generator L is the angular velocity bivector Ω. Letting Y_{ref} be the position vector within the reference copy of the body, R be the 4D rotor taking these positions to the actual position in 4D space, and defining $\Omega_B = \tilde{R} \Omega R$ as usual, we have that

$$\dot{Y} = Y \cdot \Omega = R Y_{\text{ref}} \cdot \Omega_B \tilde{R} \tag{18.16}$$

We thus have

$$\frac{\dot{Y}_i}{Y_i^2} = \frac{(Y_{\text{ref}i} \cdot \Omega_B)^2}{Y_{\text{ref}i}^2} = -\Omega_B \cdot \left(\hat{Y}_{\text{ref}i} \wedge (\hat{Y}_{\text{ref}i} \cdot \Omega_B)\right) \tag{18.17}$$

In this form, we recognise that we are back on the familiar territory of Chap. 1 and can now write the inertia tensor in 4D as

$$I(\Omega_B) = \sum_i m_i \frac{\lambda^2}{4} \left(\hat{Y}_{\text{ref}i} \wedge (\hat{Y}_{\text{ref}i} \cdot \Omega_B)\right) \tag{18.18}$$

We note that transforming this to the space frame correctly gives a version of the angular momentum in this case (with the correct units):

$$R I(\Omega_B) \tilde{R} = \sum_i m_i \frac{\lambda^2}{4} \hat{Y}_i \wedge \dot{\hat{Y}}_i \tag{18.19}$$

From this point, we can now copy quite a bit from what we have already done in the Euclidean 2-up approach in Chap. 1, but now adapted to the case where ψ is a 4D rather than 5D spinor. This means that there are fewer degrees of freedom to fix in order to turn ψ into a rotor, and we can use scalar rather than vector Lagrangian multipliers in the Lagrangian, as follows:

$$\mathcal{L} = \left\langle -\frac{1}{2}\Omega_B \cdot I(\Omega_B) - \mu(\psi\tilde{\psi} - 1) - \nu I \psi\tilde{\psi} \right\rangle$$

Here μ and ν are scalars, ψ a general spinor in 4D, and Ω_B is again a bivector function of ψ defined by

$$\Omega_B = -\tilde{\psi}\dot{\psi} + \dot{\tilde{\psi}}\psi \tag{18.20}$$

The equations of motion derived from this Lagrangian are similar to those of Chap. 1, and we obtain

$$I(\dot{\Omega}_B) - \Omega_B I(\Omega_B) = \mu + \nu I \tag{18.21}$$

and projecting onto the various grades, we have

$$-\Omega_B \cdot I(\Omega_B) = \mu \quad \text{twice the kinetic energy}$$
$$I(\dot{\Omega}_B) - \Omega_B \times I(\Omega_B) = 0 \quad \text{Euler equation} \tag{18.22}$$
$$-\Omega_B \wedge I(\Omega_B) = \nu I \quad \text{grade-4 part}$$

As in Chap. 1, the grade-4 part is somewhat mysterious, and in the examples we have tried vanishes. The identification of the Lagrangian multiplier μ with twice the K.E. is a consequence of the above discussion, but of course we will need to show that it reduces to something sensible in the limit as the curvature scale $\lambda \to \infty$ and also that it is conserved.

This is indeed true, but for now we want to consider the main equation of motion (the 'Euler equation') and will seek to derive an explicit solution for a case of interest, namely the same 'dumbbell' model as considered in Chap. 1, but this time moving in curved space.

18.2.1.3 The Dumbbell Motion

A dumbbell provides the simplest example of an extended rigid body system, and by studying its motion in curved space we can uncover some interesting features which will also apply to more complicated objects.

We assume that our dumbbell consists of equal masses m positioned at 3D space coordinates $(\pm x_0, 0, 0)$ in the reference copy of the body. For simplicity, we consider (as in the flat space cases) motion which takes place just in the x, y plane in 3D. This means that the only parts of the inertia tensor we need consider are its actions on the rotation bivector $e_1 e_2$ and the translation bivectors $e_1 e$ and $e_2 e$. Let us call these bivectors R_3, T_1 and T_2 respectively. Calculating the moment of inertia tensor on these, we find that $I(B)$ as given by (18.18) is purely diagonal with moments of inertia defined by:

$$-R_3 \cdot I(R_3) = i_3 = \frac{2m\lambda^4 x_0^2}{(\lambda^2 + x_0^2)^2}$$

$$-T_1 \cdot I(T_1) = i_1 = \frac{m\lambda^2}{2} \qquad (18.23)$$

$$-T_2 \cdot I(T_2) = i_2 = \frac{m\lambda^2(\lambda^2 - x_0^2)^2}{2(\lambda^2 + x_0^2)^2}$$

We see that in general these are unequal but that if the size of the rigid body ($\sim x_0$) is small compared to the curvature scale of the space ($\sim \lambda$), then the two translation moments of inertia are approximately equal, and we have a 3D symmetric top! Moreover we see that under these conditions, $i_3 \ll i_1 \approx i_2$, i.e. it is a highly prolate top. (Note that this 'prolateness' is nothing to do with the prolateness of the dumbbell model—the third direction here is not e_3 but e.)

The fact that for most applications of interest we will have (to a very good approximation) a symmetric top in the 3D space spanned by e_1, e_2 and e, all of which have positive square, means that the rotor solution for a symmetric top in ordinary 3D Euclidean space, discussed back in Sect. 1.2 of Chap. 1, can be directly applied here. However, we first continue with attempting an exact solution. Since this still maps onto a 3D Euclidean case, viz the 3D asymmetric top, this is a well-worked area analytically, and at least at the level of solving the Euler equations for the angular velocity bivector, we can get an analytical solution.

Let us parameterise Ω_B as

$$\Omega_B = f(t)R_3 + g(t)T_1 + h(t)T_2 \qquad (18.24)$$

Here f, g and h are scalar functions of time, and this is the most general form we need for our assumed form of motion (restricted to the (x, y) plane in 3D real space). Inserting Ω_B into the Euler equation and using the moments of inertia in (18.23), we get the surprisingly simple results

$$\dot{f} = gh, \quad \dot{g} = fh\frac{(\lambda^2 + 2\lambda x_0 - x_0^2)(\lambda^2 - 2\lambda x_0 - x_0^2)}{(\lambda^2 + x_0^2)^2}, \quad \dot{h} = -fg \quad (18.25)$$

The f and h derivatives tell us that $f^2 + h^2$ is constant, suggesting a parameterisation of the form

$$f = \omega \cos(\varepsilon(t)), \quad g = -\dot{\varepsilon}(t), \quad h = \omega \sin(\varepsilon(t)) \qquad (18.26)$$

for some function $\varepsilon(t)$ and scalar constant ω (called this since it is in fact the ordinary angular velocity of the body in 3D space).

If we insert this, then the remaining equation to solve is

$$\ddot{\varepsilon} + \alpha \sin(2\varepsilon(t)) = 0 \qquad (18.27)$$

where

$$\alpha = \frac{\omega^2(\lambda^2 + 2\lambda x_0 - x_0^2)(\lambda^2 - 2\lambda x_0 - x_0^2)}{2(\lambda^2 + x_0^2)^2} \tag{18.28}$$

is a constant.

We can see that our ε parameter (or more specifically $\varepsilon/2$) satisfies the same equation as the amplitude of an exact, or 'spherical', pendulum. Not surprisingly, therefore, the solution can expressed in terms of the Jacobi amplitude function $\mathrm{am}(u|m)$. A convenient form, incorporating two constants of integration, is

$$\varepsilon(t) = \mathrm{am}\left(a(t - t_0) \middle| \frac{\sqrt{2\alpha}}{a} \right) \tag{18.29}$$

This solves (18.27) for arbitrary t_0 and a. We can fix these values of course from the initial conditions.

With $\varepsilon(t)$ found, f, g and h then follow immediately, and for example the constancy of $f^2 + h^2$ can be understood via the elliptic function identity

$$\mathrm{cn}^2(u) + \mathrm{sn}^2(u) = 1 \tag{18.30}$$

Having found an analytic form for Ω_B, the question arises as to whether we can get an analytic expression for the 4D 'attitude spinor' ψ. If $\psi\tilde{\psi} = 1$, then we can rewrite (18.20) as

$$\dot{\psi} = -\frac{1}{2}\psi\Omega_B \tag{18.31}$$

which is therefore the dynamical equation for ψ. We note, moreover, that this equation implies that $\psi\tilde{\psi}$ is constant, so that ψ is maintained as a rotor by the equations if it starts as one. Examining the literature, it is not clear if this equation has been solved analytically in the case we are currently interested in, which (in conventional terms) corresponds to finding the attitude spinor of a general non-symmetric top in 3D. Of course, conventionally, this would be phrased not in terms of an attitude spinor, but probably via rotation matrices—the question still stands, however, has an analytic solution been found for this level of the motion?

Leaving this as a pending question, we can proceed either exactly, by integrating the equations numerically, or by making approximations. As already discussed shortly after Eq. (18.23), an obvious approximation to make, applicable to the case where the linear dimensions of the rigid body are much smaller than the curvature scale of the space, is to treat the moments of inertia as those of a symmetric top, rather than the general non-symmetric case. We already have the full analytic solution for ψ for this case, which is given by Eq. (1.11) in Chap. 1, i.e. we just

Fig. 18.2 Actual motion of centre of mass of the dumbbell in the numerical example used (*small circle*) versus geodesic motion with equivalent starting conditions (*great circle*)

use the ordinary GA solution in 3D space! For convenience, we repeat the solution here:

$$R(t) = \exp\left(-\frac{1}{2}i_1^{-1}Lt\right)R(0)\exp\left(-\frac{1}{2}\omega_3(1 - i_3/i_1)Ie_3 t\right) \qquad (18.32)$$

Of course there are some details we need to go through as to exactly how to perform the approximations required—i_3 is unambiguous, but given that in fact i_1 is not exactly equal to i_2, what should we take for i_1 in this formula, and what should we take for ω_3 and L?

Once having made these choices, which are not critical, we can integrate the rotor equations both numerically and via the analytical approximation, and very good agreement is found. Now we get to a key observation, regarding the motion of the centre of mass of the dumbbell. Naively, we would expect this centre of mass motion to be along a geodesic in the curved space.

However, this is *not* what happens! In Fig. 18.2 we show the actual motion of the centre of mass of the dumbbell in our numerical example, versus the geodesic motion that would correspond to the same starting conditions. The expected geodesic motion is of course a great circle spanning the equator of the sphere. The actual motion is (to a good approximation) a *small circle* going through the same starting point and tangent to the great circle. How 'tight' this circle is, is a function of how fast the object is spinning and of the ratio of the length of dumbbell to the total radius of curvature of the space.

Given our analytic approximation to the rotor solution, it is easy to find out what is happening. The centre of mass of the dumbbell in 2D real space is determined by where the vector e is rotated to by Ψ. How does this manage to acquire a y component, given that the dumbbell is spinning in the (x, y) plane and projected with an x velocity only? The answer, of course, is *precession*! The motion of the transformed e vector about a small circle is just exactly the expected precession behaviour of a symmetric top, where the 3-axis of the top, as we know, will trace out a cone in space. The novel thing here, is that all *real*-space motion is in the (x, y) plane only, and the 3-axis of our top is actually moving in the conformal space, where the direction it points is indicating *position* in 2D space!

The observed manifestation of this precession behaviour in 2D real space is that there is an apparent y-directed component of force on the spinning object, when it is projected in the x-direction. We interpret this as a non-relativistic analogue of the force on spinning objects in general relativity, first described by Papapetrou [6]. Using our analytic approximation, it is possible to work out the magnitude of this force in our approach and indeed to generalise it to a general velocity and angular momentum. We find, in conventional 3D vector algebra notation,

$$F = \frac{4}{\lambda^2} \ell \times v \qquad (18.33)$$

Here v is the object velocity, and ℓ is its 3D angular momentum. A clear exposition of the relativistic form of coupling expected can be found in [1], where it is shown that the expected contribution to the 'geodesic equation' is

$$v \cdot \mathscr{R}(S) \qquad (18.34)$$

where v is the 4D velocity of the object, S is its 'spin bivector', and $\mathscr{R}(B)$ is the Riemann tensor of the space (which in a geometric algebra formulation, maps bivectors to bivectors, see Chap. 19).

Now the Riemann tensor for a space of constant curvature, of the type we have set up, is

$$\mathscr{R}(B) = -\frac{4}{\lambda^2} B \qquad (18.35)$$

(see e.g. [3]). Thus we can see that the two expressions match up precisely (in fact one can show that even the overall sign agrees).

As a final interesting point, we can think about the case where this force is weak, and the centre of mass of the body moves approximately along the expected geodesic, the great circle path. We can now understand that even this case is an example of precession. In fact the precession is even more extreme than in the preceding case! Technically, it would probably be called 'fast precession' and corresponds to the case where the top has more or less tipped over by 90°, and the small circle of precession is almost a great circle.

So, in this picture, *the translational motion of a rigid body actually corresponds to fast precession in the 1D up conformal space!* This is quite a surprising picture and seems a long way removed from what we may have expected at the start, but is nevertheless quite appealing as a new way of thinking about motion.

18.3 Comparison with Charles Gunn's Work on Euclidean Rigid Body Motion

In Chap. 15, Charles Gunn discusses a homogeneous model of Euclidean Geometry and applies the techniques developed to Rigid Body Dynamics. His work is also embedded in a curved space framework, see Chap. 15, Sect. 15.11. We are considering it here in the context of using a unified rotor treatment of rigid body dynamics in a '1D-up' space.

Here we discuss an alternative '1D-up' approach to Euclidean geometry within the framework of CGA, noting similarities and differences to Gunn's approach.

One essential aspect of the Gunn approach is to use *two* homogeneous (i.e. 1D up) spaces, instead of one. These are denoted W^* and W. Their formal definitions are that W is the projectivised exterior algebra of $\mathbb{R}^{m+1}$, and W^* is the projectivised exterior algebra of the dual space $\mathbb{R}^{m+1*}$. The latter, W^*, is 'Cliffordised' by equipping it with the metric $(m, 0, 1)$, i.e. m basis vectors with positive square, none with negative, and one with zero square.

Specialising to the case of 3D Euclidean geometry, positions are represented via the trivectors of W^*, and transformations on them via the rotors of W^*. The reasons for having both spaces W^* and W available are discussed in Sect. 15.2.2. One uses W^* to calculate the meet operation; while W is used for the join operator. Unlike in non-degenerate metrics, the join of two points also provides a way to calculate the distance between the two points. In the model presented below, the point of having the additional space W available is because without it various quantities which we want to form in W^*, such as the Euclidean distance between points, or various wedge products, would end up as 0, due to the zero norm vector present in the basis.

By using an operator J, which transforms between the two spaces, one is able to carry out the operations in W instead, where they reduce to just an outer product, and then use J to transform the (non-zero) result back again.

The operator J plays a crucial role in this approach and is defined in Chap. 15 in a non-metric way. Section 15.3.3 of Chap. 15 contrasts this non-metric approach to duality with the more familiar metric approach using pseudoscalar multiplication. Our approach, described below, also involves pseudoscalar multiplication, but maps from one space to another rather than one space to itself, hence we call it $\hat{J}$.

18.3.1 Translation into CGA

Here we describe two algebras, $\hat{W}$ and $\hat{W}^*$, which play roles analogous to the algebras W and W^* in the Gunn approach. The vector corresponding to the origin in the CGA is a multiple of $\bar{n}$. Since $\hat{W}^*$ is the space where dual Euclidean points are meant to live, and $\hat{W}$ is its dual, we assume that the zero square vector of $\hat{W}$, denoted e^0 in Chap. 15, is in fact $\bar{n}/2$, the $\frac{1}{2}$ being for ease of future normalisation. Concentrating on representing 3D Euclidean geometry, so working with $m = 3$, the full basis of vectors for $\hat{W}$ is thus $e^0 = \bar{n}/2$, e^1, e^2 and e^3, with the e^i being our usual basis vectors with positive square, i.e. $e^i = e_i$, $i = 1, 2, 3$. We define the pseudoscalar of this space as $I^u = e^0 e^1 e^2 e^3$.

We now assign $\hat{W}^*$ a basis which is the *reciprocal frame* for the e^μ, $\mu = 0, \ldots, 3$, just defined, befitting its role as a dual space. This reciprocal frame e_μ satisfies

$$e_\mu \cdot e^\nu = \delta_\mu^\nu \tag{18.36}$$

The e_i are already a (self-)reciprocal frame, so we just need to find e_0. The specific assignment which works is $e_0 = n_\infty$. So in this approach, $\hat{W}^*$ just differs from $\hat{W}$ by using the point at infinity, n_∞ as its zero vector, rather than the origin $\bar{n}/2$.

We define the pseudoscalar in $\hat{W}^*$ as $I_d = e_3 e_2 e_1 e_0$. This sign choice yields $I^u \cdot I_d = 1$. We note that both pseudoscalars square to 0, and this is the basic origin for why we need the two spaces to work with. The two spaces in the current model are treated on an equal footing (indeed they are part of the same 5D space with metric coming from that). This contrasts with the Gunn approach, where it is found that it is only possible to attach a metric to W^*, and not W (see Sect. 15.3.2).

Equipped with these two pseudoscalars, we can give an explicit construction for the operator $\hat{J}$, which maps between the spaces. Generalising temporarily to the m-dimensional case, this operation is

$$\hat{J}(A_r^*) = \langle A_r^* I^u \rangle_{m+1-r}$$
$$\hat{J}(A_r) = \langle A_r I_d \rangle_{m+1-r}$$

(18.37)

where A_r^* is a homogeneous grade-r element of $\hat{W}^*$, and A_r a homogeneous grade-r element of $\hat{W}$. We can see that combining with the pseudoscalar of the opposite space converts any entity into a member of the opposite space, dualising it in the process.

Applying $\hat{J}$ twice, we find

$$\hat{J}^2(A_r^*) = A_r^*$$
$$\hat{J}^2(A_r) = A_r$$

(18.38)

or in shorthand, $\hat{J}^2 = 1$. This means that $\hat{J}$ satisfies the same set of properties as J, although $\hat{J}$ is defined metrically while J is defined non-metrically.

The application to enable products in $\hat{W}^*$ to be carried out instead in $\hat{W}$ is summarised in the definition of the *join*: if A and B are two homogeneous grade objects in $\hat{W}^*$, we define

$$A \vee B = \hat{J}\left(\hat{J}(A) \wedge \hat{J}(B)\right)$$

(18.39)

This produces the same result as obtained in Chap. 15, Eq. (15.3).

18.3.2 Applications to the Euclidean Model and Rigid Bodies

With these elements of the translation in place, which we can see involve the full 5D CGA (since we need both of the extra basis vectors e and $\bar{e}$, in the combinations n_∞ and $\bar{n}$), we can now seek to understand the relationship with the treatment of the Euclidean model and rigid body dynamics in the CGA.

Firstly, we can see that since the Euclidean points are represented by their duals (in the Gunn approach), then the object in $\hat{W}^*$ representing the origin will be $\hat{J}(\bar{n}/2) = e_1 e_2 e_3$, which we can denote as I_3.

The rotors used for moving points around in $\hat{W}^*$ are exactly the same as used in Chap. 1 in the five-dimensional approach, i.e. a combination of the usual spatial rotors, built out of even combinations of 1 and the e_i, and translation rotors of the form $R_t = 1 - \frac{1}{2}tn_\infty$, where t is the 3D spatial vector through which we translate. Starting from the 'origin', therefore, and applying a translation, we reach:

$$
R_t I_3 \tilde{R}_t = R_t \hat{J}\left(e^0\right)\tilde{R}_t = R_t \left\langle \frac{1}{2}\bar{n}I_d \right\rangle_3 \tilde{R}_t
$$

$$
= R_t \frac{1}{2}\bar{n}\cdot I_d \tilde{R}_t = \left(R_t \frac{1}{2}\bar{n}\tilde{R}_t \right)\cdot I_d \tag{18.40}
$$

with the last equality following since I_d, containing a factor n_∞, is left invariant under R_t.

We can now recognise that (up to a sign) we have found out that *the representative of a Euclidean point in the Gunn approach is the usual 5D CGA null vector representative, dotted with the grade-4 object I_d*. This follows since if we let t be the 3D position vector x, then we are translating minus the null vector representing the origin in the CGA, to position x, thereby achieving the vector representative (in our usual notation) $-X$. The trivector representing Euclidean position in the Gunn approach is thus $-X \cdot I_d$. Since I_d contains a factor n_∞, this kills off the $x^2 n_\infty$ part of X, and we get a dualised form of $\frac{1}{2}\bar{n} - x$. It is not immediately clear that it is useful to lose the quadratic part of X, but since I_d is invariant under rotations and translations, dotting it with X is at least a covariant operation (as long as we are not concerned with dilations, which for rigid body motion we are not).

If we have two 3D points p and q, then we know that in the CGA, the Euclidean distance between them is got simply from

$$
d(p,q) = -\frac{1}{2}P \cdot Q \tag{18.41}
$$

To recover this distance once we have lost the quadratic part is more complicated but can be done. Forming $-P \cdot I_d$ and taking $\hat{J}$ of this yields $\frac{1}{2}\bar{n} - p$. Doing the same with Q yields $\frac{1}{2}\bar{n} - q$, and so the outer product of these is

$$
\hat{J}(-P \cdot I_d) \wedge \hat{J}(-Q \cdot I_d) = p \wedge q - (p-q)\frac{1}{2}\bar{n} \tag{18.42}
$$

If we take $\hat{J}$ of this again, thus forming overall the 'join' $(-P \cdot I_d) \vee (-Q \cdot I_d)$, this will make the $p \wedge q$ part of the product null, and the $p - q$ part non-null, meaning that the square is just proportional to $(p-q)^2$, recovering the distance.

The remaining concept we need before passing to rigid body motion is the notion of a 'free vector'. In Chap. 1 it was argued that this corresponds to a 'boundary point', in which we lose the $\bar{n}$ part of a 5D null vector. Here therefore we would expect we should lose both the $\bar{n}$ and n_∞ parts, so that a free vector is just the dualised version of the 3D vector, i.e. of the form $\hat{J}(p)$, with p a 3D vector. A typical free vector would be the velocity of a particle, $\dot{x}$. In Chap. 15, the crucial construction for rigid body dynamics is the generalised momentum of a particle, defined by $mR \vee \dot{R}$, where R is the dualised (therefore trivector) representation of the Euclidean position x. Adopting our translation, this is just the dual of

$$L_{\text{hp}} = x \wedge (m\dot{x}) - \frac{1}{2}\bar{n}m\dot{x} \tag{18.43}$$

which, as we can see, does indeed encode both the linear and angular momentum (the 'hp' subscript is meant to indicate the 'homogeneous space' particle angular momentum).

18.3.3 Comparison with the Curved Space Approach

Finally, we look at the comparison between the '1D up' curved space approach, discussed in the bulk of this chapter, with the approach by Gunn. To apply the curved space approach to motion in Euclidean space, we need to take the limit as the radius of curvature of the space tends to infinity. As a point of comparison, we therefore look at the expansion of the curved space point particle generalised angular momentum

$$L_{\text{cp}} = m\frac{\lambda^2}{4}\hat{Y}\wedge\dot{\hat{Y}} \tag{18.44}$$

as an asymptotic expansion in λ. This yields

$$L_{\text{cp}} \approx \ell_p - \frac{\lambda}{2}pe \tag{18.45}$$

where ℓ_p is the ordinary Euclidean space angular momentum of the particle, and p its 3-space momentum. Comparing to L_{hp} above, we note two things: (i) the homogeneous space version above is in fact dimensionally inhomogeneous, whereas L_{cp} has the correct units of angular momentum throughout, and (ii) the role of $\bar{n}$ in the L_{hp} expression is taken over by e in L_{cp}. This of course is why we do not need to have recourse to a 'join' operation (which necessitates using the full 5D algebra) in working in the curved space approach—we do not have to deal with a basis element squaring to zero.

We see that in the flat limit of the curved approach, we have a sensible version of generalised angular momentum, provided that we keep a factor of λ in place to provide dimensional homogeneity. Once we have taken the limit of course, this λ

could in principle take any length value we wish—in numerical work it would not be necessary to 'unbalance' the dynamical scale of different parts of expressions by using a very large value for λ. However, the same apparently does not apply to translation rotors: we can approximate these (in the flat space limit) via

$$R_t \approx 1 + \frac{1}{2\lambda} te \qquad (18.46)$$

and these will have approximately the right effect in moving points, but only if the ratio $|t|/\lambda$ is small. An obvious way of removing this limitation, whilst still being compatible with having λ appearing in the expression for generalised angular momentum, is to have a rule that terms which are *second order* in the expansion parameter $1/\lambda$ are set to zero. This then means that the above approximation to R_t translates e through exactly t. Applying this systematically to all products, we will be able to represent Euclidean motions exactly in 4D and also be able to deal with Euclidean rigid body dynamics in the same setting. This method appears to coincide with that proposed elsewhere in Chap. 17 by Mullineux and Simpson, although the full correspondence needs further investigation.

18.4 Conclusions

We have considered a 'one dimension up' approach to conformal geometric algebra, where the advantages of the unification of translations and spatial rotation into a common rotor structure are maintained, but where, particularly with regard to rigid body motion, we have a better match between the degrees of freedom and the elements of the algebra than occurs in the 'two dimensions up' approach of the full CGA.

This necessitates working in a curved space, and this prompted an examination of the motion of a spinning, translating rigid body in such a space, where it was shown that an extra 'force' occurs as compared to motion in flat space, perpendicular to both the angular momentum and the velocity, and affording a simple non-relativistic derivation of what is called the Papapetrou force in General Relativity. As the radius of curvature of the space becomes larger, this force goes to zero, but there is still a non-trivial effect of working within a curved space, in that ordinary translational motion is revealed as *precessional motion* in the higher space. In 2D this is an exact equivalence, in that the standard 3D solutions for a precessing top may be used to find the motion of a spinning rigid body in a 2D curved space.

We then carried out a comparison with the approach to Euclidean geometry and rigid body motion discussed by Charles Gunn in this volume. This revealed that an analogous treatment was possible in CGA, and explicit analogues were given for the crucial 'J' operation, and for the definition of the 'join'. The full (5D) CGA is necessary to realise this analogy, in which, corresponding to Gunn's W and W^*, we constructed two spaces $\hat{W}$ and $\hat{W}^*$, each one just ordinary 3D Euclidean space, supplemented by either the CGA origin ($\bar{n}/2$) or point at infinity (n_∞) respectively.

We emphasise that this is only a *possible* translation of Gunn's work into CGA, motivated by the apparent correspondence of the results obtained in this way, and that establishing the exact relationship between the two approaches remains a topic for further work.

The use of the two partial spaces, rather than the full CGA, makes the construction of certain things more difficult, such as the Euclidean distance between two points, or the generalised angular momentum, but it appears the same information is present, albeit packaged in a different way. This would not apply in an obvious way to various other desirable features of the full CGA, such as ability to deal with the full conformal group, and to be able to construct and intersect geometric objects transcending simple lines and planes; however these lie outside what rigid body dynamics strictly needs.

Finally, the flat space (large curvature scale) limit of the curved space approach was discussed and compared briefly with both the Gunn work and the approach of Mullineux and Simpson. For the latter, it is likely that retaining only first-order terms in an expansion in the reciprocal curvature scale makes the methods coincide, although the full correspondence needs further investigation.

18.5 Exercises

18.1 Let $I = e_1 e_2 e_3 e$ be the pseudoscalar for 4D 'spherical' space. Show that $I^2 = +1$ and that it commutes with all vectors in the space.

Now define the two objects

$$P_+ = \frac{1}{2}(1 + I), \quad P_- = \frac{1}{2}(1 - I)$$

Show that these are *orthogonal projectors*, i.e. satisfy

$$P_+^2 = P_+, \quad P_-^2 = P_-, \quad P_+ P_- = P_- P_+ = 0$$

Using these, we can define a basis of bivectors in the 4D space as

$$\xi_1 = e_2 e_3 P_+, \quad \eta_1 = e_2 e_3 P_-$$
$$\xi_2 = e_3 e_1 P_+, \quad \eta_2 = e_3 e_1 P_-$$
$$\xi_3 = e_1 e_2 P_+, \quad \eta_3 = e_1 e_2 P_-$$

Show that ξ_i and η_i both have 'structure constant' relations of the form

$$[\xi_i, \xi_j] = \varepsilon_{ijk}\xi_k$$

where $[a, b]$ is the commutator $ab - ba$. Show also that $[\xi_i, \eta_j] = 0$ (all i, j).

By expanding each of the ξ_i and η_i explicitly, show that their exponentials correspond to *screw* motions in which there is a rotation accompanied by a translation about the same axis, and give an interpretation of each geometrically.

18.2 For spherical space, what do the *d-lines* (geodesics) look like in 3D?

Show that if $\hat{Y}_1$ and $\hat{Y}_2$ are the 4D representatives of two points on the d-line L, then $Y_3 = \alpha\hat{Y}_1 + \beta\hat{Y}_2$, where α and β are real scalars, also lies on it.

Use this to show why $L \cdot \hat{Y}_1$ (suitably normalised) can be interpreted as the unit tangent vector to the line at $\hat{Y}_1$.

18.3 Show, using the CGA translation of the Gunn approach, that the norm of a free vector with $\hat{W}^*$ representative $\mathbf{V}$ is given by the norm of $\mathbf{V} \vee \mathbf{P}$ for any normalised Euclidean point $\mathbf{P}$. Why can't we just take $\|\mathbf{V}\|$?

References

1. Doran, C.J.L., Lasenby, A.N., Challinor, A.D., Gull, S.F.: Effects of spin-torsion in gauge theory gravity. J. Math. Phys. **39**(6), 3303–3321 (1998)
2. Heath, R.S.: On the dynamics of a rigid body in elliptic space. Philos. Trans. R. Soc. Lond. **175**, 281–324 (1884)
3. Lasenby, A.N.: Conformal geometry and the Universe. Unpublished paper available on the site http://www.mrao.cam.ac.uk/~clifford/publications (2003)
4. Lasenby, A.N.: Recent applications of conformal geometric algebra. In: Li, H., Olver, P.J., Sommer, G. (eds.) Computer Algebra and Geometric Algebra with Applications. Lecture Notes in Computer Science, p. 298. Springer, Berlin (2005)
5. Lasenby, A.N.: Some results in the conformal geometry approach to the Dirac equation, electromagnetism and gravity (2011, in preparation)
6. Papapetrou, A.: Spinning test-particles in general relativity. I. Philos. Trans. R. Soc. Lond. A **209**, 248–258 (1951)

Part VII
Towards Coordinate-Free Differential Geometry

Differential geometry is an obvious target for geometric algebra. In its classical description by means of coordinate charts, its structure easily gets hidden in notation, and that limits its applications to specialized fields. Geometric algebra should be able to do better, especially if combined with modern insights in the system of geometrical invariants.

The Shape of Differential Geometry in Geometric Calculus

19

David Hestenes

Abstract

We review the foundations for coordinate-free differential geometry in *Geometric Calculus*. In particular, we see how both extrinsic and intrinsic geometry of a manifold can be characterized by a single bivector-valued one-form called the *Shape Operator*. The challenge is to adapt this formalism to *Conformal Geometric Algebra* for wide application in computer science and engineering.

19.1 Introduction

Geometric Algebra (GA) enabled the development of several new methods for *coordinate-free differential geometry* on manifolds of any dimension in [8]. In the most innovative of these methods, both extrinsic and intrinsic geometry of a manifold are characterized by a single bivector-valued one-form called the *shape operator*, which is essentially the derivative of the tangent space pseudoscalar as it slides along the manifold. I regard creation of this approach to differential geometry as some of my best work, so I am rather disappointed that, apart from one fine application [15], it has not been further exploited by me or anyone else.

As abundantly demonstrated in this volume and elsewhere [2, 7], *Conformal Geometric Algebra* (CGA) has recently emerged as an ideal tool for computational geometry in computer science and engineering. My purpose here is to prepare the way for integrating the Shape Operator into the CGA tool kit for routine applications of differential geometry. I hope this will stimulate others to deal with practical implementation and applications.

D. Hestenes (✉)
Arizona State University, Tempe, AZ, USA
e-mail: hestenes@asu.edu

L. Dorst, J. Lasenby (eds.), *Guide to Geometric Algebra in Practice*,
DOI 10.1007/978-0-85729-811-9_19, © Springer-Verlag London Limited 2011

19.2   Geometric Calculus—Basic Concepts

Geometric Algebra is essential to formulate the basic concepts of "vector derivative" and "directed integral." Their initial formulations in [3] raised questions about
relations to the Cartan's concept of "differential forms" [5]. That stimulated development of the *Geometric Calculus* (GC) in Chaps. 4–7 of [8].

To elucidate the structure of Geometric Calculus, its basic concepts are listed
here, and their unique features are described in subsequent sections. The purpose is
to explain how GC enables differential geometry without coordinates.

- *Universal Geometric Algebra*—arbitrary dimension and signature
- *Vector manifolds*—for representing any manifold
- *Directed integrals* and *differential forms*
- *Vector derivative* and the *fundamental theorem of calculus*
- *Differentials* and *codifferentials* for mappings and fields
- *Shape* and *curvature* for differential geometry

19.3   Differentiable Manifolds as Vector Manifolds

A (differentiable) manifold $\mathcal{M}^m$ of dimension m is a set on which differential and
integral calculus is well defined. The standard definition requires covering the manifold with overlapping charts of local coordinates. Calculus is then done indirectly
by local mappings to $\mathbb{R}^m = \mathbb{R} \otimes \mathbb{R} \otimes \cdots \otimes \mathbb{R}$. Proofs are then required to establish
that results are independent of coordinates.

In contrast, a *vector manifold* $\mathcal{M}^m = \{x\}$ of dimension m is defined as a set of
vectors (called points) in GA that generates at each point x a *tangent space* with
unit *pseudoscalar* $I_m(x)$. Any other manifold can then be defined as a set that is
isomorphic to a vector manifold.

Thus, GC enables a concept of manifold that is manifestly coordinate-free. As
we shall see, calculus can then be done directly with algebraic operations on points,
and geometry is completely determined by derivatives of the pseudoscalar. It should
be noted that a vector manifold can be defined without assuming that it is embedded
in a vector space of specified dimension, though embedding theorems can no doubt
be proved therefrom.

Though GC enables a coordinate-free approach to manifolds, it also provides a
very efficient formalism for handling coordinates. That is worth reviewing briefly,
because it facilitates direct connection to the standard literature and, of course, use
of coordinates when appropriate.

The vector-valued function $x = x(x^1, x^2, \ldots, x^m)$ represents a patch of $\mathcal{M}^m$ parameterized by *scalar coordinates* (Fig. 19.1). The inverse mapping into $\mathbb{R}^m$ is given
by *coordinate functions* $x^\mu = x^\mu(x)$. A *coordinate frame* $\{e_\mu = e_\mu(x)\}$ is defined
by

$$e_\mu = \partial_\mu x = \frac{\partial x}{\partial x^\mu} = \lim \frac{\Delta x}{\Delta x^\mu}$$

Fig. 19.1 Coordinate curves

with pseudoscalar

$$e_{(m)} = e_1 \wedge e_2 \wedge \cdots \wedge e_m = |e_{(m)}| I_m.$$

It is interesting to note that Elie Cartan used the expression $e_\mu = \partial_\mu x$ in an intuitive way at the foundation of his approach to differential geometry. Thus, GC provides the means to give it a more rigorous formulation.

Calculations with frames are greatly facilitated by employing a *reciprocal frame* $\{x^\mu\}$, often defined implicitly by the equations $e^\mu \cdot e_\nu = \delta^\mu_\nu$, which have the solution

$$e^\mu = \left(e_1 \wedge \cdots ()_\mu \wedge \cdots \wedge e_m \right) e_{(m)}^{-1},$$

where the μth vector is omitted from the product. This can be used for a coordinate definition of the *vector derivative*, that is, the derivative with respect to the point x:

$$\partial = \partial_x = e^\mu \partial_\mu \quad \text{where} \quad \partial_\mu = e_\mu \cdot \partial = \frac{\partial}{\partial x^\mu}. \tag{19.1}$$

Consequently, the reciprocal vectors can be expressed as *gradients*:

$$e^\mu = \partial x^\mu.$$

The question remains: How can the vector derivative be defined without coordinates? The answer is given by first defining integration on vector manifolds, to which we now turn.

19.4   Directed Integrals and the Fundamental Theorem

Let $F = F(x)$ be a multivector-valued function on the manifold $\mathcal{M} = \mathcal{M}^m$ (Fig. 19.2) with a *directed measure* $d^m x = |d^m x| I_m(x)$. The measure can be expressed in terms of coordinates by

$$d^m x = d_1 x \wedge d_2 x \wedge \cdots \wedge d_m x = e_1 \wedge e_2 \wedge \cdots \wedge e_m \, dx^1 dx^2 \ldots dx^m,$$

where $d_\mu x = e_\mu(x) d^\mu x$ (no sum). Accordingly, the usual scalar-valued volume element of integration is given by

$$|d^m x| = |e_{(m)}| dx^1 dx^2 \ldots dx^m.$$

Fig. 19.2 Vector manifold

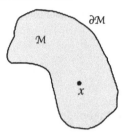

The directed integral of F can now be expressed as a standard multiple integral:

$$\int_{\mathcal{M}} d^m x\, F = \int_{\mathcal{M}} e_{(m)}\, dx^1\, dx^2 \dots dx^m.$$

This establishes contact with standard integration theory. It is worth mentioning that there are many practical and theoretical advantages to defining and evaluating the directed integral without reducing it to a multiple integral with scalar coordinates, though that cannot be addressed here.

Now we are equipped to formulate the *fundamental theorem of calculus* in the powerful general form that GC makes possible. We shall see that this leads us to a coordinate-free definition of the vector derivative in terms of the directed integral that reduces proof of the fundamental theorem to a near triviality. In addition, it generalizes the definition of derivative through (19.1) to apply to discontinuous functions (such as occur at the boundaries of material media in physics).

It is enlightening to begin with the important special case of a manifold embedded in a vector space: $\mathcal{M} = \mathcal{M}^m \subset \mathcal{V}^n$. Let $\nabla = \nabla_x$ denote the derivative of a point in the vector space $\mathcal{V}^n$. The derivative of any field $F = F(x)$ can then be decomposed algebraically into

$$\nabla F = \nabla \cdot F + \nabla \wedge F.$$

Thus GC unifies the familiar concepts of *"divergence"* and *"curl"* into a single vector derivative, which could well be dubbed the *"gradient,"* as it reduces to the usual gradient when the field is scalar-valued.

Now we can formulate the first generalization of the fundamental theorem of calculus made possible by GC:

$$\int_{\mathcal{M}} \left(d^m x\right) \cdot \nabla F = \int_{\partial \mathcal{M}} d^{m-1} x\, F. \tag{19.2}$$

As explained in [3] when this was first written down, all the integral formulas of standard vector calculus (including those attributed to Gauss, Stokes, and Green) are included as special cases of this formula.

I was puzzled for a while by the role of the inner product on the left side of (19.2). Then I realized that its function is to project the derivative ∇ to a derivative on the submanifold $\mathcal{M}$, as expressed by

$$\partial = \partial_x = I_m^{-1}(I_m \cdot \nabla).$$

Hence, one can write $d^m x \, \partial = (d^m x) \cdot \partial = (d^m x) \cdot \nabla$, so the theorem (19.2) can be written in the form:

$$\int_{\mathcal{M}} d^m x \, \partial F = \int_{\partial \mathcal{M}} d^{m-1} x F, \tag{19.3}$$

which has no explicit reference to the embedding space. That observation inspired the following *coordinate-free definition for the vector derivative with respect to x in $\mathcal{M}$ without reference to any embedding space*:

$$\partial F = \lim_{d\omega \to 0} \frac{1}{d\omega} \oint d\sigma F, \tag{19.4}$$

where $d\omega = d^m x$ and $d\sigma = d^{m-1} x$. I called this the *"tangential derivative,"* when I first proposed it, to emphasize that it is determined by the restriction of the variable x to $\mathcal{M}$. One consequence of that is that the operator $\partial = \partial_x$ is itself a function of x, so, for example, the theorem $\nabla \wedge \nabla = 0$, which holds for derivatives on a vector space (a "flat manifold"), does not apply for derivatives on a curved manifold, where $\partial \wedge \partial \neq 0$ in general. That property is essential for the formulation of differential geometry with GC, as we see below.

This is not the place to discuss limit processes for defining the vector derivative (19.4) and proving the fundamental theorem (19.3). However, the method of simplices in [16] deserves mention, because it provides a practical approach to finite element approximations.

Now we are prepared to explain how GC generalizes Cartan's theory of differential forms. For $k \leq m$, a *differential k-form* $L = L(d^k x, x)$ on a manifold $\mathcal{M}^m$ is a multivector-valued k-form, that is, it is a linear function of the k-vector $d^k x$ at each point x. The simplest example is the volume element $d^k x$, which is a k-vector-valued k-form. Another example is the $(m-1)$-form $d^{m-1} x F(x)$ on the right side of (19.3).

In Cartan's terminology, the *exterior differential* of the k-form L is a $(k+1)$-form dL defined here by

$$dL \equiv \dot{L}\big(d^{k+1} x \cdot \dot{\partial}\big) = L\big(d^{k+1} x \cdot \dot{\partial}, \dot{x}\big),$$

where the overdot indicates the variable differentiated. Cartan's abbreviated notation dL suppresses the dependence on the volume element $d^{k+1} x$ that is explicit in this definition. Note that the term "differential" as used here refers to the fact that the form is a linear function of a "volume element" intended to reside under an integral sign. In the next section we use the term "differential" in a different sense related to transformations. However, the two senses are intertwined when the transformation is applied to a form, which is just a particular kind of tensor.

Now we can express the *Fundamental Theorem of Geometric Calculus* in its most general form by the equation

$$\int_{\mathcal{M}} dL = \oint_{\partial\mathcal{M}} L.$$ (19.5)

This looks identical to the "*Generalized Stokes' Theorem*" in Cartan's calculus of differential forms. However, Cartan's forms are limited by being scalar-valued and lacking the complete algebraic structure of GC. More specifically, Cartan's theory is limited to functions of the form $L = \langle d^k x F(x)\rangle = (d^k x) \cdot F(x)$, where the angular brackets indicate scalar part, and the center-dot applies if F is k-vector-valued. Accordingly, the exterior differential becomes

$$dL = \langle d^{k+1}x \partial F(x)\rangle = \langle d^{k+1}x \partial \wedge F\rangle = (d^{k+1}x) \cdot (\partial \wedge F).$$

Thus, Cartan's exterior differential is equivalent to the curl in GC, though its use in applications is more limited. When it is used to formulate Maxwell's equations, for example, the implicit volume element is just excess baggage, except when integration is intended.

The GC generalization to multivector-valued differential forms has profound applications. For example, it follows immediately from (19.3) that, with simple provisos,

$$\partial F = 0 \quad \Longleftrightarrow \quad \int_{\partial\mathcal{M}} d^{m-1}x F = 0.$$

For $m = 2$, this can be recognized as *Cauchy's Theorem* for complex variables, so it gives a straight-forward generalization of that theorem to higher dimensions. Similarly, a generalization of the justly famous *Cauchy Integral formula* can easily be derived from (19.5), as explained elsewhere [5, 8].

19.5 Mappings and Transformations

With the concept of vector derivative in hand, we are prepared to see how Geometric Calculus enables coordinate-free *transformations* of multivector fields on a given manifold or in *mappings* from one manifold to another. The power of this formalism is amply demonstrated in an elegant new approach to General Relativity called *Gauge Theory Gravity* [1, 11].

Let f be a an invertible diffeomorphism from one region of a given manifold to another, as expressed by

$$f : x \to x' = f(x), \quad \text{so that} \quad x = f^{-1}(x').$$

Fig. 19.3 Induced
transformations of vector
fields

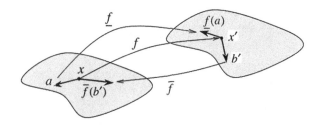

This transformation induces a linear transformation $\underline{f}$ of the tangent space at each point called the *differential* of f. Accordingly, each vector field $a = a(x)$ undergoes a transformation (Fig. 19.3) defined by

$$\underline{f} : a = a(x) \to a' = \underline{f}(a) \equiv a \cdot \partial f, \quad \text{so that} \quad a = \underline{f}^{-1}(a').$$

The *adjoint* $\overline{f}$ of the transformation is an induced linear transformation (Fig. 19.3) in the reverse direction:

$$\overline{f} : b' = b'(x') \to b = \overline{f}(b') \equiv \partial_x f(x) \cdot b'.$$

For applications, one needs the theorem that the adjoint of the inverse transformation is the inverse of the adjoint:

$$\overline{f^{-1}} = \overline{f}^{-1} : b(x) \to b'(x') = \overline{f^{-1}}[b(f(x'))].$$

Note the complementary roles of directional derivative and gradient in the definitions of differential and adjoint. Also note that no notion of "differential as infinitesimal displacement" is involved.

To relate the GC approach to standard tensor calculus, consider a *rank-2 tensor* (field) $T(a, b')$ that is a linear function of vector fields a and b'. If these fields transform according to the differential and adjoint laws respectively, the tensor is said to be *contravariant* in the first argument and *covariant* in the second.

The unique power of GC is manifest in the concept of *outermorphism*: the unique extension of a linear transformation defined on a vector space to *a linear transformation that preserves the outer product* and hence defined on the entire GA generated by the vector space [4, 8]. For a pair of vector fields, the outermorphism of the differential gives us

$$\underline{f} : a \wedge b \to \underline{f}(a \wedge b) = \underline{f}(a) \wedge \underline{f}(b),$$

as illustrated in Fig. 19.4. By linearity, this property generalizes easily to the outermorphism of any multivector field [8]. In particular, it follows that the outermorphism of the pseudoscalar $I = I(x)$ is

$$\underline{f} : I \to \underline{f}(I) = J_f I \quad \text{so that} \quad J_f = \det \underline{f} = I^{-1} \underline{f}(I),$$

Fig. 19.4 Outermorphism of
the differential

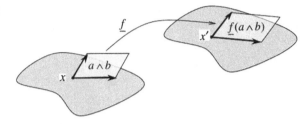

which shows that the Jacobian of the transformation, J_f, is just a scale factor induced by the outermorphism of the pseudoscalar.

A generalization of the familiar *chain rule* for differentiation is given by the transformation law for the vector derivative:

$$\overline{f} : \partial' \to \partial = \overline{f}(\partial') \quad \text{or} \quad \partial_x = \overline{f}(\partial_{x'}).$$

This implies invariance of the directional derivative:

$$a \cdot \partial = a \cdot \overline{f}(\partial') = \underline{f}(a) \cdot \partial' = a' \cdot \partial'. \tag{19.6}$$

Note that this applies whether a is a vector field or just a single vector in a given tangent space. For example, if $x = x(\tau)$ is a curve with tangent $\dot{x} = dx/d\tau$, then (19.6) implies invariance of the chain rule for differentiating fields:

$$\frac{d}{d\tau} = \dot{x} \cdot \partial_x = \dot{x} \cdot \overline{f}(\partial_{x'}) = \underline{f}(\dot{x}) \cdot \partial_{x'} = \dot{x}' \cdot \partial_{x'}.$$

Now we have all the necessary tools in hand for addressing the main subject of this chapter: *coordinate-free differential geometry*.

19.6   Shape and Curvature

My purpose here is to explain how *the differential geometry of a given vector manifold* $\mathcal{M} = \{x\}$ *is completely determined by properties of its pseudoscalar* $I = I(x)$. For an *oriented manifold*, the pseudoscalar is a single-valued field defined on the manifold. It can be visualized at each point (Fig. 19.5) as the tangent space (which it determines). Actually, as we have seen, it is a defining property of the manifold. For an unoriented manifold like a Möbius strip, the pseudoscalar is double-valued, as the orientation (algebraic sign) can be reversed by sliding it smoothly along a closed curve. But that is a minor point that will not concern us.

Let $a = a(x)$ be a *vector-valued function* defined on the manifold. We say that it is a *vector field* if its values lie in the tangent space at each point of the manifold. This property is definitively determined by the pseudoscalar. Thus *projection* into the tangent space is a linear function defined by $P(a) \equiv (a \cdot I)I^{-1} \equiv a_{\parallel}$, while

Fig. 19.5 Manifold
pseudoscalar

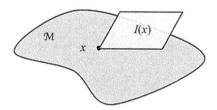

rejection from the tangent space is defined by $P_\perp(a) \equiv (a \wedge I)I^{-1} \equiv a_\perp$. Whence
we derive the obvious result

$$a = (a \cdot I + a \wedge I)I^{-1} = P(a) + P_\perp(a) = a_\| + a_\perp.$$

Of course, a vector field has the tangency properties $a \wedge I = 0$ and $a = P(a)$.

The differential $P_b(a)$ of the manifold projection operator is given by straightforward differentiation:

$$P_b(a) \equiv b \cdot \dot\partial \dot P(a) = b \cdot \partial P(a) - P(b \cdot \partial a).$$

Note that $b \cdot \partial = [P(b)] \cdot \partial$, that is, the inner product of any vector with the vector
derivative projects that vector into the tangent space, so differentials are always
taken with respect to tangent vectors or vector fields.

As we are not interested in the specific vector direction b, we can differentiate it
out to get the *shape tensor*:

$$S(a) \equiv \dot\partial \dot P(a) = \partial_b P_b(a) = \dot\partial \wedge \dot P(a) + \dot\partial \cdot \dot P(a).$$

It is easy to prove the following theorems:

$$\dot\partial \wedge \dot P(a) = S(a_\|) \quad \Rightarrow \quad \dot\partial \wedge \dot P(a_\perp) = 0,$$
$$\dot\partial \cdot \dot P(a) = S(a_\perp) \quad \Rightarrow \quad \dot\partial \wedge \dot P(a_\|) = 0.$$

Consequently, we can decompose the shape tensor into bivector and scalar parts:

$$S(a) = S_a + N \cdot a,$$

where

$$S_a \equiv \dot\partial \wedge \dot P(a) = S[P(a)] = S(a_\|) \tag{19.7}$$

and

$$N \equiv \dot P(\dot\partial) = \partial_a S_a. \tag{19.8}$$

Fig. 19.6 Shape and spur

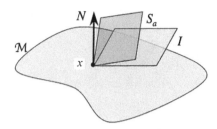

By virtue of (19.7), the *shape bivector* S_a could well be called the *curl* of the manifold $\mathcal{M}$. It follows that $P(S_a) = 0$, so the bivector-valued tensor S_a is not a field, as its values are not in the tangent algebra of $\mathcal{M}$.

The vector N is called the *spur* (of $\mathcal{M}$), see Fig. 19.6. It follows from (19.8) that $N \cdot P(a) = N \cdot a_{\parallel} = 0$, so N is not a vector field, as it is everywhere orthogonal to the tangent algebra of $\mathcal{M}$. As far as I know, the spur was not identified as a significant geometrical concept until it was first formulated in GC. We will not pursue it here. Rather, we aim to see how the shape tensor relates to standard concepts of differential geometry.

The shape bivector has a simple geometric interpretation with great intuitive appeal. It is easy to prove from its definition above that *the shape bivector is the rotational velocity of the pseudoscalar as it slides along the manifold*; formally,

$$S_a = I^{-1} a \cdot \partial I.$$

Alternatively,

$$\partial I = I S_a = I \times S_a,$$

where the symbol $\times$ denotes the commutator product, and the last inequality is a consequence of $I \cdot S_a = 0$ and $I \wedge S_a = 0$.

The *curvature* of the manifold is given by the shape commutator, defined for vectors a and b by

$$C(a \wedge b) \equiv S_a \times S_b = P(S_a \times S_b) + P_{\perp}(S_a \times S_b). \tag{19.9}$$

The right side shows that the full curvature decomposes into distinct intrinsic and extrinsic parts. It can be proved that the intrinsic part is the usual *Riemann curvature*, which can accordingly be defined by

$$R(a \wedge b) \equiv P(S_a \times S_b).$$

Readers may be surprised that this simple expression does not involve the usual "coefficients of connexion." The moral is that the treatment of intrinsic geometry can be simplified by coordinating it with extrinsic geometry!—a striking claim that surely deserves close scrutiny. Supported by the power of GC, the shape tensor provides the means for investigating this claim.

Extension of the derivative concept to "covariant derivative" is at the heart of standard differential geometry. GC generalizes this to extension of the vector derivative $\partial = \partial_x$ to a *coderivative* $D = D_x$ defined, for action on any multivector-valued function $A = A(x)$, by

$$DA \equiv P(\partial A) = D \wedge A + D \cdot A.$$

It follows that

$$\partial A = DA + S(A),$$

where the *shape tensor* for A is given by

$$S(A) \equiv \dot{\partial}\dot{P}(A) = \dot{\partial} \wedge \dot{P}(A) + \dot{\partial} \cdot \dot{P}(A) = S(A_{\parallel}) + S(A_{\perp}).$$

For any tangent field $A = P(A) = A(x)$, the *cocurl* is given by

$$D \wedge A = P(\partial \wedge A) = \partial \wedge A - S(A),$$

while the *codivergence* is given by

$$\partial \cdot A = D \cdot A = \left[D \wedge (AI) \right] I^{-1}.$$

Many valuable differential identities can be derived from these definitions, such as

$$D \wedge D \wedge A = P(\partial \wedge \partial \wedge A) = 0,$$

$$D \cdot (D \cdot A) = \partial \cdot (\partial \cdot A) = 0.$$

The equivalent of the covariant derivative is the *directional coderivative* (or *codifferential*) defined by

$$\delta_a A \equiv a \cdot DA \equiv P(a \cdot \partial A).$$

The commutator of codifferentials is determined by the intrinsic curvature:

$$(\delta_a \delta_b - \delta_b \delta_a)A = A \times R(a \wedge b).$$

Fig. 19.7 The normal and its differential

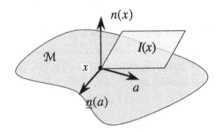

The Riemann curvature is a bivector function of a bivector variable with the following (mostly well-known) properties of great importance in General Relativity Theory:

$$\text{Symmetry:} \quad (a \wedge b) \cdot R(c \wedge d) = (c \wedge d) \cdot R(a \wedge b)$$

$$\text{Ricci Identity:} \quad a \cdot R(b \wedge c) + b \cdot R(c \wedge a) + c \cdot R(a \wedge b) = 0$$

$$\text{Ricci tensor:} \quad R(b) \equiv \partial_a R(a \wedge b) = -D \cdot S_b$$

$$\text{Scalar curvature:} \quad R \equiv \partial_b R(b) = \partial \cdot N$$

$$\text{Bianchi identity:} \quad \dot{D} \wedge \dot{R}(a \wedge b) = 0$$

$$\text{Einstein tensor:} \quad G(a) \equiv R(a) - \frac{1}{2} a R$$

This should suffice to clarify how the shape tensor relates to conventional formulations of differential geometry. For important examples of manifold geometry, we turn to the special case of manifolds that are hypersurfaces in a given manifold.

19.7 Hypersurfaces and Classical Geometry

The shape tensor generalizes the original approach to classical differential geometry developed by Gauss, who characterized *surfaces* (2D manifolds) in terms of their normals. The straightforward generalization of his approach to hypersurfaces of any dimension has been formulated in modern terms by [10]. Let us see how to do it with GC.

Let $\mathcal{M} = \mathcal{M}_m$ be an m-dimensional hypersurface in $\mathbb{E}_{m+1}$. Let $i = \langle i \rangle_{m+1} =$ constant be the unit pseudoscalar for $\mathbb{E}_{m+1}$. Then the pseudoscalar for $\mathcal{M}$ is given by $I = ni$, where $n = n(x)$ is the *unit normal*. The function $n = n(x)$ is often called the "Gauss map" to support the intuition that it is a mapping of the manifold onto a unit sphere. Instead, we describe the normal sliding on the hypersurface in terms of its differential $\underline{n}(a) = a \cdot \partial n$. Then the shape of the hypersurface reduces to a function of the normal and its differential, see Fig. 19.7:

$$S_a = I^{-1} a \cdot \partial I = n\underline{n}(a) = n \wedge \underline{n}(a).$$

Fig. 19.8 Ellipsoidal
equipotentials

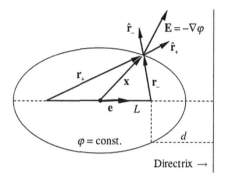

It is now straightforward to reduce geometric quantities in the previous section to functions of the normal and its differential:

$$\text{Curvature:} \quad R(a \wedge b) = P(S_a \times S_b) = \underline{n}(a) \wedge \underline{n}(b) = \underline{n}(a \wedge b)$$

$$\text{Mean Curvature:} \quad H \equiv \frac{1}{m} \partial_a \underline{n}(a), \text{ with } \partial_a \underline{n}(a) = \text{tr}\,\underline{n} = \partial \cdot n = -n \cdot N$$

$$\text{Scalar Curvature:} \quad R = (\partial_b \wedge \partial_a) \cdot \underline{n}(a) \wedge \underline{n}(b) = (\text{tr}\,\underline{n})^2 - \text{tr}\,\underline{n}^2$$

$$\text{Gaussian Curvature:} \quad \kappa = I^{-1} \cdot \underline{n}(I)$$

For $m = 2$, the Gaussian curvature can be written $\kappa = I^{-1} \cdot R(I)$, so it is equivalent to the Riemann curvature, which has only one component. It is worth noting that the extrinsic component of the curvature (19.9) gives the classical Codazzi–Mainardi equations for extrinsic geometry of a hypersurface, but we will no go into that.

The rest of this section is devoted to surfaces in $\mathbb{E}_3$, since that is the case of greatest practical interest to engineering and computer graphics. Interested readers are invited to compare the present approach to the classical treatment in [17] using vector calculus. Rather than review examples in [8], let me discuss an important classical example with a new twist.

According to Coulomb's law, the electric potential of a finite line charge of length L and charge density λ is given by the integral

$$V(\mathbf{x}) = \int \frac{k\,dq(s)}{|\mathbf{x} - \mathbf{x}'(s)|} = \int_{-L/2}^{L/2} \frac{k\lambda\,ds}{|\mathbf{x} - s\mathbf{e}|} \equiv k\lambda\varphi(\mathbf{x}).$$

Remarkably, a simple expression for the value of this integral was overlooked until recently when Rowley [14] discovered

$$\varphi(\mathbf{x}) = \ln\left(\frac{r_+ + r_- + L}{r_+ + r_- - L}\right), \quad \text{where} \quad r_\pm = |\mathbf{r}_\pm| = \left|\mathbf{x} \pm \frac{1}{2}L\mathbf{e}\right|.$$

The equipotentials compose a family of *confocal ellipsoids* (Fig. 19.8). The *eccentricity* ε and *directrix* d of each ellipsoid is given by $\varepsilon \equiv \tanh(\varphi/2) < 1$, so that

$$\frac{1+\varepsilon}{1-\varepsilon} = e^{\varphi} = \frac{r_+ + r_- + L}{r_+ + r_- - L},$$

$$d = \frac{(r_+ + r_-)^2 - L^2}{2L} = (\varepsilon^{-2} - 1)\frac{L}{2},$$

$$r_+ + r_- = \frac{L}{\varepsilon}.$$

The electric field (for $k\lambda = 1$) is given by

$$\mathbf{E} = -\nabla\varphi = \frac{\hat{\mathbf{r}}_+ + \hat{\mathbf{r}}_-}{d}.$$

Here is the surprising new geometric fact that Rowley discovered: The unit normal $\mathbf{n}$ at each point of an ellipse or ellipsoid is given by

$$\hat{\mathbf{r}}_+ + \hat{\mathbf{r}}_- = \nabla(r_+ + r_-) = \Lambda\mathbf{n}, \qquad \Lambda^2 = (\hat{\mathbf{r}}_+ + \hat{\mathbf{r}}_-)^2 = 2(1 + \hat{\mathbf{r}}_+ \cdot \hat{\mathbf{r}}_-).$$

Of course, the difference of the unit coradius vectors is a tangent vector $\hat{\mathbf{r}}_+ - \hat{\mathbf{r}}_- \equiv 2\mathbf{t}$. All this gives us a simple and perspicuous expression for the differential of the normal: For any tangent vector $\mathbf{a} = P(\mathbf{a})$,

$$\underline{\mathbf{n}}(\mathbf{a}) \equiv \mathbf{a} \cdot \nabla\mathbf{n} = \lambda\big[\mathbf{a} - (\mathbf{a} \cdot \mathbf{t})\mathbf{t}\big] \quad \text{with} \quad \lambda = \left(\frac{1}{r_+} + \frac{1}{r_-}\right)\frac{1}{\Lambda} = \frac{L}{\varepsilon r_+ r_- \Lambda}.$$

Note that the tangent vector $\mathbf{t}$ is an eigenvector of the differential $\underline{\mathbf{n}}$.

It is now a simple matter to compute all geometric properties of an ellipsoid of revolution using the apparatus developed above. Since ellipsoids have many practical applications and the present approach is new, it is worth recording the main results for future reference. For an m-dimensional ellipsoid of revolution, we find

$$\text{tr}\,\underline{\mathbf{n}} = \lambda\big(m - \mathbf{t}^2\big),$$

$$\underline{\mathbf{n}}^2(\mathbf{a}) = \lambda\big[\mathbf{n}(\mathbf{a}) - \mathbf{t}\mathbf{t} \cdot \mathbf{n}(\mathbf{a})\big] = \lambda^2\big[\mathbf{a} + (\mathbf{a} \cdot \mathbf{t})(\mathbf{t}^2 - 2)\mathbf{t}\big],$$

$$\text{tr}\,\underline{\mathbf{n}}^2 = \lambda^2\big[m + (\mathbf{t}^2 - 2)\mathbf{t}^2\big] = \frac{\lambda^2}{4}\big(4m - 3 - 5\hat{\mathbf{r}}_+ \cdot \hat{\mathbf{r}}_- + 2(\hat{\mathbf{r}}_+ \cdot \hat{\mathbf{r}}_-)^2\big).$$

Therefore:

Shape: $\quad S_{\mathbf{a}} = \underline{\mathbf{n}}\underline{\mathbf{n}}(\mathbf{a}) = \lambda\mathbf{n}\big[\mathbf{a} - (\mathbf{a} \cdot \mathbf{t})\mathbf{t}\big]$

Curvature: $\quad R(\mathbf{a} \wedge \mathbf{b}) = \underline{\mathbf{n}}(\mathbf{a}) \wedge \underline{\mathbf{n}}(\mathbf{b}) = \underline{\mathbf{n}}(\mathbf{a} \wedge \mathbf{b})$

$$= \lambda^2\big(\mathbf{a} \wedge \mathbf{b} - \mathbf{t} \wedge \big[\mathbf{t} \cdot (\mathbf{a} \wedge \mathbf{b})\big]\big)$$

Mean Curvature: $\quad H \equiv \frac{1}{m}\text{tr}\,\underline{\mathbf{n}} = \frac{\lambda}{m}\big(m - \mathbf{t}^2\big).$

For the case $m = 2$, we have the particular results:

$$\text{Curvature:} \quad R(\mathbf{a} \wedge \mathbf{b}) = \kappa \mathbf{a} \wedge \mathbf{b}$$

$$\text{Gaussian Curvature:} \quad \kappa = \lambda^2(1 - \mathbf{t}^2) = \frac{1}{2}\lambda^2(1 + \hat{\mathbf{r}}_+ \cdot \hat{\mathbf{r}}_-)$$

$$\text{Mean Curvature:} \quad H = \frac{1}{2}\operatorname{tr}\underline{\mathbf{n}} = \lambda\left(1 - \frac{1}{2}\mathbf{t}^2\right) = \frac{1}{2}\lambda\left(\frac{3}{2} + \hat{\mathbf{r}}_+ \cdot \hat{\mathbf{r}}_-\right).$$

All this has some obvious generalizations, for example, to a general ellipsoid with an orthonormal set of tangent vectors, which, like $\mathbf{t}$, are eigenvectors of the differential $\underline{\mathbf{n}}$.

Now let us turn to a general question of great interest and utility: *What is the "shape" of a curve embedded in a manifold?* Shape and curvature are not defined for a curve, because it is a one-dimensional manifold. Instead, shape and curvature bivectors are replaced by the Darboux bivector [8], which completely characterizes the geometry of the curve. Let us address our question for curves in $\mathbb{E}_3$ embedded in some surface, since that is the case of greatest practical interest. The GC apparatus we are using makes generalization to higher dimensions (and even mixed signature) fairly straightforward. In deference to that possibility, we drop the convention of boldface type for vectors in Euclidean space.

Let $x = x(s)$ be a curve with arc length s. Then its "velocity" is a unit tangent vector $v = dx/ds \equiv \dot{x}$. All derivatives of v are determined by the *Darboux bivector* Ω_v. In particular, the acceleration is given by the Frenet equation

$$\dot{v} = \Omega_v \cdot v.$$

Its magnitude is called the *first curvature* $\kappa_\tau = |\dot{v}|$.

The condition that the curve is embedded in a surface with normal $n = n(x)$ is that $v = P(v)$ is a tangent vector and

$$S_v = P_\perp(\Omega_v) = n\underline{n}(v) = n \wedge \underline{n}(v).$$

This decomposes the "Darboux" into two parts:

$$\Omega_v = S_v + \omega_v = P_\perp(\Omega_v) + P(\Omega_v),$$

where $\omega_v = P(\Omega_v)$ is the rotation rate of the curve within the surface. This decomposition can be characterized by two *bending invariants* [17], the *normal curvature* $\kappa_v = v \cdot \underline{n}(v)$ and the *geodesic (tangential) curvature* $\kappa_g = -I \cdot \omega_v = u \cdot \omega_v \cdot v = u \cdot \dot{v}$, where $u = Iv$ is a unit vector orthogonal to v. Obviously, $\kappa_g^2 = -S_v^2$ and $\kappa_g^2 = -\omega_v^2$. This completes our answer to the question about the "shape" of an embedded curve.

Fig. 19.9 Triangular domain
for the Gauss–Bonnet
formula

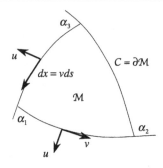

We can use what we have just learned to understand the beautiful and profound
Gauss–Bonnet Formula:

$$\int_{\mathcal{M}} \kappa \, dA + \oint_{\mathcal{C}} \kappa_g \, ds + \sum_i \alpha_i = 2\pi. \tag{19.10}$$

This applies to any simply connected surface $\mathcal{M}$ bounded by a piecewise differentiable closed curve $\mathcal{C} = \partial \mathcal{M}$ with outer normal $u = Iv$ and exterior angles α_i,
as illustrated in Fig. 19.9. As before, $\kappa = I^{-1}R(I)$ is the Gaussian curvature, and
$\kappa_g = u \cdot \dot{v} = -\dot{u} \cdot \dot{x}$ is the geodesic curvature. In GC terms, using the Riemann curvature $R(d^2x) = \kappa \, d^2x$ with directed area element $d^2x = I \, dA$, the formula can be
written

$$\int_{\mathcal{M}} I^{-1}R(d^2x) - \oint_{\mathcal{C}} \dot{u} \cdot dx + \sum_i \alpha_i = 2\pi. \tag{19.11}$$

Proof of the Gauss–Bonnet formula is a nice application of the Fundamental Theorem [17]. Generalization of the formula to higher dimensions is highly nontrivial
[8], and it involves the Riemann curvature in the way it appears in (19.11). No doubt,
there is more to be learned about this generalization and variations on the theme.

Now consider an important special case. The bounding curves are geodesics if
$\kappa_g = 0$, and the figure in Fig. 19.9 is a *geodesic triangle*. For a sphere of radius r,
the Gaussian curvature is r^2; whence the first term in (19.10) and (19.11) is the
solid angle subtended by the region $\mathcal{M}$. An elegant expression for this solid angle
in terms of the vectorial endpoints is derived in [8], which uses GA to describe the
geometry of human body movement. Therein is discussed the amazing fact that the
human eye has learned to implement this theorem to keep the retinal image upright
in saccadic motion. That is the import of the psychophysical discovery known as
Listing's Law.

19.8   Challenges

Let me conclude this review with a few challenges for further development of the
theory and applications.

- *Extension to Conformal Geometric Algebra.* The concept of vector manifold is so general that there should be no problem in applying it to the case where all points are null vectors as required for CGA. I would recommend concentrating first on the geometry of hypersurfaces using the conformal split [7] with the normal at each point x given by the unit bivector $E = x \wedge e_\infty$.
- *Finite Element Differential Geometry.* There is an abundant literature on this subject with many examples worth translating into GA and CGA. Reference [16] should be especially helpful for discrete versions of the vector derivative and fundamental theorem. *Regge Calculus* is an elegant approach to discretizing Riemannian geometry developed for applications to General Relativity [12, 13]. Translation and adaptation to GC should be fairly easy and enlightening. Applications to engineering and computer science as well as physics look promising.
- *Geometry of Movement.* Using CGA to rework and extend the approach in [6] has great potential for robotics as well as biomechanics.
- *Elasticity.* The geometry of material media, including constitutive relations as well as stresses, strains and deformations should be a fertile domain for GC applications.
- *Tangent cones for discontinuities.* So far our approach to differential geometry has ignored discontinuities and singularities of all kinds. GC is well suited to handle such issues, especially in concert with the finite element approach to geometry proposed above. But here is another approach worth investigating. My father developed the concept of *tangent cone* as a portion of the tangent space at a point wherein convergence to a limit obtains, and he applied it with great success to rigorous treatment of singularities in calculus of variations [9]. We have characterized the geometry of a manifold by properties of the pseudoscalar for the tangent space. My suggestion is to meld this notion with the tangent cone idea by using a more general multivector to describe limit structure in the tangent space at points that lie on creases, edges, corners and other discontinuities. I regard this as a hard problem, because it is not well defined and I do not really know how to approach it.

19.9   Exercises

19.1 Suppose that the geodesic triangle in Fig. 19.9 lies on a unit sphere with vertices at $\mathbf{a}, \mathbf{b}, \mathbf{c}$, so $\mathbf{a}^2 = \mathbf{b}^2 = \mathbf{c}^2 = 1$. Parallel transfer of a vector $\mathbf{p}$ around the triangle can be calculated as follows: The tangent vector $\mathbf{p}$ at $\mathbf{a}$ is transferred to a tangent vector $A\mathbf{p}A^{-1}$ at $\mathbf{b}$ by the spinor $A = 1 + \mathbf{ba}$. It can subsequently be transferred to the point $\mathbf{c}$ by $B = 1 + \mathbf{cb}$ and back to $\mathbf{a}$ by $C = 1 + \mathbf{ac}$. The net result is rotation by a spinor $T = CBA$. Show that

$$\frac{1}{2}T = 1 + \mathbf{a} \cdot \mathbf{b} + \mathbf{b} \cdot \mathbf{c} + \mathbf{c} \cdot \mathbf{a} + \mathbf{a}(\mathbf{c} \wedge \mathbf{b} \wedge \mathbf{a}),$$

so $\mathbf{p}$ is rotated about the axis $\mathbf{a}$ through an angle ϕ given by

$$\tan\left(\frac{1}{2}\phi\right) = \frac{\mathbf{a}\cdot(\mathbf{b}\times\mathbf{c})}{1+\mathbf{a}\cdot\mathbf{b}+\mathbf{b}\cdot\mathbf{c}+\mathbf{c}\cdot\mathbf{a}}.$$

How does this angle relate to the area of the triangle?

19.2 Use the result of the previous exercise to explain how the eye must rotate during saccades in order to keep the image on the retina erect. See [6] for details.

19.3 Generalize Rowley's potential $\phi(\mathbf{x})$ for an ellipsoid of revolution to an ellipsoid with axes $\mathbf{a}$, $\mathbf{b}$, $\mathbf{c}$. Calculate the shape and curvature tensors.

19.4 Find an explicit expression for the Darboux bivector of a geodesic on an ellipsoid. Calculate its normal and geodesic curvatures. How do these relate to the curvatures of the ellipsoid?

19.5 Apply the Fundamental Theorem of Geometric Calculus to prove the Gauss–Bonnet Formula (19.10).

References

1. Doran, C., Lasenby, A.: Geometric Algebra for Physicists. The University Press, Cambridge (2003)
2. Dorst, L., Fontijne, D., Mann, S.: Geometric Algebra for Computer Science. Morgan Kaufmann, San Francisco (2007)
3. Hestenes, D.: Space–Time Algebra. Gordon and Breach, New York (1966)
4. Hestenes, D.: The design of linear algebra and geometry. Acta Appl. Math. **23**, 65–93 (1991)
5. Hestenes, D.: Differential forms in geometric calculus. In: Brackx, F. et al. (eds.) Clifford Algebras and Their Applications in Mathematical Physics, pp. 269–285. Kluwer, Dordrecht (1993)
6. Hestenes, D.: Invariant body kinematics: I. Saccadic and compensatory eye movements and II. Reaching and neurogeometry. Neural Netw. **7**, 65–88 (1994)
7. Hestenes, D.: New tools for computational geometry and rejuvenation of screw theory. In: Bayro-Corrochano, E., Scheuermann, G. (eds.) Geometric Algebra Computing for Engineering and Computer Science. Springer, London (2009)
8. Hestenes, D., Sobczyk, G.: Clifford Algebra to Geometric Calculus, a Unified Language for Mathematics and Physics, 4th printing 1999. Kluwer, Dordrecht (1984)
9. Hestenes, M.R.: Calculus of Variations and Optimal Control Theory. Wiley, New York (1966)
10. Hicks, N.: Notes on Differential Geometry. Van Nostrand, New York (1965)
11. Lasenby, A., Doran, C., Gull, S.: Gravity, gauge theories and geometric algebra. Philos. Trans. R. Soc. Lond. A **356**, 161 (2000)
12. Miller, W.: The geometrodynamic content of the Regge equations as illuminated by the boundary of a boundary principle. Found. Phys. **16**(2), 143–169 (1986)
13. Regge, T.: General relativity without coordinates. Nuovo Cimento **19**, 558–571 (1961)
14. Rowley, R.: Finite line of charge. Am. J. Phys. **74**, 1120–1125 (2006)
15. Sobczyk, G.: Killing vectors and embedding of exact solutions in general relativity. In: Chisholm, J., Common, A. (eds.) Clifford Algebras and Their Applications in Mathematical Physics, pp. 227–244. Reidel, Dordrecht (1986)
16. Sobczyk, G.: Simplicial calculus with geometric algebra. In: Micali, A., Boudet, R., Helmstetter, J. (eds.) Clifford Algebras and Their Applications in Mathematical Physics, pp. 227–244. Kluwer, Dordrecht (1992)
17. Struik, D.: Lectures on Classical Differential Geometry. Addison Wesley, Reading (1961)

On the Modern Notion of a Moving Frame

20

Elizabeth Mansfield and Jun Zhao

Abstract

A tutorial on the modern definition and application of moving frames, with a variety of examples and exercises, is given. We show three types of invariants; differential, joint and integral, and the running example is the linear action of $SL(2)$ on smooth surfaces, on sets of points in the plane, and path integrals over curves in the plane. We also give details of moving frames for the group of rotations and translations acting on smooth curves, and on discrete sets of points, in the conformal geometric algebra.

20.1 Introduction

This chapter gives a tutorial on the modern definition of moving frames and details a range of examples. The mathematical context is that of Lie group actions and their invariants. On spaces of smooth curves and surfaces, we obtain differential invariants and invariant differential operators. More generally, we can speak of integral invariants, difference invariants, differential-difference invariants, and so on.

The notion of a moving frame is associated with Élie Cartan [2], who used it to solve equivalence problems in differential geometry, relativity, and so on. Moving frames were further developed and applied in a substantial body of work, in particular to differential geometry and (exterior) differential systems, see for example papers by Green [7] and Griffiths [10]. From the point of view of symbolic computation, a breakthrough in the understanding of Cartan's methods came in a series of papers by Fels and Olver [4, 5], Olver [22, 23], Hubert [11–13], Hubert and Ko-

E. Mansfield (✉) · J. Zhao
School of Mathematics, Statistics and Actuarial Science, University of Kent, Canterbury CT2 7NF, UK
e-mail: E.L.Mansfield@kent.ac.uk

J. Zhao
e-mail: J.Zhao-73@kent.ac.uk

L. Dorst, J. Lasenby (eds.), *Guide to Geometric Algebra in Practice*,
DOI 10.1007/978-0-85729-811-9_20, © Springer-Verlag London Limited 2011

gan [14, 15], which provide a coherent, rigorous and constructive moving frame method free from any particular application, and hence applicable to a huge range of examples, from classical invariant theory to numerical schemes.

The definition of moving frames we use is frameless in the sense that the examples are not restricted to the classical examples where the group actions were applied to frames of vectors on curves and surfaces. However we show how the modern definition applies to the classical examples.

One of the main results of the Fels and Olver papers is the derivation of symbolic formulae for differential invariants and their invariant differentiation. The textbook [21] contains a detailed exposition of the calculations for the resulting symbolic invariant calculus. Applications are to the integration of Lie group invariant differential equations, and in the Calculus of Variations to the computation of invariant Euler–Lagrange equations and the conservation laws guaranteed by Noether's Theorem (see also [9, 19]).

Integral invariants are conjectured to play a significant role in computer vision and graphics. For example, recent software for handwriting recognition [8] uses integral invariants developed in [6] to "quotient out" the effects of translations, rotations and shear. Indeed, it is not hard to see that trying to study pixellated images via difference analogues of Euclidean curvature (for example) leads to processes that are highly sensitive to noise. Even the differential invariants computed by Gaussian derivatives in the scale space approach, while less sensitive, are usually limited to third order because of noise problems. Integrals, on the other hand, are far more stable under small fluctuations in (digitised) curves and surfaces.

Finally, moving frames are being used to incorporate known symmetries of differential systems into numerical schemes designed to integrate them [3, 16–18]. These methods are claimed to give better results regarding singularities and "blow up" of solutions, as well as being able to incorporate first integrals.

20.2 Invariants

We take for our running example the linear action of $SL(2)$ on the (u, v) plane, given by[1]

$$g \cdot \begin{pmatrix} u \\ v \end{pmatrix} = \begin{pmatrix} au + bv \\ cu + dv \end{pmatrix}, \tag{20.1}$$

where

$$g = \begin{pmatrix} a & b \\ c & d \end{pmatrix}, \quad ad - bc = 1.$$

---

[1] *Editorial note*: In this chapter only, the '·' does not denote the dot product, but function composition; this can also be used for a function 'acting on' its argument.

As given, there are no invariants which is unsurprising as we have a three-parameter group acting on a two-dimensional space. In the applications, the action is extended in several different ways, and the actions on the extended spaces do have invariants: differential, integral and joint.

20.2.1 Differential Invariants and Their Syzygies

The first extension of the action in (20.1) is to declare that $u = u(s,t)$, $v = v(s,t)$ are smooth functions of the parameters (s,t), so that we are really considering the action of $SL(2)$ on two-dimensional smooth surfaces, and we use the chain rule to induce an action on derivatives, so that if K is a multi-index and

$$u_K = \frac{\partial^{|K|} u}{\partial s^{K_1} \partial t^{K_2}},$$

then

$$g \cdot \begin{pmatrix} u_K \\ v_K \end{pmatrix} = \begin{pmatrix} au_K + bv_K \\ cu_K + dv_K \end{pmatrix}. \tag{20.2}$$

Historically known invariants of this extended action include

$$uv_s - vu_s, \qquad uv_t - vu_t, \qquad uv_{ss} - vu_{ss}.$$

Indeed, for any two multi-indices, $u_K v_J - u_J v_K$ is invariant. For example,

$$\begin{aligned}
g \cdot (uv_s - vu_s) &= (g \cdot u)(g \cdot v_s) - (g \cdot u_s)(g \cdot v) \\
&= (au + bv)(cu_s + dv_s) - (cu + dv)(au_s + bv_s) \\
&= (ad - bc)(uv_s - vu_s) \\
&= (uv_s - vu_s)
\end{aligned}$$

and similarly for the others. Implicitly, s and t are both invariant, and hence the operators $\partial/\partial s$ and $\partial/\partial t$ map differential invariants to differential invariants, for example

$$\frac{\partial}{\partial s}(uv_s - vu_s) = uv_{ss} - vu_{ss}.$$

Further there are differential relations or *syzygies* between these invariants. If we set $I_{K,J} = u_K v_J - u_J v_K = -I_{J,K}$, then we have, for example,

$$\frac{\partial}{\partial s} I_{0,t} - \frac{\partial}{\partial t} I_{0,s} - 2I_{s,t} = 0,$$

which is easily verified.

In order to ease computations that are naturally expressed in terms of these invariants and manipulate them in a symbolic computation environment, it is natural to ask questions such as:

given a Lie group action, how to compute, algorithmically, a (small) finite set of generators of the differential algebra of differential invariants and of the module of their syzygies?

20.2.2 Integral Invariants

Not all applications use differential invariants; integral invariants on curves are far less subject to noise and are more stable under small perturbations of curves. Thus, a second extension is to consider curves $t \mapsto (u(t), v(t))$ in the plane that satisfy, say, $(u(0), v(0)) = (0, 0)$. The action can be induced on integrals of the form

$$\int u^m v^n \, du := \int_0^t u^m v^n u_\tau \, d\tau, \qquad \int u^m v^n \, dv,$$

and by extension to their sums with polynomial (in u and v) coefficients. Using integration by parts,

$$\int u^m v^n \, du = \frac{1}{m+1} \left(u^{m+1} v^m - n \int u^{m+1} v^{n-1} \, dv \right), \tag{20.3}$$

so it is only necessary to consider integrals with respect to dv. The induced action is $g \cdot (u^m v^n) = (g \cdot u)^m (g \cdot v)^n$ and

$$g \cdot \int u^m v^n \, dv = \int_0^t (g \cdot u)^m (g \cdot v)^n (g \cdot v)_\tau \, d\tau.$$

Some simple invariants are the well-known area integral

$$I_1 = \int u \, dv - \frac{1}{2} uv \tag{20.4}$$

and also

$$I_2 = u \int uv \, dv - \frac{1}{2} v \int u^2 \, dv - \frac{1}{6} u^2 v^2. \tag{20.5}$$

There are a countably infinite set of integral invariants for this action [6].

To show that the area integral I_1 in (20.4) is invariant, we need to show that $g \cdot I_1 = I_1$:

$$g \cdot \left(\int u \, dv - \frac{1}{2} uv \right)$$

$$= \int \left((au + bv)(cu_t + dv_t) \right) dt - \frac{1}{2}(au + bv)(cu + dv)$$

$$= ac \left(\int u \, du - \frac{1}{2}u^2 \right) + bd \left(\int v \, dv - \frac{1}{2}v^2 \right) + ad \left(\int u \, dv - \frac{1}{2}uv \right)$$

$$+ bc \left(\int v \, du - \frac{1}{2}uv \right)$$

$$= ad \left(\int u \, dv - \frac{1}{2}uv \right) + bc \left(\frac{1}{2}uv - \int u \, dv \right)$$

$$= (ad - bc) \left(\int u \, dv - \frac{1}{2}uv \right)$$

$$= \int u \, dv - \frac{1}{2}uv.$$

In this case, a natural question is:

What is a generating set of integral invariants, in the sense that any invariant involving such integrals can be written as a sum (product, quotient, ...) of invariants in my generating set?

20.2.3 Joint Invariants

A third extension of the action (20.1) is to N copies of the plane or, equivalently, to (ordered) sets of N points in the (u, v) plane. The action is extended component wise: if $z_i = (u_i, v_i)$, then we define the *joint action* to be

$$g \cdot (z_1, z_2, \ldots, z_N) = (g \cdot z_1, g \cdot z_2, \ldots, g \cdot z_N). \tag{20.6}$$

It is not hard to see that the quantities

$$u_j v_k - u_k v_j$$

are invariant. Indeed,

$$g \cdot (u_j v_k - u_k v_j) = (au_j + bv_j)(cu_k + dv_k) - (au_k + bv_k)(cu_j + dv_j)$$

$$= (ad - bc)(u_j v_k - u_k v_j)$$

$$= u_j v_k - u_k v_j.$$

Further, it is not hard to imagine mix-and-match invariants such as joint differential invariants and joint integral invariants.

In the following, we will show how moving frames allow one to write down generating sets of invariants in all these situations.

20.3 Moving Frames

For many applications, what is wanted is:

> Given the Lie group action, derive the invariants and their relationships algorithmically, that is, without prior knowledge of 100 years of differential geometry, and with minimal effort.

The modern definition of a moving frame, as used and developed by Fels and Olver and further developed by Hubert, provides a conceptually simple path to solving the problem of finding sets of generators of Lie group actions. The one catch is that the Lie group action must be *regular* and *free* in an open domain in the space on which the Lie group acts. For many actions, this is not as restrictive as it sounds, as there are ways and means of extending the space on which the group acts to achieve regularity and freeness, most notably by prolongation to higher and higher order derivatives (see §9 of [5], where conditions on the action are given that guarantee that a sufficiently high prolongation of the action will be free and regular on dense open subsets of the relevant jet bundle), or by taking multiple copies [1].

20.3.1 The Definition of a Moving Frame

Our starting point is a (left) action by a Lie group G on some space M. This is given as a map

$$G \times M \to M, \quad (g, z) \mapsto g \cdot z$$

which satisfies[2]

$$g \cdot (h \cdot z) = (gh) \cdot z.$$

A moving frame can be defined where the group action is regular and free; see Fig. 20.1. This means that:
- the group orbits foliate the space on which the group acts,
- given a surface $\mathcal{K}$ transverse to the orbits, the intersection of $\mathcal{K}$ with an orbit is a unique point, and
- given z lying in the orbit $\mathcal{O}(z)$, such that $k = \mathcal{O}(z) \cap \mathcal{K}$, there is a unique element h of the group such that $k = h \cdot z$.

Typically, moving frames exist only locally, that is, in some open domain in M.

Definition For the group action $G \times M \to M$, a *moving frame* is an equivariant map $\rho : M \to G$; for a left action, the equivariance of a right frame is given by

$$\rho(g \cdot z) = \rho(z) g^{-1}. \tag{20.7}$$

---

[2]A right action satisfies $g \cdot (h \cdot z) = (hg) \cdot z$. The moving frame theory for right actions is entirely equivalent.

Fig. 20.1 The picture for a free and regular group action

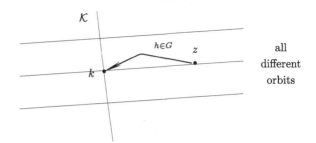

Fig. 20.2 A moving frame is an equivariant map, $\rho(g \cdot z) = \rho(z)g^{-1}$

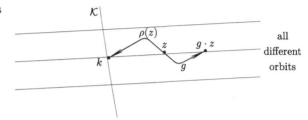

Fig. 20.3 The classical idea of a moving frame: at each point on the curve, there is a "frame" of vectors, (e_1, e_2) which has the same information as a translation and a rotation

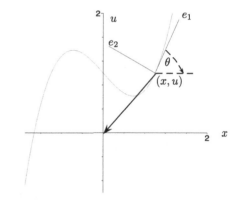

Consider Fig. 20.2. For $z \in M$, let h be the unique element of G that satisfies $h \cdot z = k \in \mathcal{K}$, and define $\rho(z) = h$. Since we have a left action, hg^{-1} takes $g \cdot z$ to k; indeed, $(hg^{-1}) \cdot (g \cdot z) = (hg^{-1}g) \cdot z = h \cdot z = k$. In other words, $\rho(g \cdot z) = \rho(z)g^{-1}$.

If instead we took the inverse h^{-1} to be the frame, we would obtain a so-called left frame, for which the equivariance is $\rho(g \cdot z) = g\rho(z)$. However, the method used to calculate a frame given in Sect. 20.3.2 yields a right frame, so we stick with that.

To see how this modern definition of a moving frame includes the classical examples, consider Fig. 20.3. The space M is the set of smooth curves in the plane, and the group G is the three-parameter group of rotations and translations in the plane. The action is extended to curves pointwise, and to tangent vectors and higher-order derivatives via the chain rule. We think of M as having coordinates $(x, u, u_x, u_{xx}, u_{xxx}, \ldots)$. The classical frame consists, at any given point of the curve, of the unit tangent and unit normal vectors of the curve. These are given as

$$e_1 = \left(\frac{1}{\sqrt{1+u_x^2}}, \frac{u_x}{\sqrt{1+u_x^2}} \right), \qquad e_2 = \left(-\frac{u_x}{\sqrt{1+u_x^2}}, \frac{1}{\sqrt{1+u_x^2}} \right).$$

While it is difficult to visualise the orbits in such a high-dimensional space, we can easily visualise a transverse section $\mathcal{K}$, which is the plane given as the locus of the equations, $x = 0$, $u = 0$, $u_x = 0$. Thus the group element taking $(x, u, u_x, u_{xx}, u_{xxx}, \dots)$ to $\mathcal{K}$ is the translation taking (x, u) to the origin, followed by the rotation that takes the unit vector e_1 to $(1, 0)$.[3] In parameter form, the angle and the vector of translation are

$$\rho(x, u, u_x) = \bigl(-\arctan(u_x), -(x, u) \bigr).$$

20.3.2 The Calculation of a Moving Frame

In practice, one specifies $\mathcal{K}$, the cross-section, as the locus of a system of equations $\Phi(z) = 0$. These are known as the *normalisation equations*. Then the frame $h = \rho(z)$ solves $\Phi(h \cdot z) = 0$. If the Lie group depends on r independent parameters, one has r equations

$$\phi_j(g \cdot z) = 0, \quad j = 1, \dots, r = \dim(G)$$

for the r independent parameters describing g. The requirements that the group action be free and regular amount to requiring the hypotheses of the Implicit Function Theorem (IFT) to hold.[4] Since the solution is guaranteed by the IFT to be unique, it must be that

$$\rho(g \cdot z) = \rho(z) \cdot g^{-1}$$

since both solve $\Phi(h \cdot (g \cdot z)) = 0$ for h. Thus, this method solves for a right frame, as in (20.7). Invoking the Implicit Function Theorem means that the moving frame holds only locally. For researchers committed to symbolic, algebraic processes, this is a non-constructive step. In response to this, Hubert and Kogan [14, 15] formulated a way of constructing a moving frame for algebraic group actions acting rationally. In this development, the Implicit Function Theorem is replaced by an equally powerful result in commutative algebra.

We now demonstrate a moving frame for two of the three extended $SL(2)$ actions detailed above. Integral invariants will be discussed in Sect. 20.4.3.

---

[3] *Editorial note*: To relate to a standard term in robotics, in Cartan's examples moving frames simplify to 'the group element that sends the frame of vectors at a point to a *reference frame* of vectors at the origin'.

[4] For any $z \in \mathcal{K}$, if one stacks the tangent vectors to $\mathcal{K}$ at z and the tangent vectors of the orbit at z as columns in a matrix, then the matrix must have $n = \dim M$ columns and have full rank. The tangent vectors to the orbits can be obtained by differentiating $g \cdot z$ with respect to the group parameters at $g = e$, the identity of the group.

20.3.2.1 A Frame for the Action on Derivatives

Consider the action (20.1) of $SL(2)$, as prolonged to (20.2). If we take the equations for the transverse cross-section to the orbits to be

$$g \cdot u = 1, \qquad g \cdot v = 0, \qquad g \cdot u_x = 0,$$

then solving these for (a, b, c), the independent parameters of the group, yields

$$a = \frac{-v_x}{u_x v - u v_x}, \qquad b = \frac{u_x}{u_x v - u v_x}, \qquad c = -v. \tag{20.8}$$

In matrix form, we have

$$\rho(u, v, u_x, v_x) = \begin{pmatrix} \frac{-v_x}{u_x v - u v_x} & \frac{u_x}{u_x v - u v_x} \\ -v & u \end{pmatrix}.$$

To demonstrate the equivariance, we have

$$\rho(g \cdot u, g \cdot v, g \cdot u_x, g \cdot v_x)$$

$$= \begin{pmatrix} \frac{-g \cdot v_x}{g \cdot u_x \, g \cdot v - g \cdot u \, g \cdot v_x} & \frac{g \cdot u_x}{g \cdot u_x \, g \cdot v - g \cdot u g \cdot v_x} \\ -g \cdot v & g \cdot u \end{pmatrix}$$

$$= \begin{pmatrix} \frac{-(cu_x + dv_x)}{u_x v - u v_x} & \frac{au_x + bv_x}{u_x v - u v_x} \\ -(cu + dv) & (au + bv) \end{pmatrix}$$

$$= \begin{pmatrix} \frac{-v_x}{u_x v - u v_x} & \frac{u_x}{u_x v - u v_x} \\ -v & u \end{pmatrix} \begin{pmatrix} d & -b \\ -c & a \end{pmatrix}$$

$$= \rho(u, v, u_x, v_x) g^{-1},$$

since the denominators appearing in ρ are invariant.

20.3.2.2 A Frame for the Joint Action

Consider the action (20.6). If we take the normalisation equations to be

$$g \cdot \begin{pmatrix} u_1 \\ v_1 \end{pmatrix} = \begin{pmatrix} 1 \\ 0 \end{pmatrix}, \qquad g \cdot \begin{pmatrix} u_2 \\ v_2 \end{pmatrix} = \begin{pmatrix} 0 \\ \square \end{pmatrix},$$

where $\square$ means undetermined (by the normalisation equations, we only need three equations for the three independent parameters), then we have

$$a = \frac{v_2}{u_1 v_2 - v_1 u_2}, \qquad b = \frac{-u_2}{u_1 v_2 - v_1 u_2}, \qquad c = -v_1, \qquad d = u_1 \tag{20.9}$$

in a calculation strongly resembling that for (20.8). The proof of equivariance is also similar.

20.4   Invariants via Moving Frames

By definition, a function F which is constant on orbits, $F(g \cdot z) = F(z)$, is an *invariant* of the group action. Considering Fig. 20.1, an invariant is determined by its values on any transverse cross-section $\mathcal{K}$. Indeed, if $F(k) = c$, then $F(z) = c$ for every z on the orbit of G through k. Since we have $k = \rho(z) \cdot z$ for all z on the orbit through k in the domain U of the frame, we can see that any invariant on U is a function of $\rho(z) \cdot z$.

Conversely, we obtain the following result:

Theorem 20.1 *If $z \in M$ is in the domain of a moving frame ρ, then the components of $I(z) = \rho(z) \cdot z$ are invariant.*

Proof The proof that $I(z)$ is invariant looks simple:

$$I(g \cdot z) = \rho(g \cdot z) \cdot (g \cdot z) = \rho(z)g^{-1}g \cdot z = \rho(z) \cdot z.$$

In practice, the invariance of expressions obtained via this method can look miraculous, as the examples show.   □

Given any function F on the domain U of a moving frame ρ in M, we say $F(\rho(z) \cdot z)$ is the *invariantisation* of F on U. If $z = (z_1, \ldots, z_n)$ in co-ordinates, let $I(z) = (I(z_1), \ldots, I(z_n))$ define the invariantised co-ordinate functions $I(z_i)$. As a rule, invariantisation of co-ordinates via a moving frame leads to complete sets of invariants. The reason is the following result.

Theorem 20.2 (Replacement Theorem) *If F is an invariant function defined in the domain U of a moving frame ρ, then*

$$F(z_1, \ldots, z_n) = F\big(I(z_1), I(z_2), \ldots, I(z_n)\big).$$

Proof

$$\begin{aligned} F(z_1, \ldots, z_n) &= F(g \cdot z_1, \ldots g \cdot z_n) \quad \text{for all } g \\ &= F(g \cdot z_1, \ldots g \cdot z_n)|_{g=\text{frame}} \\ &= F(I(z_1), \ldots, I(z_n)). \end{aligned}$$ □

The Replacement Theorem implies that the $I(z_i)$ form a *generating* set of invariants, in the sense that any invariant is a function of these. Note that the invariantisation of the normalisation equations give rise to functional relations between the $I(z_i)$, so the $I(z_i)$ are not functionally independent.

A second use of the Replacement Theorem is in cases where we cannot solve for the frame, but invariants are known either historically or from geometrical considerations. In this case, it may be possible to obtain enough information to solve for the distinguished symbolic invariants $I(z_j)$ in terms of known invariants. This can be important, especially in the case of differential invariants, since a great deal is known about symbolic calculation with the $I(z_j)$, which has been coded in software such as Maple.

We now turn to the examples to illustrate these results.

20.4.1 Joint Invariants via Moving Frames

Recall that in Sect. 20.2.3 we defined the action on N copies of the plane (or equivalently, sets of N points in the plane), given in (20.6). Our running example is the group $SL(2)$ acting linearly, as given in (20.1), for which a frame ρ was defined in (20.9). The frame maps (u_1, v_1) to $(1, 0)$ and u_2 to 0, so that $I(u_1) = 1$, $I(v_1) = 0$ and $I(u_2) = 0$. We then have $I(v_2) = (cu_2 + dv_2)|_\rho = -v_1 u_2 + u_1 v_2$, and indeed, for all j, we have

$$I(u_j) = (au_j + bv_j)|_\rho = \frac{u_j v_2 - v_j u_2}{u_1 v_2 - v_1 u_2},$$
$$I(v_j) = (cu_j + dv_j)|_\rho = -v_1 u_j + u_1 v_j.$$

To verify the Replacement rule, we have for the invariant $u_j v_k - v_k u_j$ that

$$I(u_j)I(v_k) - I(u_k)I(v_j)$$
$$= \frac{u_j v_2 - v_j u_2}{u_1 v_2 - v_1 u_2}(-v_1 u_k + u_1 v_k) - \frac{u_k v_2 - v_k u_2}{u_1 v_2 - v_1 u_2}(-v_1 u_j + u_1 v_j)$$
$$= \frac{(u_j v_k - u_k v_j)}{-v_1 u_j + u_1 v_j}(-v_1 u_j + u_1 v_j)$$
$$= u_j v_k - u_k v_j,$$

as required.

It can be seen that if $\mathscr{S}$ is the shift operator, $\mathscr{S}(j) = j + 1$, $\mathscr{S}u_j = u_{j+1}$, and so on, then

$$\mathscr{S}\left(I(u_j)\right) = \frac{u_{j+1}v_3 - v_{j+1}u_3}{u_2 v_3 - v_2 u_3} \neq I(\mathscr{S}u_j) = \frac{u_{j+1}v_2 - v_{j+1}u_2}{u_1 v_2 - v_1 u_2}.$$

Therefore applications of moving frames to the study of finite difference schemes for differential systems with a Lie group symmetry have involved defining a frame ρ_n for each index n, see for example [3, 17].

20.4.2 Differential Invariants via Moving Frames

When considering prolonged actions on derivatives, the co-ordinates of the space are $(x_i, u^\alpha, u_x^\alpha, \ldots, u_K^\alpha, \ldots)$. In this case, the components of $I(z)$ are

$$g \cdot x_i|_{g=\text{frame}} = I(x_i), \qquad g \cdot u_K^\alpha|_{g=\text{frame}} = I_K^\alpha.$$

Other notations in use are $\iota(u_K^\alpha)$ and $\bar{\iota} u_K^\alpha$. Since in general $I(u_x) \neq \partial I(u)/\partial x$, an equivalent numerical multi-index for K is often used, for example $I(u_{xxy}) = I_{112}^u$, and so on.

All differential invariants are functions of the I_K^α by the version of the Replacement Theorem adapted to this case:

Theorem 20.3 (Replacement—differential case) *If $F(x_i, u^\alpha, \ldots, u_K^\alpha, \ldots)$ is an invariant, then*

$$
\begin{aligned}
F(x_i, u^\alpha, \ldots, u_K^\alpha, \ldots) &= F(g \cdot x_i, g \cdot u^\alpha, \ldots, g \cdot u_K^\alpha, \ldots) \quad \text{for all } g \\
&= F(g \cdot x_i, g \cdot u^\alpha, \ldots, g \cdot u_K^\alpha, \ldots)\big|_{g=\text{frame}} \\
&= F(I(x_i), I^\alpha, I_1^\alpha, \ldots, I_K^u, \ldots).
\end{aligned}
$$

To illustrate the generation of the invariants and the Replacement Theorem in the running example, we consider the linear $SL(2)$ action (20.1) prolonged to derivatives in (20.2). A frame for this action was calculated in (20.8) using the normalisation equations

$$g \cdot u = 1, \qquad g \cdot v = 0, \qquad g \cdot u_x = 0.$$

We obtain, for example,

$$
\begin{aligned}
I(v_x) = I_1^v &= (cu_x + dv_x)|_{\text{frame}}, \\
&= -vu_x + uv_x \\
I(u_{xx}) = I_{11}^u &= (au_{xx} + bv_{xx})|_{\text{frame}} \\
&= \frac{-v_x}{u_x v - u v_x} u_{xx} + \frac{u_x}{u_x v - u v_x} v_{xx} \\
&= \frac{-v_x u_{xx} + u_x v_{xx}}{u_x v - u v_x},
\end{aligned}
$$

while in general,

$$
\begin{aligned}
I_K^v &= -vu_K + uv_K, \\
I_K^u &= \frac{-v_x u_K + u_x v_K}{u_x v - u v_x}.
\end{aligned}
$$

We then note that, for example, using the replacement rule on I_K^u yields

$$
I(u_K) = \frac{-I(v_x)I(u_K) + I(u_x)I(v_K)}{I(u_x)I(v) - I(u)I(v_x)} = \frac{-I(v_x) \cdot I(u_K) + 0 \cdot I(v_K)}{0 \cdot 0 - 1 \cdot I(v_x)},
$$

as required, while for the general invariant,

$$
I_K^u I_J^v - I_J^u I_K^v = \frac{-v_x u_K + u_x v_K}{u_x v - u v_x}(-v u_J + u v_J)
$$

$$
- \frac{-v_x u_J + u_x v_J}{u_x v - u v_x}(-v u_K + u v_K)
$$

$$
= \frac{u_K v_J - u_J v_K}{u_x v - u v_x}(u_x v - u v_x)
$$

$$
= u_K v_J - u_J v_K,
$$

as expected.

Differential invariants are related to each other using invariant differentiation. A maximal set of distinguished invariant differential operators may be defined by the same process of invariantisation, namely

$$
\mathcal{D}_j = \frac{\partial}{\partial(g \cdot x_j)}\bigg|_{g=\text{frame}} = \sum_j \frac{\partial x_k}{\partial(g \cdot x_j)}\bigg|_{g=\text{frame}} \frac{\partial}{\partial x_k}.
$$

In our examples in this chapter, we take both independent variables to be invariant, and thus $\mathcal{D}_x = \partial/\partial x$ and $\mathcal{D}_y = \partial/\partial y$. In general, the $\mathcal{D}_j$ are linear operators with non-constant coefficients. While they may not commute, we have, however,

$$
[\mathcal{D}_j, \mathcal{D}_k] = \sum_\ell \mathcal{A}_{jk}^\ell \mathcal{D}_\ell
$$

where the $\mathcal{A}_{jk}^\ell$ are invariants. Formulae for the $\mathcal{A}_{jk}^\ell$ in terms of the normalisation equations and the group action appear in [5].

Perhaps the most striking result by Fels and Olver in [5] is the proof that there exist "correction", or "error" terms, M_{Kj}^α such that

$$
\mathcal{D}_j I_K^\alpha = I_{Kj}^\alpha + M_{Kj}^\alpha; \tag{20.10}
$$

moreover, formulae for their calculation are given.[5] *Indeed, the M_{Kj}^α can be obtained, as expressions in the I_K^α, simply from knowing the normalisation equations for the frame and the group action: it is not necessary to have solved for the frame.* This result means that a symbolic invariant calculus is possible, since one can obtain all the relations, both functional and differential, between the distinguished invariants $\{I(x_i), I^\alpha, I_K^\alpha\}$, even when these are only known symbolically.

---

[5]The formulae are fully explained in terms of undergraduate multi-variable calculus in [21], while the Fels and Olver papers use (nontrivial) exterior calculus.

Fig. 20.4 In standard differential algebra, the entire set of derivative terms $\{u_K^\alpha\}$ is generated by u^α, shown here for two independent variables

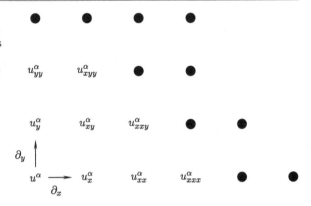

In Fig. 20.4 is shown the standard structure of the set of derivative terms and how they are related via differentiation. In this case, the partial derivative operators commute, there are no functional relations relating the u_K^α, and all differential equations connecting them are in terms of the trivial relations, $\partial_K u_J^\alpha = \partial_J u_K^\alpha = u_{JK}^\alpha$.

By contrast, when we invariantise, a radically different picture emerges. In Fig. 20.5, we see the diagrams for the invariantised derivatives $I(u_K) = I_K^u$ and I_K^v, which stand, in general, for complex differential expressions. Equation (20.10) shows that the invariant derivatives of I_K^α involve I_{Kj}^α as well as additional terms, so that arrows in these diagrams indicate merely, for example, that $\mathscr{D}_j I_K^u$ "involves" I_{Kj}^u rather than "equals" I_{Kj}^u.

Next, we note that if I_K^α has been normalised to a constant, say, then it is not necessarily the case that I_{Kj}^α is zero, merely that the error term M_{Kj}^α will cancel the I_{Kj}^α term.

Finally, the existence of non-zero error terms means that eliminating[6] $I_{KJ}^\alpha = I_{JK}^\alpha$ in the equations

$$\mathscr{D}_K I_J^\alpha = I_{KJ}^\alpha + M_{JK}^\alpha,$$

$$\mathscr{D}_J I_K^\alpha = I_{KJ}^\alpha + M_{KJ}^\alpha$$

yields the differential relation or *syzygy*,

$$\mathscr{D}_K I_J^\alpha - \mathscr{D}_J I_K^\alpha = M_{JK}^\alpha - M_{KJ}^\alpha.$$

In general, $M_{JK}^\alpha \neq M_{KJ}^\alpha$ so these syzygies will be non-trivial.

When the normalisation equations are relatively simple, diagrams such as Fig. 20.5 can be used to locate the most important syzygies. For example, it can be seen that $I_2^u = I(u_y)$ can be differentiated twice with respect to x, producing an expression with I_{112}^u and additional terms, similarly I_{11}^u can be differentiated once

---

[6]We have $I_{KJ}^\alpha = I_{JK}^\alpha$ since they are equal to the invariantisation of $u_{KJ}^\alpha = u_{JK}^\alpha$ respectively.

Fig. 20.5 In invariantised differential algebra, more than one generator may be needed to obtain the complete set of invariants I_K^α for each dependent variable u^α. The picture here is for the running example with a linear $SL(2)$ action, see (20.2), with frame determined by the normalisation equations $I^u = 1$, $I_1^u = 0$, $I^v = 0$

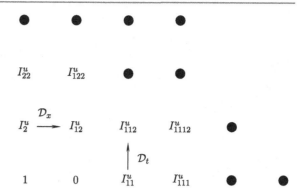

(i) The diagram of invariants for the dependent variable u

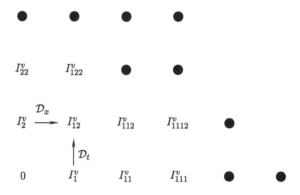

(ii) The diagram of invariants for the dependent variable v

with respect to y to produce an expression with I_{112}^u in it. Subtracting, and rewriting the result in terms of the generating invariants

$$\kappa_1 = I_{11}^u, \qquad \kappa_2 = I_2^u, \qquad \sigma_1 = I_1^v, \qquad \sigma_2 = I_2^v$$

to ease the appearance of the formula yields

$$\mathscr{D}_x^2 \kappa_2 - \mathscr{D}_y \kappa_1 = \frac{1}{\sigma_1^2}(2\kappa_1 \sigma_2 \mathscr{D}_x \sigma_1 - \sigma_1 \sigma_2 \mathscr{D}_x \kappa_1 - 2\sigma_1 \kappa_1 \mathscr{D}_x \sigma_2 + \sigma_1 \kappa_2 \mathscr{D}_x \sigma_1).$$

(This expression was obtained using the Maple package, Indiff that encodes the invariant differentiation formulae [20].) In this case, we know the invariants explicitly, and this result can be verified by direct calculation. The basic syzygy for v is much simpler, it is $\mathscr{D}_y I_1^v - \mathscr{D}_x I_2^v = 2I_1^v I_2^u$ or

$$\mathscr{D}_y \sigma_1 - \mathscr{D}_x \sigma_1 = \sigma_1 \kappa_2.$$

For more general kinds of normalisation equations, the complete result concerning which syzygies generate the set of all syzygies was obtained by Hubert [12].

To summarise the main differences of invariantised differential algebra from the standard differential algebra, we have:

- More than one generator may be needed to obtain the complete set of the I_K^α under invariant differentiation.
- There exist functional relations between the I_K^α, namely the normalisation equations.
- There exist non-trivial differential syzygies between the I_K^α.
- The $\mathscr{D}_j$ are non-commuting in general.

20.4.3 Moving Frames for Integral Invariants

We first consider the group $SO(2) \subset SL(2)$ which has only one parameter, the angle θ of rotation, acting linearly on (u, v) as

$$\theta \cdot u = \cos\theta u - \sin\theta v, \qquad \theta \cdot v = \sin\theta u + \cos\theta v.$$

We assume our set of curves all begin at $(0,0)$. Bearing in mind that $\int u^n \, du = u^{n+1}/(n+1)$ and the integration by parts formula, (20.3), we take the co-ordinates of our space to be $w(0,0), w(0,1), \ldots, z(1,0), z(1,1), \ldots$, where

$$w(m,n) = u^m v^n, \qquad z(m,n) = \int u^m v^n \, dv.$$

We include the monomials in our co-ordinates since they appear naturally in the calculations. The action on the monomials is the standard action induced on functions:

$$\theta \cdot w(m,n) = (\cos\theta\, u - \sin\theta\, v)^m (\sin\theta\, u + \cos\theta\, v)^n,$$

the right-hand side of which expands to a sum in the $w(i,j)$. The action on the integrals is obtained as follows:

$$\theta \cdot z(1,0) = \int (\cos\theta\, u - \sin\theta\, v)(\sin\theta\, du + \cos\theta\, dv)$$
$$= \cos\theta \sin\theta \left(\int u \, du - \int v \, dv \right) - \sin^2\theta \int v \, du + \cos^2\theta \int u \, dv$$
$$= \frac{1}{2}\cos\theta \sin\theta (u^2 - v^2) - \sin^2\theta \left(uv - \int u \, dv \right) + \cos^2\theta \int u \, dv$$
$$= \frac{1}{2}\cos\theta \sin\theta (w(2,0) - w(0,2)) - \sin^2\theta\, w(1,1) + z(1,0),$$

and similarly for the other $z(i,j)$. In words, we apply the action to the integral in the standard way, expand and put into "normal form" by performing integrations where possible and applying integration by parts to remove any integrations with respect to u. It is helpful to write a procedure in Maple (for example) to do this.

We need only one normalisation equation to obtain the frame, and we take $g \cdot v = 0$. Thus the frame is $\rho = \arctan(-v/u)$ or more helpfully,

$$\cos\theta = \frac{u}{\sqrt{u^2 + v^2}}, \qquad \sin\theta = \frac{-v}{\sqrt{u^2 + v^2}}.$$

Applying the frame to the $w(i, j)$ leads either to 0 or 1; however, we are interested in the integral invariants. Applying the frame to $z(1, 0)$ yields

$$I\big(z(1,0)\big) = -\frac{1}{2}\frac{uv(u^2 - v^2)}{u^2 + v^2} - \frac{v^2(uv)}{u^2 + v^2} + z(1,0) = -\frac{1}{2}uv + z(1,0),$$

thus recovering the well-known area invariant,

$$I\big(z(1,0)\big) = \int u \, dv - \frac{1}{2}uv,$$

which is the area between the curve and the diagonal $t \mapsto (u(t), v(t))$ in the (u, v) plane, on the t-interval implicit in the integration. Applying the method to the $z(i, j)$, $i \neq 0$ in turn yields a countably infinite set of integral invariants. For example,

$$I\big(z(1,1)\big) = -\frac{1}{6}\frac{u^2v^2}{\sqrt{u^2 + v^2}} - \frac{1}{2}\frac{v}{\sqrt{u^2 + v^2}}\int u^2 \, dv + \frac{u}{\sqrt{u^2 + v^2}}\int uv \, dv.$$

Turning now to the full linear $SL(2)$ action, we have that the action on the monomials is

$$g \cdot u^m v^n = (au + bv)^m (cu + dv)^n,$$

while for integrals, we begin with

$$g \cdot z(m, n) = \int (au + bv)^m (cu + dv)^n (c \, du + d \, dv)$$

and then expand the right-hand side, performing integrations where possible and integration by parts to remove all integrals with respect to u. The use of symbolic software to perform the calculations is recommended.

Since there are now three independent parameters in the group, we need three normalisation equations. As above, we can take $g \cdot u = 1$, $g \cdot v = 0$, and then we need one other, so we can take $g \cdot z(2, 0) = 0$. It is not possible to normalise with an equation for $z(1, 0)$, as the action is not free in $(u, v, z(1, 0))$ space. The frame obtained is then

$$a = \frac{2uv^2 - 6z(1, 1)}{u^2v^2 + 3z(2, 0) - 6uz(1, 1)}, \qquad b = \frac{-u^2v + 3z(2, 0)}{u^2v^2 + 3z(2, 0) - 6uz(1, 1)},$$
$$c = -v, \qquad d = u,$$

where we have written the expressions in terms of the $z(i, j)$ for ease of reading.

Applying the frame to the $z(i, j)$ yields a countably infinite set of integral invariants, for example, $I(z(1, 1)) = -\frac{1}{6}u^2v^2 - \frac{1}{2}v\, z(2, 0) + u\, z(1, 1)$ or

$$I\big(z(1, 1)\big) = -\frac{1}{6}u^2v^2 - \frac{1}{2}v\int u^2\, dv + u\int uv\, dv.$$

The invariants rapidly become complex with increasing powers of u and v in the integrand, but are nevertheless straightforward to obtain.

20.5 Moving Frames for the $SE(3)$ Action in Conformal Geometric Algebra

Finally we consider the standard linear action of $SE(3)$, the group of rotations and translations in three dimensions. We discuss the standard Serret–Frenet frame acting on vectors in $\mathbb{R}^3$, since this is the classically defined, well-known frame in common use for the standard action of $SE(3)$ on curves. We show how one may write this frame in the language of conformal geometric algebra. However, we conjecture that a different frame, more "native" to CGA, will exist that offers improved computational ease and better geometric properties than the Serret–Frenet frame for calculations in CGA.

Given two non-parallel vectors $\mathbf{a}$ and $\mathbf{b}$, which we can assume have been normalised to have unit length, at a point $\mathbf{P}$ in $\mathbb{R}^3$, the standard Serret–Frenet frame is the group action that translates $\mathbf{P}$ to the origin, followed by the rotation that takes both $\mathbf{a}$ to the x-axis and $\mathbf{b}$ to the (x, y)-plane. It is not hard to see that the rotation matrix given by

$$R_{\mathbf{a},\mathbf{b}} = \begin{pmatrix} \mathbf{a} \\ (\mathbf{a} \times \mathbf{b}) \times \mathbf{a} \\ \mathbf{a} \times \mathbf{b} \end{pmatrix}$$

satisfies the two requirements; further, it is overtly equivariant. The quaternionic form of this rotation matrix seems complex, nor is it obviously equivariant; the axis of rotation of $R = R_{\mathbf{a},\mathbf{b}}$ is

$$e_1 \times e_2 + R(e_1 \times e_2) + e_1 \times R(e_2) + R(e_1) \times e_2,$$

where e_1 and e_2 are the unit vectors on the x and y axes respectively, and while the angle of rotation is easily calculated, the formula seems uninformative. Perhaps a different frame, defined by a different cross-section to the group orbits, might be more natural in the context of conformal geometric algebra; however, in what follows, we show that we can write the Serret–Frenet frame in a natural way as a product.

The two standard applications are the differential case, where we wish to construct this frame relative to the vectors $\mathbf{a} = \mathbf{x}'(t)$ and $\mathbf{b} = \mathbf{x}''(t)$ at the point $\mathbf{x}(t)$, so that the frame depends on t, and the discrete case, where we are given three non-collinear points p, p_1 and p_2 so that $\mathbf{a} = p_1 - p$ and $\mathbf{b} = p_2 - p$. In what follows we restrict ourselves to the discrete case, as the differential case follows mutatis mutandis.

20.5.1 The Serret–Frenet Frame in CGA

In this section, we will use the notation

$$g \cdot z = z'$$

to ease the appearance of the formulae.

In conformal geometric algebra, the generic elements are points

$$p_i = \mathbf{p}_i + \frac{1}{2}\mathbf{p}_i^2 n_\infty + n_o,$$

where $\mathbf{p}_i$ denotes the position vector of the ith point. Consider the group $SE(3)$ acting as

$$p_i{}' = g \cdot p_i = (RT)p_i(RT)^\sim = \mathbf{p}_i' + \frac{1}{2}(\mathbf{p}_i')^2 n_\infty + n_o, \tag{20.11}$$

where T and R are the unit translator and rotor, and $(RT)^\sim$ denotes the reverse of RT.

We construct the moving frame for the action on three non-collinear points, or equivalently two vectors defined at one point, in space. We illustrate the construction by taking the three points related to a sphere as follows. Let a sphere be given as

$$S = p - \frac{1}{2}\rho^2 n_\infty = \mathbf{p} + \frac{1}{2}(\mathbf{p}^2 - \rho^2)n_\infty + n_o.$$

From this expression we obtain a point p (centre) and an invariant ρ (radius). After the group action, we obtain:

$$S' = p' - \frac{1}{2}\rho^2 n_\infty = \mathbf{p}' + \frac{1}{2}(\mathbf{p}'^2 - \rho^2)n_\infty + n_o.$$

We move the centre of the sphere to the origin, and this gives three normalisation equations for the translation part of the group action. We will need two more points in order to determine all the six parameters in $SE(3)$. Let the first point be a point p_1 on the surface of the sphere, and the second point p_2 on the end of any one unit tangent vector to the sphere at p_1. Now define three vectors,

$$E_2 = \mathbf{p}_1 - \mathbf{p}, \qquad E_3 = (\mathbf{p}_1 - \mathbf{p}) \times (\mathbf{p}_2 - \mathbf{p}), \qquad E_1 = E_2 \times E_3.$$

After the translation T, the group element $R_2 R_1 = R \in SO(3)$ will act on our sphere and will take both E_2 to $\|E_2\|e_2$, and the plane whose normal is E_3 to the xy-plane, whose normal is e_3.

Firstly we apply R_1 to the sphere which takes the plane spanned by $\mathbf{p}_1 - \mathbf{p}$ and $\mathbf{p}_2 - \mathbf{p}$ to the $e_2 e_1$ plane (xy-plane). We then apply R_2, which is a rotation in the $e_2 e_1$ plane and will take $g \cdot E_2 = E_2'$ to e_2 (y-axis). Thus,

$$R_1 = \cos\left(\frac{\theta}{2}\right) + \sin\left(\frac{\theta}{2}\right)l_1, \qquad R_2 = \cos\left(\frac{\alpha}{2}\right) + \sin\left(\frac{\alpha}{2}\right)l_2,$$

where

$$l_1 = \frac{E_3 \wedge e_3}{\|E_3 \wedge e_3\|}, \qquad \theta = \cos^{-1}\left(\frac{\langle E_3, e_3 \rangle}{\|E_3\|}\right)$$

are the rotation axis and angle of R_1; and

$$l_2 = \frac{\hat{E}_2 \wedge e_2}{\|\hat{E}_2 \wedge e_2\|}, \qquad \alpha = \cos^{-1}\left(\frac{\langle \hat{E}_2, e_2 \rangle}{\|\hat{E}_2\|}\right)$$

are the rotation axis and angle of R_2, where $\hat{E}_2 = R_1 E_2 \tilde{R}_1$, with $\tilde{R}_1$ the reverse of R_1, and $\langle \cdot, \cdot \rangle$ represents the inner product between two vectors.

Hence our normalisation equations are

$$E_2' = \|E_2\|e_2, \qquad E_3' = \|E_3\|e_3.$$

To show an example, let a sphere be given as

$$S = \left(e_1 + e_2 + e_3 + \frac{\sqrt{3}}{2}n_\infty + n_o\right) - \frac{1}{2}n_\infty,$$

with centre at $p = e_1 + e_2 + e_3 + (\sqrt{3}/2)n_\infty + n_o$ and radius 1. Choose the other two points as $p_1 = e_1 + e_2 + 2e_3 + (\sqrt{6}/2)n_\infty + n_o$ and $p_2 = 2e_1 + 2e_2 + 2e_3 + \sqrt{3}n_\infty + n_o$. The parameters of the frame can be then calculated as follows:

$$E_2 = e_3, \qquad E_3 = e_2 - e_1, \qquad l_1 = \frac{1}{\sqrt{2}}(e_2 e_3 - e_1 e_3),$$

$$\theta = \frac{\pi}{2}, \qquad l_2 = e_1 e_2, \qquad \alpha = \frac{\pi}{4}.$$

Hence the translator and the rotors are:

$$T = 1 + \frac{\mathbf{p}}{2}n_\infty, \qquad R_1 = \cos\left(\frac{\theta}{2}\right) + \sin\left(\frac{\theta}{2}\right)l_1, \qquad R_2 = \cos\left(\frac{\alpha}{2}\right) + \sin\left(\frac{\alpha}{2}\right)l_2.$$

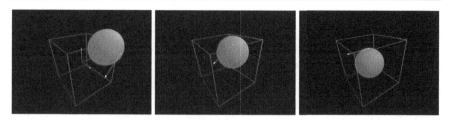

Fig. 20.6 An animation of the action of the moving frame taking the sphere to its normalised position, $t = 0$ (*left*), $t = 0.5$ (*middle*) and $t = 1$ (*right*)

Figure 20.6 shows that the moving frames of $SE(3)$ act on a sphere at the time of $t = 0$, $t = 0.5$ and $t = 1$ in an animation of the action of the moving frame taking the sphere to its normalised position.

20.5.2 Going Co-ordinate Free

By taking more general normalisation equations for the frame, we can see how the frame can be defined in a more co-ordinate free way. Let two points be $v_1, v_2 \in \mathbb{R}^{4,1}$, and two other points be $w_1, w_2 \in \mathbb{R}^{4,1}$. The group element $g \in SO(3)$ acts on our points as

$$g \cdot v_i = (R_2 R_1) v_i (R_2 R_1)^{\sim}.$$

Now let

$$E_2 = \mathbf{v}_1, \qquad E_3 = \mathbf{v}_1 \times \mathbf{v}_2, \qquad E_1 = E_2 \times E_3,$$

and

$$F_2 = \mathbf{w}_1, \qquad F_3 = \mathbf{w}_1 \times \mathbf{w}_2, \qquad F_1 = F_2 \times F_3.$$

Our normalisation equations are:

$$E_2' = \frac{\|E_2\|}{\|F_2\|} F_2, \qquad E_3' = \frac{\|E_3\|}{\|F_3\|} F_3.$$

Then the parameters of the moving frame can be calculated without using any co-ordinates:

$$R_1 = \cos\left(\frac{\theta}{2}\right) + \sin\left(\frac{\theta}{2}\right) l_1, \qquad R_2 = \cos\left(\frac{\alpha}{2}\right) + \sin\left(\frac{\alpha}{2}\right) l_2,$$

where

$$l_1 = \frac{E_3 \times F_3}{\|E_3 \times F_3\|}, \qquad \theta = \cos^{-1}\left(\frac{\langle E_3, F_3 \rangle}{\|E_3\| \|F_3\|}\right)$$

are rotation axis and rotation angle for R_1, and

$$l_2 = \frac{\hat{E}_2 \times F_2}{\|\hat{E}_2 \times F_2\|}, \qquad \theta = \cos^{-1}\left(\frac{\langle \hat{E}_2, F_2 \rangle}{\|\hat{E}_2\| \|F_2\|}\right),$$

where $\hat{E}_2 = R_1 E_2 \tilde{R}_1$, are the rotation axis and rotation angle for R_2.

20.6 Exercises

20.1 Verify that the integral invariant I_2 in (20.5) is indeed an invariant under the action of $SL(2)$ given in (20.2). Using the method of Sect. 20.4.3, verify the calculations of $I(z(1, 1))$ for the given frames and find $I(z(2, 0))$ for both the $SO(2)$ and $SL(2)$ actions.

20.2 (This exercise is preliminary to the next, in that the calculations required are similar but for a much simpler action.) Let the group action of $\mathbb{R}^+ \ltimes \mathbb{R}$ on smooth curves in the plane parameterised as $(x, u(x))$ be given by

$$(\lambda, a) \cdot (x, u(x)) = (x, \lambda u(x) + a).$$

Show that the frame defined by $u' = 0$, $u_x' = 1$ is

$$\rho(x, u, u_x) = \left(\frac{1}{u_x}, -\frac{u}{u_x}\right),$$

which is valid, provided that $u_x > 0$. If we set $u_n = d^n u/dx^n$, show that $I_n = u_n'|_\rho = u_n/u_x$ and that $d/dx(I_n) = I_{n+1} - I_n I_2$. If the group is represented as

$$(\lambda, a) \mapsto \begin{pmatrix} \lambda & a \\ 0 & 1 \end{pmatrix},$$

show that

$$\rho_x \rho^{-1} = \begin{pmatrix} -I_2 & -1 \\ 0 & 0 \end{pmatrix}. \tag{20.12}$$

Note: equations for ρ in the form of (20.12) can always be obtained from the normalisation equations, even when one cannot solve these for the frame explicitly; see [21], Chap. 5, with an application to solving Lie group invariant ODEs in Chap. 6.

20.3 If M is the set of smooth curves in a CGA parameterised by s (say), and $\mathbb{H}$ the group of quaternions with the standard action on M, define a right frame $\rho : \mathbb{H} \times M \to M$. Show that $\rho_s \rho^{-1}$ is an element of the Lie algebra of $\mathbb{H}$ with invariant components. What is the geometric meaning of the invariants appearing in the components of $\rho_s \rho^{-1}$? (Note: for the Serret–Frenet frame ρ_{SF}, the non-zero components of $\rho_{SF,s} \rho_{SF}^{-1}$ are the curvature and torsion of the curve.)

References

1. Boutin, M.: On orbit dimensions under a simultaneous Lie group action on n copies of a manifold. J. Lie Theory **12**, 191–203 (2002)
2. Cartan, E.: Oeuvres complètes. Gauthier-Villars, Paris (1952–1955)
3. Chhay, M., Hamdouni, A.: A new construction for invariant numerical schemes using moving frames. C. R. Acad. Sci. Mec. **338**, 97–101 (2010)
4. Fels, M., Olver, P.J.: Moving coframes I. Acta Appl. Math. **51**, 161–213 (1998)
5. Fels, M., Olver, P.J.: Moving coframes II. Acta Appl. Math. **55**, 127–208 (1999)
6. Feng, S., Kogan, I., Krim, H.: Classification of curves in 2D and 3D via affine integral signatures. Acta Appl. Math. (2010). doi:10.1007/s10440-008-9353-9
7. Green, M.L.: The moving frame, differential invariants and rigidity theorems for curves in homogeneous spaces. Duke Math. J. **45**, 735–779 (1978)
8. Golubitsky, O., Mazalov, V., Watt, S.M.: Toward affine recognition of handwritten mathematical characters. In: Proc. International Workshop on Document Analysis Systems (DAS 2010), Boston, USA, June 9–11 2010, pp. 35–42. ACM, New York (2010)
9. Gonçalves, T.M.N., Mansfield, E.L.: On moving frames and Noether's conservation laws. arxiv.org/abs/1006.4660 (2010)
10. Griffiths, P.: On Cartan's methods of Lie groups and moving frames as applied to uniqueness and existence questions in differential geometry. Duke Math. J. **41**, 775–814 (1974)
11. Hubert, E.: Differential algebra for derivations with nontrivial commutation rules. J. Pure Appl. Algebra **200**(1–2), 163–190 (2005)
12. Hubert, E.: Differential invariants of a Lie group action: syzygies on a generating set. J. Symb. Comput. **44**(4), 382–416 (2009)
13. Hubert, E.: Generation properties of Maurer–Cartan invariants, preprint [hal:inria-00194528] (2009)
14. Hubert, E., Kogan, I.A.: Smooth and algebraic invariants of a group action. Local and global constructions. Found. Comput. Math. **7**(4), 345–383 (2007)
15. Hubert, E., Kogan, I.A.: Rational invariants of a group action. Construction and rewriting. J. Symb. Comput. **42**(1–2), 203–217 (2007)
16. Kim, P., Olver, P.J.: Geometric integration via multi-space. Regul. Chaotic Dyn. **9**(3), 213–226 (2004)
17. Kim, P.: Invariantization of numerical schemes using moving frames. BIT Numer. Math. **47**(3), 525 (2007)
18. Kim, P.: Invariantization of the Crank–Nicolson method for Burgers' equation. Physica D: Nonlinear Phenomena **237**(2), 243 (2008)
19. Kogan, I.A., Olver, P.J.: Invariant Euler–Lagrange equations and the invariant variational bicomplex. Acta Appl. Math. **76**, 137–193 (2003)
20. Mansfield, E.L.: Indiff a Maple package to calculate with differential expressions referred to a moving frame. Available from http://www.kent.ac.uk/ims/personal/elm2 (2001)
21. Mansfield, E.L.: A Practical Guide to the Invariant Calculus. Cambridge University Press, Cambridge (2010)
22. Olver, P.J.: Joint invariant signatures. Found. Comput. Math. **1**, 3–67 (2001)

434434434

434434434

23. Olver, P.J.: Moving frames—in geometry, algebra, computer vision, and numerical analysis. In: DeVore, R., Iserles, A., Suli, E. (eds.) Foundations of Computational Mathematics. London Math. Soc. Lecture Note Series, vol. 284, pp. 267–297. Cambridge University Press, Cambridge (2001)

Tutorial Appendix: Structure Preserving Representation of Euclidean Motions Through Conformal Geometric Algebra

21

Leo Dorst

Abstract

Using conformal geometric algebra, Euclidean motions in n-D are represented as orthogonal transformations of a representational space of two extra dimensions, and a well-chosen metric. Orthogonal transformations are representable as multiple reflections, and by means of the geometric product this takes an efficient and structure preserving form as a 'sandwiching product'. The antisymmetric part of the geometric product produces a spanning operation that permits the construction of lines, planes, spheres and tangents from vectors, and since the sandwiching operation distributes over this construction, 'objects' are fully integrated with 'motions'. Duality and the logarithms complete the computational techniques.

The resulting geometric algebra incorporates general conformal transformations and can be implemented to run almost as efficiently as classical homogeneous coordinates. It thus becomes a high-level programming language which naturally integrates quantitative computation with the automatic administration of geometric data structures.

This appendix provides a concise introduction to these ideas and techniques. *Editorial note*: This appendix is a slightly improved version of (Dorst in: Bayro-Corrochano, E., Scheuermann, G. (eds.) Geometric Algebra Computing for Engineering and Computer Science, pp. 457–476, 2011). We provide it to make this book more self-contained.

21.1 Introduction

"Doing geometry" in computer science or engineering requires at least the following ingredients in a practical computational framework:

- *descriptive primitives*: such as points, lines, planes, circles, spheres, tangents

L. Dorst (✉)
Intelligent Systems Laboratory, University of Amsterdam, Amsterdam, The Netherlands
e-mail: l.dorst@uva.nl

L. Dorst, J. Lasenby (eds.), *Guide to Geometric Algebra in Practice*,
DOI 10.1007/978-0-85729-811-9_21, © Springer-Verlag London Limited 2011

- *basic constructions*: connections, intersections, parametric specification
- *motions*: translation, rotation, reflection, projection
- *properties*: size, location, orientation
- *practical numerics*: approximation, estimation, interpolation, linearization

These ingredients should interweave seamlessly. Notably, the framework should be structure preserving, in the sense that *constructions and properties of primitives should be covariant under motions*. For instance, when moving a circle determined by three points, it should not be necessary to decompose the circle back into the points, move those, and then reconstruct; rather, the circle should be a basic element of computation with an associated motion operator (which should moreover preferably be identical to that for points). Also, all ingredients should be specifiable in a sufficiently high-level programming language, which avoids coordinates as specification level though it may revert to them when executing the operations. The usual linear algebra tools have neither of these desirable properties, not even when using homogeneous coordinates. Yet a practical computational framework exists that can do all of the above. It is called "conformal geometric algebra" (CGA), and this appendix briefly exposes its essential structure. We will explain all elements of Fig. 21.1, and more.

21.2 Conformal Geometric Algebra

21.2.1 Trick 1: Representing Euclidean Points in Minkowski Space

Let us focus on a 3D space in which we want to perform Euclidean rigid body motions. We can consider it as a 3D vector space and use a position vector $\mathbf{x}$ to denote a point X relative to an (arbitrary) origin. This is naive practice, and not very convenient, since Euclidean motions are then not even linear transformations. A commonly used improvement is the *homogeneous model*, in which the space is augmented with an extra dimension e_o, and the point X at $\mathbf{x}$ represented as $e_o + \mathbf{x}$. Now Euclidean motions are linear transformations but still not structure preserving. More is required.

In CGA, we introduce *two extra dimensions* for representational purposes, thus constructing a five-dimensional space. We introduce two basis vectors for these extra dimensions, n_o and n_∞, and the specific metric given below. As we will see, the null vectors in this extended space (i.e., the vectors x satisfying $x \cdot x = 0$ in the chosen metric) represent weighted points in the Euclidean space (though one usually employs unit weight points satisfying $x \cdot n_\infty = -1$). Such vectors representing points have algebraic properties which enable to construct other elements in a coordinate-free, invariant manner.

In this introductory appendix, we simply use an explicit expression for such a vector x representing a point X, relating it to the "classical" Euclidean position vector $\mathbf{x}$ of the point relative to the chosen origin through

$$x = n_o + \mathbf{x} + \frac{1}{2}\|\mathbf{x}\|^2 n_\infty. \tag{21.1}$$

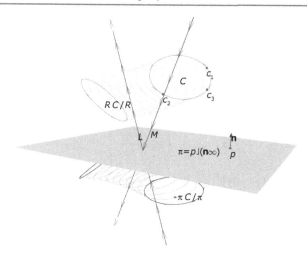

Fig. 21.1 An example of the ease of CGA (from [3]). A circle C is generated from three points c_1, c_2, c_3 as $C = c_1 \wedge c_2 \wedge c_3$. A line is given as a 3-blade L. The circle C is to be rotated around the line L, producing RCR^{-1}, with R specified as $R = \exp(L^*\phi/2)$. The rotation is interpolated in k steps using $R^{1/k}$. Then the whole scene is reflected in the plane π given by a normal vector $\mathbf{n}$ and a point p on it as $\pi = p \cdot (\mathbf{n} \wedge n_\infty)$; any element X is reflected as $X \mapsto (-1)^{\text{grade}(X)} \pi X \pi^{-1}$. In appropriate software such as [4], these coordinate-free formulas are the literal specification of a computer program producing the scene

If we ignore the component for n_∞, we recognize in $n_o + \mathbf{x}$ just the homogeneous model trick. In that model, the extra dimension e_o represents *the point at the origin*; and the same interpretation holds for n_o in CGA (set $\mathbf{x} = \mathbf{0}$). We see that the term with n_∞ dominates as $\mathbf{x}$ gets large. In fact, n_∞ can be interpreted consistently as *the point at infinity* which is used in mathematics to "compactify" Euclidean space to remove special cases from its algebra.

The Big Trick of CGA is the choice of a specific metric for the 5D representational space. We extend the dot product $\mathbf{x} \cdot \mathbf{x}$ for Euclidean vectors to the new dimensions according to the multiplication table

$\cdot$	n_o	$\mathbf{x}$	n_∞
n_o	0	0	-1
$\mathbf{x}$	0	$\mathbf{x} \cdot \mathbf{x}$	0
n_∞	-1	0	0

where the bold elements are purely Euclidean and borrow the 3D Euclidean dot product. This table shows that the usual Euclidean metric holds for the bold vectors, but a strange metric applies to the two additional dimensions n_o and n_∞, which are moreover "orthogonal" to the Euclidean part of the representational space since they have dot product zero with Euclidean vectors. (In fact, the full 5D space now has a Minkowski metric, as can be seen by considering the alternative basis vectors

$\sigma_+ = n_o - \frac{1}{2}n_\infty$ and $\sigma_- = n_o + \frac{1}{2}n_\infty$ that have squared norms of $+1$ and -1, respectively.) Since there are thus four orthogonal basis vectors squaring to $+1$, and one basis vector that squares to -1, we will denote this space by $\mathbb{R}^{4,1}$, and its geometric algebra by $\mathbb{R}_{4,1}$.

This metric is introduced to give a sensible real world meaning to the dot product of two point representatives x and y:

$$
\begin{aligned}
x \cdot y &= \left(n_o + \mathbf{x} + \frac{1}{2}\|\mathbf{x}\|^2 n_\infty\right) \cdot \left(n_o + \mathbf{y} + \frac{1}{2}\|\mathbf{y}\|^2 n_\infty\right) \\
&= \left(0 + 0 - \frac{1}{2}\|\mathbf{y}\|^2\right) + (0 + \mathbf{x} \cdot \mathbf{y} + 0) + \left(-\frac{1}{2}\|\mathbf{x}\|^2 + 0 + 0\right) \\
&= -\frac{1}{2}(\mathbf{x} - \mathbf{y}) \cdot (\mathbf{x} - \mathbf{y}) \\
&= -\frac{1}{2}\|\mathbf{x} - \mathbf{y}\|^2.
\end{aligned}
\tag{21.2}
$$

The dot product in conformal space therefore encodes the (squared) Euclidean distance of the original points! Since points have distance zero to themselves, they are represented by null vectors. Since Euclidean rigid body motions should preserve the inter-point distance, they should preserve the dot product.[1]

Euclidean transformations are represented as orthogonal transformations

This is more specific than their representation as a certain strange class of linear transformations in the usual homogeneous model, and it permits us to design a more effective computational framework tailored to this property. Matrices are actually not that great for representing orthogonal transformations, but fortunately there is something better, as we will see in the next section.

First, let us determine what the vectors in the 5D representation space signify geometrically. Suppose that we want to represent a sphere with center C and radius ρ in Euclidean space. A point X on such a sphere would satisfy $\|\mathbf{x} - \mathbf{c}\|^2 = \rho^2$ (using Euclidean vectors). Using (21.2), this can be written in terms of the dot product of the representative vectors x and c as $x \cdot c = -\frac{1}{2}\rho^2$. Using $-n_\infty \cdot x = 1$, we can even group into $x \cdot (c - \frac{1}{2}\rho^2 n_\infty) = 0$. The vector $\sigma = \alpha(c - \frac{1}{2}\rho^2 n_\infty)$, with $\alpha \in \mathbb{R}$, is the most general vector we can make in the conformal space (it has five parameters), and written in this form, we recognize it as representing a sphere with center c, radius ρ and "weight" α through the equation $x \cdot \sigma = 0$. You may verify that $\|\sigma\|^2 = \alpha^2 \rho^2$ (even "imaginary spheres" with $\rho^2 < 0$ are included) and that a point is just a sphere with radius zero, represented by a null vector (for which

---

[1] We have simplified slightly; the general representation of a point at $\mathbf{x}$ in CGA is a scalar multiple of x in (21.1); the scalar factor is the scalar $-n_\infty \cdot x$ (as you may verify), and this can be consistently interpreted as the *weight* of the point. The squared distance between weighted points is computed by normalizing first as $(x/(-n_\infty \cdot x)) \cdot (y/(-n_\infty \cdot y))$. Euclidean transformations should then not affect this formula; this implies that they are the specific orthogonal transformations that preserve the special vector n_∞.

$\|x\|^2 = 0$). A plane is the degenerate case of a sphere, and it is represented by a vector of the form $\pi = \alpha(\mathbf{n} + \delta n_\infty)$ (which has no n_o-component and therefore satisfies $n_\infty \cdot \pi = 0$). Here $\mathbf{n}$ is the unit normal vector of the plane, δ is its oriented distance from the origin, and α a weight. So:

the vectors in conformal space represent weighted spheres and planes.

In this tutorial, we will mostly use unit weights, focusing on the merely geometrical aspects of the representation. In our *notation*, we will use bold for the elements of the conformal model that are in its n-D Euclidean subspace and nonbold for elements residing in the full $(n + 2)$-D representational space or its geometric algebra. Since there is a clear correspondence between elements of Euclidean geometry and their conformal representation, we will drop the distinction between X and x, and talk about a point x at location $\mathbf{x}$.

21.2.2 Trick 2: Orthogonal Transformations as Multiple Reflections in a Sandwiching Representation

In mathematics, the Cartan–Dieudonné theorem states that all orthogonal transformations can be represented as multiple reflections. In linear algebra, this fact is not used much, since reflections are represented awkwardly and therefore unsuitable as atoms of representation. If we want to reflect a Euclidean vector $\mathbf{x}$ in a plane through the origin with normal vector $\mathbf{a}$, this is the linear transformation

$$\mathbf{x} \mapsto \mathbf{x} - 2(\mathbf{x} \cdot \mathbf{a})\mathbf{a}/(\mathbf{a} \cdot \mathbf{a}). \tag{21.3}$$

It does not look elementary at all, and within linear algebra the dot products cannot be simplified.

We now introduce a clever trick: we consider the dot product (which is symmetric) as merely the symmetrical part of a more fundamental product between vectors. That product (invented by Clifford in 1872) is called the *geometric product* and denoted by a space. So we rewrite:

$$\mathbf{a} \cdot \mathbf{x} = \frac{1}{2}(\mathbf{a}\mathbf{x} + \mathbf{x}\mathbf{a}). \tag{21.4}$$

This more fundamental product is defined to be bilinear and associative but not necessarily commutative. We see that $\|\mathbf{x}\|^2 = \mathbf{x} \cdot \mathbf{x} = \mathbf{x}\mathbf{x} = \mathbf{x}^2$, so that the square of a vector under the geometric product is a scalar. We extend the geometric product to scalars (and later to other elements). Scalars commute under the geometric product, so $\alpha\mathbf{x} = \mathbf{x}\alpha$ for vector $\mathbf{x}$ and scalar α. A vector $\mathbf{x}$ has a unique inverse $\mathbf{x}^{-1}$ under the geometric product, defined through $\mathbf{x}\mathbf{x}^{-1} = 1 = \mathbf{x}^{-1}\mathbf{x}$ and therefore found explicitly as

inverse of a vector: $\mathbf{x}^{-1} = \mathbf{x}/(\mathbf{x}^2)$.

Now we see how this simplifies the reflection representation:

$$\textit{reflection in origin hyperplane with normal } \mathbf{a}: \quad \mathbf{x} \mapsto \mathbf{x} - 2(\mathbf{x} \cdot \mathbf{a})\mathbf{a}/(\mathbf{a} \cdot \mathbf{a})$$
$$= \mathbf{x} - (\mathbf{xa} + \mathbf{ax})\mathbf{a}^{-1}$$
$$= -\mathbf{axa}^{-1}. \qquad (21.5)$$

The reflection of $\mathbf{x}$ in the origin hyperplane with normal vector $\mathbf{a}$ is therefore simply a "sandwiching" of $\mathbf{x}$ by $\mathbf{a}$ and $\mathbf{a}^{-1}$ (with a minus sign). In this form, the fundamental nature of reflections for the representation of transformations is more obvious than it was in (21.3).

You may rightly object that we have not really reflected a point x, but only its Euclidean part $\mathbf{x}$. Let us try to extend the formula to the point x, using the explicit representation (21.1). Postulating distributivity of the geometric product, we get $-axa^{-1} = -\mathbf{a}(n_o + \mathbf{x} + \frac{1}{2}\|\mathbf{x}\|^2 n_\infty)\mathbf{a}^{-1} = -\mathbf{a}n_o\mathbf{a}^{-1} - \mathbf{axa}^{-1} - \frac{1}{2}\|\mathbf{x}\|^2\mathbf{a}n_\infty\mathbf{a}^{-1}$. Evaluating this requires computing what $-\mathbf{a}n_o\mathbf{a}^{-1}$ and $-\mathbf{a}n_\infty\mathbf{a}^{-1}$ are. We realize from the definition of (21.4) and the dot product table that $-\mathbf{a}n_o\mathbf{a}^{-1} = -(\mathbf{a}n_o)\mathbf{a}^{-1} = -(2\mathbf{a} \cdot n_o - n_o\mathbf{a})\mathbf{a}^{-1} = 0 + n_o\mathbf{a}\mathbf{a}^{-1} = n_o$. Of course, you would expect this geometrically: the point at the origin does not change after the reflection. Similarly for n_∞, as you may verify. Further realize that $\|-\mathbf{axa}^{-1}\|^2 = (-\mathbf{axa}^{-1})(-\mathbf{axa}^{-1}) = \mathbf{axxa}^{-1} = \mathbf{x}^2(\mathbf{aa}^{-1}) = \|\mathbf{x}^2\|$—obviously, since reflection is an orthogonal transformation. Combining all this, we find $-axa^{-1} = n_o - \mathbf{axa}^{-1} + \frac{1}{2}\|-\mathbf{axa}^{-1}\|^2 n_\infty$, which is precisely the representation of a point at the reflected location. Therefore a point x is reflected by transfer of the Euclidean formula (21.5), as $x \mapsto -axa^{-1}$. This structural principle may be illustrated as the commutative diagram

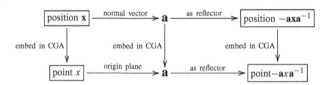

If we perform a second reflection in another origin hyperplane, with normal vector $\mathbf{b}$, this should be the mapping

$$x \mapsto = -\mathbf{b}(-axa^{-1})\mathbf{b}^{-1} = (\mathbf{ba})x(\mathbf{ba})^{-1},$$

using the associativity of the geometric product in the rewriting. Geometrically, a double reflection is a rotation (see Fig. 21.2), so the operator $(\mathbf{ba})$ represents a rotation operator (in an axis through the origin, determined as the intersection of the planes $\mathbf{a}$ and $\mathbf{b}$). In this manner, we can generate all orthogonal transformations as sandwiching products by elements that are themselves the geometric product of vectors. These elements are called *versors*, or (when normalized) *rotors*. A delightful property of versors is that they do not only apply to vectors, but also directly to

Fig. 21.2 A reflection in two successive planes is equivalent to a rotation over double their separating angle, around the line of their intersection (in 3D)

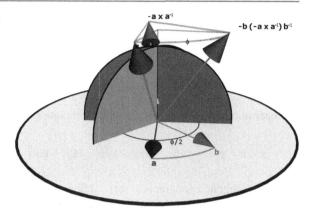

other geometric elements like lines and circles. Let us first make those geometric elements part of our algebra.

21.2.3 Trick 3: Constructing Elements by Anti-symmetry

When we introduced the geometric product for vectors, we used only its symmetric part (that was the dot product). But of course there is an anti-symmetric part as well. Let us denote that by $\wedge$ and call it the *outer product*. For vectors, it is defined as

$$\mathbf{x} \wedge \mathbf{a} = \frac{1}{2}(\mathbf{xa} - \mathbf{ax}).$$

It is clear that $\mathbf{x} \wedge \mathbf{a} = -\mathbf{a} \wedge \mathbf{x}$, so that $\mathbf{x} \wedge \mathbf{x} = 0$.

To interpret this new element $\mathbf{x} \wedge \mathbf{a}$ geometrically, let us use some classical linear algebra and take $\mathbf{x}$ and $\mathbf{a}$ as direction vectors. If we take an orthonormal basis $\{\mathbf{e}_1, \mathbf{e}_2\}$ in the plane spanned by $\mathbf{x}$ and $\mathbf{a}$, and choose it such that $\mathbf{x} = \|\mathbf{x}\|\mathbf{e}_1$, then $\mathbf{a}$ can be written as $\mathbf{a} = \|\mathbf{a}\|(\cos(\phi)\mathbf{e}_1 + \sin(\phi)\mathbf{e}_2)$ with ϕ the angle from $\mathbf{x}$ to $\mathbf{a}$. We evaluate:

$$\begin{aligned}
\mathbf{x} \wedge \mathbf{a} &= \big(\|\mathbf{x}\|\mathbf{e}_1\big) \wedge \big(\|\mathbf{a}\|(\cos(\phi)\mathbf{e}_1 + \sin(\phi)\mathbf{e}_2)\big) \\
&= \|\mathbf{x}\|\,\|\mathbf{a}\|\big(\cos(\phi)\mathbf{e}_1 \wedge \mathbf{e}_1 + \sin(\phi)\mathbf{e}_1 \wedge \mathbf{e}_2\big) \\
&= \|\mathbf{x}\|\,\|\mathbf{a}\|\sin(\phi)\mathbf{e}_1 \wedge \mathbf{e}_2,
\end{aligned}$$

for, being the sum of two bilinear products, the outer product is itself bilinear. We recognize in $\|\mathbf{x}\|\,\|\mathbf{a}\|\sin(\phi)$ the signed area of the oriented parallelogram spanned by $\mathbf{x}$ and $\mathbf{a}$ (in that order) and can therefore interpret $\mathbf{e}_1 \wedge \mathbf{e}_2$ as the algebraic specification of the unit area element in the $(\mathbf{e}_1, \mathbf{e}_2)$-plane. We call this a *unit 2-blade*. We then interpret the 2-blade $\mathbf{x} \wedge \mathbf{a}$ of direction vectors as the full specification of the geometric area element spanned by $\mathbf{x}$ and $\mathbf{a}$ (in that order) in terms of its magnitude, orientation and geometrical attitude (i.e., spatial stance). Only the shape of the element is not determined uniquely, for you can easily verify that, for instance,

$\mathbf{x} \wedge (\mathbf{a} + \lambda \mathbf{x}) = \mathbf{x} \wedge \mathbf{a}$ so that $\mathbf{x}$ and $\mathbf{a} + \lambda \mathbf{x}$ span the same element as $\mathbf{x}$ and $\mathbf{a}$. For parallel direction vectors $\mathbf{x}$ and $\mathbf{a}$, the outer product $\mathbf{x} \wedge \mathbf{a}$ is zero, so the commutativity relationship $\mathbf{xa} = \mathbf{ax}$ is the algebraic way of expressing parallelness of vectors. Orthogonality of vectors is expressed as $\mathbf{xa} = -\mathbf{ax}$, or $\mathbf{x} \cdot \mathbf{a} = 0$.

The outer product can be extended over more vector terms, always as the antisymmetric sum. This is done by permuting the geometric products and endowing even permutations with a plus and odd permutations with a minus. For instance:

$$\mathbf{a} \wedge \mathbf{b} \wedge \mathbf{c} = \frac{1}{3!}(\mathbf{abc} - \mathbf{bac} + \mathbf{bca} - \mathbf{cba} + \mathbf{cab} - \mathbf{acb}) \qquad (21.6)$$

(but this algebraic equation is a very inefficient way of computing the value of the outer product; the equivalent $\mathbf{a} \wedge \mathbf{b} \wedge \mathbf{c} = \frac{1}{2}(\mathbf{abc} - \mathbf{cba})$ is already better). It can be shown that the outer product thus defined is associative and multilinear. To make it fully defined over all elements, we can extend it to scalars simply by defining $\alpha \wedge \mathbf{a} = \alpha \mathbf{a}$ for scalar α and vector $\mathbf{a}$.

The outer product of k vector factors is called a k-*blade*, and the number of vector factors k is called its *grade*. Geometrically, a k-blade is a quantitative representation of a weighted, oriented k-dimensional subspace of the space its vectors reside in, and its signed magnitude is an oriented hypervolume. For instance, if you would compute the outer product of three direction vectors in 3D space, you would find that the coordinates of the vectors combine to a familiar signed scalar multiple of the unit volume: $\mathbf{a} \wedge \mathbf{b} \wedge \mathbf{c} = \det([\![\mathbf{abc}]\!])\mathbf{e}_1 \wedge \mathbf{e}_2 \wedge \mathbf{e}_3$. This volume is zero when the vectors are co-planar, and therefore $\mathbf{x} \wedge (\mathbf{a} \wedge \mathbf{b}) = 0$ can be solved for $\mathbf{x}$ as $\mathbf{x} = \lambda \mathbf{a} + \mu \mathbf{b}$. Again, the 2-blade $\mathbf{a} \wedge \mathbf{b}$ is seen to be a single computational element representing the plane spanned by the direction vectors $\mathbf{a}$ and $\mathbf{b}$.

In the conformal model, the outer product of vectors representing points a and b takes on a different geometric interpretation, even though its algebra is the same. In CGA, the blade $a \wedge b$ *represents an oriented point-pair*, in the sense that the set of points x satisfying $x \wedge a \wedge b = 0$ is either $x = a$ or $x = b$. (Comparing to the derivation just given, we do get $x = \lambda a + \mu b$, as before. Yet to be a point in CGA, x has to satisfy $x \cdot x = 0$ by (21.2), as do a and b. Some algebra then leads to $\lambda \mu (a \cdot b) = 0$, and this implies $\lambda = 0$ and/or $\mu = 0$.) Similarly, $a \wedge b \wedge c$ *represents the oriented circle through the points a, b and c*, and the outer product of four points $a \wedge b \wedge c \wedge d$ represents an oriented *sphere*. We call these elements *rounds*. If the points are in degenerate positions, or if one of them is the point at infinity n_∞, an oriented *flat* results (in 3D, these are: a line $a \wedge b \wedge n_\infty$, a plane $a \wedge b \wedge c \wedge n_\infty$, or a "flat point" $a \wedge n_\infty$). Showing these facts without too much computation requires the technique of dual representation, introduced next.

21.2.4 Trick 4: Dual Specification of Elements Permits Intersection

A subspace can be characterized by the outer product, but it is often convenient to take a "dual" approach, not specifying the vectors in it but instead the vectors orthogonal to it. We have already seen this for spheres: the orthogonality demand

$x \cdot (c - \frac{1}{2}\rho^2 n_\infty) = 0$ is equivalent to x lying on a sphere with center c and radius ρ. Duality is a fundamental concept of geometric algebra and requires no more than complementation relative to the volume of the vector space, through division.[2]

An n-dimensional vector space cannot have nonzero blades of a grade exceeding n. A nonzero blade of the maximum grade n is called a *pseudoscalar* for the space. It is common to normalize this to a unit pseudoscalar and to denote it by $\mathbf{I}_n$ or I_n. The choice of the sign of the unit pseudoscalar amounts to choosing a reference orientation for the space. In a 3D Euclidean space of direction vectors with an orthonormal basis, $\mathbf{I}_3 = \mathbf{e}_1 \wedge \mathbf{e}_2 \wedge \mathbf{e}_3 (= \mathbf{e}_1 \mathbf{e}_2 \mathbf{e}_3)$ picks the standard "right-handed" orientation. In the conformal model space, a suitable pseudoscalar is $I_{4,1} = n_o \wedge \mathbf{I}_3 \wedge n_\infty$. The inverse of the unit pseudoscalar in 3D Euclidean space is $\mathbf{I}_3^{-1} = -\mathbf{I}_3$ (verify that $\mathbf{I}_3 \mathbf{I}_3^{-1} = 1$!). In the conformal space, $I_{4,1}^{-1} = n_o \wedge \mathbf{I}_3^{-1} \wedge n_\infty = -I_{4,1}$.

One can find the blade representing the orthogonal complement of any subspace through right-dividing its blade A by the pseudoscalar, as $A I_n^{-1}$. This is called the *dual* of A and denoted A^*:

$$\text{dualization}: \quad A^* = A I_n^{-1}. \tag{21.7}$$

For instance, the dual of the 2-blade $\mathbf{e}_1 \wedge \mathbf{e}_2$ in 3D-space is $(\mathbf{e}_1 \wedge \mathbf{e}_2)(\mathbf{e}_1 \wedge \mathbf{e}_2 \wedge \mathbf{e}_3)^{-1} = -(\mathbf{e}_1 \mathbf{e}_2)(\mathbf{e}_1 \mathbf{e}_2 \mathbf{e}_3) = \mathbf{e}_3$. This is indeed the normal vector of the $(\mathbf{e}_1, \mathbf{e}_2)$-plane using the right-hand rule. The familiar 3D cross product of vectors can be made in CGA as $\mathbf{x} \times \mathbf{a} = (\mathbf{x} \wedge \mathbf{a})\mathbf{I}_3^{-1}$, though its use should be avoided.

Duality permits us to intersect subspaces. Let us denote the *intersection* (or meet) of blades A and B as $A \cap B$; then we can define it in terms of outer product and dual as

$$\text{dual specification of meet}: \quad (A \cap B)^* = B^* \wedge A^*, \tag{21.8}$$

where the duality is to be taken relative to the smallest-grade blade containing both A and B (this is known as their join, and the intersection as their meet). If one simply takes duality relative to the full space, a meet can become zero in degenerate situations.

An extension of the inner product beyond vector arguments can be developed as a product in its own right, with its own set of algebraic rules. When done properly, it is consistent with the rest of the framework in the sense that

$$\text{extension of inner product}: \quad (A \cdot B)^* \equiv (A \wedge B^*), \tag{21.9}$$

with duality relative to a blade containing the join (one usually takes the pseudoscalar I_n).[3] This inner product has properties like

---

[2]*Editorial note*: Since duality sometimes plays nonmetric roles, others prefer to introduce it differently, see for instance Chap. 14 and Chap. 15.

[3]This inner product is called the *left contraction* and denoted "$\rfloor$" in [3] and some chapters of this book. It differs in details from the inner product used in [1].

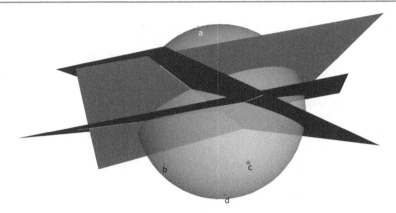

Fig. 21.3 The proof that $a \wedge b \wedge c \wedge d$ represents a sphere involves the intersection of the midplanes $b - a, c - a$ and $d - a$

$$x \cdot (a \wedge b) = (x \cdot a)b - (x \cdot b)a. \tag{21.10}$$

When the arguments of the inner product are blades of the same grade, the result coincides with the *scalar product* $A * B$. For general arguments, that is defined as the scalar part of the geometric product AB and denoted $\langle AB \rangle$; it is symmetric.

The inner product is especially convenient to define orthogonal projection of subspaces as

orthogonal projection of X onto B: $\quad X \mapsto (X \cdot B^{-1}) \cdot B.$

For flats, this corresponds to the usual orthogonal projection, but it is more general: for instance, in CGA projecting a line onto a sphere produces a great circle.

Knowing duality also permits us to interpret elements like $\mathbf{a} \wedge \mathbf{b}$. In CGA, $\mathbf{a}$ and $\mathbf{b}$ are the dual representations of planes through the origin, for the points on these planes satisfy $x \cdot \mathbf{a} = 0$ and $x \cdot \mathbf{b} = 0$. Therefore by (21.8), the 2-blade $\mathbf{a} \wedge \mathbf{b}$ should be the dual representation of their intersection line. Points x on that line then satisfy $x \cdot (\mathbf{a} \wedge \mathbf{b}) = 0$, and the point at infinity n_∞ also is on the line (see the exercises in Sect. 21.9).

We now have enough to show that in CGA, $S = a \wedge b \wedge c \wedge d$ represents the sphere through the four points a, b, c, d. The geometry is illustrated in Fig. 21.3. By antisymmetry of $\wedge$, we can subtract any factor from the others without changing the value of S. We pick a and produce $S = a \wedge (b - a) \wedge (c - a) \wedge (d - a)$. To find out what $(b - a)$ represents, solve $x \cdot (b - a) = 0$. This evaluates to $x \cdot a = x \cdot b$, and because of (21.2), this means that x has the same distance to a and b. So $(b - a)$ is the dual representation of the midplane between a and b. Therefore $(b - a) \wedge (c - a) \wedge (d - a)$ is the dual representation of the intersection of three midplanes. These planes intersect in two points: the center of the sphere m and the point at infinity n_∞, so $(b - a) \wedge (c - a) \wedge (d - a)$ is proportional to $(m \wedge n_\infty)^*$. Then we find $S \propto a \wedge (m \wedge n_\infty)^* = (a \cdot (m \wedge n_\infty))^* = (m - \frac{1}{2}\rho^2 n_\infty)^*$ with $a \cdot m = -\frac{1}{2}\rho^2$.

So indeed S is the dual of a dual sphere representation and therefore a sphere. This also gives a very compact way to compute center and radius of a sphere given by four points: they are simply the appropriate components of $(a \wedge b \wedge c \wedge d)^*$.

21.3   Bonus: The Elements of Euclidean Geometry as Blades

Closure of the operations of outer product and duality produces a suite of blades representing recognizable elements of Euclidean geometry. We have seen many examples of this already, and the full list is given in this table from [3] (where n is the dimension of the Euclidean space, $\mathbf{E}$ a purely Euclidean element of appropriate grade, $\mathbf{E}^{\star}$ denotes the Euclidean dual $\mathbf{EI}_n^{-1}$, and $T_{\mathbf{p}}$ denotes the translation versor over $\mathbf{p}$, see (21.12)). Care has been taken to orient the blades and their duals consistently.

Element	Standard form X	Defining properties
Direction	$\mathbf{E} \wedge n_\infty$	$n_\infty \wedge X = 0; n_\infty \cdot X = 0$
Dual direction	$-\mathbf{E}^{\star} \wedge n_\infty$	$n_\infty \wedge X = 0; n_\infty \cdot X = 0$
Flat	$T_{\mathbf{p}}(n_o \wedge \mathbf{E} \wedge n_\infty)T_{\mathbf{p}}^{-1}$	$n_\infty \wedge X = 0; n_\infty \cdot X \neq 0$
Dual flat	$T_{\mathbf{p}}(\mathbf{E}^{\star}(-1)^{n-\text{grade}(\mathbf{E})})T_{\mathbf{p}}^{-1}$	$n_\infty \wedge X \neq 0; n_\infty \cdot X = 0$
Tangent	$T_{\mathbf{p}}(n_o \wedge \mathbf{E})T_{\mathbf{p}}^{-1}$	$n_\infty \wedge X \neq 0; n_\infty \cdot X \neq 0; X^2 = 0$
Dual tangent	$T_{\mathbf{p}}(n_o \wedge \mathbf{E}^{\star}(-1)^n)T_{\mathbf{p}}^{-1}$	$n_\infty \wedge X \neq 0; n_\infty \cdot X \neq 0; X^2 = 0$
Round	$T_{\mathbf{p}}((n_o + \frac{1}{2}\rho^2 n_\infty) \wedge \mathbf{E})T_{\mathbf{p}}^{-1}$	$n_\infty \wedge X \neq 0; n_\infty \cdot X \neq 0; X^2 \neq 0$
Dual round	$T_{\mathbf{p}}((n_o - \frac{1}{2}\rho^2 n_\infty) \wedge \mathbf{E}^{\star}(-1)^n)T_{\mathbf{p}}^{-1}$	$n_\infty \wedge X \neq 0; n_\infty \cdot X \neq 0; X^2 \neq 0$

The square of a normalized round gives its radius squared, and this may be negative. Such "imaginary rounds" occur naturally, for instance when intersecting two spheres that are further apart than the sum of their radii. Because only the squared radius occurs in the conformal model, these elements are tractable in a real algebra. Tangents are in fact rounds of zero radius, indicative of their infinitesimal size. A tangent 2-blade occurs for instance as the grade 3 element that is the meet of two touching spheres. In this context, a weighted point may be viewed as a localized tangent scalar.

It is especially notable that the various uses and meanings of "vector with direction $\mathbf{u}$" from applied linear algebra get their own "algebraic data structures":

- *a point at location* $\mathbf{u}$ is represented by the CGA vector $n_o + \mathbf{u} + \frac{1}{2}\mathbf{u}^2 n_\infty$
- *a free vector* is represented by the translation invariant 2-blade $\mathbf{u} \wedge n_\infty$
- *a normal vector* is the vector $p \cdot (\mathbf{u} \wedge n_\infty)$ and can move on a localized plane
- *a force vector* is represented by the 3-blade $p \wedge \mathbf{u} \wedge n_\infty$ and can move along a line
- *a tangent vector* $\mathbf{u}$ at p is the localized 2-blade $p \cdot (p \wedge \mathbf{u} \wedge n_\infty)$

All these automatically move appropriately under Euclidean versors, without a programmer needing to specify that they should (by explicitly giving them their own "classes" and "methods", as is required in common practice in classical software, even when based on homogeneous coordinates).

21.4   Bonus: Rigid Body Motions Through Sandwiching

We have seen how all orthogonal transformations can be made as multiple reflections and that a single reflection is represented by an invertible vector $\mathbf{a}$ as the transformation $x \mapsto -\mathbf{a}x\mathbf{a}^{-1}$. Now that we know what the vectors in the conformal model represent, we can easily generate the versors for common motions. Euclidean motions are generated by multiple reflections in planes, and we have seen that those are dually represented by vectors of the form $\pi = \mathbf{n} + \delta n_\infty$; they satisfy $n_\infty \cdot \pi = 0$.

- *Rotation in a plane through the origin*: If we take two unit dual planes at the origin $\mathbf{n}_1$ and $\mathbf{n}_2$ with a relative angle of $\phi/2$ from $\mathbf{n}_1$ to $\mathbf{n}_2$, the double reflection first in $\mathbf{n}_1$ and then in $\mathbf{n}_2$ is represented as

$$R_{\mathbf{I}\phi} = \mathbf{n}_2\mathbf{n}_1 = \mathbf{n}_2 \cdot \mathbf{n}_1 + \mathbf{n}_2 \wedge \mathbf{n}_1 = \cos(\phi/2) - \mathbf{I}\sin(\phi/2). \tag{21.11}$$

When used in a sandwiching operation, this is a rotation over the angle ϕ around the dual line given by the unit 2-blade $\mathbf{I}$ (proportional to $\mathbf{n}_1 \wedge \mathbf{n}_2$). Such a 2-blade has the property $\mathbf{I}^2 = -1$. To show this, introduce an orthonormal basis $\{\mathbf{e}_1, \mathbf{e}_2\}$, write $\mathbf{I} = \mathbf{e}_1 \wedge \mathbf{e}_2 = \mathbf{e}_1\mathbf{e}_2$ and compute using the associativity property of the geometric product: $(\mathbf{e}_1 \wedge \mathbf{e}_2)(\mathbf{e}_1 \wedge \mathbf{e}_2) = (\mathbf{e}_1\mathbf{e}_2)(\mathbf{e}_1\mathbf{e}_2) = -\mathbf{e}_2\mathbf{e}_1\mathbf{e}_1\mathbf{e}_2 = -\mathbf{e}_2\mathbf{e}_2 = -1$. In this real geometric algebra, we therefore naturally get elements that square to -1. In 3D, there is a basis for 2-blades consisting of the elements $\mathbf{I} = \mathbf{e}_1 \wedge \mathbf{e}_2$, $\mathbf{J} = \mathbf{e}_2 \wedge \mathbf{e}_3$ and $\mathbf{K} = \mathbf{e}_3 \wedge \mathbf{e}_1$, each squaring to -1 and having multiplicative relationships like $\mathbf{IJ} = -\mathbf{JI} = -\mathbf{K}$. These are of course isomorphic to the elementary quaternions which have proven so useful for 3D rotation computations. In geometric algebra, they are introduced in a real manner as products of vectors, fully integrated with the real elements they operate on. We will soon see that they can rotate any element and derive the versor for a rotation around a general line in Sect. 21.6.

- *Translation*: A translation over a vector $\mathbf{t}$ is generated by reflection in two dual planes separated by a vector $\mathbf{t}/2$, resulting in the element: $(\mathbf{t} + \frac{1}{2}\mathbf{t} \cdot \mathbf{t}n_\infty)\mathbf{t} = \mathbf{t}^2(1 - \mathbf{t}n_\infty/2)$. Since a scalar multiple generates the same motion in the sandwiching product with the inverse, we prefer to define

$$\textit{versor for translation over } \mathbf{t}: \quad T_{\mathbf{t}} \equiv 1 - \frac{1}{2}\mathbf{t}n_\infty. \tag{21.12}$$

You can check that the point representation (21.1) is indeed related to the point at the origin n_o by translation over $\mathbf{x}$, since $x = T_{\mathbf{x}}n_o T_{\mathbf{x}}^{-1}$.

- *General rigid body motion*: A general rigid body motion can be constructed in the usual manner as a rotation followed by a translation. The resulting versors are often called 'motors' rather than 'rotors'. In CGA, an alternative is to make it directly as the reflection in four planes. When two of those are chosen perpendicular to the other two, this produces Chasles' screw motion representation (see [3], and Chap. 5 and Chap. 6 in this book).
- *Uniform scaling*: Although not strictly a rigid body motion, the Euclidean similarity transformation of uniform scaling can be made by subsequent reflection in two dual spheres at the origin such as $n_o - \frac{1}{2}\rho_1^2 n_\infty$ and $n_o - \frac{1}{2}\rho_2^2 n_\infty$. After some simplification, the scaling versor for a uniform scaling by e^γ is found to be

$$S_\gamma \equiv \cosh(\gamma/2) + \sinh(\gamma/2) n_o \wedge n_\infty.$$

More versors can be generated by reflection in spheres. Since these reflections can change n_∞ to a finite point, they can convert between lines and circles, and between planes and spheres. Ultimately, all even conformal transformations can be generated from a combination of translation, rotation, scaling and *transversion* (reflection in two touching equal radius spheres)—details may be found elsewhere [3].

21.5 Bonus: Structure Preservation and the Transfer Principle

All constructions of elements were based on the linear combinations of geometric products, since the other products are ultimately expressible in that manner. Therefore, when we act on them with a versor V in the sandwiching product, all constructions transform covariantly. For the outer product, this means that equations hold like the following:

$$V(a \wedge b)V^{-1} = \left(VaV^{-1}\right) \wedge \left(VbV^{-1}\right).$$

The same structure-preserving property holds for *all* operations we introduced, be they spanning, inner product or duality (relative to a transformed pseudoscalar). In words: "*the transformation of a construction equals the construction of the transformed elements*". This fact is very convenient, for it implies that we can simply construct something at the origin and then move it into place to find the general form (hence our preference for origin-based specification in the table above). And composite elements move by the same versor as points do: the translation versor T_t universally translates points, lines, planes, spheres or tangent elements. As we mentioned, there is no longer any need for data structures distinguishing between "position vectors" which feel translations, and "direction vectors" which do not; all is automatically administrated in the algebraic behavior of the corresponding elements. This is an enormous advantage relative to the classical homogeneous model for the development of structural code, either by hand or using a code generator [4].

This principle is also extremely useful in derivations. Let us for instance use it to prove the general formula for the reflection of a line Λ in a dual plane π as $\Lambda \mapsto -\pi \Lambda \pi^{-1}$, simply from the 1D direction reflection formula (21.5). A line Λ_0

with direction $\mathbf{u}$ through the origin is given as $\Lambda_0 = n_o \wedge \mathbf{u} \wedge n_\infty$, and a dual plane π_0 through the origin with normal vector $\mathbf{n}$ as $\pi_0 = \mathbf{n}$. The reflection of the direction $\mathbf{u}$ is affected by (21.5) as $\mathbf{u} \mapsto \mathbf{u}' \equiv -\mathbf{n}\mathbf{u}\mathbf{n}^{-1} = -\pi_0 \mathbf{u}\pi_0^{-1}$. The reflected line is then $\Lambda_0' = n_o \wedge \mathbf{u}' \wedge n_\infty$. Now we note that due to the algebraic commutation (i.e., the geometric orthogonality) of the bold Euclidean and the nonbold extra dimensions n_o and n_∞, we have $-\pi_0 n_o \pi_0^{-1} = -\mathbf{n} n_o \mathbf{n}^{-1} = n_o$ and $-\pi_0 n_\infty \pi_0^{-1} = -\mathbf{n} n_\infty \mathbf{n}^{-1} = n_\infty$. Therefore we can "pull out" the reflection operator to act on the whole line Λ_0 by (21.5):

$$\Lambda_0' = \left(-\pi_0 n_o \pi_0^{-1}\right) \wedge \left(-\pi_0 \mathbf{u}\pi_0^{-1}\right) \wedge \left(-\pi_0 n_\infty \pi_0^{-1}\right)$$
$$= -\pi_0 (n_o \wedge \mathbf{u} \wedge n_\infty)\pi_0^{-1} = -\pi_0 \Lambda_0 \pi_0^{-1}.$$

This is still only true at the origin, but we can move this construction by a motion versor V to an arbitrary location. All elements change to their general form $\pi = V\pi_0 V^{-1}$, $\Lambda = V\Lambda_0 V^{-1}$, and the reflection transformation preserves its structure since $V\Lambda_0' V^{-1} = (V\pi_0 V^{-1})(V\Lambda_0 V^{-1})(V\pi_0 V^{-1}) = \pi\Lambda\pi^{-1}$. Therefore the general reflection formula of a line in a plane is simply:

reflection of a line Λ in the dual plane π: $\Lambda \mapsto -\pi\Lambda\pi^{-1}$.

This includes all aspects of location, direction and orientation. Note that this computation reflects a general line in a general plane without computing its intersection point—try doing that using linear algebra! (If you need the intersection point of line and plane, it is $\pi \cdot \Lambda$, by straightforward application of the universal meet operation (21.8) and duality (21.7), (21.9).)

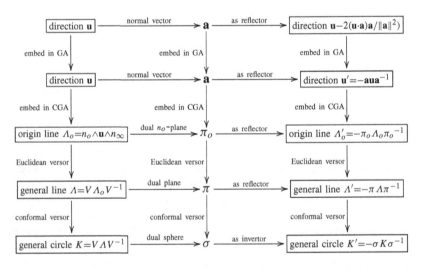

We can even apply an arbitrary conformal versor and change the reflecting dual plane π into a dual sphere σ, and the line L into a circle K; the result is a spherical inversion operation. (As a further extension, another application of the structure preservation property shows that the reflection in σ of a general element X is $X \mapsto (-1)^{\text{grade}(X)} \sigma X \sigma^{-1}$.)

The conformal model renders all transitions trivial in this transfer, all the way from a reflection of a Euclidean direction vector at the origin to the inversion of a general circle in a general sphere. Such is the power of a structure-preserving framework!

21.6   Trick 5: Exponential Representation of Versors

Even-graded versors, made by an even number of reflections, represent motions that can be performed continuously and in small amounts. In Euclidean and Minkowski spaces, *all even-graded versors can be written as the exponentials of bivectors*. The bivector specification of an even versor is often more directly related to the geometry of the situation than the "product of vectors" method.

As an example of the exponential rewriting, take the rotation $R_{\mathbf{I}\phi}$ over the angle ϕ, parallel to the $\mathbf{I}$-plane as treated in (21.11):

$$R_{\mathbf{I}\phi} = \cos\left(\frac{1}{2}\phi\right) - \sin\left(\frac{1}{2}\phi\right)\mathbf{I} = e^{-\mathbf{I}\phi/2}.$$

It is the property $\mathbf{I}^2 = -1$ that makes the exponential rewriting permitted:

$$e^{-\mathbf{I}\phi/2} = 1 + \frac{1}{1!}\left(-\frac{\mathbf{I}\phi}{2}\right)^1 + \frac{1}{2!}\left(-\frac{\mathbf{I}\phi}{2}\right)^2 + \cdots$$

$$= \left(1 - \frac{1}{2!}\left(\frac{\phi}{2}\right)^2 + \cdots\right) + \left(\frac{1}{1!}\left(\frac{\phi}{2}\right)^1 - \frac{1}{3!}\left(\frac{\phi}{2}\right)^3 + \cdots\right)\mathbf{I}$$

$$= \cos\left(\frac{1}{2}\phi\right) - \sin\left(\frac{1}{2}\phi\right)\mathbf{I}.$$

The translation versor of (21.12) can also be written in this exponential form; but since it involves the bivector $\mathbf{t} \wedge n_\infty$, the expansion truncates after two terms (fundamentally due to $n_\infty^2 = 0$):

$$T_{\mathbf{t}} = 1 - \frac{1}{2}\mathbf{t} \wedge n_\infty = e^{-\mathbf{t} \wedge n_\infty/2}.$$

A rotation around a general 3D unit line Λ over ϕ is now generated by the versor:

rotation around Λ over ϕ: $R_{\Lambda,\phi} = e^{\Lambda^*\phi/2}.$

Proof This follows from the simply derived structural property

$$V \exp(B) V^{-1} = \exp(V B V^{-1}),$$

and the transfer property applied as follows. First recognize that the rotation axis of the origin rotation $R_{\mathbf{I}\phi}$ is the line $\Lambda_0 = \mathbf{I}^*$, so the origin rotation is $\exp(\Lambda_0{}^*\phi/2)$. Then transfer this by a translation T to the actual location of the desired axis Λ, which changes $\Lambda_0{}^*$ to $T(\Lambda_0{}^*)T^{-1} = (T\Lambda_0 T^{-1})/(T I_{4,1} T^{-1}) = \Lambda^*$ since the pseudoscalar $I_{4,1}$ involved in the dualization is translation invariant. Done.   □

General rigid body motions can of course also be made, for instance by the usual method of combining an origin rotation with a translation. You find that the result can be written as the exponential of a general conformal bivector on the basis $\{\mathbf{e}_1 \wedge \mathbf{e}_2, \mathbf{e}_2 \wedge \mathbf{e}_3, \mathbf{e}_3 \wedge \mathbf{e}_1, \mathbf{e}_1 \wedge n_\infty, \mathbf{e}_2 \wedge n_\infty, \mathbf{e}_3 \wedge n_\infty\}$, giving the six degrees of freedom required. Since this space of bivectors is linear, it can be used for motion interpolation. To interpolate between two poses characterized by the versors M_0 and M_1, find their bivectors $B_0 = \log(M_0)$ and $B_1 = \log(M_1)$. Now apply a standard vector interpolation method to smoothly change B_0 into B_1 through intermediate bivectors B_i; then use the versors $\exp(B_i)$ to generate the interpolated poses. To execute this procedure, one needs to find the bivector corresponding to a given versor; such "versor logarithms" are treated in Chap. 5. Effective use of the bivectors for motion tracking may be found in Chap. 6. The linear space of motors (versors of rigid body motions) is studied in great detail in Chap. 2.

21.7   Trick 6: Geometric Calculus

Linearization of versor motions for extrapolation or estimation is also possible and requires geometric calculus. When performed (see [1]), the first-order change in an element X that is moved by a changing versor $V(\tau)$ from a standard element X_0 as $X(\tau) = V(\tau)X_0 V(\tau)^{-1}$ is

$$X(\tau + \varepsilon) = X(\tau) + \big(\Omega(\tau)X(\tau) - X(\tau)\Omega(\tau)\big)\varepsilon$$

$$\text{with } \Omega(\tau) = \left(\frac{d}{d\tau}V(\tau)\right)V(\tau)^{-1}.$$

If V is normalized, Ω is a bivector, and its 'commutator product' with $X(\tau)$ (half the second term above) preserves the grade. This linearization of geometrical perturbations is very useful in applications, see Fig. 21.4.

The full geometric calculus is truly powerful, and one can differentiate relative to an arbitrary element of the algebra (such as a blade or a versor). The main difference with classical differentiation is that such geometric differentiation, denoted by ∂_X, also has the scalar aspects of a "derivative operator", but that these are now tied to directions in space. The geometric product makes the scalar aspects commute, but not their geometric factors. Because of the noncommutativity of the geometric product, we cannot always move the geometric differentiation operator next to the element it acts on. This leads to additional refinement of the classical results in the

Fig. 21.4 The mirror plane Π rotates an angle ϕ round a line Λ, and a line X is reflected in it. Using a local first-order linearization of the reflection versor, one can derive the perturbation of the reflected line to second order (in *black*) to be the rotation with versor $\exp(-\phi((\Lambda \cdot \Pi)/\Pi)^*)$, i.e. around the projection of Λ onto Π with angle $2\phi \cos(\Pi, \Lambda)$. For details, see [3]

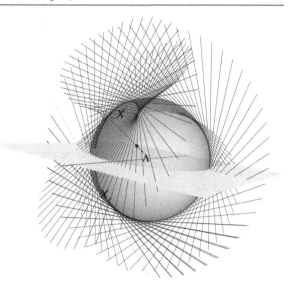

derivative results, with interpretable geometrical significance. It has become convention to denote the differentiation action by an overdot, giving expressions like

$$\partial_X AB = \dot{\partial}_X \dot{A}B + \dot{\partial}_X A \dot{B}$$

for the product rule of geometric differentiation.

We give a few derivatives that come up in this book (notably in Chaps. 1, 2 and 7). In typical applications, one probes a parameterized multivector by a variation in one of its parameters. The two elements need not lie in the same unrestricted space; usually X is on a manifold in some embedding space from which A can take its values. We denote by P[·] the projection from the embedding space to the local tangent space of the X-manifold, and assume that to be m-dimensional. Then:

$$(A * \partial_X)X = \mathsf{P}[A], \tag{21.13}$$

$$\partial_X X = m, \tag{21.14}$$

$$\partial_X \langle XA \rangle = \mathsf{P}[A]. \tag{21.15}$$

For a vector $\mathbf{x}$ and an α-blade $\mathbf{A}$, a result required in Chap. 5 is

$$\partial_{\mathbf{x}}(\mathbf{A}\mathbf{x}) = (-1)^{\alpha}(m - 2\alpha)\mathsf{P}[\mathbf{A}] \quad \text{with } \alpha = \text{grade}(\mathbf{A}). \tag{21.16}$$

We cannot treat more on differentiation here, and the reader is referred to introductions like [1] and [3]. According to David Hestenes, geometric algebra can provide compact new tools for practical coordinate-free differential geometry; he gives hints and tips in Chap. 19 of this volume.

21.8   Trick 7: Sparse Implementation at Compiler Level

Implementation of CGA may seem to be expensive. After all, to treat a 3D space, we embed into a 5D representational space $\mathbb{R}^{4,1}$ and use the geometric algebra $\mathbb{R}_{4,1}$ of that, which involves a 2^5-D basis of constructible elements of all grades. Yet the use we make of this space is restricted, and the elements are therefore somehow sparse.

Ultimately, the main purpose of the algebraic organization is to keep track automatically of the administration of the meaning of the coordinates of points, lines, planes, spheres, etc., simultaneous with performing the quantitative computations. That is in a sense a Boolean selection task of the algebra, which one would intuitively expect not to be too expensive. Indeed it has proved possible to limit the overhead of the use of CGA to about 10% relative to the best available coordinate code programmed classically. For the computer science techniques that achieve this, consult [3] and [4]. A warning: before you start using CGA in commercial applications, be aware that it is covered by a *US patent* [5].

21.9   Exercises

21.1 Show that the point at infinity n_∞ is preserved by a reflection in a plane through the origin characterized by normal vector **a**. Then show it is preserved by reflection in any plane.

21.2 Compute the reflection of the point at infinity n_∞ in a (dual) sphere $\sigma = c - \frac{1}{2}\rho^2 n_\infty$. (Hint: what do you expect it to be?)

21.3 Show that $\mathbf{a} \wedge \mathbf{b} \wedge \mathbf{c} = \frac{1}{2}(\mathbf{abc} - \mathbf{cba})$, with the outer product defined through (21.6).

21.4 In 4D space, show that the bivector $\mathbf{e}_1 \wedge \mathbf{e}_2 + \mathbf{e}_3 \wedge \mathbf{e}_4$ is not a 2-blade.

21.5 With **a** and **b** Euclidean direction vectors in 3D, $(\mathbf{a} \wedge \mathbf{b})^*$ is a line. Investigate which line, by finding all points x such that $x \cdot (\mathbf{a} \wedge \mathbf{b}) = 0$. Show that the point at infinity also satisfies this equation. (Hint: use (21.10).)

21.6 Check that the point representation (21.1) is indeed related to the point at the origin n_o by the translation versor $T_{\mathbf{x}}$, defined as in (21.12).

21.7 Show that the intersection point of a general line Λ with the dual plane π is $\pi \cdot \Lambda$. Interpret its weight geometrically.

21.8 Many of the structural tricks that make CGA work can be used in other models of geometry. Chapter 15 presents an alternative "homogeneous model". Identify which CGA tricks are similar, and which are resolved differently.

References

1. Doran, C., Lasenby, A.: Geometric Algebra for Physicists. Cambridge University Press, Cambridge (2000)
2. Dorst, L.: Tutorial: Structure preserving representation of Euclidean motions through conformal geometric algebra. In: Bayro-Corrochano, E., Scheuermann, G. (eds.) Geometric Algebra Computing for Engineering and Computer Science, pp. 457–476. Springer, Berlin (2011)
3. Dorst, L., Fontijne, D., Mann, S.: Geometric Algebra for Computer Science: An Object-Oriented Approach to Geometry. Morgan Kaufman, San Mateo (2007/2009). See www.geometricalgebra.net
4. Fontijne, D.: Efficient Implementation of Geometric Algebra. University of Amsterdam, Amsterdam (2007), ISBN: 13 978-90-889-10-142, online at www.science.uva.nl/~fontijne
5. Hestenes, D., Rockwood, A., Li, H.: System for encoding and manipulating models of objects. U.S. Patent 6,853,964, February 8, 2005

Index

L. Dorst, J. Lasenby (eds.), *Guide to Geometric Algebra in Practice*,
DOI 10.1007/978-0-85729-811-9, © Springer-Verlag London Limited 2011